Handbook of Spectroscopy

Edited by G. Gauglitz and T. Vo-Dinh

Related Titles from WILEY-VCH

H. Günzler and H.-U. Gremlich (eds)
IR Spectroscopy

2002. ca. 361 pages.
Hardcover. ISBN 3-527-28896-1

H. W. Siesler, Y. Ozaki, S. Kawata and H. M. Heise (eds)
Near-Infrared Spectroscopy
Principles, Instruments, Applications

2001. ca. 348 pages.
Hardcover. ISBN 3-527-30149-6

H. Günzler and A. Williams (eds)
Handbook of Analytical Techniques

2001. 2 Volumes, 1182 pages.
Hardcover. ISBN 3-527-30165-8

J. F. Haw (ed)
In-situ Spectroscopy in Heterogeneous Catalysis

2002. ca. 276 pages.
Hardcover. ISBN 3-527-30248-4

Handbook of Spectroscopy

Edited by G. Gauglitz and T. Vo-Dinh

Volume 2

WILEY-VCH GmbH & Co. KGaA

Prof. Dr. Guenter Gauglitz
Institute for Physical and Theoretical
Chemistry
University of Tübingen
Auf der Morgenstelle 8
72976 Tübingen
Germany

Prof. Dr. Tuan Vo-Dinh
Advanced Biomedical Science
and Technology Group
Oak Ridge National Laboratory
P. O. Box 2008
Oak Ridge, Tennessee 37831-6101
USA

■ This book was carefully produced. Nevertheless, editors, authors and publisher do not warrant the information contained therein to be free of errors. Readers are advised to keep in mind that statements, data, illustrations, procedural details or other items may inadvertently be inaccurate.

Library of Congress Card No.: applied for
A catalogue record for this book is available from the British Library.

Bibliographic information published by Die Deutsche Bibliothek
Die Deutsche Bibliothek lists this publication in the Deutsche Nationalbibliografie; detailed bibliographic data is available in the Internet at http://dnb.ddb.de.

© 2003 WILEY-VCH Verlag GmbH & Co. KGaA, Weinheim

All rights reserved (including those of translation in other languages). No part of this book may be reproduced in any form – by photoprinting, microfilm, or any other means – nor transmitted or translated into machine language without written permission from the publishers. Registered names, trademarks, etc. used in this book, even when not specifically marked as such, are not to be considered unprotected by law.

Printed in the Federal Republic of Germany.

Printed on acid-free paper.

Typesetting Hagedorn Kommunikation, Viernheim
Printing Strauss Offsetdruck GmbH, Mörlenbach
Bookbinding J. Schäffer GmbH & Co. KG, Grünstadt

ISBN 3-527-29782-0

Contents

Volume 1

    Preface XXVIII

    List of Contributors

Section I Sample Preparation and Sample Pretreatment 1

Introduction 3

1	**Collection and Preparation of Gaseous Samples** 4	
1.1	Introduction 4	
1.2	Sampling considerations 5	
1.3	Active vs. Passive Sampling 8	
1.3.1	Active Air Collection Methods 8	
1.3.1.1	Sorbents 9	
1.3.1.2	Bags 11	
1.3.1.3	Canisters 11	
1.3.1.4	Bubblers 12	
1.3.1.5	Mist Chambers 13	
1.3.1.6	Cryogenic Trapping 13	
1.3.2	Passive Sampling 13	
1.4	Extraction and Preparation of Samples 14	
1.5	Summary 15	
2	**Sample Collection and Preparation of Liquid and Solids** 17	
2.1	Introduction 17	
2.2	Collection of a Representative Sample 17	
2.2.1	Statistics of Sampling 18	
2.2.2	How Many Samples Should be Obtained? 21	

2.2.3	Sampling 22
2.2.3.1	Liquids 22
2.2.3.2	Solids 23
2.3	Preparation of Samples for Analysis 24
2.3.1	Solid Samples 24
2.3.1.1	Sample Preparation for Inorganic Analysis 25
2.3.1.2	Decomposition of Organics 28
2.3.2	Liquid Samples 29
2.3.2.1	Extraction/Separation and Preconcentration 29
2.3.2.2	Chromatographic Separation 31

Section II Methods 1: Optical Spectroscopy 37

3	**Basics of Optical Spectroscopy** 39
3.1	Absorption of Light 39
3.2	Infrared Spectroscopy 41
3.3	Raman Spectroscopy 43
3.4	UV/VIS Absorption and Luminescence 44

4	**Instrumentation** 48
4.1	MIR Spectrometers 48
4.1.1	Dispersive Spectrometers 49
4.1.2	Fourier-Transform Spectrometers 50
4.1.2.1	Detectors 53
4.1.2.2	Step-scan Operation 53
4.1.2.3	Combined Techniques 54
4.2	NIR Spectrometers 54
4.2.1	FT-NIR Spectrometers 55
4.2.2	Scanning-Grating Spectrometers 55
4.2.3	Diode Array Spectrometers 56
4.2.4	Filter Spectrometers 56
4.2.5	LED Spectrometers 56
4.2.6	AOTF Spectrometers 56
4.3	Raman Spectrometers 57
4.3.1	Raman Grating Spectrometer with Single Channel Detector 57
4.3.1.1	Detectors 59
4.3.1.2	Calibration 60
4.3.2	FT-Raman Spectrometers with Near-Infrared Excitation 61
4.3.3	Raman Grating Polychromator with Multichannel Detector 61
4.4	UV/VIS Spectrometers 63
4.4.1	Sources 64
4.4.2	Monochromators 64
4.4.3	Detectors 64
4.5	Fluorescence Spectrometers 66

5	**Measurement Techniques** 70	
5.1	Transmission Measurements 71	
5.2	Reflection Measurements 73	
5.2.1	External Reflection 73	
5.2.2	Reflection Absorption 75	
5.2.3	Attenuated Total Reflection (ATR) 75	
5.2.4	Reflection at Thin Films 77	
5.2.5	Diffuse Reflection 78	
5.3	Spectroscopy with Polarized Light 81	
5.3.1	Optical Rotatory Dispersion 81	
5.3.2	Circular Dichroism (CD) 82	
5.4	Photoacoustic Measurements 83	
5.5	Microscopic Measurements 84	
5.5.1	Infrared Microscopes 85	
5.5.2	Confocal Microscopes 85	
5.5.3	Near-field Microscopes 86	
6	**Applications** 89	
6.1	Mid-Infrared (MIR) Spectroscopy 89	
6.1.1	Sample Preparation and Measurement 89	
6.1.1.1	Gases 90	
6.1.1.2	Solutions and Neat Liquids 91	
6.1.1.3	Pellets and Mulls 92	
6.1.1.4	Neat Solid Samples 94	
6.1.1.5	Reflection–Absorption Sampling Technique 94	
6.1.1.6	Sampling with the ATR Technique 95	
6.1.1.7	Thin Samples 96	
6.1.1.8	Diffuse Reflection Sampling Technique 97	
6.1.1.9	Sampling by Photoacoustic Detection 97	
6.1.1.10	Microsampling 98	
6.1.2	Structural Analysis 98	
6.1.2.1	The Region from 4000 to 1400 cm^{-1} 102	
6.1.2.2	The Region 1400–900 cm^{-1} 102	
6.1.2.3	The Region from 900 to 400 cm^{-1} 102	
6.1.3	Special Applications 103	
6.2	Near-Infrared Spectroscopy 104	
6.2.1	Sample Preparation and Measurement 105	
6.2.2	Applications of NIR Spectroscopy 110	
6.3	Raman Spectroscopy 112	
6.3.1	Sample Preparation and Measurements 112	
6.3.1.1	Sample Illumination and Light Collection 113	
6.3.1.2	Polarization Measurements 118	
6.3.1.3	Enhanced Raman Scattering 119	
6.3.2	Special Applications 120	
6.4	UV/VIS Spectroscopy 125	

6.4.1	Sample Preparation	125
6.4.2	Structural Analysis	129
6.4.3	Special Applications	132
6.5	Fluorescence Spectroscopy	135
6.5.1	Sample Preparation and Measurements	138
6.5.1.1	Fluorescence Quantum Yield and Lifetime	138
6.5.1.2	Fluorescence Quencher	139
6.5.1.3	Solvent Relaxation	144
6.5.1.4	Polarized Fluorescence	148
6.5.2	Special Applications	152

Section III   Methods 2: Nuclear Magnetic Resonance Spectroscopy   169

Introduction   *171*

7	**An Introduction to Solution, Solid-State, and Imaging NMR Spectroscopy** *177*	
7.1	Introduction *177*	
7.2	Solution-state ^{1}H NMR *179*	
7.3	Solid-state NMR *187*	
7.3.1	Dipolar Interaction *188*	
7.3.2	Chemical Shift Anisotropy *190*	
7.3.3	Quadrupolar Interaction *191*	
7.3.4	Magic Angle Spinning (MAS) NMR *194*	
7.3.5	T_1 and $T_{1\rho}$ Relaxation *195*	
7.3.6	Dynamics *198*	
7.4	Imaging *199*	
7.5	3D NMR: The HNCA Pulse Sequence *204*	
7.6	Conclusion *207*	

8	**Solution NMR Spectroscopy** *209*	
8.1	Introduction *209*	
8.2	1D (One-dimensional) NMR Methods *210*	
8.2.1	Proton Spin Decoupling Experiments *211*	
8.2.2	Proton Decoupled Difference Spectroscopy *212*	
8.2.3	Nuclear Overhauser Effect (NOE) Difference Spectroscopy *212*	
8.2.4	Selective Population Transfer (SPT) *213*	
8.2.5	*J*-Modulated Spin Echo Experiments *213*	
8.2.5.1	INEPT (Insensitive Nucleus Enhancement by Polarization Transfer) *214*	
8.2.5.2	DEPT (Distortionless Enhancement Polarization Transfer) *215*	
8.2.6	Off-Resonance Decoupling *216*	
8.2.7	Relaxation Measurements *217*	

8.3	Two-dimensional NMR Experiments	218
8.3.1	2D J-Resolved NMR Experiments	219
8.3.2	Homonuclear 2D NMR Spectroscopy	223
8.3.2.1	COSY, Homonuclear Correlated Spectroscopy	223
8.3.2.2	Homonuclear TOCSY, Total Correlated Spectroscopy	226
8.3.2.3	NOESY, Nuclear Overhauser Enhancement Spectroscopy	228
8.3.2.4	ROESY, Rotating Frame Overhauser Enhanced Spectroscopy	230
8.3.2.5	NOESY vs. ROESY	231
8.3.2.6	Other Homonuclear Autocorrelation Experiments	231
8.3.3	Gradient Homonuclear 2D NMR Experiments	232
8.3.4	Heteronuclear Shift Correlation	234
8.3.5	Direct Heteronuclear Chemical Shift Correlation Methods	234
8.3.5.1	HMQC, Heteronuclear Multiple Quantum Coherence	234
8.3.6	HSQC, Heteronuclear Single Quantum Coherence Chemical Shift Correlation Techniques	236
8.3.6.1	Multiplicity-edited Heteronuclear Shift Correlation Experiments	237
8.3.6.2	Accordion-optimized Direct Heteronuclear Shift Correlation Experiments	239
8.3.7	Long-range Heteronuclear Chemical Shift Correlation	240
8.3.7.1	HMBC, Heteronuclear Multiple Bond Correlation	242
8.3.7.2	Variants of the Basic HMBC Experiment	243
8.3.7.3	Accordion-optimized Long-range Heteronuclear Shift Correlation Methods.	244
8.3.7.4	$^2J^3J$-HMBC	248
8.3.7.5	Relative Sensitivity of Long-range Heteronuclear Shift Correlation Experiments	251
8.3.7.6	Applications of Accordion-optimized Long-range Heteronuclear Shift Correlation Experiments	252
8.3.8	Hyphenated-2D NMR Experiments	252
8.3.9	One-dimensional Analogues of 2D NMR Experiments	255
8.3.10	Gradient 1D NOESY	255
8.3.11	Selective 1D Long-range Heteronuclear Shift Correlation Experiments	257
8.3.12	Small Sample NMR Studies	257
8.4	Conclusions	262
9	**Solid-State NMR**	269
9.1	Introduction	269
9.2	Solid-state NMR Lineshapes	272
9.2.1	The Orientational Dependence of the NMR Resonance Frequency	272
9.2.2	Single-crystal NMR	273
9.2.3	Powder Spectra	275
9.2.4	One-dimensional ^{2}H NMR	278
9.3	Magic-angle Spinning	280
9.3.1	CP MAS NMR	281

9.3.2	¹H Solid-State NMR *285*
9.4	Recoupling Methods *287*
9.4.1	Heteronuclear Dipolar-coupled Spins: REDOR *287*
9.4.2	Homonuclear Dipolar-coupled Spins *290*
9.4.3	The CSA: CODEX *291*
9.5	Homonuclear Two-dimensional Experiments *292*
9.5.1	Establishing the Backbone Connectivity in an Organic Molecule *293*
9.5.2	Dipolar-mediated Double-quantum Spectroscopy *295*
9.5.3	High-resolution ¹H Solid-state NMR *298*
9.5.4	Anisotropic – Isotropic Correlation: The Measurement of CSAs *300*
9.5.5	The Investigation of Slow Dynamics: 2D Exchange *303*
9.5.6	¹H–¹H DQ MAS Spinning-sideband Patterns *305*
9.6	Heteronuclear Two-dimensional Experiments *307*
9.6.1	Heteronuclear Correlation *307*
9.6.2	The Quantitative Determination of Heteronuclear Dipolar Couplings *310*
9.6.3	Torsional Angles *312*
9.6.4	Oriented Samples *313*
9.7	Half-integer Quadrupole Nuclei *315*
9.8	Summary *319*

Section IV   Methods 3: Mass Spectrometry   *327*

10	**Mass Spectrometry** *329*
10.1	Introduction: Principles of Mass Spectrometry *329*
10.1.1	Application of Mass Spectrometry to Biopolymer Analysis *330*
10.2	Techniques and Instrumentation of Mass Spectrometry *331*
10.2.1	Sample Introduction and Ionisation Methods *331*
10.2.1.1	Pre-conditions *331*
10.2.1.2	Gas Phase ("Hard") Ionisation Methods *331*
10.2.1.3	"Soft" Ionisation Techniques *332*
10.2.2	Mass Spectrometric Analysers *335*
10.2.2.1	Magnetic Sector Mass Analysers *335*
10.2.2.2	Quadrupole Mass Analysers *337*
10.2.2.3	Time-of-Flight Mass Analysers *338*
10.2.2.4	Trapped-Ion Mass Analysers *339*
10.2.2.5	Hybrid Instruments *340*
10.2.3	Ion Detection and Spectra Acquisition *340*
10.2.4	High Resolution Fourier Transform Ion Cyclotron Resonance (ICR) Mass Spectrometry *341*
10.2.5	Sample Preparation and Handling in Bioanalytical Applications *344*
10.2.5.1	Liquid–Liquid Extraction (LLE) *344*
10.2.5.2	Solid Phase Extraction (SPE) *345*
10.2.5.3	Immunoaffinity Extraction (IAE) *345*

10.2.5.4	Solid-phase Microextraction 345
10.2.5.5	Supercritical-Fluid Extraction (SFE) 346
10.2.6	Coupling of Mass Spectrometry with Microseparation Methods 346
10.2.6.1	Liquid Chromatography-Mass Spectrometry Coupling (LC-MS) 347
10.2.6.2	Capillary Electrophoresis (CE)-Mass Spectrometry 348
10.3	Applications of Mass Spectrometry to Biopolymer Analysis 349
10.3.1	Introduction 349
10.3.2	Analysis of Peptide and Protein Primary Structures and Post-Translational Structure Modifications 349
10.3.3	Tertiary Structure Characterisation by Chemical Modification and Mass Spectrometry 353
10.3.4	Characterisation of Non-Covalent Supramolecular Complexes 354
10.3.5	Mass Spectrometric Proteome Analysis 356

Section V Methods 4: Elemental Analysis 363

11	**X-ray Fluorescence Analysis** 365
11.1	Introduction 365
11.2	Basic Principles 367
11.2.1	X-ray Wavelength and Energy Scales 367
11.2.2	Interaction of X-rays with Matter 367
11.2.3	Photoelectric Effect 369
11.2.4	Scattering 371
11.2.5	Bremsstrahlung 372
11.2.6	Selection Rules, Characteristic Lines and X-ray Spectra 373
11.2.7	Figures-of-merit for XRF Spectrometers 376
11.2.7.1	Analytical Sensitivity 376
11.2.7.2	Detection and Determination Limits 377
11.3	Instrumentation 380
11.3.1	X-ray Sources 380
11.3.2	X-ray Detectors 384
11.3.3	Wavelength-dispersive XRF 390
11.3.4	Energy-dispersive XRF 393
11.3.5	Radioisotope XRF 397
11.3.6	Total Reflection XRF 398
11.3.7	Microscopic XRF 399
11.4	Matrix Effects 401
11.4.1	Thin and Thick Samples 401
11.4.2	Primary and Secondary Absorption, Direct and Third Element Enhancement 403
11.5	Data Treatment 404
11.5.1	Counting Statistics 404
11.5.2	Spectrum Evaluation Techniques 405
11.5.2.1	Data Extraction in WDXRF 406

11.5.2.2	Data Extraction in EDXRF: Simple Case, No Peak Overlap	407
11.5.2.3	Data Extraction in EDXRF, Multiple Peak Overlap	408
11.5.3	Quantitative Calibration Procedures	409
11.5.3.1	Single-element Techniques	412
11.5.3.2	Multiple-element Techniques	413
11.5.4	Error Sources in X-ray Fluorescence Analysis	415
11.5.5	Specimen Preparation for X-ray Fluorescence	416
11.6	Advantages and Limitations	417
11.6.1	Qualitative Analysis	417
11.6.2	Detection Limits	418
11.6.3	Quantitative Reliability	418
11.7	Summary	419
12	**Atomic Absorption Spectrometry (AAS) and Atomic Emission Spectrometry (AES)**	**421**
12.1	Introduction	421
12.2	Theory of Atomic Spectroscopy	421
12.2.1	Basic Principles	421
12.2.2	Fundamentals of Absorption and Emission	426
12.2.2.1	Absorption	429
12.2.2.2	Line Broadening	430
12.2.2.3	Self-absorption	431
12.2.2.4	Ionisation	432
12.2.2.5	Dissociation	434
12.2.2.6	Radiation Sources and Atom Reservoirs	434
12.3	Atomic Absorption Spectrometry (AAS)	436
12.3.1	Introduction	436
12.3.2	Instrumentation	436
12.3.2.1	Radiation Sources	437
12.3.2.2	Atomisers	440
12.3.2.3	Optical Set-up and Components of Atomic Absorption Instruments	453
12.3.3	Spectral Interference	454
12.3.3.1	Origin of Spectral Interference	454
12.3.3.2	Methods for Correcting for Spectral Interference	455
12.3.4	Chemical Interferences	462
12.3.4.1	The Formation of Compounds of Low Volatility	463
12.3.4.2	Influence on Dissociation Equilibria	463
12.3.4.3	Ionisation in Flames	464
12.3.5	Data Treatment	465
12.3.5.1	Quantitative Analysis	465
12.3.6	Hyphenated Techniques	466
12.3.6.1	Gas Chromatography-Atomic Absorption Spectrometry	467
12.3.6.2	Liquid Chromatography-Atomic Absorption Spectrometry	469
12.3.7	Conclusion and Future Directions	470

12.4	Atomic Emission Spectrometry (AES) 471
12.4.1	Introduction 471
12.4.2	Instrumentation 471
12.4.2.1	Atomisation Devices 471
12.4.2.2	Optical Set-up and Detection 480
12.4.2.3	Instrumentation for Solid Sample Introduction 483
12.4.3	Matrix Effects and Interference 486
12.4.3.1	Spectral Interferences 486
12.4.3.2	Matrix Effects and Chemical Interferences 487
12.4.4	Quantitative and Qualitative Analysis 488
12.4.5	Advantages and Limitations 491
12.4.5.1	Absolute and Relative Sensitivity 491
12.4.5.2	Hyphenated Techniques 491
12.5	Summary 493

Section VI   Methods 5: Surface Analysis Techniques 497

13	**Surface Analysis Techniques** 499
13.1	Introduction 499
13.2	Definition of the Surface 501
13.3	Selection of Method 501
13.4	Individual Techniques 506
13.4.1	Angle Resolved Ultraviolet Photoelectron Spectroscopy 506
13.4.1.1	Introduction 507
13.4.1.2	Instrumentation 507
13.4.1.3	Sample 507
13.4.1.4	Analytical Information 507
13.4.1.5	Performance Criteria 507
13.4.1.6	Applications 508
13.4.1.7	Other Techniques 508
13.4.2	Appearance Potential Spectroscopy 508
13.4.2.1	Introduction 508
13.4.2.2	Instrumentation 508
13.4.2.3	Sample 509
13.4.2.4	Analytical Information 509
13.4.2.5	Performance Criteria 509
13.4.2.6	Applications 509
13.4.2.7	Other Techniques 510
13.4.3	Atom Probe Field Ion Microscopy 510
13.4.3.1	Introduction 510
13.4.3.2	Instrumentation 510
13.4.3.3	Analytical Information 510
13.4.3.4	Performance Criteria 510
13.4.3.5	Applications 510

13.4.4	Attenuated Total Reflection Spectroscopy	*511*
13.4.4.1	Introduction	*511*
13.4.4.2	Instrumentation	*511*
13.4.4.3	Analytical Information	*511*
13.4.4.4	Performance Criteria	*511*
13.4.4.5	Applications	*512*
13.4.5	Auger Electron Spectroscopy	*512*
13.4.5.1	Introduction	*512*
13.4.5.2	Instrumentation	*512*
13.4.5.3	Sample	*513*
13.4.5.4	Analytical Information	*513*
13.4.5.5	Performance Criteria	*513*
13.4.5.6	Applications	*514*
13.4.5.7	Other Techniques	*514*
13.4.6	Auger Photoelectron Coincidence Spectroscopy	*514*
13.4.6.1	Introduction	*514*
13.4.6.2	Instrumentation	*515*
13.4.6.3	Sample	*515*
13.4.6.4	Analytical Information	*515*
13.4.6.5	Performance Criteria	*515*
13.4.6.6	Applications	*516*
13.4.6.7	Other Techniques	*516*
13.4.7	Charge Particle Activation Analysis	*516*
13.4.7.1	Introduction	*516*
13.4.7.2	Instrumentation	*516*
13.4.7.3	Sample	*517*
13.4.7.4	Analytical Information	*517*
13.4.7.5	Performance Criteria	*517*
13.4.7.6	Application	*518*
13.4.7.7	Other Technique	*518*
13.4.8	Diffuse Reflection Spectroscopy	*518*
13.4.8.1	Introduction	*518*
13.4.8.2	Instrumentation	*518*
13.4.8.3	Analytical Information	*519*
13.4.8.4	Performance Criteria	*519*
13.4.8.5	Applications	*519*
13.4.9	Elastic Recoil Detection Analysis	*520*
13.4.9.1	Introduction	*520*
13.4.9.2	Instrumentation	*520*
13.4.9.3	Sample	*520*
13.4.9.4	Analytical Information	*520*
13.4.9.5	Performance Criteria	*521*
13.4.9.6	Applications	*522*
13.4.9.7	Other Techniques	*522*
13.4.10	Electron Momentum Spectroscopy	*522*

13.4.10.1 Introduction *523*
13.4.10.2 Instrumentation *523*
13.4.10.3 Sample *523*
13.4.10.4 Analytical Information *523*
13.4.10.5 Performance Criteria *523*
13.4.10.6 Applications *523*
13.4.11 Electron Probe Microanalysis *524*
13.4.11.1 Introduction *524*
13.4.11.2 Instrumentation *524*
13.4.11.3 Sample *524*
13.4.11.4 Analytical Information *524*
13.4.11.5 Performance Criteria *525*
13.4.11.6 Applications *525*
13.4.12 Electron Stimulated Desorption *525*
13.4.12.1 Introduction *525*
13.4.12.2 Instrumentation *525*
13.4.12.3 Sample *526*
13.4.12.4 Analytical Information *526*
13.4.12.5 Performance Criteria *526*
13.4.12.6 Applications *526*
13.4.13 Electron Stimulated Desorption Ion Angular Distributions *526*
13.4.13.1 Introduction *526*
13.4.13.2 Instrumentation *527*
13.4.13.3 Sample *527*
13.4.13.4 Analytical Information *527*
13.4.13.5 Performance Criteria *527*
13.4.13.6 Applications *527*
13.4.14 Ellipsometry *528*
13.4.14.1 Introduction *528*
13.4.14.2 Instrumentation *528*
13.4.14.3 Sample *528*
13.4.14.4 Analytical Information *528*
13.4.14.5 Performance Criteria *529*
13.4.14.6 Applications *529*
13.4.15 Extended Energy Loss Fine Structure *529*
13.4.15.1 Introduction *529*
13.4.15.2 Instrumentation *530*
13.4.15.3 Analytical Information *530*
13.4.15.4 Performance Criteria *530*
13.4.15.5 Applications *530*
13.4.15.6 Other Techniques *530*
13.4.16 Evanescent Wave Cavity Ring-down Spectroscopy *530*
13.4.16.1 Introduction *531*
13.4.16.2 Instrumentation *531*
13.4.16.3 Performance Criteria *531*

13.4.16.4 Applications *531*
13.4.17   Glow Discharge Optical Emission Spectrometry *531*
13.4.17.1 Introduction *531*
13.4.17.2 Instrumentation *532*
13.4.17.3 Sample *532*
13.4.17.4 Analytical Information *532*
13.4.17.5 Performance Criteria *532*
13.4.17.6 Application *533*
13.4.17.7 Other Techniques *533*
13.4.18   High Resolution Electron Energy Loss Spectroscopy *533*
13.4.18.1 Introduction *533*
13.4.18.2 Instrumentation *533*
13.4.18.3 Sample *534*
13.4.18.4 Analytical Information *534*
13.4.18.5 Performance Criteria *534*
13.4.18.6 Applications *535*
13.4.18.7 Other Techniques *535*
13.4.19   Inelastic Electron Tunneling Spectroscopy *535*
13.4.19.1 Introduction *535*
13.4.19.2 Instrumentation *536*
13.4.19.3 Sample *536*
13.4.19.4 Analytical Information *536*
13.4.19.5 Performance Criteria *536*
13.4.19.6 Applications *536*
13.4.20   Inverse Photoelectron Spectroscopy *536*
13.4.20.1 Introduction *536*
13.4.20.2 Instrumentation *537*
13.4.20.3 Sample *537*
13.4.20.4 Analytical Information *537*
13.4.20.5 Performance Criteria *538*
13.4.20.6 Applications *538*
13.4.21   Ion Neutralization Spectroscopy *538*
13.4.21.1 Introduction *538*
13.4.21.2 Instrumentation *538*
13.4.21.3 Sample *539*
13.4.21.4 Analytical Information *539*
13.4.21.5 Performance Criteria *539*
13.4.21.6 Applications *539*
13.4.21.7 Other Techniques *539*
13.4.22   Ion Probe Microanalysis *539*
13.4.22.1 Introduction *540*
13.4.22.2 Instrumentation *540*
13.4.22.3 Sample *540*
13.4.22.4 Analytical Information *540*
13.4.22.5 Performance Criteria *541*

13.4.22.6	Application	*541*
13.4.22.7	Other Techniques	*541*
13.4.23	Low-energy Ion Scattering Spectrometry	*542*
13.4.23.1	Introduction	*542*
13.4.23.2	Instrumentation	*542*
13.4.23.3	Sample	*542*
13.4.23.4	Analytical Information	*542*
13.4.23.5	Performance Criteria	*543*
13.4.23.6	Application	*543*
13.4.23.7	Other Technique	*543*
13.4.24	Near Edge X-ray Absorption Spectroscopy	*544*
13.4.24.1	Introduction	*544*
13.4.24.2	Instrumentation	*544*
13.4.24.3	Sample	*544*
13.4.24.4	Analytical Information	*544*
13.4.24.5	Performance Criteria	*544*
13.4.24.6	Applications	*545*
13.4.24.7	Other Techniques	*545*
13.4.25	Neutron Depth Profiling	*545*
13.4.25.1	Introduction	*545*
13.4.25.2	Instrumentation	*545*
13.4.25.3	Sample	*545*
13.4.25.4	Analytical Information	*545*
13.4.25.5	Performance Criteria	*546*
13.4.25.6	Application	*546*
13.4.26	Particle Induced Gamma Ray Emission	*546*
13.4.26.1	Introduction	*547*
13.4.26.2	Instrumentation	*547*
13.4.26.3	Sample	*547*
13.4.26.4	Analytical Information	*547*
13.4.26.5	Performance Criteria	*547*
13.4.26.6	Applications	*548*
13.4.27	Particle Induced X-ray Emission	*548*
13.4.27.1	Introduction	*548*
13.4.27.2	Instrumentation	*548*
13.4.27.3	Sample	*549*
13.4.27.4	Spectrum	*549*
13.4.27.5	Analytical Information	*549*
13.4.27.6	Performance Criteria	*550*
13.4.27.7	Application	*550*
13.4.27.8	Other Techniques	*550*
13.4.28	Penning Ionisation Electron Spectroscopy	*551*
13.4.28.1	Introduction	*551*
13.4.28.2	Instrumentation	*551*
13.4.28.3	Sample	*551*

13.4.28.4 Analytical Information *551*
13.4.28.5 Performance Criteria *552*
13.4.28.6 Applications *552*
13.4.28.7 Other Techniques *552*
13.4.29 Photoacoustic Spectroscopy *552*
13.4.29.1 Introduction *552*
13.4.29.2 Instrumentation *553*
13.4.29.3 Analytical Information *553*
13.4.29.4 Performance Criteria *553*
13.4.29.5 Application *553*
13.4.30 Photoemission Electron Microscopy *553*
13.4.30.1 Introduction *554*
13.4.30.2 Instrumentation *554*
13.4.30.3 Sample *554*
13.4.30.4 Analytical Information *554*
13.4.30.5 Performance Criteria *554*
13.4.30.6 Applications *555*
13.4.31 Positron Annihilation Auger Electron Spectroscopy *555*
13.4.31.1 Introduction *555*
13.4.31.2 Instrumentation *555*
13.4.31.3 Sample *556*
13.4.31.4 Analytical Information *556*
13.4.31.5 Performance Criteria *556*
13.4.31.6 Applications *557*
13.4.31.7 Other Techniques *557*
13.4.32 Raman Spectroscopy *557*
13.4.32.1 Introduction *557*
13.4.32.2 Instrumentation *557*
13.4.32.3 Sample *557*
13.4.32.4 Analytical Information *558*
13.4.32.5 Performance Criteria *558*
13.4.32.6 Application *558*
13.4.33 Reflection-absorption Spectroscopy *559*
13.4.33.1 Introduction *559*
13.4.33.2 Instrumentation *559*
13.4.33.3 Sample *559*
13.4.33.4 Analytical Information *560*
13.4.33.5 Performance Criteria *560*
13.4.33.6 Limitations *560*
13.4.33.7 Applications *560*
13.4.33.8 Other techniques *561*
13.4.34 Reflection Electron Energy Loss Spectroscopy *561*
13.4.34.1 Introduction *561*
13.4.34.2 Instrumentation *561*
13.4.34.3 Sample *561*

13.4.34.4	Analytical Information	562
13.4.34.5	Performance Criteria	562
13.4.34.6	Applications	562
13.4.34.7	Other Techniques	562
13.4.35	Resonant Nuclear Reaction Analysis	563
13.4.35.1	Introduction	563
13.4.35.2	Instrumentation	563
13.4.35.3	Sample	564
13.4.35.4	Analytical Information	564
13.4.35.5	Performance Criteria	564
13.4.35.6	Application	564
13.4.35.7	Other Techniques	565
13.4.36	Rutherford Backscattering Spectrometry	565
13.4.36.1	Introduction	565
13.4.36.2	Instrumentation	565
13.4.36.3	Sample	565
13.4.36.4	Analytical Information	565
13.4.36.5	Performance Criteria	566
13.4.36.6	Applications	567
13.4.36.7	Other Techniques	567
13.4.37	Scanning Electron Microscopy	567
13.4.37.1	Introduction	568
13.4.37.2	Instrumentation	568
13.4.37.3	Sample	569
13.4.37.4	Analytical Information	569
13.4.37.5	Performance Criteria	569
13.4.37.6	Applications	570
13.4.38	Scanning Tunneling Spectroscopy	570
13.4.38.1	Introduction	570
13.4.38.2	Instrumentation	570
13.4.38.3	Sample	571
13.4.38.4	Analytical Information	571
13.4.38.5	Performance Criteria	571
13.4.38.6	Applications	571
13.4.38.7	Other Techniques	571
13.4.39	Secondary Ion Mass Spectrometry	571
13.4.39.1	Introduction	571
13.4.39.2	Instrumentation	572
13.4.39.3	Sample	572
13.4.39.4	Analytical Information	572
13.4.39.5	Performance Criteria	573
13.4.39.6	Application	573
13.4.39.7	Other Techniques	573
13.4.40	Spectroscopy of Surface Electromagnetic Waves	574
13.4.40.1	Introduction	574

13.4.40.2 Instrumentation *574*
13.4.40.3 Performance Criteria *574*
13.4.40.4 Applications *574*
13.4.41 Spin Polarized Electron Energy Loss Spectroscopy *575*
13.4.41.1 Introduction *575*
13.4.41.2 Instrumentation *575*
13.4.41.3 Sample *575*
13.4.41.4 Analytical Information *575*
13.4.41.5 Performance Criteria *575*
13.4.41.6 Applications *576*
13.4.41.7 Other Techniques *576*
13.4.42 Spin Polarized Ultraviolet Photoelectron Spectroscopy *576*
13.4.42.1 Introduction *576*
13.4.42.2 Instrumentation *576*
13.4.42.3 Sample *577*
13.4.42.4 Analytical Information *577*
13.4.42.5 Performance Criteria *577*
13.4.42.6 Applications *577*
13.4.43 Sum-Frequency Generation Vibrational Spectroscopy *578*
13.4.43.1 Introduction *578*
13.4.43.2 Instrumentation *578*
13.4.43.3 Analytical Information *578*
13.4.43.4 Performance Criteria *578*
13.4.43.5 Applications *579*
13.4.43.6 Other Methods *579*
13.4.44 Surface Plasmon Resonance Spectroscopy *579*
13.4.44.1 Introduction *579*
13.4.44.2 Instrumentation *579*
13.4.44.3 Analytical Information *579*
13.4.44.4 Performance Criteria *580*
13.4.44.5 Applications *580*
13.4.45 Total Reflection X-ray Fluorescence Spectroscopy *580*
13.4.45.1 Introduction *580*
13.4.45.2 Instrumentation *580*
13.4.45.3 Sample *581*
13.4.45.4 Analytical Information *581*
13.4.45.5 Performance Criteria *581*
13.4.45.6 Applications *582*
13.4.46 Transmission Spectroscopy *582*
13.4.46.1 Introduction *582*
13.4.46.2 Instrumentation *582*
13.4.46.3 Performance Criteria *582*
13.4.46.4 Applications *582*
13.4.47 Ultraviolet Photoelectron Spectroscopy *583*
13.4.47.1 Introduction *583*

13.4.47.2	Instrumentation	583
13.4.47.3	Sample	583
13.4.47.4	Analytical Information	583
13.4.47.5	Performance Criteria	583
13.4.47.6	Applications	584
13.4.47.7	Other Techniques	584
13.4.48	X-ray Absorption Fine Structure	584
13.4.48.1	Introduction	584
13.4.48.2	Instrumentation	585
13.4.48.3	Analytical Information	585
13.4.48.4	Performance Criteria	585
13.4.48.5	Applications	585
13.4.48.6	Other Techniques	586
13.4.49	X-ray Photoelectron Diffraction	586
13.4.49.1	Introduction	586
13.4.49.2	Instrumentation	586
13.4.49.3	Sample	587
13.4.49.4	Analytical Information	587
13.4.49.5	Performance Criteria	587
13.4.49.6	Applications	587
13.4.50	X-ray Photoelectron Spectroscopy	587
13.4.50.1	Introduction	588
13.4.50.2	Instrumentation	588
13.4.50.3	Sample	589
13.4.50.4	Analytical Information	589
13.4.50.5	Performance Criteria	590
13.4.50.6	Applications	590
13.4.50.7	Other Techniques	591
13.4.51	X-ray Standing Wave	591
13.4.51.1	Introduction	591
13.4.51.2	Instrumentation	591
13.4.51.3	Sample	592
13.4.51.4	Analytical Information	592
13.4.51.5	Performance Criteria	592
13.4.51.6	Applications	593
13.5	Further Information	593
13.6	Appendix: List of Acronyms Related to Surface Analysis	594

Volume 2

Section VII Applications 1: Bioanalysis 1

14 **Bioanalysis** 3
14.1 General Introduction 3
14.1.1 Spectroscopy in the Biosensor and Genomics Age 3
14.1.2 Genomics, Proteomics and Drug Discovery 4
14.1.3 Biosensor Technologies 5
14.1.4 Biomolecular Structure Determination 6
14.1.5 Bioinformatics 6
14.2 Optical Spectroscopy in Bioanalysis 7
14.2.1 Introduction 7
14.2.2 VIS/NIR Fluorescence Spectroscopy in DNA Sequencing and Immunoassay 10
14.2.2.1 Introduction 10
14.2.2.2 Chemistry of VIS/NIR Dyes 28
14.2.2.3 Bioanalytical Applications of NIR and Visible Fluorescent Dyes 36
14.2.2.4 Fluorescence Polarisation Methods 54
14.2.2.5 Time-resolved Fluorescence 55
14.2.2.6 Fluorescence Excitation Transfer 56
14.2.2.7 Bioanalytical Applications of Fluorescent Proteins 57
14.2.3 Bioanalytical Applications of Multi-photon Fluorescence Excitation (MPE) 58
14.2.3.1 Introduction 58
14.2.3.2 MPE Fluorescence Dyes 59
14.2.3.3 Two-photon Excitation Immunoassays 61
14.2.3.4 MPE in Gel and Capillary Electrophoresis 61
14.2.3.5 MPE in Tissue Imaging 63
14.2.3.6 Future Prospects of MPE Fluorescence Spectroscopy 63
14.2.4 Bioluminescence, Chemiluminescence and Electrochemiluminescence 65
14.2.5 Bioanalytical Applications of NIR Absorption Spectroscopy 68
14.2.6 Bulk Optical Sensing Techniques 69
14.2.7 Evanescent Wave Spectroscopy and Sensors 71
14.2.7.1 Introduction 71
14.2.7.2 Theory of Total Internal Reflection 72
14.2.7.3 Measurement Configurations 81
14.2.7.4 Surface Plasmon Resonance (SPR) 85
14.2.7.5 Reflectometric Interference Spectroscopy (RIfS) 89
14.2.7.6 Total Internal Reflection Fluorescence (TIRF) and Surface Enhanced Fluorescence 91
14.2.8 Infrared and Raman Spectroscopy in Bioanalysis 92

14.2.8.1	FTIR, FTIR Microscopy and ATR-FTIR 92
14.2.8.2	Raman Spectroscopy 92
14.2.8.3	Surface Enhanced Raman Spectroscopy (SERS) 93
14.2.9	Circular Dichroism 93
14.3	NMR Spectroscopy of Proteins 94
14.3.1	Introduction 94
14.3.2	Protein Sample 95
14.3.2.1	Solubility and Stability 95
14.3.2.2	Isotope Labeling 97
14.3.2.3	Dilute Liquid Crystals 99
14.3.3	Proton NMR Experiments 102
14.3.3.1	One-dimensional NMR Experiment 103
14.3.3.2	Correlation Experiments 105
14.3.3.3	Cross-relaxation Experiments 110
14.3.4	Heteronuclear NMR Experiments 113
14.3.4.1	Basic Heteronuclear Correlation Experiments 113
14.3.4.2	Edited and Filtered Experiments 117
14.3.4.3	Triple Resonance Experiments 119
14.4	Bioanalytical Mass Spectroscopy 122
14.4.1	Introduction 122
14.4.2	MALDI-TOF 122
14.4.3	Electrospray Methods (ESI-MS) 123
14.4.4	Tandem-MS 124
14.4.5	TOF-SIMS 125
14.4.6	MS in Protein Analysis 126
14.4.7	MS in Nucleic Acid Analysis 130
14.5	Conclusions 130

Section VIII Applications 2: Environmental Analysis 149

Introduction *151*

15	**LC-MS in Environmental Analysis** *152*
15.1	Introduction *152*
15.1.1	Historical Survey of the Development of LC-MS *152*
15.1.2	First Applications of LC-MS *153*
15.2	Applications of LC-MS Interfaces in Environmental Analyses *155*
15.2.1	Moving Belt Interface (MBI) *156*
15.2.2	Direct Liquid Introduction (DLI) *156*
15.2.3	Particle Beam Interface (PBI) *157*
15.2.4	Fast Atom Bombardment (FAB) and Continuous Flow FAB (CF-FAB) *160*
15.3	LC-MS Interfaces Applied in Environmental Analysis During the Last Decade *163*

15.3.1	Achievements and Obstacles 163
15.3.2	Soft Ionisation Interfaces (TSP, APCI and ESI) 168
15.3.3	The Applications of Soft Ionising Interfaces 172
15.3.3.1	Applications Using Thermospray Ionization Interface (TSP) 172
15.3.3.2	Atmospheric Pressure Ionization Interfaces (API) 183
15.4	Conclusions 226

16 **Gas Chromatography/Ion Trap Mass Spectrometry (GC/ITMS) for Environmental Analysis** 244

16.1	Introduction 244
16.2	Practical Aspects of GC/ITMS 245
16.2.1	Historical survey 245
16.2.2	Principles of Operation 245
16.2.3	Ionization and Scanning Modes 247
16.2.3.1	Electron Ionization 247
16.2.3.2	Chemical ionization 249
16.2.3.3	Full Scan Versus Selected-Ion Monitoring 251
16.2.4	Advances in GC/ITMS 251
16.2.4.1	Methods for Improving Performances: Increasing the Signal-to-Background Ratio 252
16.2.4.2	External Ion Sources 252
16.2.4.3	GC/MS/MS 253
16.3	Examples of Applications of GC/ITMS 254
16.3.1	Requirements for Environmental Analysis 254
16.3.2	Determination of Volatile Organic Compounds in Drinking Water; EPA Methods 256
16.3.3	Detection of Dioxins and Furans 257
16.3.4	Other Examples 258
16.4	Future Prospects in GC/Chemical Ionization-ITMS 260
16.4.1	Chemical Ionization in Environmental Analysis 260
16.4.2	Examples of Unusual Reagents for Chemical Ionization 261
16.4.3	Ion Attachment Mass Spectrometry 262
16.4.3.1	Principle 262
16.4.3.2	Sodium Ion Attachment Reactions with GC/ITMS 263
16.5	Conclusion 265
16.6	Appendix: List of Main Manufacturers and Representative Products for GC/ITMS 266

Section IX   Application 3: Process Control   268

Introduction   269

17	**Optical Spectroscopy** 279	
17.1	Introduction 279	
17.2	Mid-infrared 281	
17.3	Non-dispersive Infrared Analysers 281	
17.4	Near-infrared Spectroscopy 282	
17.5	Ultraviolet/Visible Spectroscopy 286	
17.6	Raman Spectroscopy 287	
17.7	Laser Diode Techniques 291	
17.8	Fluorescence 293	
17.9	Chemiluminescence 293	
17.10	Optical Sensors 294	
17.11	Cavity Ringdown Spectroscopy 294	
18	**NMR** 297	
18.1	Introduction 297	
18.2	Motivations for Using NMR in Process Control 297	
18.3	Broadline NMR 301	
18.4	FT-NMR 307	
18.5	Conclusion 314	
19	**Process Mass Spectrometry** 316	
19.1	Introduction 316	
19.2	Hardware Technology 317	
19.2.1	Sample Collection and Conditioning 319	
19.2.2	Sample Inlet 319	
19.2.2.1	Direct Capillary Inlets 320	
19.2.2.2	Membrane Inlets 320	
19.2.2.3	Gas Chromatography (GC) 320	
19.2.3	Ionization 321	
19.2.4	Mass Analyzers 322	
19.2.4.1	Sector Mass Analyzers 322	
19.2.4.2	Quadrupole Mass Analyzers 323	
19.2.4.3	Choice of Analyzer 324	
19.2.5	Detectors 325	
19.2.6	Vacuum System 325	
19.2.7	Data Analysis and Output 325	
19.2.8	Calibration System 327	
19.2.9	Gas Cylinders 328	
19.2.10	Permeation Devices 328	
19.2.11	Sample Loops 329	
19.2.12	Maintenance Requirements 329	

19.2.13	Modes of Operation 329
19.3	Applications 330
19.3.1	Example Application: Fermentation Off-gas Analysis 331
19.4	Summary 334

20	**Elemental Analysis** 336
20.1	Applications of Atomic Spectrometry in Process Analysis 336
20.1.1	Catalyst Control 337
20.1.2	Corrosion Monitoring 339
20.1.3	Reducing Environmental Impact 341
20.1.4	Troubleshooting Process Problems 342
20.2	On-stream/at-line Analysis 343
20.2.1	X-ray Fluorescence (XRF) 344
20.2.1.1	Liquid Process Streams 348
20.2.1.2	Trace Analysis and Corrosion Monitoring 351
20.2.1.3	Analysis of Slurries and Powders 352
20.2.1.4	Direct Analysis 354
20.2.2	Atomic Emission Spectrometry 356
20.2.2.1	Plasma Spectrometry 356
20.2.2.2	Laser Based Techniques 362
20.3	Conclusions 368

Section X   Hyphenated Techniques   377

Introduction   379

21	**Hyphenated Techniques for Chromatographic Detection** 381
21.1	Introduction 381
21.2	Electronic Spectral Detection 383
21.3	MS Detection 400
21.4	NMR Detection 412
21.5	FTIR Detection 415
21.6	Atomic Spectrometric Detection 421
21.7	Other Types of Detection 428
21.8	Serial or Parallel Multiple Detection 430

Section XI   General Data Treatment: Data Bases/Spectral Libaries   437

Introduction   439

22	**Optical Spectroscopy** 441
22.1	Introduction 441
22.2	Basic Operations 442

22.2.1	Centering	442
22.2.2	Standardization (Autoscaling)	443
22.3	Evaluation of Spectra	444
22.3.1	Introduction	444
22.3.2	Qualitative Evaluation of Spectra	446
22.3.2.1	Spectral Data Banks	446
22.3.2.2	Data Banks Containing Spectroscopic Information	452
22.3.2.3	Interpretation of Spectra by Means of Group Frequencies and of Characteristic Bands	452
22.3.2.4	PCA (Principal Component Analysis)	452
22.3.2.5	Cluster Analysis	455
22.3.2.6	Discriminant analysis	455
22.3.2.7	SIMCA Soft Independent Modeling of Class Analogy (SIMCA)	455
22.3.3	Quantitative Evaluation of Spectra	455
22.3.3.1	Univariate Methods	456
22.3.3.2	Multivariate Methods	459
23	**Nuclear Magnetic Resonance Spectroscopy**	**469**
23.1	Introduction	469
23.2	Comparison of NMR-Spectroscopy with IR and MS	470
23.3	Methods in NMR Spectroscopy	471
23.4	Spectral Similarity Search Techniques	471
23.5	Spectrum Estimation, Techniques	473
23.6	Spectrum Prediction, Quality Consideration	474
23.7	Spectrum Prediction and Quality Control, Examples	475
23.8	Spectrum Interpretation and Isomer Generation	481
23.9	Ranking of Candidate Structures	484
23.10	Conclusions	484
24	**Mass spectrometry**	**488**
24.1	Introduction	488
24.2	Mass Spectrometry Databases	489
24.2.1	NIST/EPA/NIH Mass Spectral Library	490
24.2.2	Wiley Registry of Mass Spectral Data	491
24.2.3	SpecInfo/SpecData	491
24.2.4	SDBS, Integrated Spectra Data Base System for Organic Compounds	492
24.2.5	Other Smaller Collections	492
24.2.5.1	Pfleger/Maurer/Weber: Mass Spectral and GC Data of Drugs, Poisons, Pesticides, Pollutants and Their Metabolites	494
24.2.5.2	Ehrenstorfer	494
24.2.5.3	Wiley-SIMS	494
24.2.5.4	American Academy of Forensic Sciences, Toxicology Section, Mass Spectrometry Database Committee	494
24.2.5.5	The International Association of Forensic Toxicologists (TIAFT)	494

24.3	Mass Spectrometry Search Software 495
24.3.1	INCOS 496
24.3.2	Probability Based Matching (PBM) 496
24.3.3	MassLib/SISCOM 497
24.3.4	AMDIS 498
24.3.5	Mass Frontier 499
24.3.6	The WebBook 500
24.3.7	General Spectroscopy Packages 501
24.4	Biological Mass Spectrometry and General Works 502

Index 505

Section VII
Applications 1: Bioanalysis

14
Bioanalysis

Willem M. Albers, Arto Annila, Nicholas J. Goddard, Gabor Patonay, and Erkki Soini

14.1
General Introduction

14.1.1
Spectroscopy in the Biosensor and Genomics Age

Bioanalysis is presently in a new breakthrough stage, in which much recently developed technology and knowledge is revolutionising the way bioanalysis is practised. Microarray platforms have become commercially available during the last ten years and microfluidics and micro-electrophoresis systems will also be making their entry on the market very soon. Due to advances in microsystems in the 1990s (MST and MEMS), the realisation of complete systems for chemical analysis and synthesis within chips or within compact analysers is gradually becoming a reality. Such systems may be directly linked to or integrated into desktop or laptop computers (e.g. as a disk drive). It can thus be expected that within a few years the appearance of a bioanalytical laboratory will be very different. Large, costly facilities and hand labour will be largely replaced by miniaturised, autonomous, high-throughput analysis systems. It has been particularly the *Human Genome Project* that has accelerated the advances in this miniaturisation process, but the methods are now also crossing over to other bioanalysis areas.

The 1990s also witnessed the rapid commercialisation of biosensor technology. Although a large part of the biosensor literature lies in the field of electroanalysis, much work is concerned with new optical detection techniques, suitable for coverage in a handbook dedicated to spectroscopy. Advances in quite a number of science disciplines, most notably materials science, have contributed to the development of biosensors, but the major force was the successful marriage between electronics and biotechnology. Presently, there are many new programmes that modify and integrate biosensors, producing devices that have better sensitivity, specificity, stability and decreased manufacturing costs and are cast in array format, enabling fast "multidetection". Biosensors can play an important role in applications where rapid detection and continuous use are important. Presently, the advances in biosensors are quickly crossing over to (or merging with) other bioanalysis areas. It can be expected that during the first five years of the new millennium genomics and proteomics techniques will be effectively combined with biosensors and this will particularly revolutionise the practice of clinical diagnostics and drug discovery.

In many cases, however, novel technology is still rooted in conventional chemical analysis concepts, in which spectroscopy still holds a firm lead. In many cases the stages of protein analysis still comprise sample handling, separation, detection and (to an ever increasing degree) data processing and interpretation. For detection, spectroscopic techniques hold the advantage of rapidity, high information content and high sensitivity for quite a number of tasks to be performed in the multifarious bioanalysis field. The scope of the present chapter, therefore, lies predominantly in describing the basic methods and chemistry that are likely to preserve their significance also in the future analytical device technologies.

14.1.2
Genomics, Proteomics and Drug Discovery

The field of *genomics* aims at characterising all the genes of an organism, starting from the DNA sequence to its structure and polymorphism. The most ambitious enterprise within genomics has been the Human Genome Project, which is dedicated to the sequencing of the 3 billion base pairs of the whole human genome. This endeavour required new, advanced tools for DNA replication, sequencing and analysis. Although the techniques for DNA analysis are still being further developed, the Human Genome Project actually ran 4 years ahead of schedule, when the first working draft of the human genome was completed in June 2000. Also the genomes of some other relevant organisms have been completely mapped. As a logical consequence of the results in genomics, the year 2000 witnessed the birth of the *proteomics* field [1]. The term 'proteomics' was coined in 1995 and refers to experimental tools for locating, isolating and characterising all proteins that are produced by a genome [2]. Both genomics and proteomics rely heavily on gel electrophoresis (GE), capillary electrophoresis (CE) and, more recently, capillary array electrophoresis (CAE) [3, 4].

The term 'functional genomics' is presently used for the elucidation of the function of genes (regulation, interactions and products) [5, 6]. Functional genomics relies on much the same methods as employed in proteomics. From a spectroscopic point of view the most interesting methods for detection of DNA involve fluorescence detection [7], particularly laser-induced fluorescence (LIF) in the visible and near-infrared (VIS/NIR) region (400–1100 nm). Apart from fluorescence detection for DNA, other spectroscopic methods have gained importance in genomics, particularly mass spectroscopy [8].

Presently, attention has also been moving to 'functional proteomics' in which specific functions of proteins are mapped. Functional proteomics is evolving fast and various novel tools have been developed for the study of protein–protein interactions, quantification and comparisons of protein expression, and advanced protein function mapping [9]. Also annotated proteomic databases have been set up. All these advances will have a large impact on diagnostic and therapeutic product development and the identification of important biomarkers and novel drug targets. Functional proteomics and drug discovery are closely related fields. Drug discovery is mainly concerned with the screening of small organic compounds against a range of receptors relevant in disease control, but also involves the screening of peptides, either produced by organisms or by combinatorial chemistry methods. Completely new approaches for screening very large numbers of chemicals have been devised in the 1980s and 1990s [10].

14.1.3
Biosensor Technologies

Biosensors are chemical sensors that utilise a biomolecule for the determination of an analyte by intimate coupling of the biomolecule to a suitable detection device (optical, electrochemical, piezoelectric, calorimetric) [11, 12]. Frequently, additional filters are used to provide some initial selection of analytes, to shield the sensing device from fouling and to protect the sensor as a whole from mechanical damage. An ideal biosensor should provide a real-time readout of the concentration of an analyte in its natural environment and follow the changes in concentrations reversibly. This ideal has been realised to date only for some types of catalytic sensors (based on enzymatic action). For many other types of devices, however, this is not yet realised, such as for biosensors that rely on affinity interactions, for which the term "single-use biosensor" or "dosimeter" is more appropriate. The minimum requirement for a device to be called a biosensor is that the device should provide a quick readout of the concentration without the need for adding reagents [13].

Biosensors comprise a rapidly growing technology field that continuously reshapes bioanalysis, particularly diagnostics and functional proteomics, but also various other chemical monitoring fields (environmental monitoring, drug discovery, process industry, food analysis). The first biosensor system was already commercialised in the early 1970s by Yellow Springs Instrument Co. (YSI) and was used for the determination of glucose in blood, urine and bioprocesses. In the 1980s relatively little commercial activity was yet observed, but new sensor types were developed at a rapid pace in many research laboratories. There were many successful demonstrations of biosensors based on electrochemical, optical, microcalorimetric and piezoelectric transducers. In the 1990s a number of new sensor technologies moved successfully from the research laboratory into the marketplace.

Advances in optical biosensors have contributed significantly to the speed of bioanalysis by supplying real-time monitoring of binding reactions, without the need for labelling of biomolecules (real-time BioInteraction Analysis or BIA). Some of the methods in biosensor technology are based on optical phenomena and can therefore be placed in a spectroscopic context (see Section 14.2.7). Optical biosensors have traditionally been used in a number of proteomics subfields, such as antibody screening and epitope mapping, in which the dissociation and association rates and affinity constants can be directly determined.

In vivo monitoring, particularly of glucose for the care of diabetics, still receives much attention in R&D worldwide. This is not surprising, since home glucose monitoring forms the largest single-analyte market. This market is still expanding: a growth from $2.6 billion in 1997 to an estimated $5.9 billion in 2002 has been projected. Although various electrochemical sensors for *in vivo* monitoring have been developed, non-invasive monitoring by direct spectroscopic methods is still receiving attention. More than 20 firms and institutes have notified that they are working on such systems, and a few have released equipment for direct measurement of blood glucose, although these instruments are still bulky and expensive. The reliability of the glucose measurements is, however, still under critical evaluation.

14.1.4
Biomolecular Structure Determination

The elucidation of three-dimensional biomolecular structure is still heavily dominated by X-ray crystallography, but for the analysis of biomolecular structure in aqueous solution NMR spectroscopy has become the method of choice, particularly when dynamic molecular information is also needed. Therefore, a rather large part of this chapter will be dedicated to biomolecular NMR, focussing on the analysis of proteins (see Section 14.3). Apart from the determination of primary and tertiary structure, the determination of the secondary structure of proteins is still important in bioanalysis. Secondary structure features (mainly α-helix and β-sheet content) are still dominated by IR spectroscopy and circular dichroism techniques (see Section 14.2.8). Secondary structural features of proteins and conformational stability will probably grow in significance through the advent of diseases like transmissible spongiform encephalopathy (TSE), which are caused by conformational changes in Prion proteins. Furthermore, due to the increase in the production and use of chiral compounds in the pharmaceutical sector, the demand for analysis by circular dichroism has also seen a steady growth.

14.1.5
Bioinformatics

Scientists become more and more dependent on extensive data banks to access structural information on proteins and nucleic acids. With the constant growth of these public databases, there has emerged a large need for robust analytical data handling software that is able to partly elucidate the significance of the data. For instance, geneticists use special software to analyse the hybridisation patterns obtained from DNA chips. Bioinformatics is the branch of science that enables scientists, researchers and physicians to manage large amounts of data via powerful computational tools, with which the large amounts of data can be classified and organised [14, 15]. The application areas of bioinformatics lie in drug design, gene research, and advanced medical procedures. Computational approaches have been applied to biology and medicine, databases and search tools, to genome and proteome analysis, and mapping of the human brain.

An important factor in the progress of bioinformatics has been the constant increase in computer speed and memory capacity of desktop computers and the increasing sophistication of data processing techniques. The computation power of common personal computers has increased within 12 years approximately 100-fold in processor speed, 250-fold in RAM memory space and 500-fold or more in hard disk space, while the price has nearly halved. This enables acquisition, transformation, visualisation and interpretation of large amounts of data at a fraction of the cost compared to 12 years ago. Presently, bioanalytical databases are also growing quickly in size and many databases are directly accessible via the Internet. One of the first chemical databases to be placed on the Internet was the Brookhaven protein data bank, which contains very valuable three-dimensional structural data of proteins. The primary resource for proteomics is the ExPASy (**Ex**pert **Pro**tein **A**nalysis **Sy**stem) database, which is dedicated to the analysis of protein sequences and structures and contains a rapidly growing index of 2D-gel electrophoresis maps. Some primary biomolecular database resources compiled from spectroscopic data are given in Tab. 14.1.

14.2
Optical Spectroscopy in Bioanalysis

14.2.1
Introduction

Applications of optical spectroscopy have advanced rapidly in bioanalysis and can be considered as a primary tool for detection in DNA analysis and immunoassays as well as in medicinal and pharmaceutical analysis where rapid data acquisition, sensitivity and reproducibility are crucial. It is especially fluorescence spectroscopy that has flourished in bioanalysis, where a plethora of interesting approaches are competing in the scientific and commercial sector. Besides synthetic dyes, the intrinsic fluorescence of enzymes [16, 17] and fluorescent proteins (see Section 14.2.2.7) is increasingly used, because some of these biomolecules have highly optimised photochemical properties. The following paragraphs intend to give an overview of the most important trends and emerging techniques in optical bioanalysis methods, the emphasis being on the underlying synthetic and bioconjugation issues.

Table 14.1 Overview of some major biomolecular databases on the internet.

Database name	Site	Provider	Contents	Charge
Map of the human genome	http://www.ncbi.nlm.nih.gov/genemap99/	National Center for Biotechnology Information (NCBI), a division of the National Library of Medicine (NLM) at the National Institutes of Health (NIH).	One of the starting links for retrieving human genome sequence data and data of other genomes.	Free access via the internet
Brookhaven Protein Data Bank	http://www.rcsb.org/pdb/	Research Collaboratory for Structural Bioinformatics (RCSB), non-profit consortium, CA, USA	World-wide repository for the processing and distribution of 3-D biological macromolecular structure data, X-ray crystallographic and NMR data.	Freely downloadable structure files. Display of structures via free programs of other vendors (e.g. Chime, Rasmol, WebLab Viewer)
Cambridge Crystallographic Database	http://www.ccdc.cam.ac.uk/index.html	Cambridge Crystallographic Data Centre, Cambridge, UK	Computerised database containing comprehensive data for organic and metal-organic compounds studied by X-ray and neutron diffraction	Paid subscription on database and programs for search and display
ExPASy	http://www.expasy.ch/	Swiss Institute of Bioinformatics (SIB), Geneva, Switzerland	Major proteomics database comprising interpreted 2DE gels of various organisms (Liver, plasma, HepG2, HepG2SP, RBC, lymphoma, CSF, macrophage-CL, erythroleukemia-CL, platelet, yeast, E.coli, colorectal, kidney, muscle, macrophage-like-CL, Pancreatic Islets, Epididymus, dictyostelium).	Freely downloadable database and program for searching and displaying (Melanie Viewer)
Argonne Protein Mapping Group	http://www.anl.gov/BIO/PMG/	Argonne National Laboratory, Center for Mechanistic Biology and Biotechnology, University of Chicago	2DE proteome database of Mouse liver, human breast cell lines, and Pyrococcus furiosus	Free access via the internet

Danish Centre for Human Genome Research	http://biobase.dk/cgi-bin/celis/	Danish Centre for Human Genome Research, University of Aarhus	2DE proteome database of human (primary keratinocytes, epithelial, hematopoietic, mesenchymal, hematopoietic, tumors, urothelium, amnion fluid, serum, urine, proteasomes, ribosomes, phosphorylations) and mpuse (epithelial and new born) proteomes	Free access via the internet
HOMSTRAD (HOMologous STRucture Alignment Database)	http://www.cryst.bioc.cam.ac.uk/~homstrad/	Crystallography and Bioinformatics Department of Biochemistry University of Cambridge	HOMSTRAD is a derived database of structure-based alignments for homologous protein families. Known protein structure (from the PDB database) are clustered into homologous families (i.e., common ancestry), and the sequences of representative members of each family can be aligned on the basis of their 3D structures using various programs.	Freely downloadable structure files and display programs
LIPIDAT	http://www.lipidat.chemistry.ohio-state.edu/	Ohio State University, Ohio, USA	relational database of thermodynamic and associated information on lipid mesophase and crystal polymorphic transitions. There are 19,957 records in the database. The database includes lipid molecular structures.	Free access via the internet
CarbBank, CCSD and CCRC	http://bssv01.lancs.ac.uk/gig/pages/gag/carbbank.htm http://www.ccrc.uga.edu/	Georgia State University, Athens, Georgia, USA	The CCSD is a database containing complex carbohydrate structures and associated text. The information is derived from scientific publications and submissions by authors. CarbBank is the computer program to allow access to the information in the CCSD database files. CCRC is the web version of the database.	Free access after registration

14.2.2
VIS/NIR Fluorescence Spectroscopy in DNA Sequencing and Immunoassay

14.2.2.1 Introduction

Fluorescence and absorption techniques are of prime importance in bioscience today. Most of the recently developed separation techniques, including gel electrophoresis (GE), capillary electrophoresis (CE) and high-performance liquid chromatography (HPLC) as well as gene probes and immunoassays, rely on detection in the visible and near-infrared (VIS/NIR) region through the use of fluorescent or absorbing probes [7, 18–21]. Fluorescent probes are highly versatile and can be used in very many types of assays. An overview of the use of fluorescent probes in bioanalytical applications is presented in Tab. 14.2.

The large majority of fluorescent probes absorb in the near-UV region (300–400 nm) and emit in the visible light region (400–700 nm), but the utilisation of the far-red (700–900 nm) and particularly the near-infrared region (900–2000 nm) is more advantageous and profitable. Optical light sources and detectors in this spectral region become continuously cheaper through their large-scale use in the telecommunications field. Recently developed photodiodes and diode lasers have been applied in quite novel techniques for biomolecule detection. The near-IR region is also promising due to the lower interference from the environment and the low scattering in turbid media, which makes biomedical sensing applications, in particular, very attractive. For fiber optic sensing the near-infrared is also favourable, because of the lower losses in quartz fibers around 1400 nm. Another advantage provided by VIS/NIR spectroscopic detection is that relatively low energy transitions are used which do not readily affect the structural integrity of the biological analytes or media. Thus, recent advances in spectroscopic methods utilising VIS/NIR fluorophores and chromophores in biomolecular analysis of nucleic acids and immunological assays will be the larger focus of this section.

The development of new absorbing or fluorescent probe molecules for the near-IR region is a relatively new field. The main problem with near-IR chromophores lies in the broadness of the absorption peaks, the small Stokes' shifts and the low quantum yields observed with the more conventional dyes. Recently, novel chromophores have been designed and synthesised with much more favourable properties and many types of probes are commercially available. These dyes generally comprise xanthenes, rhodamines, polymethines and phthalocyanines (Pc). There are also important applications of rare earth elements, such as europium, osmium and ruthenium in time-resolved fluorescence. With visible probes that absorb and fluoresce in the 400–700 nm region there is the danger of high background absorption or fluorescence of the bio-matrix. Thus, background absorbance and fluorescence needs to be eliminated, which may be problematic and time consuming. Probes in the 700–1100 nm region provide an advantage over visible fluorophores and chromophores due to a significantly reduced interference from the bio-matrix and thus increase the detection sensitivity. Also scattering in the NIR region is greatly reduced in comparison to the visible region due to the

Table 14.2 Fluorescent probes and their applications [compiled from Ref. 21 with permission].

Application area	Subfield or technology	Typical dyes or assay system used	Description	References
Ion indicator dyes	pH	Near-neutral pH: fluorescein diacetate, carboxyfluorescein and Its esters, 5-sulfofluorescein diacetate, BCECF Acidic pH: di- and trifluorofluoresceins (Oregon Green) and dichlorofluorescein, 9-amino-6-chloro-2-methoxyacridine, 8-hydroxypyrene-1,3,6-trisulfonic acid (HPTS) LysoSensor probes	Physiological detection of pH is presently performed with various new fluorescent dyes. BCECF and its membrane-permeant ester have become the most widely used fluorescent indicators for estimating intracellular pH. Studies related to: role of intracellular pH in diverse physiological and pathological processes, cell proliferation, apoptosis, muscle contraction, malignancy, multidrug resistance, ion transport and homeostasis, endocytosis and Alzheimer's disease.	22
	Na^+ and K^+	Crown ethers conjugated to the benzofuranyl fluorophore (SBFI and PBFI), sodium green, CoroNa Red,	Detection of physiological concentrations of Na^+ and K^+ in the presence of other monovalent cations. Applications are: • Estimation of Na^+ gradients in isolated mitochondria, • Measurement of intracellular Na^+ levels or Na^+ efflux in cells from a variety of tissues: Blood (platelets, monocytes and lymphocytes), Brain (astrocytes, neurons, and presynaptic terminals), Muscle (perfused heart, cardiomyocytes and smooth muscle), secretory epithelium cells.	23

Table 14.2 (continuing)

Application area	Subfield or technology	Typical dyes or assay system used	Description	References
			Correlation of changes in intracellular Na$^+$ with Ca^{2+} concentrations, intracellular pH and membrane potential has been used in combination with other fluorescent indicators.	23
	Ca^{2+}, Mg^{2+}, Al^{3+}	Calcein, fura-2, BTC, various der. of Calcium Green, fluo-3 AM and fluo-4 AM and some of their dextran conjugates	Measurement of intracellular and extracellular, Ca^{2+} concentrations (fluorescence microscopy, flow cytometry and fluorescence spectroscopy). Conjugates with dextrans are used to confine the indicator to the cytosol. Major applications to the study of calcium regulation and transport.	24
		EGTA, APTRA and BAPTA		
		Aequorin (bioluminescent indicator for Ca^{2+})		
		Calcein, Mag-fura-2, Magnesium Green and others	Mg^{2+} detection is important for studies on enzymatic reactions, DNA synthesis, hormonal secretion and muscular contraction. E.g. mag-fura-2 was first used to detect Mg^{2+} fluctuations in embryonic chicken heart cells.	

Zinc and other metal ions	Newport Green DCF and PDX, FuraZin, FluoZin and TSQ indicators (Zn^{2+}). Calcein, Phen Green FL and SK (Fe^{2+}, Cu^{2+}, Cu^+, Hg^{2+}, Pb^{2+}, Cd^{2+} and Ni^{2+}) 2,3-Diaminonaphthalene (selenium)	Zinc is an important divalent cation in biological systems, influencing DNA synthesis, microtubule polymerization, gene expression, apoptosis, immune system function and the activity of enzymes such as carbonic anhydrase and matrix metalloproteinases (MMP). Zn^{2+} is functionally active in synaptic transmission and is a contributory factor in neurological disorders including epilepsy and Alzheimer's disease.	25
F^-, Cl^-, Br^-, I^-	Calcein, 6-methoxyquinolinium derivatives (SPQ), MQAE, MEQ and cell-permeant DiH-MEQ, lucigenin,	Measurement of intracellular Cl^- has renewed relevance to research on cystic fibrosis, and can be performed with a variety of probes. Conventional applications of Cl^- detection are: • Membrane chloride transport (e.g. sodium-dependent transport, Renal brush-border, $GABA_A$ receptor), • Intracellular chloride activity • Quenching of Al^{3+}-calcein complex fluorescence has been used as the basis of method for fluoride determination with a detection limit of 0.2 ng mL^{-1}. • Detection of Cl^-, Br^- and I^- are also much performed via chemiluminescence.	26

Table 14.2 (continuing)

Application area	Subfield or technology	Typical dyes or assay system used	Description	References
	CN^-	o-Phthaldialdehyde and naphthalene-2,3-dicarboxaldehyde	Determination of cyanide in blood, urine and other samples via reaction of fluorophore in presence of primary amine.	27
	Sulfide	5,5'-Dithiobis-(2-nitrobenzoic acid) (DTNB or Ellman's reagent), monobromobimane,	Sulfide has been determined for histochemical studies, probing of dynamic changes of red cell membrane thiol groups, subcellular location of cathepsin D, and studies related to glucose-6-phosphate dehydrogenase, glutathione, glutathione transferase, 6-mercaptopurine e.o.	28
	Phosphate and pyrophosphate	DCIP, NBT, Amplex Red/resorufin	Detection of free phosphate in solution through the formation of the fluorescent products (e.g. resorufin via maltose phosphorylase, glucose oxidase and HRP resorufin)	29
	Nitric oxides	4-Amino-5-methylamino-2',7'-difluorofluorescein (DAF), 2,3-Diaminonaphthalene and N-methyl-4-hydrazino-7-nitrobenzofurazan	Role of nitric oxide in signal transduction has been recently realised, yielding new fluorimetric assays for NO^{2-} and NO.	30

Gasses	Oxygen, peroxide, etc.	fluorophores that generate or detect reactive oxygen species, such as including singlet oxygen, superoxide, hydroxyl radical, peroxides. E.g.: merocyanine 540, Rose Bengal diacetate; xanthene dyes e.o. detection of peroxide via HRP.	Various applications in bioscience: oxydation/peroxydation of lipids, fatty acids and cholesterols, NADH, NADPH, dopamines, ascorbic acid, histidine, tryptophan, tyrosine, cysteine, glutathione, proteins and nucleic acids. Relevance to research of Alzheimers disease.	31
General biomolecule detection	Functional group labelng and detection with single dye molecules.	Nearly all known fluorescent dye conjugates of drugs, haptens, proteins, immunoglobulins, and nucleic acid: coumarin's, fluorescein's cyanines, rhodamines. Specialised dyes: BODIPY & Alexa Fluor, various extended cyanine dyes	Main application in diagnostics, laboratory assays, immunoassays, histochemistry, immunohistochemistry, flow cytometry and cellular diagnostics.	7, 18–21
	Labeling and detection with fluorescent proteins or other fluorescent biomolecules	Phycobiliproteins, green fluorescent proteins	Diagnostic assay use, histochemical staining, imaging of live organisms and their functions.	32
	Functional group labeling and detection with fluorescent latex particles	Fluorescent dye loaded micro- and nanoparticles.	Diagnostic assay use (particularly in immunochromatography tests) and flow cytometry.	33
	Enzyme-labeled fluorescence (ELF)	Various phosphate-labeled fluorescent dyes, ELF 97	Uses phosphatase-based signal amplification. Applicable to: immunohistochemical and cytological staining, mRNA *in situ* hybridization, detection of endogenous phosphatase activity, and blot analyses.	34

Table 14.2 (continuing)

Application area	Subfield or technology	Typical dyes or assay system used	Description	References
	Tyramide signal amplification (TSA) Technology	Various tyramide-labeled dyes	Utilizes the catalytic activity of horseradish peroxidase (HRP) to generate high-density labeling of a target (also *in situ*). Involves coupling of tyramide-fluorescent dye conjugate to protein tyrosine sidechains via peroxidase-mediated formation of an O,O'-dityrosine adduct. Applied to: immunohistochemical staining and *in situ* hybridization.	35
Protein detection	Protein quantitation in solution in gels, in CE and 2DE, and on blots	SYTO, SYBR, fluorescamine and o-phthaldialdehyde, novel dyes: BODIPY, NanoOrange, CBQCA and SYPRO protein gel stains	In addition to the conventional dyes for protein staining (Coomassie Blue, colloidal gold), many novel fluorescent stains have been developed that allow highly sensitive protein quantitation in solution and in gels, particularly the SYPRO protein gel stains.	36
	Peptide analysis, sequencing and synthesis	All fluorescent dye labels with reactive groups for amine, dansyl chloride, dabsyl chloride. Fluorescent isothiocyanates.	N-Terminal amino acid analysis, peptide sequencing, peptide synthesis, labeling peptides in solution, solid-phase synthesis of labeled peptides	37

	Detection of cytoskeletal proteins (actin, tubulin).	Fluoresceine, Alexa Fluor, BODIPY and other conjugates of actin, tubulin and phallotoxins	Study of *in vivo* cytoskeleton dynamics. Labels of phalloidin and phallacidin are used for selectively labeling F-actin. Tubulin conjugates are used for observation of cell cycle-dependent microtubule dynamics, mitotic spindle morphogenesis and visualisation of tubulin transport in neurons.	38
Nucleic acid analysis	Nucleic acid labelling, detection and quantitation with single fluorescent dye molecules	Cyanine dyes and phenanthridine dyes, ethidium bromide and propidium iodide DAPI, "Hoechst dyes", acridine orange, 7-AAD and hydroxystilbamidine. New dyes have been developed for nucleic acid detection in solution and staining in gels Cell membrane-impermeant dyes, incl. stains for dead cells (SYTOX Dyes), high affinity stains (the cyanine dimers TOTO, YOYO etc.), and counterstains (cyanine monomers, YO-PRO, TO-PRO e.o.), SYTO cyanine cell-permeant nucleic acid dye Dyes for ultrasensitive solution quantitation (PicoGreen for dsDNA, OliGreen for ssDNA and RiboGreen for RNA),		

Table 14.2 (continuing)

Application area	Subfield or technology	Typical dyes or assay system used	Description	References
		SYBR dyes for sensitive detection in gels and blots,	The three classes of "classic" nucleic acid stains are:	39
		Chemically reactive SYBR dyes for bioconjugates.	Intercalating dyes (ethidium bromide and propidium iodide),	
			Minor-groove binders, (DAPI and the "Hoechst dyes"),	
			Miscellaneous nucleic acid stains with special properties (acridine orange, 7-AAD and hydroxystilbamidine).	
	Hybridization detection incl. FISH (fluorescence in situ hybridization)	ChromaTide nucleotides	Fluorophore- and hapten-labeled nucleotides for enzymatic incorporation into DNA or RNA probes for FISH (fluorescence in situ hybridization), for DNA arrays and microarrays and for other hybridization techniques	40
		SYBR Green	Real-time quantitative nucleic acid gel stain used in PCR	41
		Universal Linkage System (ULS)	Platinum-based chemistry for producing bright, fluorophore-labeled hybridization probes	42
		Enzyme-Labeled Fluorescence (ELF)	Phosphatase-based signal amplification assay also applicable to nucleic acid labeling (see description under proteins)	43

	Tyramide Signal Amplification (TSA) Technology	HRP-based assay with tyramide-fluorescent dye conjugate also applicable to nucleic acid labeling (see description under proteins)	44
Membrane and lipid research	General lipid and cell membrane labeling	Lipid and fatty acid derivatives of BODIPY, nitrobenzoxadiazole (NBD), pyrene, perylene, 9-anthroyloxy and dansyl fluorophores, cis-parinaric acid.	Fluorescence labeled lipids and fatty acids are generally used to: investigate lipid traffic, e.g. fluorescence recovery after photobleaching (FRAP) and other techniques examine lipid-lipid and lipid-protein interactions via FRET measurements characterize lipid domains by fluorescence correlation spectroscopy (FCS) characterize lipid domains by near-field scanning optical microscopy. detect phospholipase activity investigate the cellular uptake of lipids investigate lipid metabolism and signalling study membrane fusion (pyrene eximer formation) and structural dynamics Detect lipid peroxidation	45
	Membrane potential-sensitive probes	Various styryl dyes (ANEP, RH) and Indo-, thia- and oxa-carbocyanines, merocyanines, bisoxonols and rhodamines.	Dynamic optical detection and imaging of membrane potential changes.	46

Table 14.2 (continuing)

Application area	Subfield or technology	Typical dyes or assay system used	Description	References
Detection of enzymatic action	Glycosidases, glucuronidases etc.	A large variety of β-galactosidase, β-glucuronidase, amylase, neuraminidase, sialidase, chitinase, cellulase substrates	Endogenous glycosidase activity is frequently used to characterize strains of microorganisms and to selectively label organelles of mammalian cells. Defects in glycosidase activity are characteristic of several diseases. β-D-glucuronide activity is used primarily for contamination detection of E-coli. β-amylase levels in various fluids of the human body are of clinical importance. Plant and microbial β-amylases are important industrial enzymes. Cellulase enzymes are relevant in food, fuel, animal feed and clothing applications	47
	Proteases and peptidases	7-Aminocoumarins, rhodamine 110, fluorescein casein	Peptidases and proteases play essential roles in protein activation, cell regulation and signaling, in the generation of amino acids for protein synthesis or utilization in other metabolic pathways.	48
	Metabolism of phosphates and Polyphosphates, ATP e.o. • phosphatases	Phosphatase substrates (classic chromogenic sustrates: BCIP and NBT. Novel fluorescent substrates: Fluorescein diphosphate, Dimethyl-acridinone phosphate, methylumbelliferyl Phosphates, ELF 97 e.o.); bioluminescent determination of ATP;	detection of phospholipases phosphodiester-ases, alkaline phosphates, ATPases, GTPases, DNA and RNA polymerases	49

	Oxidoreductases	Fluorogenic reagents for H_2O_2 in conjunction with peroxidases (HRP) and catalase. (Amplex Red/resorufin etc.). Electrochemical and chemiluminescent approaches are also much used Nitroreductase/Nitrate reductase	Systems are used in a wide variety of bioassays by linking various oxidases to HRP via hydrogen peroxide. Determination of glucose, cholesterol, choline glutamate, xantine, uric acid, galactose etc.	50
	Miscellaneous enzymes	Various probes for microsomal dealkylase, lipase, acetylcholinesterase, acetyltransferase and carbonic anhydrase		51
Carbohydrate analysis	Lectins	Fluorescent Concanavalin A, wheat germ agglutinin (WGA) fluorescein conjugates, lectins from *Griffonia simplicifolia*, *Phaseolus vulgaris*, *Arachis hypogaea*, *Helix pomatia*, *Glycine max*, Cholera Toxin Subunits A and B	Lectins and other carbohydrate-binding proteins bind to specific configurations of sugar molecules can serve to identify cell types or cellular components, making them versatile primary detection reagents in histochemical applications and flow cytometry. Fluorescent derivatives of carbohydrate-binding proteins have been used to detect cell-surface and intracellular glycoconjugates by microscopy and flow cytometry, to localize glycoproteins in gels and on protein blots, to precipitate glycoproteins in solution and to cause agglutination of specific cell types.	52

Table 14.2 (continuing)

Application area	Subfield or technology	Typical dyes or assay system used	Description	References
Cellular biochemistry	Probes for mitochondria	Rhodamines and rosamines, Carbocyanines, stryryl dyes, lipophilic acridine orange, lucigenin, MitoTracker and MitoFluor Probes	Mitochondrion-selective reagents for assessment of mitochondrial activity, localization and abundance, monitoring effects of pharmacological agents, such as anesthetics that alter mitochondrial function. Important role in apoptosis	53
	Probes for the endoplasmic reticulum and Golgi apparatus	The flattened membranous sacs of the endoplasmic reticulum (ER) and the Golgi apparatus can be stained with a variety of lipophilic probes and then distinguished on the basis of their morphology. Probes for the Golgi apparatus are fluorescently labeled ceramides and sphingolipids, which tend to associate preferentially with the trans-Golgi.	The endoplasmic reticulum (ER) and Golgi apparatus are responsible for the proper sorting of lipids and proteins in cells. Cell-permeant probes for these organelles are lipids or chemicals that affect protein movement. Enzymes in the ER are involved in synthesis of cholesterol and membranes and in the detoxification of hydrophobic drugs through the cytochrome P-450 system. Several enzymes in the Golgi glycosylate lipids and proteins, resulting in some fluorescent lectins being useful markers for this organelle.	54

Calcium regulation	Cyclic adenosine 5'-diphosphate ribose (cADP-ribose), 3-deaza-cADP-ribose, Ryanodine derivatives, BODIPY FL thapsigargin etc. fluorescently labeled calmodulin	Studies of calcium release. Calmodulin mediates many of the regulatory functions of calcium ions. Fluorescently labeled calmodulin is used to study the in vivo behavior of the protein (e.g. in the mammalian mitotic spindle) In vitro, fluorescently labeled calmodulin has been used for: • Following the binding of protein kinase substrates by calmodulin • Studying the interactions of myogenic basic helix-loop-helix transcription factors with calmodulin • Investigating the molecular mechanisms for calmodulin trapping by calcium/calmodulin-dependent protein kinase I, • Characterizing inhibitors of calmodulin activation of MLCK-catalyzed phosphorylation of the smooth-muscle regulatory chain

Table 14.2 (continuing)

Application area	Subfield or technology	Typical dyes or assay system used	Description	References
	Blood coagulation	Fluorescent heparin conjugates	Study of heparin binding to thrombin, low-density lipoproteins, lipoprotein lipase, circulatory serine proteases, proteinase inhibitors, heparin-binding growth factors, blood vessel–associated proteins (fibronectin and laminin) and binding to cells and tissues. Study of anticoagulant activity and the modulation of the structure, function and metabolism of many proteins and enzymes.	56
	Protein kinases, protein phosphatases and nucleotide-binding proteins	Bisindolylmaleimides, fluorescent polymyxin B analogs, hypericin and hypocrellins, blue-fluorescent N-methylanthraniloyl (MANT) analog of cGMP, fluorescent forskolin	Studies on protein kinase inhibitors and activators, cyclic nucleotides, adenylate cyclase, etc.	57

Receptor binding	Various acetylcholine receptors (probes to α-bungarotoxin probes, fluorescent derivatives of pirenzepine), adrenergic receptors (derivatives of prazosi and CGP 12177), $GABA_A$ receptor (muscimol conjugates), neurokinin receptors (conjugates of substance P), neuromedin C receptors, angiotensin II receptor, opioid receptors (naloxone and naltrexone probes)... Ca^{2+} channels: α-conotoxin probes, fluorescent dihydropyridines and verapamil ryanodine etc. Probes for the Na^+ channel and the Na^+/H^+ antiporter: amiloride analogs Na^+/K^+-ATPase: ouabain probes Probes for K^+ channels and carriers: glibenclamide conjugates for the ATP-dependent K^+ channel, apamin probes for small-conductance Ca^{2+}-activated K^+ channels, Probes for Cl^- channels and carriers: Ivermectin probes for glutamate-gated Cl^- channels, stilbene disulfonates: (anion-transport Inhibitors),	Fluorescent receptor ligands can provide a sensitive means of identifying and localizing some of the most pivotal molecules in cell biology. Many types of fluorescently labeled and unlabeled ligands exist for various cellular receptors, ion channels and ion carriers. Many of these site-selective fluorescent probes may be used on live or fixed cells, as well as in cell-free extracts. Many new dyes provide extremely sensitive detection, which enables measurement of low-abundance receptors. Various methods for further amplifying detection of these receptors have been reported. A variety of probes for Ca^{2+}, Na^+, K^+ and Cl^- ion channels and carriers have been described. Ion flux that affects the cell membrane potential, can be indicated with potential-sensitive probes.	58

Table 14.2 (continuing)

Application area	Subfield or technology	Typical dyes or assay system used	Description	References
	Phagocytosis	Dichlorodihydrofluorescein diacetate, OxyBURST technology	Monitoring of the oxidative "burst" produced by activation of an NADPH oxidase in a chain of events starting with the binding of surface-bound IgG immune complexes interact with Fc receptors.	59
	Apoptosis	Various fluorescence based kits, e.g. those using annexin V conjugates Caspase protease activity	Apoptosis is a genetically controlled cell death. Various approaches are used in studying apoptosis, and to distinguish live cells from early and late apoptotic cells and from necrotic cells. Apoptosis assays may be based on: • Nucleic acid stains, • Annexin V conjugates • Protease activity • Mitochondrial stains • Free radicals • Ion indicators • Esterase activity • ATP:ADP ratio	60

Table 14.3 Comparison of noise levels in the NIR in visible regions.

Noise source	NIR region	Visible region
Detector	low	high
Scatter (Rayleigh/Raman)	reduced	6 × greater at 250 nm than 820 nm
Autofluorescence	mostly absent	autofluorescence of biomolecules

wavelength dependence of Raman scattering. These benefits are shown in Tab. 14.3.

Excitation in the visible part of the electromagnetic spectrum (400–700 nm) can be accomplished with the 488 nm argon, 546 nm mercury-arc, 633 nm HeNe, and 647 nm Kr laser lines. Longer wavelength electronic transitions can be induced by the near-infrared gallium–aluminum-arsenide (GaAlAs) laser diode with an output at 785 nm. The advantages of using laser diodes instead of argon lasers are listed in Tab. 14.4. Fluorescence detection in the NIR region utilising avalanche photodiodes (APD) instead of conventional photo multiplier tubes (PMT) provides additional benefits as shown in Tab. 14.5. In addition, APDs not only offer greater quantum efficiency but also show lower power consumption in comparison to PMTs. The complementary advantages of diode laser excitation and APD signal transduction are especially applicable to bioanalytical techniques.

Table 14.4 Comparison of diode lasers and argon ion lasers as excitation sources.

Parameter	Laser diode	Argon laser
Wavelength (nm)	785	488
Lifespan (h)	100 000	3000
Power output (W)	0.02	~ 5
Power output	0.15	1800
Replacement ($)	10	5000

Table 14.5 Comparison of avalanche photodiodes an photomultilpier tubes as signal transducers.

Parameter	APD	PMT
Quantum effeciency at 820 (%)	80	0.3
Internal amplification	low	high
Size	mm	cm
Power Consumption	very low	low
Lifetime (h)	100 000	10 000
Replacement cost ($)	50	500

14.2.2.2 Chemistry of VIS/NIR Dyes

The chemistry of VIS/NIR-absorbing dyes is well established in a variety of bioanalysis applications including DNA analysis and immunological assays [7]. Numerous approaches have been applied to the development of dyes for these specific fields. The major tools for designing new dyes are (1) to apply structure–color correlation rules or (2) to approximate the absorption range of the target dye by quantum chemical calculation. A bathochromic or hypsochromic shift can be affected by simple structural modifications, applying general guidelines of the structure–color correlation. These comprise: (1) increasing the overall electron density, (2) increasing the strength of the donor and acceptor groups in the dye, (3) increasing the conjugation length (longer oscillator), e.g. by polymerisation and (4) increasing the "dimensionality" of the molecule. Semiempirical quantum chemical methods are still quite often used to assess the absorption wavelengths of candidate dyes prior to synthesis. Older methods were based on the Pariser–Parr–Pople molecular orbital (PPP MO) method [61] or XNDO/S method [62]. Presently, INDO methods in conjunction with more extensive configuration interaction schemes are frequently used [63]. These methods are nowadays applicable to quite large molecules and even on a desktop computer quite large structures (up to 500 atoms) can be readily evaluated. DFT methods, however, yield very much improved accuracy of calculations, particularly when metals are also involved [64], but are only applicable to smaller molecules.

In addition to the desired absorption maxima, other factors must be considered to tailor the chromophore for its desired function. Such customisation leads to incorporation of proper functionalities for covalent coupling to biomolecules and organic or water solubility. Furthermore, the fluorescence maximum in the desired region, the quantum yield (φ_f), and the fluorescence lifetime (τ) should also be taken into consideration in the design of dye molecules. These alterations can generally be achieved by structural modification of known chromophore structures. For example, the quantum yield of a fluorophore can be estimated by comparison with existing fluorophores with known quantum yields. This estimation can be achieved simply by comparing the wavelength-integrated intensity of the unknown to that of the standard using the following equation where Q is the quantum yield, I is the integrated intensity, OD is the optical density, and n is the refractive index.

$$Q = Q_R \frac{I}{I_R} \frac{OD_R}{OD} \frac{n^2}{n_R^2} \tag{1}$$

Visible-absorbing dyes

Rhodamines and fluoresceins

The most commonly used visible fluorophores are the rhodamine and fluorescein analogues. These dyes can be derivatized with either an isothiocyanato group (–NCS) or an N-hydroxysuccinimidyl (NHS) ester functionality for covalent labelling at the amino group of proteins or amino-functionalised nucleotides (Fig. 14.1). A number of dyes containing reactive groups are commercially available and they are as a rule inexpensive.

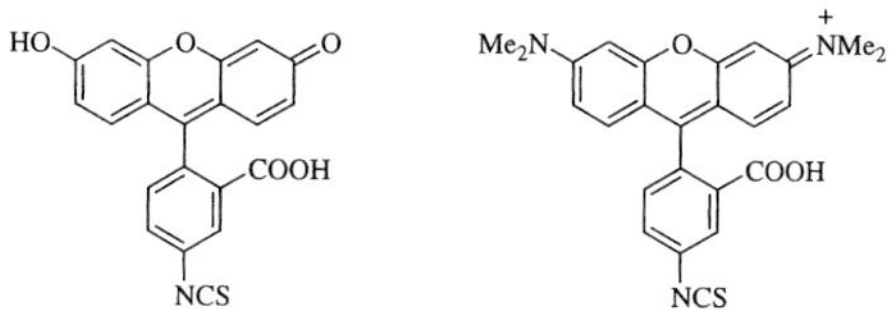

Fig. 14.1 Structure of isothiocyanato-funtionalized fluorescein (**1**) and tetramethylrhodamine (**2**).

Fluorescein is a hydroxylated xanthene (fluorone) with intense green fluorescence. This well-known dye has been used as a tracer in underground streams and water supplies, as well as a marker for sea-rescue proceedings [65]. The biolabel FITC, fluorescein isothiocyanate (**1**, Fig. 14.1), was first introduced for use as an antibody label for rapid identification of pathogens in 1958 by Riggs et al. [66]. Since then, this dye has been an established marking tool in DNA synthesisers, crime-scene analysis, and immunoassay. Other visible dyes, exemplified by tetramethylrhodamine isothiocyanate (**2**, Fig. 14.1), are similar in structure to fluorescein except they possess amino moieties ($-NR_2$) in place of the peripheral hydroxy/oxo groups. These cationic xanthene dyes exhibit many of the desirable properties for biological labeling such as a high fluorescence quantum yield and good water solubility [67].

Recently, fluorescein and tetraethylrhodamine derivatives were utilized in the evaluation of biotin–dye conjugates for use in an HPLC assay to assess relative binding of biotin derivatives with avidin (Av) and streptavidin (SAv) [68]. These derivatives are targeted for radiotherapy of cancer in which the biotin derivative carries a radionucleide to cancer cells. These biotin–dye conjugates incorporated a 4,7,10-trioxatridecane-1,13-diamino linker as a 17 Å spacer to assess the changes in rates of association and dissociation with Av or SAv.

Thiazole orange
Thiazole orange (TO, Fig. 14.2) is a cyanine dye that shows a relatively large emission enhancement upon intercalation with double-stranded DNA (dsDNA) and a moderate DNA binding affinity ($K_d = 10^{-5}$ M). These specific features make thiazole orange derivatives ideal candidates for fluorescent DNA-binding probes. Thompson et al. synthesized a carboxylic acid-functionalized thiazole orange **9** and conjugated it to the amino-terminal zinc finger of the glucocorticoid receptor DNA binding domain (GR-DBD) for DNA sequence specificity binding studies [69]. Photooxidation of the thiazole orange intercalation complex cleaves DNA, which provides a convenient assay for determining the preferred binding site, a 5'-TGTTCT-3' sequence.

Synthesis of TOTO dimers is illustrated in Fig. 14.3. The initially formed iodoalkyl derivatives **12** were treated with *N,N,N',N'*-tetramethyl-1,3-diaminopropane in anhydrous MeOH to furnish the symmetrical dimers **13**. As shown by Jacobsen

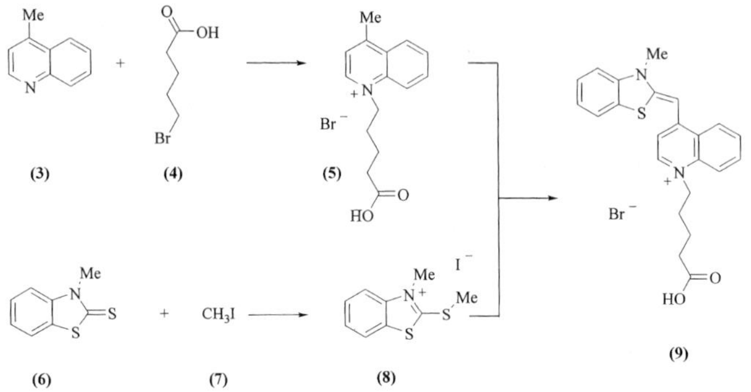

Fig. 14.2 Synthetic route to carboxylic acid-functionalised thiazole orange **9**.

and his co-workers, these bis-cyanines **13** interact with dsDNA in a sequence-selective manner via noncovalent bis-intercalation [70]. Upon complex formation with the 5'-CTAG-3' sequence of dsDNA, these chromophores exhibit an enhanced fluorescence emission several thousand orders of magnitude greater than the free TOTO dimers.

The final TOTO chromophores were evaluated as sequence-selective bis-intercalators via fluorescence and ^{1}H–^{1}H NOESY NMR spectroscopy. The TOTO chromophores were found to form a complex with dsDNA, where each separate chromophoric unit is sandwiched between two base pairs in a (5'-CpT-3'):(5'-ApG-3') site while the linker lies in the minor groove. Fluorescence dsDNA binding studies provided a correlation between the linker length of the TOTO dimers and the binding strength. The results indicated that the fluorescence quantum yield of the dimers increases with the length of the linker upon binding to the CTAG sequence. However, this effect does not correlate with the binding selectivity.

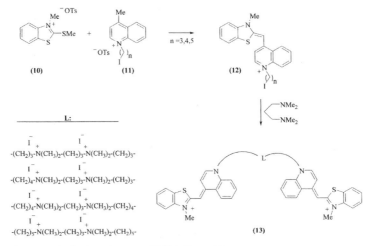

Fig. 14.3 Synthetic pathways to TOTO dimers **13**.

Cy^{TM} Dyes

The Cy™ dyes were developed by Waggoner and his group at Carnegie Mellon University in Pittsburgh, PA. The dyes, with reactive moieties such as a carboxylic acid group or an N-hydroxysuccinimidyl (NHS) ester (**16, 17**, Fig. 14.4), are commercially available from Molecular Probes (Eugene, OR, USA) and Amersham Life Science (Pittsburgh, PA, USA). The dyes contain one or more peripheral sulfonate groups to improve their aqueous solubility. Bio-conjugates, such as Cy5-dUTP (a pentamethine derivative), are commercially available or can be readily prepared. The Cy3™ series consist of indolium trimethine cyanine dyes that absorb in the 535–555 nm range, which is compatible with the 546 nm mercury-arc line. Emission of this series of cyanines occurs in the 570–605 nm range. The Cy5™ dyes are vinyl analogues of the Cy3™ dyes and give a bathochromic shift with absorption in the 620–650 nm region and emission in the 665–725 nm region. This absorption in the orange part of the electromagnetic spectrum allows the pentamethines to be excited via the 633 nm HeNe or 647 nm Kr laser lines. Both series have large extinction coefficients in the range of 130,000 to 250,000 M cm^{-1}.

Recently, the active N-hydroxysuccinimidyl (NHS) esters of Cy3 and Cy3NOS (**17**, Fig. 14.4) were used to label oligonucleotides by Randolph et al. [71]. While Cy3 contains a bis-sulfonate functionality for improved water-solubility, Cy3NOS is mono-sulfonated. The reaction of **14** and **15** produced the carboxy-substituted dye **16**, which was further treated with N,N'-disuccinimidyl carbonate in a mixture of DMF/pyridine to give the nonsymmetric N-hydroxysuccinimidyl substituted Cy™ dyes **17**. Dyes **17** were then covalently linked via a short or long tether at the C-5 position of deoxyuridine to multiply-label DNA. The correlation study between the labeling density and the sensitivity of the DNA fluorescent probe demonstrated that labeling at every 6th base pair with Cy3 showed the optimal fluorescence.

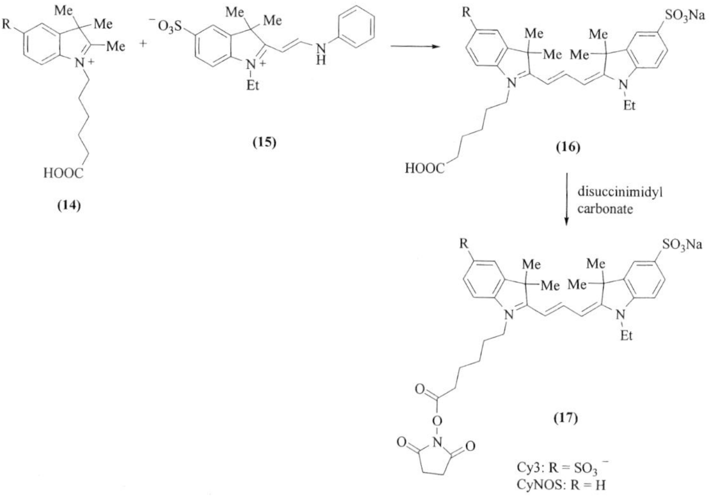

Fig. 14.4 Synthetic route to nonsymmetric dyes Cy3 and Cy3NOS.

NIR-absorbing dyes

The recent availability of inexpensive semiconductor laser diodes such as gallium-aluminum-arsenide (GaAlAs) with an emission wavelength at 785 nm has led to an increase in the use of NIR fluorophores for bioanalytical applications. As mentioned earlier, the use of a NIR fluorophore can lower the background interference from the biological sample. The IRD™ dyes are commercially available from LI-COR Inc. They are indolium heptamethine cyanine dyes whose absorption coincides with the output of the GaAlAs laser diode. These dyes contain isothiocyanate or phosphoramidite functionalities for labeling amino residues of antibodies or synthetically modified nucleotides, respectively. The general commercial availability of functionalized NIR absorbing chromophores is partially limited and the dyes are rather expensive due to the laborious purification procedures. The most commonly used NIR chromophores belong to the cyanine-type dyes including carbocyanines, squaryliums, and phthalocyanines (Pc), and naphthalocyanines (NPc).

Carbocyanine dyes

The classical synthetic route to symmetrical indolium heptamethine cyanine dyes by the Strekowski approach [72] is represented in Fig. 14.5. Introduction of an isothiocyanato (-NCS) functionality into the dye is accomplished by facile nucleophilic displacement of the meso-chloro substituent in the intermediate product **22**. This method provides amino-reactive cyanines with absorption in the 770–800 nm region.

A similar approach also supplies heptamethine cyanine dyes with altered spectroscopic properties for more specific applications. For example, Flanagan et al. incorporated fluorine and heavy atoms such as I, Br, and Cl into polymethine dyes, as shown in Fig. 14.6, to alter the fluorescence lifetime of the dye without disturbing the other spectroscopic properties [73]. The fluorine or heavy-atom modified

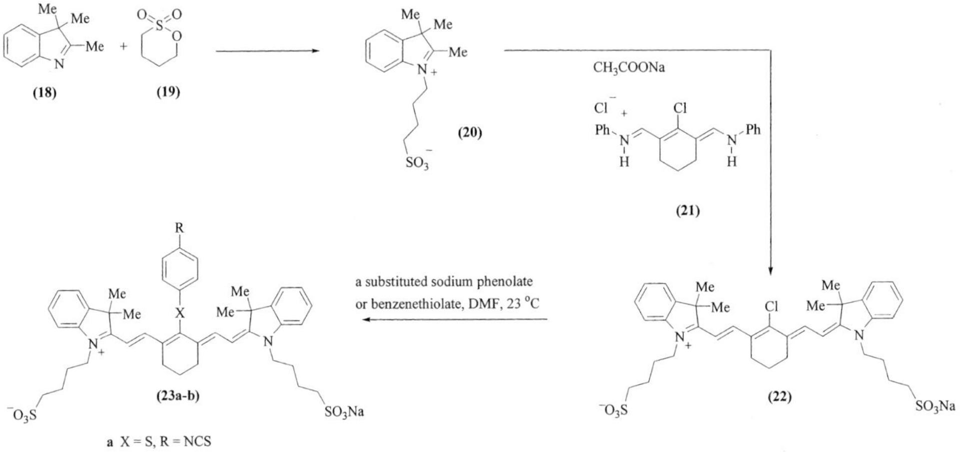

Fig. 14.5 Synthetic route to symmetrical indolium heptamethine dyes.

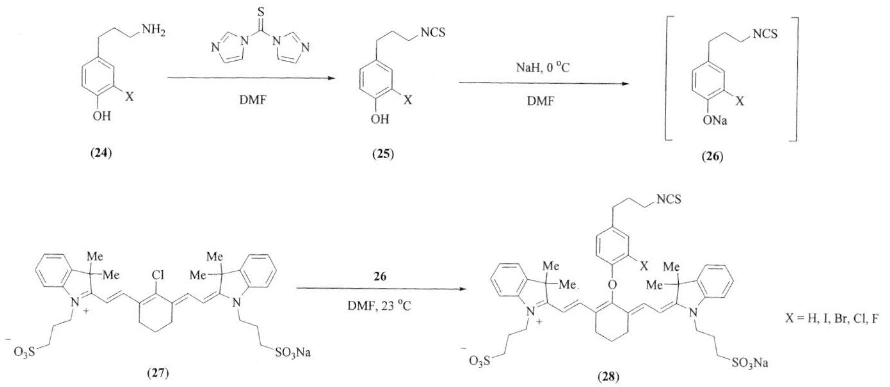

Fig. 14.6 Synthetic route to near-infrared heavy atom modified isothiocyanate dyes.

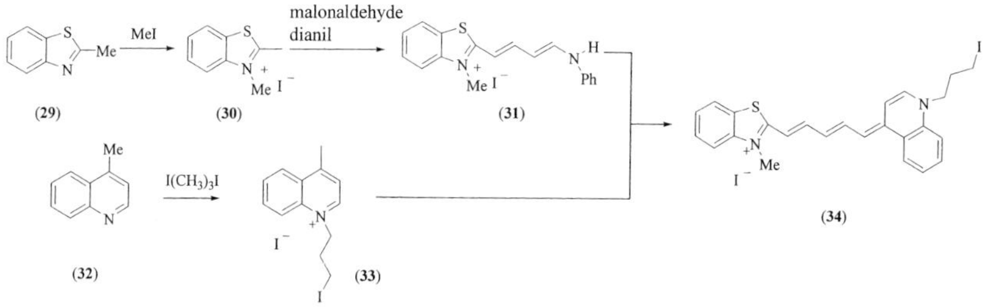

Fig. 14.7 Synthesis of thiazole green **34** (TAG) used for low-level detection of DNA restriction fragment by CE.

isothiocyanato phenol **25** was obtained from the reaction of the appropriate substituted tyramine **24** with 1,1-thiocarbonyldiimidazole in DMF. The sodium phenolate of **24** was further treated with meso-chloro dye **27** to afford the final dye **28**.

Thiazole green represents another class of carbocyanines, which was recently utilized in the low-level detection of DNA restriction fragments by Soper et al. [74]. This dye is comprised of a quinolinium and a thiazolium nucleus joined by a pentamethine chain. The synthetic route to the thiazole green **34** is shown in Fig. 14.7. 2-Methylbenzothiazole **29** was allowed to react with methyl iodide to give the N-methyl derivative **30** which was further treated with malonaldehyde-dianil hydrochloride to afford the intermediate product **31**. The intermediate product **31** was then allowed to react with N-(3-iodopropyl)lepidinium iodide **33**, which had been obtained from the reaction of lepidine with 1,3-diiodopropane, to finally afford the asymmetrical thiazole green **34**.

Recently, Chairs analyzed the oxazolium pentamethine cyanine dye **35** (DODC, Fig. 14.8) in DNA binding studies utilizing his competitive dialysis method [75]. The nucleic acid structures included single-stranded, duplex, triplex, and tetraplex forms. The comparative assay demonstrated that DODC binds to triplex DNA more selectively than any of the tetraplex forms included in the assay. In addition, the

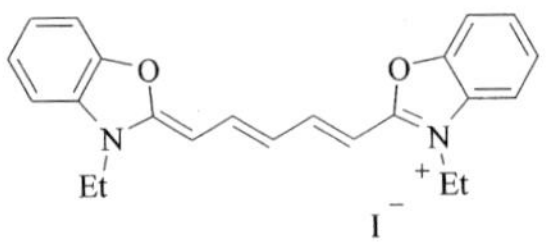

Fig. 14.8 Structure of 3,3'-diethyloxadicarbocyanine (DODC) studied by Ren et al. [75].

absorbance spectrum of DODC is red-shifted and a signal centered at 610 nm is observed in the CD spectrum upon binding.

A new class of long-wavelength (> 1000 nm) NIR dyes suitable for bio-conjugation **37** (Fig. 14.9) has been developed by Strekowski et al. [76]. These long wavelength NIR chromophores provide biolabels with increased sensitivity and detection limits. The isothiocyanato-substituted benz[c,d]indolium heptamethine cyanine dyes **37** were obtained by nucleophilic displacement of the meso-chlorine atom of the commercially available dye IR-1048 (**36**). This nucleophilic displacement is suggested to proceed through an $S_{RN}1$ pathway that includes a cationic radical dye intermediate [77].

Squaraine and croconine dyes
Squaraine dyes and the structurally related croconine dyes have also been described with intense absorptions in the NIR region (ε > 200,000 M.cm^{-1}) [78]. In the synthetic field there is presently much activity to find squaraine dyes with very high absorption wavelengths. For instance, Meier and Petermann described novel NIR dyes with absorption wavelengths in excess of 900 nm by coupling ferrocenes to the squarine moiety [79] and squaraine and croconine chromoinophores compatible with lipophilic matrices have been described by Citterio et al. [80]. Presently, a variety of squarine dyes and croconine dyes are commercially available from H.W. Sands Corporation (Jupiter, Florida, USA).

Of the squaraine dyes N-succinimidyl ester-derivatized indolium-squaraine dyes have been specifically developed for conjugation to biomolecules, because this class of dyes has been shown to exhibit high photostability, a long fluorescence lifetime and spectral properties (absorption, emission, and fluorescence lifetime) which are

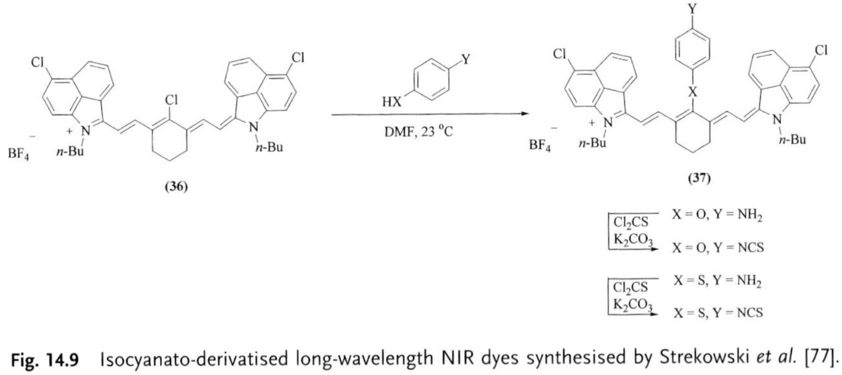

Fig. 14.9 Isocyanato-derivatised long-wavelength NIR dyes synthesised by Strekowski et al. [77].

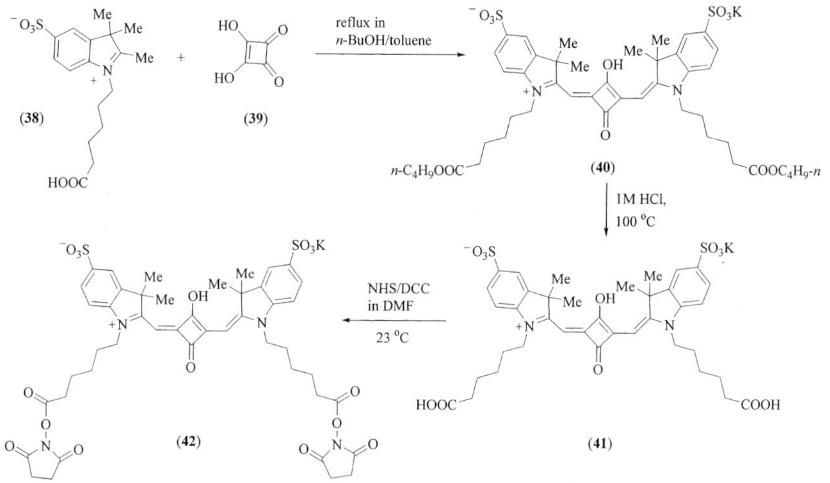

Fig. 14.10 Synthetic route to Sq635 squaraine dye **42** containing N-succinimidyl ester moieties [81].

independent of pH at physiological conditions (pH 6–9) [81]. The water-soluble squaraine dyes exhibit a four- to five-fold fluorescence quantum yield enhancement when covalently bound to proteins [82]. Recently, Oswald et al. described the synthesis of indolium Sq635 (**42**, Fig. 14.10) and the benz[e]indolium Sq660 NHS esters for use in a fluorescence resonance energy transfer (FRET) immunoassay [83]. In the competitive immunoassay, the donor Sq635-HSA has spectral overlap with the acceptor Sq660-anti-HSA which results in a high R_o of 70 and leads to detection limits of 10^{-7} M.

Phthalocyanins and naphthalocyanins
The NIR absorbing phthalocyanines (Pcs) and naphthalocyanines (Npcs, Fig. 14.11) are planar chromophores comprising four 1,3-diiminoisoindolenine subunits with an 18-electron cavity. These cyanines can exist as the metal-free (Pc and NPc) or metal-complexed form (MPc and MNPc). The absorbance, emission, fluorescence lifetime, and other photophysical properties are strongly influenced by the central metal ion. Specifically, metal complexation with transition metals provides a selective method to alter the quantum yield (ϕ_f) and the fluorescence lifetime (τ) of the photoexcited triplet state, which allows MPcs and MNPcs to be strong candidates for use in the photodynamic therapy (PDT) of cancer [84]. These dyes are obtained by cyclotetramerization of the appropriate precursors including di-imino-isoindoline, phthalic anhydride, phthalimide, or benzene dicarbonitrile [85]. The metal-free Pcs or NPcs can then be altered to give the metal-functionalized dye via an insertion reaction with appropriate metal salt. However, most synthetic chemists rely on the direct method to obtain the metal-ion Pcs and NPcs which involves heating the precursors with the appropriate metal salt in a high

boiling solvent such as quinoline, o-dichlorobenzene, or tetrahydronaphthalene. Although the photophysical properties can be fine-tuned by the central metal substituent, other structural modifications of these dyes are difficult to accomplish in comparison to carbocyanines or squaraines due to the formation of numerous isomeric products. In addition, these molecules are extremely insoluble, especially in the aqueous solutions that are used in bioconjugation protocols. On the other hand, phthalocyanines and naphthalocyanines are extremely stable to strong bases and acids, heat, and direct exposure to light.

In order to overcome the limited solubility of MNPcs, Brasseur et al. recently synthesized and evaluated the PDT (photodynamic therapy) activity of bis(alkylsiloxysilyl) complexes of naphthalocyanines (45, Fig. 14.11) [86, 87]. It has been established that axial substitution of the central metal ion diminishes the tendency of these dyes to form H-aggregates in solution and the diaxial substituted SiNPcs follow this trend with an enhanced stability to photooxidation in comparison to Al- and Zn-NPcs [88, 89]. In the synthesis of 45, benz[f]diimino-isoindolenine **43** undergoes cyclotetramerization with silicon tetrachloride to yield the intermediate dichloride. The axial ligands of the dichloro-intermediate are then hydrolyzed to furnish silicon(IV) 2,3-naphthalocyanine dihydroxide **44**. The bis(alkysiloxy) ligands are then introduced via substitution of the axial dihydroxide moieties with the appropriate chlorosilane derivatives to provide the photosensitizing SiNPcs **45**. The photodynamic properties of bis-substituted SiNPcs were evaluated against the EMT-6 tumor in Balb/c mice. Although the bis(alkylsiloxy) SiNPcs showed some phototoxicity in vitro, they gave excellent results *in vivo*, especially the derivative **45d**.

14.2.2.3 Bioanalytical Applications of NIR and Visible Fluorescent Dyes

DNA sequencing

As mentioned above, the Human Genome Project, which was initiated in the late 1980s, has spurred the development of technology necessary to facilitate the formidable task of sequencing and decoding the human genome. The dideoxy

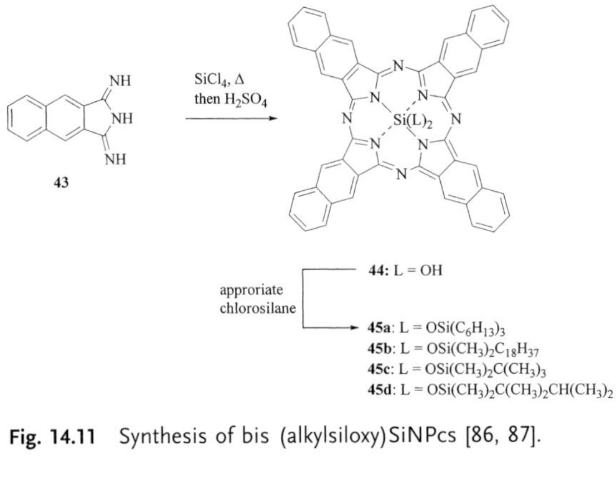

Fig. 14.11 Synthesis of bis (alkylsiloxy)SiNPcs [86, 87].

chain termination method of DNA sequencing, developed by Sanger and coworkers in 1977 is the primary method used in DNA sequencing [90]. Generally, DNA sequencing involves three stages of fragmentation, beginning with the chromosome and ending up with segments of DNA four to five hundred base pairs in length (Fig. 14.12) [91–94]. The starting point of the Sanger method is the M13 vector containing the foreign DNA insert. A short strand of DNA complementary to a portion of the M13 vector is synthesized. This strand of DNA, referred to as a primer, will bind to the complementary portion of the M13 vector. The M13 vector and the primer are placed in a solution containing the enzyme DNA polymerease. DNA polymerase will catalyze the synthesis of DNA from the 3' end of the primer. The newly synthesized DNA will be complementary to the sequence of the foreign or inserted DNA. DNA synthesis, catalyzed by DNA polymerase, requires the presence of all of the deoxynucleotide bases. The Sanger method utilizes modified bases, called dideoxynucleotides, which lack the 3'-hydroxy on the sugar residue that normal nucleotides have. When DNA polymerase incorporates a dideoxynucleotide into a growing strand of DNA, the strand terminates immediately thereafter. Chain termination occurs because the dideoxy nucleotide lacks the 3'-hydroxy

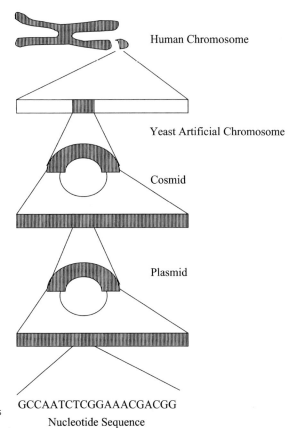

Fig. 14.12 Schematic illustration of the fragmentation steps in the sequencing of DNA.

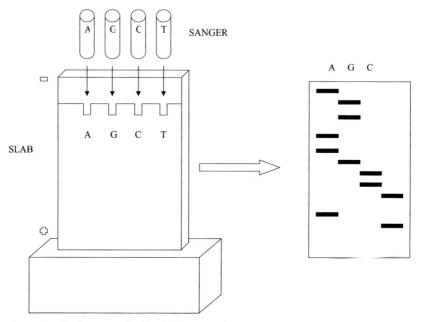

Fig. 14.13 The Sanger method for DNA sequencing.

functionality necessary to form the phosphodiester linkage with the next nucleotide. Four reactions are carried out in the Sanger method. In each case, a mixture of the primer, inserted DNA or template, all four deoxynucleotides, and one of the four dideoxynucleotides are allowed to react. Consider the reaction mixture containing the primer, template, DNA polymerase, the four deoxynucleotides, and dideoxycytidine. The newly synthesized DNA will be complementary to the template DNA. However, DNA polymerase will occasionally incorporate dideoxycytidine into the growing chain, resulting in termination of the growing chain. Dideoxycytidine is incorporated into the growing chain only when deoxyguanosine is the complementary base present in the template DNA. If the reaction was carried out properly, the products of the reaction are DNA fragments of varying lengths, which are complementary to the template DNA. All fragments terminated with the incorporation of dideoxycytidine, corresponding to the point when DNA polymerase encountered deoxyguanosine in the template DNA. The only difference in the other three reactions is that a different dideoxynucleotide is used in each. The reaction products are then separated via gel electrophoresis. In the past, isotopic labels were used to visualize components present in the mixture. The sequence is then read directly from the gel (Fig. 14.13). The sequence of the template may then be determined, since it is complementary.

One of the initial goals of the Human Genome Project was the development of technology that could increase the rate at which DNA could be sequenced. Traditionally, isotopic labeling was used for the detection of DNA fragments. Sequencing using isotopic labels was a very time consuming process. A great deal of human skill was needed for the interpretation of band patterns. Also, it was necessary to transcribe the data once they had been compiled. Furthermore, radioisotopes pose health problems and are costly. As sequencing technology progressed, the reading of audioradiographs became a limiting step in the process. The use of laser induced fluorescence (LIF) detection offered an alternative to radiolabeling. LIF detection offered the possibility of real time, automated detection. Sequence information could be stored directly by computer, eliminating possible data transcription errors. The exposure time necessary when using radiolabels is eliminated with LIF detection since detection is real-time. Furthermore, software may be written for the interpretation of data, eliminating the need for skilled personnel to read the data. DNA sequencing with LIF detection offers substantial advantages over isotopic labeling, with respect to speed and safety. Consequently, LIF detection is now the most common method of detection used for DNA sequencing.

While DNA sequencing by slab gel electrophoresis has been the primary method used in the past, sequencing by capillary gel electrophoresis is fast becoming the dominant technique. Sequencing rates in slab gel electrophoresis can be increased by operating the gels at high voltages. However, this produces excessive heat. Gel-filled capillaries allow for the use of high electric field strengths. Due to the high surface to volume ratio associated with capillaries, heat is dissipated rapidly, allowing for high separation voltages. Resolution of capillary based systems is generally better than slab gels and read lengths are up to 25 times longer when using gel-filled capillaries [95]. In order to match the throughput of DNA sequencing by capillary gel electrophoresis to that of slab gels, multiple capillary instruments must be used. The first capillary array instrument was developed by Mathies and coworkers [96, 97], and an integrated capillary array electrophoresis (CAE) system for the simultaneous processing of 96 samples in 48 electrophoresis channels has been devised [98]. Dovichi has also designed capillary-based instruments for DNA sequencing [95, 99, 100].

DNA sequencing with visible fluorophores
The use of fluorescent labels with DNA sequencing was first demonstrated in 1986 [101–103]. Ansorge and coworkers developed a DNA sequencing method that used a primer labeled with rhodamine [103]. The products of the sequencing reactions were separated on four different lanes. Excitation was accomplished from a single laser and detection was achieved through the use of photodiodes. Smith and coworkers developed a DNA sequencing method which used a primer labeled with four different fluorescent tags [101]. The sequencing products were then combined and separated on a single lane. The sequence analysis was based on the different spectral characteristics of the chosen dyes. An advantage of using four dyes is that the separation occurs in one lane. Consequently, the throughput of the method is improved four times. The use of fluorescent detection schemes in DNA sequencing also allows for longer read lengths with respect to isotopic labelling. When radioactive labels were used, the separation had to be stopped when the fastest moving fragment reached the end of the plate. As a result, resolution was poor

for the longer fragments since they did not travel as far in the gel. To accommodate this loss of resolution, smaller lengths of DNA were used. With fluorescence detection, the excitation source is fixed at the end of the capillary or gel. All fragments must traverse the length of the gel or column in order to be detected. Consequently, the resolution of the larger fragments is much better, allowing for longer read lengths with respect to isotopic labelling.

Karger and coworkers developed a capillary gel electrophoretic method for sequencing that uses primers labeled with four dyes; FAM, JOE, ROX, and TAMRA (Fig. 14.14) [104]. The dyes used by Karger are commercially available fluorescein and rhodamine derivatives (see earlier section on visible absorbing dyes). Gel-filled capillaries have been shown to have distinct advantages over conventional slab gels. They allow for the use of higher applied voltages since capillaries dissipate heat more efficiently than slab gels, resulting in rapid, high resolution separations [95, 105–106, 107, 108]. The instrument design uses two lasers, an argon ion laser at 488 nm and a helium-neon laser at 514 nm, and two detection windows (Fig. 14.15). Detection limits for the dye labelled primers are in the low picomolar range (0.7–3.5 pM). Dovichi and coworkers describe a similar method which utilizes the same four dye-labeled primer [95], however, a filter wheel design is used for detection (Fig. 14.16).

In an effort to simplify detection schemes, two dye sequencing methods have also been developed [99, 100] (Fig. 14.17). TAMRA and ROX were used to label the primers. In one Sanger reaction, TAMRA labeled primer was extended in the presence of ddATP and ddCTP. The concentrations of ddATP and ddCTP used in the reaction were adjusted to give a 3:1 peak height ratio. Consequently, the termination fragments are distinguishable on the basis of peak height. Similarly, ROX labeled primer was extended in the presence of ddGTP and ddTTP.

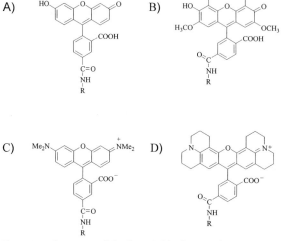

Fig. 14.14 Structures of the four visible dyes used in DNA sequencing: A) FAM, B) JOE, C) TAMRA and D) ROX. R denotes a linker between the chromophore and the nucleotide primer.

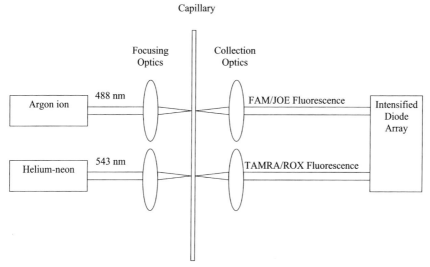

Fig. 14.15 Schematic of two-laser, two-detection window, four-dye DNA sequencing instrumentation developed by Karger and co-workers. Adapted from ref. 104.

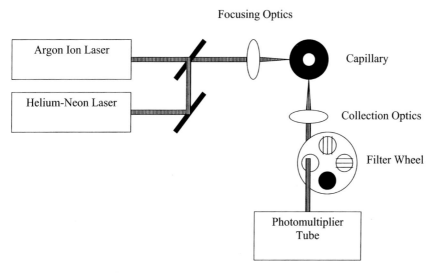

Fig. 14.16 Schematic of two-laser, filter wheel, four-dye DNA sequencing instrumentation developed by Dovichi. Adapted from ref. 93.

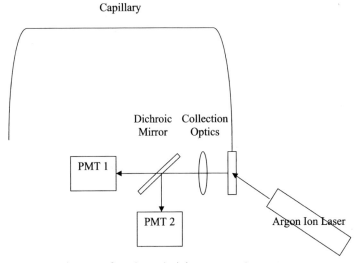

Fig. 14.17 Schematic of one-laser, dual detector, two-dye DNA sequencing instrumentation. Adapted from ref. 99 and 100.

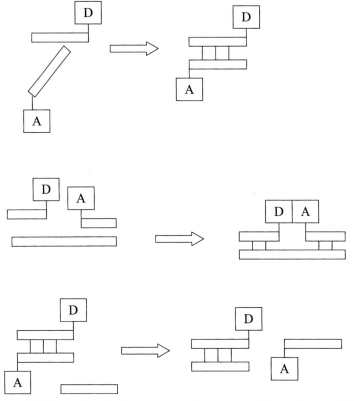

Fig. 14.18 Examples of various energy transfer schemes. D is the donor and A is the acceptor.

Again, the concentrations of the dideoxynucleotides were adjusted to give a 3:1 peak height ratio.

When using multiple dyes for DNA sequencing, it is advantageous for the dyes to have comparable emission intensities upon excitation at a common wavelength, so that the emission intensity of each dye is comparable. Furthermore, the dyes should display distinct emission spectra so that they are discernible from each other. This is difficult to accomplish. However, the use of fluorescence energy transfer allows for simultaneous optimization of both of these requirements. The use of donor-acceptor pairs allows for the equalization of emission intensity at a common excitation wavelength. The use of energy transfer fluorescent dye labeled primers offers improved sensitivity with respect to single dye labeled primers [90, 101, 109, 110]. Examples of energy transfer schemes are shown in Fig. 14.18.

Mathies and coworkers have performed work on energy transfer using FAM, a fluorescein derivative, as the donor, and FAM, JOE, TAMRA, and ROX as the acceptors (Fig. 14.19) [111]. The acceptor dyes exhibit fairly distinct emission maxima at 525, 555, 580, and 605 nm, respectively, and all energy transfer fluorescent dye

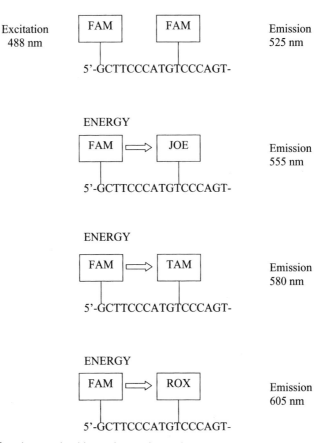

Fig. 14.19 Energy transfer scheme utilised by Mathies and coworkers.

labeled primers exhibited a strong absorbance at the 488 nm line of the argon ion laser. The group systematically adjusted the distance between the donor and acceptor pair in order to arrive at the optimal characteristics for the desired application. While fluorescein and rhodamine derivatives are the most common visible dyes used for DNA sequencing, other groups are using cyanine dyes [71], as well as DNA-intercalating dyes such as thiazole orange, to label DNA [69, 70].

Further work done by Mathies and coworkers utilized a cyanine dye energy transfer system [112]. The cyanine donor was attached at the 5' end of the primer. Fluorescein or rhodamine derivatives were used as the acceptor and were located 10 bases away from the donor. The 488 nm line of an argon ion laser was used for excitation. The fluorescence intensity of the energy transfer dyes was 1.4 to 24 times stronger than when the individual acceptor chromophores were excited at the 488 nm line. The cyanine dye used as the donor showed improved photostability and higher extinction coefficients, with respect to FAM. The cyanine dye exhibited a large absorption cross section but a low quantum yield. This increased the Stoke's shift of the acceptor dyes and minimized cross-talk between the detection channels. The system design illustrates that dyes with high absorption cross sections but low quantum yields may be used for improved energy transfer systems, thereby greatly expanding the repertoire of available donor dyes.

DNA sequencing with NIR fluorophores

While the use of visible fluorescent dyes for DNA sequencing offers multiple advantages over traditional methods of detection, these dyes absorb and fluoresce in a region of the spectrum that is prone to autofluorescence from the sample matrix, gel, or from impurities. The result of these interferences is an overall increase in noise, resulting in a loss of sensitivity. Recently, near-infrared fluorophores have been used as labels in DNA sequencing in an effort to increase sensitivity [113–115]. The NIR portion of the spectrum is generally defined as the region between 700 and 1200 nm. NIR dyes have good molar extinction coefficients ($\sim$150,000–250,000 M cm^{-1}) and quantum yields in the range 0.05 to 0.5. Since biological molecules do not posses intrinsic fluorescence in this spectral region, NIR dyes are well suited for bioanalytical applications [116–126]. Furthermore, the autofluorescence exhibited by gels, glass, and solvents is nonexistent in the NIR region. In addition, scatter noise (Rayleigh and Raman) is reduced with NIR dyes, since noise intensity is related to the wavelength of detection by 1/4. As a result of the decreased noise levels associated with the use of NIR dyes, detection is not limited by noise levels, but rather by detector performance. Consequently, laser induced fluorescence detection methods using NIR fluorophores, are potentially more sensitive than detection schemes that employ visible fluorophores.

Some of the major disadvantages associated with visible LIF detection are the cost and complexity associated with the necessary equipment. LIF detection in the NIR region requires the use of solid-state components. Semiconductor diode lasers have proven to be optimal excitation sources for NIR fluorophores. Diode lasers are inexpensive, compact, have long operating lifetimes, can be operated in continuous or pulsed mode, and provide satisfactory power [127]. Another attractive feature is that diode lasers are available at a variety of wavelengths (635, 750, 780, 810, and 830 nm).

Photomultiplier tubes are the most common signal transducers used in visible LIF detection. PMTs are ideal for work in the visible region of the spectrum. However, their quantum efficiencies deteriorate rapidly at longer wavelengths. Even red-sensitive PMTs only possess quantum efficiencies of 0.01 at 800 nm [127]. For these reasons PMTs are poor choices for signal transducers when working in the NIR region. However, photodiodes are an excellent choice as signal transducers when working in the NIR region, since silicon based semiconductor materials have here high quantum efficiencies, typically 80 % [127]. In most cases avalanche photodiodes are used since they posses internal amplification, unlike conventional photodiodes.

Soper and coworkers have developed instrumentation for highly sensitive DNA sequencing by capillary gel electrophoresis using NIR LIF detection [115]. The nucleotide sequence was determined using a single lane/single dye technique. An M13 sequencing primer was labelled at the 5' end with a tricarbocyanine dye with an isothiocyanate functionality. The molar concentrations of the dideoxynucleotides were varied such that the molar ratios were 4:2:1:0 (A:C:G:T). Base identification was based on peak intensity. A comparison was done with the NIR LIF system versus visible LIF. The 488 nm line of an argon ion laser was used as the excitation source. An M13 sequencing primer was labelled with the visible fluorescent dye TAMRA. The noise generated by the gel matrix was 20 times larger with the argon ion laser than it was with the Ti:Saphire laser. The quantum yield of the visible dye was 0.9 vs. 0.07 for the NIR dye. Despite the lower quantum yield of the NIR dye, the detection limit for the NIR dye labeled primer was 34 zmol, as opposed to 1.5 amol for the visible dye labeled primer. The significant improvement in the detection limit of the NIR dye, roughly two orders of magnitude, was attributed to the minimal background interference at the wavelength of detection.

Patonay and coworkers have done work with NIR LIF detection for DNA sequencing on slab gels [114]. Four heptamethine cyanine dyes were used in the study, although the four line, one dye method was employed for sequencing. A modified thymine base with a terminal amino linker was incorporated into the primer. All dyes possessed an isothiocyanate functionality which allowed for conjugation. Reverse phase HPLC was used to remove excess dye. Sequencing was carried out on a Li-Cor model 4000 DNA sequencer (Figs. 14.20 and 14.21). 500 bases were read with a 1 % error rate. Individual bands consisting of 0.1 fmol were routinely detected.

While spectral discrimination is the most common method of base calling used, temporal discrimination offers several advantages: (1) The lifetime of the fluorophore is independent of concentration, (2) fluorescence lifetime values may be determined with more accuracy than fluorescence intensity, (3) temporal measurements are not hindered by broad emission profiles, and (4) a single detection channel may be used. Soper has developed a method using heavy atom modified NIR dyes for base calling in DNA sequencing using temporal discrimination [74]. One

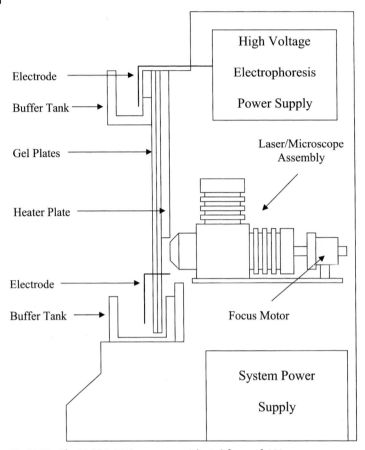

Fig. 14.20 The LI-COR DNA sequencer. Adapted from ref. 113.

of the major problems associated with lifetime determinations is the complexity of the apparatus required to carry out the measurements. Many problems associated with lifetime determinations for DNA sequencing are alleviated by using NIR fluorescence. Soper et al. attained very high precision due to the lack of interfering photons associated with the NIR region. In order to obtain fluorophores with distinct decay times, the photophysical properties of tricarbocyanine dyes were modified by the incorporation of a heavy atom. This resulted in a change in the fluorophores' singlet state photophysics (quantum yield and lifetime), due to increased inter-system crossing [128–130]. The emission and absorption maxima of the fluorophores were unchanged by the heavy atom modification. The electrophoretic mobility remained unchanged, negating the need for post-run corrections for mobility discrepancies often encountered in multiple dye approaches. The dyes developed by Soper could serve as ideal labels for DNA sequencing applications since only one excitation source is required and detection may be accomplished on a single channel.

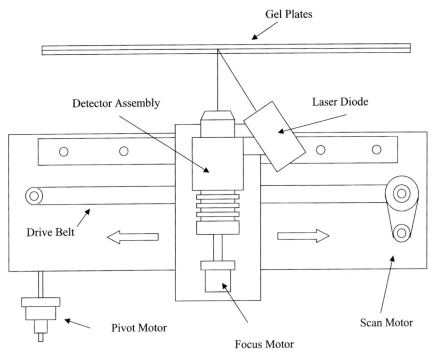

Fig. 14.21 Top view of the LI-COR DNA sequencer. Adapted form ref. 113.

Immunochemistry

Immunoassays are an analytical technique based on the highly specific interaction between an antibody and an antigen. Due to their ability to measure trace amounts of analyte in complex matrices, immunoassays are widely used in clinical, pharmaceutical, and environmental chemistry. Berson and Yalow were the first to use antibodies as an analytical tool, reporting picogram detection of insulin [131]. Immunoassays are the primary method used in diagnosing diseases such as acquired immunodeficiency syndrome (AIDS) [132], cysticerocosis [133], and schistosomiasis [134, 135]. Due to their widespread use, growth in the field has been tremendous over the past decades.

One of the advantages of the immunoassay technique is the wide variety of formats available. Generally speaking, immunoassays are carried out using either competitive or non-competitive formats. Competitive assays may be carried out by labelling either the antibody or the antigen, depending on what the analyte is. If the analyte is an antibody, then antigen is coated on a support surface. Labelled antibody is then added. Finally, unlabeled antibody from the sample is added. The labelled and unlabeled antibody bind competitively on a limited number of sites, i.e., the coated antigen. The amount of antibody present is determined by the change in signal (Fig. 14.22). Alternatively, if the analyte is an antigen, then

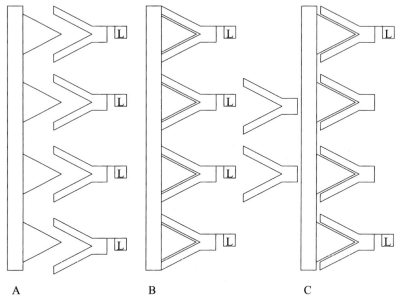

Fig. 14.22 Illustration of a competitive immunoassay procedure where the analyte is an antibody. A) Antigen is coated onto a solid support and labelled antibody is introduced. The excess of labelled antibody is removed. B) Analyte is introduced, which competes with labelled antibody for binding sites. C) The amount of label is quantified.

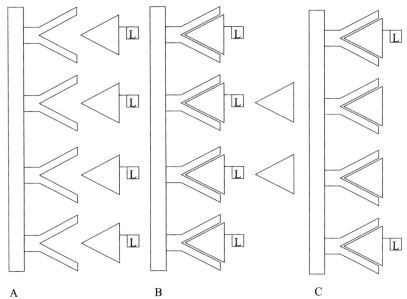

Fig. 14.23 Illustration of a competitive immunoassay procedure where the analyte is antigen. A) Antibody is coated onto a solid support and labelled antigen is introduced. The excess of labelled antigen is removed. B) Analyte is introduced, which competes with labelled antigen for binding sites. C) The amount of label is quantified.

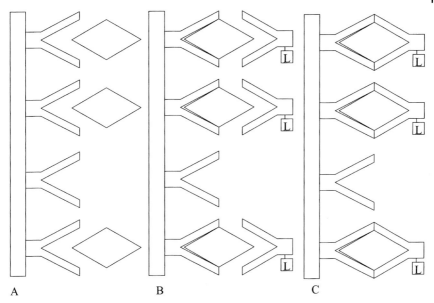

Fig. 14.24 Illustration of a non-competitive ('sandwich') immunoassay format. A) Antibody is coated onto a solid support (the 'capture' antibody) and antigen from the sample is introduced. B) A second labelled antibody (the detection antibody) is introduced. C) The amount of labelled antibody is quantified.

antibodies are fixed to a solid support. Labelled antigen is then added. Finally, unlabeled antigen from the sample is added. The labelled and unlabeled antigen compete for the binding sites on the antibody. The amount of antigen present in the sample is determined by signal change. Signal intensity decreases with increasing analyte concentration (Fig. 14.23).

Non-competitive assays operate on a different principle. The basic principle involves the binding of a limited amount of one reagent to an excess of the second reagent. Most often, the solid support is saturated with antigen. Excess antigen is removed during a washing step. The sample antibody is then added. In the final step, a second labelled antibody, specific for the primary antibody, is added. Unlike competitive assays, signal intensity increases with analyte concentration in non-competitive assays (Fig 14.24).

In the early stages of immunoassay technology, radiolabels were employed as tracers. Isotopic labels were popular due to their high sensitivity, selectivity, and unobtrusive nature. The use of radioactive labels does have drawbacks, such as health and safety issues, short shelf life, cost, long exposure times, and disposal problems. Due to these drawbacks, there has been considerable growth in the development of nonisotopic labels for use in immunoassays. Consequently, radioactive labels are now rarely used in immunoassays. Several criteria needed to be met in order for nonisotopic labels to become a viable alternative to radioactive labels: 1. Nonisotopic labels needs to match radioisotopes in terms of sensitivity; 2. the label needs to be easily coupled to the antibody or antigen; 3. the label should be nontoxic and amenable to automation; and 4. the label should not affect the antibody-antigen interaction [136].

Enzyme-linked immunosorbent assay (ELISA) is perhaps the most popular immunoassay format, due to its versatility and high sensitivity. In ELISA a solid support is coated with antibody and antigen is then added. Hereafter, a second antibody covalently bound to an enzyme is added. Following an appropriate incubation period and washing steps, enzyme substrate is added, resulting in the appearance of a colored or fluorescent product. A major advantage of this method is that with a high turnover number associated with the enzyme a very high sensitivity can be reached. A single enzyme may produce as many as 1000 product molecules. Horseradish peroxidase (HRP) and alkaline phosphatase are common enzyme labels. There are, however, disadvantages associated with the use of enzyme labels. The large size of the enzyme can affect the antibody-antigen interaction and the assays are quite sensitive to temperature and pH. Finally, the enzyme may cause nonspecific binding.

Immunochemistry with visible fluorophores
The use of fluorescent labels is a superior alternative to enzyme labels. While some enzyme labels do produce fluorescent products, most produce colored substrate products, necessitating the use of absorbance detection. Due to the high background noise associated with absorbance detection, the method is inherently less sensitive than fluorescence detection. Fluorescent labels are generally smaller than enzyme labels and therefore are less likely to interfere in antibody-antigen interactions or promote nonspecific binding. Fluorescein and rhodamine derivatives are commonly used in fluorescence immunoassays. A typical conjugation reaction involves a dye functionalized with an isothiocyanate group that is reactive toward primary amines at basic pH. Factors that affect the sensitivity of fluorescence immunoassays are autofluorescence from matrix components (such as bilirubin), scattering of excitation light, fluorescence quenching, and photobleaching processes. The instrumentation used to measure fluorescence is also a key component in method sensitivity. The use of a suitable fluorophore is critical in the development of sensitive assays.

An ideal fluorophore should have the following characteristics: a high molar adsorptivity and a high quantum yield, a large Stokes' shift in order to minimize light scatter, quick efficient coupling reactions, good solubility under physiological conditions, low non-covalent affinity for biomolecules in order to minimize nonspecific interactions, photostability, and small label size relative to the molecule to which it is to be attached. Variations of the basic fluorescence immunoassay format may be used in order to tailor photophysical properties to a specific application or to overcome some of the disadvantages associated with fluorescence detection. Examples of these variations include fluorescence polarization, time-resolved fluorescence and fluorescence energy transfer.

There is presently still much research on novel visible fluorescent labels for immunoassay. Examples are fluorinated fluoresceins [137], BODIPY dyes [138] and Alexa Fluor dyes [139]. The BODIPY fluorophores were designed as replacements for fluorescein, tetramethylrhodamine and Texas Red. The Alexa Fluor series of dyes, which presently span the whole visible and part of the near-infrared region, have several advantages, such as a high absorbance at wavelengths of maximal output of common excitation sources, efficient fluorescence and high photostability of the bioconjugates. Additionally, Alexa Fluor dyes are well soluble in water, which simplifies bioconjugation and lowers the danger of precipitation and aggregation of the conjugates. The spectra are also not affected by pH between 4 and 10.

Immunochemistry with NIR fluorophores

The use of near-infrared fluorophores as labels in fluorescence immunoassays provides several advantages over visible fluorophores. When using visible dyes, sensitivity is ultimately limited by interference resulting from light scatter, quenching effects, and matrix autofluorescence. Furthermore, visible lasers have the disadvantages of high cost, relatively short operational lifetimes, size, maintenance cost, and limited wavelength range [140]. Near-infrared fluorescence immunoassays have the ability to overcome many of the limitations imposed by visible LIF detection schemes. Detection in the NIR region of the spectrum requires the use of solid-state components. Laser diodes, characterised by long operating lifetimes, low cost, and small size, are ideal as excitation sources. Furthermore, the emission of available laser diodes is compatible with several classes of polymethine cyanine dyes. Avalanche photodiodes are used for detection in the near-infrared region. Avalanche photodiodes have long operating lifetimes, are inexpensive, have internal amplification, and have high quantum efficiencies in the near-infrared region. All of the advantages associated with detection in the near-infrared region allow for rugged, compact, and sensitive instrumentation.

Boyer and coworkers were the first to develop instrumentation for near-infrared fluorescence immunoassays [117]. Williams and coworkers also developed instrumentation for detection of near-infrared fluorescence in solid-phase immunoassays [118]. The instrument consists of a semiconductor laser coupled with a fiber-optic cable, a silicon photodiode for detection, a sample stage coupled to a motor drive, and a data acquisition device. The instrument could detect 500 pM concentrations of human immunoglobulin G (IgG) on a nitrocellulose matrix. The assay was performed in roughly two hours. The detection limits obtained on this instrument were comparable to that obtainable with ELISA. The assay developed by Williams suffers from excessive scatter generated from the membrane, nonspecific binding, and incompatibility with conventional microtiter plate immunoassay formats [140]. Patonay and coworkers developed a NIR fluorescence immunoassay apparatus that overcame many of these limitations. Baars and Patonay have evaluated a novel NIR dye NN382 (Fig. 14.25) for the ultrasensitive detection of peptides with capilary electrophoresis [141]. A solid-phase, NIR fluorescence immunoassay system was

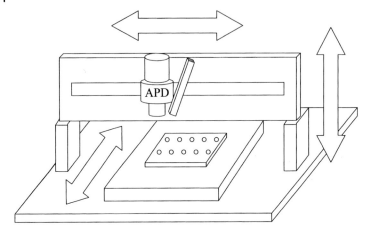

Fig. 14.25 Schematic illustration of near-infrared fluorescence immunoassay instrumentation.

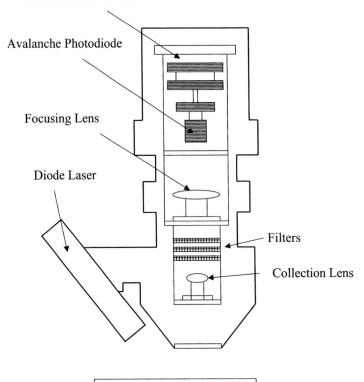

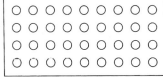

Fig. 14.26 Schematic illustration of the avalanche photodiode detector used in the NIR fluorescence immunoassay instrumentation.

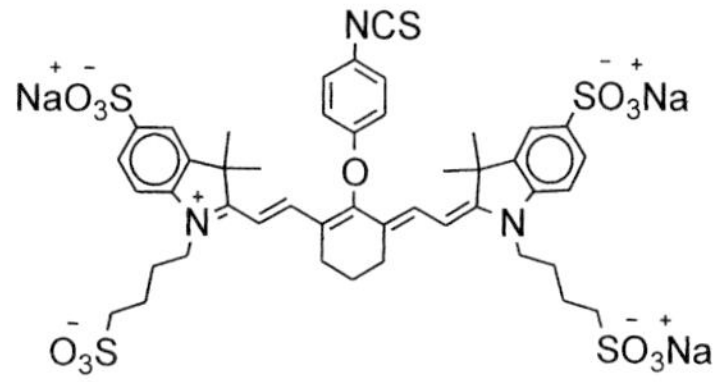

Fig. 14.27 Structure of NIR dye NN382.

developed by coupling a Li-Cor 4200 fluorescence microscope with an orthogonal scanner as illustrated in Figs. 14.26 and 14.27.

Recently several novel NIR fluorescence probes have been reported. Heptamethine cyanine dyes have previously been used for DNA sequencing, metal ion detection, protein labeling, and pH and hydrophobicity determinations [142]. The novel dye NN382 has a high molar adsorptivity (180,000) and a good quantum yield for the dye-protein conjugate (0.59) [143, 144]. The isothiocyanate functionality present on the dye is reactive toward the primary amine groups of the antibody under basic conditions. The sulfonate groups present on the dye increase its aqueous solubility and minimize nonspecific binding to the polystyrene matrix. Numerous microtiter plates were evaluated for background scatter and the one which produced the least noise was chosen for the experiments. It was necessary to optimize the conditions, i.e., temperature, pH, molar ratio, and reaction time, for conjugation of NN382 to goat anti-human IgG. Deviations from the optimal derivatisaton conditions result in deterioration of assay sensitivity. The assay developed was able to detect 20 pM human IgG, resulting in roughly an order of magnitude improvement in sensitivity as compared to traditional labels. Additionally, the assay was less time-consuming than ELISA.

NIR fluorophore-based immuno and DNA-probes

It is presently attractive to utilise NIR fluorescent dyes in optical fiber sensors, allowing for miniaturization, small analyte volumes, and the ability to carry out the analysis in remote locations, while retaining the advantages of NIR detection, such as the lower background interference. Danesvar et al. developed a NIR fiber optic immunosensor (Fig. 14.28), which was applied to the detection of human IgG [145], legionella pneumophila serogroup 1 (LPS1) [146] and the pesticide bromacil [147]. Generally, similar or slightly higher detection limits could be obtained compared to that of an ELISA assay. Fiber optic NIR-fluorescence probes have recently also been used for the assessment of heterogeneity of immobilised antibodies in a fiber optic sensor [148]. The homogeneity of antibody preparations and site-directed immobilisation of the antibodies via the Fc domain clearly yielded a lower heterogeneity.

Although NIR-fluorescence markers are now widespread in bioanalysis, further improvements are still needed to raise the power and lower the cost of the current techniques for applications in high-throughput and in vivo experiments. An array

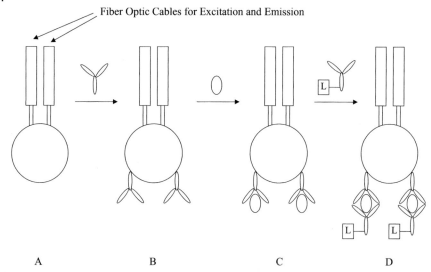

Fig. 14.28 Schematic illustration of the NIR-fluorescence immunoprobe as decribed by Danesvar et al. A) Fiber optic cables are used for remote excitation and emission collection from a sensitive terminal prepared from polymethylmethacrylate or polystyrene. B) A primary antibody is coated onto the surface of the terminal. C) Antigen from the sample is introduced. D) The labelled secondary antibody is quantified.

immunosensor based on NIR fluorescence has been described by Ligler and co-workers [149]. The system consisted of a diode laser at 635 nm and a cooled CCD detector. Samples were introduced by a fluidics system and the analysis was based on a sandwich assay using the Cy5 dye. With this set-up three different analytes could be detected simultaneously.

Another field of work is directed to DNA analysis with NIR fluorophores. Pilevar et al. have recently described a fiber optic sensor for DNA hybridisation, in which the dye IRD 41 (of LI-COR Inc., Lincoln, NE, USA) was used for the real-time hybridisation of RNA of Helicobacter Pylori at picomolar concentrations [150]. Recently optical nanosensing in which nanometer-scale probes are used for intra-cellular measurements has also been pioneered [151].

14.2.2.4 Fluorescence Polarisation Methods

Fluorescence polarisation spectroscopy is still very much used to probe the rotational dynamics of single molecules, either on surfaces or in solution [152]. In bioanalytical assays the fluorescence emission intensity is measured as a function of rotational speed. When a solution of fluorophores is excited with polarised light, the fluorophores selectively absorb those photons that are parallel to the transition moment of the fluorophore, resulting in photoselective excitation. The fluorophore molecules rotate to varying extents during the fluorophore lifetime. If the fluorescence emission is polarised, then small, rapidly rotating fluorophores will give small signals and fluorophores that rotate at slower rates will give large signals. Hence, fluorescence emission intensity is a function of rotational speed, which is correlated to size. In a fluorescence polarisation immunoassay, a labelled antigen is excited with polarised light. When the antibody binds to the antigen, the rotation of the complex is much slower than that of the labelled antigen and the emission intensity increases. It is important that the labelled antigen is small so that its signal may be differentiated from the signal of the immunocomplex. Although fluorescence polarization assays have a limited dynamic range, they are useful for identification or quantification of small molecules. With state-of-the-art optical microscopes and novel fluorescent probes attached to specific protein domains it has also recently become possible to quantify angular rotations in individual protein molecules that mediate specific functions [153].

There have been several new developments in fluorescence polarization immunoassay (FPIA). Lackowicz and Terpetschnig reviewed the use of long-lifetime metal–ligand complexes in fluorescence polarization assays [154, 155]. New complexes with Re(I) and Ru(II) were described for the highly sensitive detection of high-molecular weight analytes by FPIA. Laser-induced fluorescence polarisation detection has been used by Yatscoff and coworkers in capillary electrophoresis detection (CE-LIFP) [156]. For the analyte cyclosporin picomolar detection limits were attained.

14.2.2.5 Time-resolved Fluorescence

One of the factors, which significantly decreases sensitivity in fluorescence immunoassays, is the background fluorescence exhibited by the sample matrix. This 'autofluorescence' decreases exponentially as a function of time. Therefore, it is favourable to use fluorophores with a very long lifetime and use time-resolved fluorescence detection (TR-FIA). In TR-FIA the excitation source is pulsed and the detection is gated such that the detector does not become 'active' until the autofluorescence emission has decayed. This gives a very effective elimination of the fluorescence background. Rare earth metals (e.g. europium) complexed to organic ligands have been much used as labels, due to the very long lifetime of their excited state [157–160]. Additionally, these complexes display a high quantum yield, a large Stokes' shift and very narrow emission lines.

The lanthanide chelates have found general application in fluorometric immunoassays and also form the basis of one of the more popular immunoassay platforms: the DELFIA system by Perkin-Elmer/Wallac. To date a large number of lanthanide chelates have been synthesised and investigated as labels in TR-FIA [160]. In a first generation of reagents, the lanthanides still had to be extracted into a stable soluble complex to give an optimal signal, but by now assays have also been designed in which the extraction step can be omitted. This enables more simplified TR-FIA assays and also gives the possibility of homogeneous assays and micro-imaging applications. Already a number of studies have employed

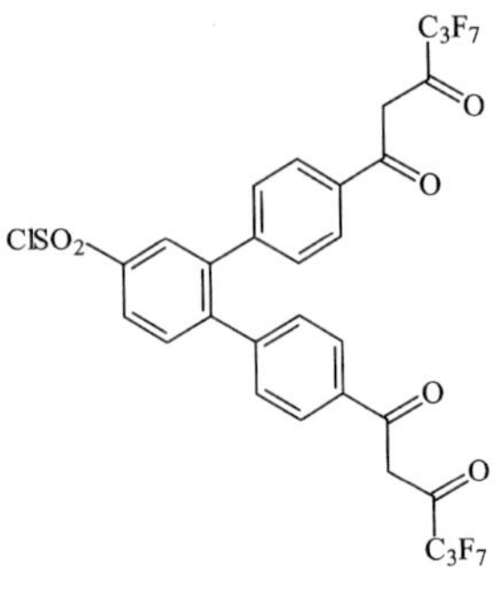

Fig. 14.29 Structure of the novel label 4,4'-bis (1", 1", 1", 2", 2", 3", 3",-heptafluoro-4",6"-hexanedion-6"-yl)-chlorosulfo-o-terphenyl (BHHCT) for TR-FIA [161]. The compound can be covalently bound to proteins and functions as a tetradendate ligand for complexation of Eu^{3+}.

the novel label BHHCT (Fig. 14.29) for TR-FIA. This dye enabled much lower detection limits in a number of assays: e.g. 3.6 pg ml^{-1} for an assay of IgE [161] and 46 fg ml^{-1} for α-fetoprotein (AFP) [162]. More recently, various other reports have appeared on improvements of TR-FIA immunoassays. Zuber et al. developed a mathematical model for the kinetics of a homogeneous immunometric assay with TR-FIA [163]. Europium chelates have been used for the determination of prostate-specific antigen on individual microparticles [164]. A dual-label TR-FIA method has been reported for the simultaneous detection of phenytoin and phenobarbital [165] and for the simultaneous determination of pregnancy-associated plasma protein A and the free β-subunit of human chorionic gonadotropin (hCG) [166].

14.2.2.6 Fluorescence Excitation Transfer

Fluorescence excitation transfer (or fluorescence resonance energy transfer), FRET, is a non-radiative process of energy transfer from one fluorophore (the donor) to another fluorophore (the acceptor), in which excitation of the donor gives rise to a fluorescence signal at the emission wavelength of the acceptor. The process of energy transfer is efficient when the emission spectrum of the donor significantly overlaps with the excitation spectrum of the acceptor [167]. Additionally, the process depends strongly on the distance between the donor and acceptor. FRET is dependent on the inverse sixth power of the intermolecular separation. Thus, fluorescence excitation transfer may be used for enhancement of the Stokes' shift and in bioassays where the distance between two binding partners should be reported.

The first obvious use of FRET is in sandwich immunoassay. A first report along this line was presented by Ullman et al. in 1976 [168]. Wei et al. investigated 10 combinatorial pairs of conventional fluorescent dyes and assessed the optimal conditions for energy transfer [169]. Oswald et al. described a fluorescence resonance energy transfer immunoassay for human serun albumin (HSA) based on the indolium dye Sq635 (Fig. 14.10) and the benz[e]indolium dye Sq660 [83]. A detection limit of 10^{-7} M of HSA was reported.

FRET has been also used to obtain fluorescent particles in which the Stokes' shift can be significantly enlarged and to some degree controlled [170]. This is the main principle behind the TransFluorSpheres of Molecular Probes Inc (Eugene, Oregon, USA). In these microparticles more than two dyes can be used, which form a fluorescence energy transfer chain. The TransFluoSpheres are used as immunofluorescent reagents, as retrograde neuronal tracers, microinjectable cell tracers and standardization reagents for flow cytometry and microscopy (see ref. 33).

14.2.2.7 Bioanalytical Applications of Fluorescent Proteins

Phycobiliproteins

Phycobiliproteins are stable and highly soluble fluorescent proteins derived from cyanobacteria and eukaryotic algae. They contain covalently linked tetrapyrroles that play a role in harvesting light in the photosynthetic reaction center. The biological process involves fluorescence resonance energy transfer from the tetrapyrrole to a pair of chlorophyll molecules [171, 172]. Because of their role in light collection, phycobiliproteins have optimized their absorption and fluorescence and reduced quenching caused either by internal energy transfer or by external factors such as changes in pH or ionic composition. For bioanalytical applications a number of phycobiliproteins are used: B-phycoerythrin (B-PE), R-phycoerythrin (R-PE) and allophycocyanin (APC). These are dyes that enable highly sensitive assay applications with the possibility for multiparameter detection detection. Quantum yields as high as 0.98 and extinction coefficients as high as 2,400,000 have been reported. These compounds are able to give five to ten times greater fluorescence signals than that of conventional fluorescein conjugates. The fluorescence properties of B-PE, R-PE and APC are compared in Tab. 14.6.

Bioanalytical applications of phycobiliproteins (and also green fluorescent proteins, see next paragraph) predominantly lie in flow cytometry and immunoassay [173, 174]. In particular, flow cytometers capable of collecting data from three or four chromophores are presently being developed, and polychromatic flow cytometry (PFC) will soon be a major tool for explorations in cell biology and immunology. The much enlarged amount of information enables characterisation of rare cell populations, allows identification and characterization of novel cell subsets, and identification of functionally homogeneous subsets of cells within the immune system. Recently, Baumgarth et al. have investigated multicolor cytometric systems for up

Table 14.6 Fluorescence data of phycobiliproteins.

Phycobiliprotein	Molecular weight	λ_{Max} (nm)	ε (cm^{-1}M^{-1})	λ_{Max} (nm)	Fluorescence quantum yield
B-Phycoerythrin	240 000	546,565	2,410,000	575	0.98
R-Phycoerythrin	240 000	480,546,565	1,960,000	578	0.82
Allophycocyanin	104 000	650	700,000	660	0.68

14.2 Optical Spectroscopy in Bioanalysis

Table 14.7 Fluorescence data of green fluorescent proteins (from Ref. 177).

Protein abbreviation/ species/colour	Molecular weight	λ_{max} (nm)	λ_{em} (nm)	ε (cm^{-1}M^{-1})	Fluorescence quantum yield
GFP/*Aequorea victoria* (yellyfish)/green	27 000	397	509	27 600	0.80
amFP486/*Anemonia majano*/green	25 400	458	486	40 000	0.24
zFP506/*Zoanthus* sp./ Yellow-green	26 100	496	506	35 600	0.63
dsRed/*Discosoma*/ Orange-Red	28 000	558	583	75 000	0.70
dsFP483/*Discosoma striata*/blue-green	26 400	443	483	23 900	0.46
cFP484/*Clavularia*/ green	30 400	456	484	35 300	0.48

to 11 distinct fluorescent signals and two scattered light parameters for characterizing single cells also using phycobiliproteins [175].

Green fluorescent proteins

Green fluorescent proteins from the jellyfish *Aequorea victoria* (GFP) and various other marine organisms like the *Anthozoa* species present in coral reefs provide interesting fluorescent proteins, which are increasingly used in bioassay work today. They are investigated as markers of gene expression and for protein localisation (Tab. 14.7) [176, 177]. Many of these proteins are presently commercially available from CLONTECH Laboratories Inc. (Palo Alto, CA, USA). Presently the orange-red fluorescent protein "DsRed" (trade name by CLONTECH) from the corallimorpharian *Discosoma* genus has received much interest, because its fluorescence in the red enables better suppression of autofluorescence and can be very well used in FRET experiments [178]. (DsRed is an excellent acceptor for excitation by the yellow fluorescent variants of GFP as a donor dye.) The fluorophore in green fluorescent proteins is formed from a tripeptide (Gln-Tyr-Gly) sequence in the wild type GFP by autocatalysis. It has been proposed by Gross et al. that a continued oxidation may lead to a further extension of the conjugated side chain leading to the chromophore in the DsRed protein molecule (Fig. 14.30) [179].

14.2.3
Bioanalytical Applications of Multi-photon Fluorescence Excitation (MPE)

14.2.3.1 Introduction
As an interesting alternative to normal fluorescence excitation, multiphoton excitation (MPE, see next paragraph) has recently appeared as a viable method in various bioanalytical applications [180, 181]. Two-photon excitation (2PE) has been used

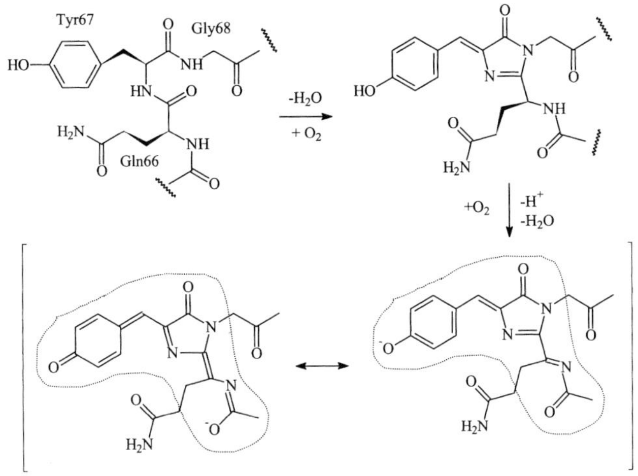

Fig. 14.30 Proposed mechanism of formation of the red fluorescent chromophore from amino acid side chains in DsRed.

most frequently, but there are also many examples of three-photon excitation (3PE) to be found in the literature. The low-background and small excitation volumes attainable with MPE suggest that multi-photon excitation is best suited for applications requiring small volumes (femtoliters or lower). Thus, MPE is presently regarded as a superior detection method in capillary electrophoresis enabling also multi-parameter measurements. As the detection limit of the direct integration measurement may reach down to a single molecule, 2PE is also applied successfully in bioaffinity assays. Additionally, due to the high transparency of tissues and other biological media in the far-red and near IR, MPE is a unique tool for three-dimensional imaging of tissues and cells. In these paragraphs the general bioconjugate chemistry and some recent bioanalytical applications are briefly introduced.

14.2.3.2 MPE Fluorescence Dyes

Various investigations on the multiphoton excitation of biological molecules, such as the aromatic aminoacids (tryptophan, tyrosine and phenylalanine), FAD, NADH, serotonin and melatonin [182–185], have appeared in the literature. There are also many synthetic dyes that have peak two-photon cross sections large enough to be useful as probes in bioassays [186]. Albota et al. recently designed novel compounds with increased MPE fluorescence [187]. The concept was to construct donor–acceptor–donor (D–A–D) or acceptor–donor–acceptor (A–D–A) dyes of sufficient length, in which the excitation causes a symmetric charge transfer from the ends of the molecule towards the center or vice versa. Excitation usually effects a charge displacement from the donor end of the molecule towards the acceptor end. Three important features of the molecules were studied: 1. the nature of the molecule (A–D–A or D–A–D), 2. the conjugation length and 3. the strength of the donor or acceptor groups. By varying these parameters the two-photon excitation could be maximised through increasing the 2PE cross section, $\delta(\omega)$, which is related to the imaginary part of the second hyperpolarizability, Im $\gamma(-\omega;\omega,\omega,-\omega)$ according to:

$$\delta(\omega) = \frac{8\pi^2 \eta w^2}{n^2 c^2} L^4 \mathrm{Im}\, \gamma(-\omega;\omega,\omega,-\omega) \tag{2}$$

Since the second hyperpolarizability is involved in the equation, symmetrical molecules of the D–A–D type and A–D–A type are suitable, but also donor-substituted porphyrins, in which charge transfer may occur in two dimensions. Table 14.8 briefly summarises MPE cross sections of some reported dyes. Some of the special dyes synthesised for very large MPE effects are presented in Fig. 14.31. As can be observed from the data, there is a reasonably good correlation between the observed and calculated excitation and cross section data. The increase in length gave the most substantial increase in the 2PE cross section and in excitation wavelength.

Table 14.8 Comparison of MPE cross sections of conventional and engineered chromophores.

Compound	λ_{max} (nm)	λ_{2PE} (nm)	δ_{2PE} (10^{-50} cm^4.s/photon)
rhodamine B		840	210
fluorescein (pH=11)		782	38
compound 1		514	12
compound 2	374	605	210
compound 3	408	730	995
compound 4	428	730	900
compound 5	456	775	1250
compound 6	472	835	1940

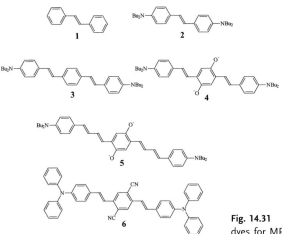

Fig. 14.31 Structures of some engineered dyes for MPE (from ref. 200).

14.2.3.3 Two-photon Excitation Immunoassays

Recently, 2PE has been applied in a variety of immunological and DNA hybridisation assays with promising results. The experiments of Soini and coworkers have shown that homogenous single step bioaffinity assays can be performed with good sensitivity and dynamic range by using two-photon fluorescence excitation [188]. The assay concept is a very promising alternative when there is a need to reduce sample volumes, because the signal is obtained from a very small focal volume, in which the signal strength is independent of the sample volume (Fig. 14.32). The sensitivity is two orders of magnitude larger in solution measurements compared to conventional fluorometric techniques. The main advantage is that the technique allows for single-step, non-separation assays, working for both immunometric and competitive binding assays (homogeneous immunoassay). Other advantages are that the sample cuvette does not contribute to the background signal, and that kinetic reaction monitoring and multiparametric measurements are possible.

The usefulness of two-photon fluorescence excitation in homogeneous bioaffinity assays has been verified by using microbeads as the solid phase [188]. Biochemically activated 3 μm polystyrene microbeads were used as solid phase to bind the analyte α-fetoprotein (AFP) (Fig. 14.33). Each microbead acts as a local concentrator of the analyte. When a fluorescent biospecific reagent molecule either attaches directly to the analyte forming a sandwich-type Ab–Ag–Ab*-complex on the microbead surface (immunometric measurement) or competes in binding to the surface of the microbead (competitive measurement), the amount of analyte molecules bound to the microbeads becomes measurable by observing the two-photon fluorescence signal from individual microbeads. In such an assay the analyte and the reagent solution, comprised of microbeads and a fluorescent tracer, can be dispensed simultaneously into a single reaction volume. The signal from the microbeads is measured after incubation directly from same reaction volume.

Due to the concentrating effect on the microbeads, the signal from the tracer bound to each microbead for the full dose of the analyte is several orders of magnitude stronger than the signal background from the free tracer. In fact, the signal of the free tracer at the zero dose level determines the lowest limit of the working range. The working range and sensitivity depend on the assay parameters such as affinity, microbead capacity, number of microbeads in an assay and tracer concentration. Theoretically, a linear working range of up to four orders of magnitude can be reached.

14.2.3.4 MPE in Gel and Capillary Electrophoresis

As discussed in Section 14.2.1, normal (one-photon) fluorescence spectroscopy (1PE) has been the method of choice in capillary electrophoresis (CE) and gel electrophoresis (GE) with visible and NIR fluorescent dyes. MPE fluorescence, however, is also quite suitable for detection in CE and GE. For instance, Song et al. fractionated coumarine dyes with capillary electrophoresis, and detected the dyes at attomole concentrations by 2PE fluorescence [189]. Generally, detection limits are comparable to those attainable by normal fluorescence. However, MPE is particu-

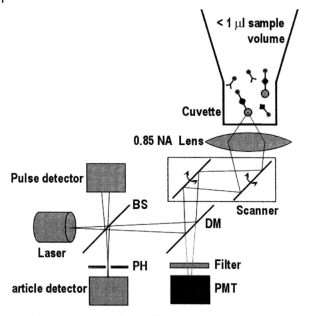

Fig. 14.32 The optical scheme of a typical 2PE set-up; BS is a beam splitter, DM is a dichroic mirror, PH is a pinhole and PMT is a photomultiplier tube.

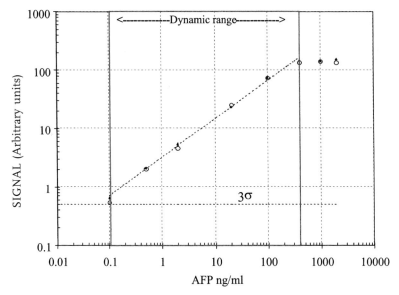

Fig. 14.33 Two standard curves for an AFP assay with microbeads with two different measurement volumes, 25 μl (◆) and 2 μl (O). The curves show that both volumes give practically same result.

larly useful in situations where multiparameter measurements are required and where the compounds to be detected have similar spectroscopic properties, but differ in their MPE cross sections. MPE is also useful when the analytes have very different spectral properties, but where the MPE gives possibilities for excitation with a single light source. 2PE and 3PE fluorescence of FAD, NADH, serotonin, melatonin and similar compounds in CE has been extensively studied by Shear and coworkers [180, 184, 190]. All these analytes had substantial MPE cross sections in the wavelength range between 710 and 750 nm. The detection limits of the CE/MPE detection scheme ranged from 350 zmols for FAD to 27 amols for serotonin [190].

14.2.3.5 MPE in Tissue Imaging

As remarked in Section 14.2.2.1, the high transparency of biological tissues in the near-IR region and the low photon energy allow high intensities to be used without the risk of photobleaching and photodamage of the specimens. Additionally, with the use of highly focussing optics, as used in confocal microscopy, very small volumes in biological tissues can be sampled with MPE fluorescence, achieving unprecedented 3-D resolution. Thus, microscopic specimens can be raster-scanned, keeping the focal point within a plane perpendicular to the laser beam (x–y direction) and collecting fluorescence and background photons from a tightly confined spatial region as a function of the x and y positions. Additionally, the specimen (or objective) can be adjusted in micrometer steps in the z direction at different planes within the sample.

An interesting demonstration of such imaging capabilities was given by Kleinfeld et al., who introduced MPE laser scanning microscopy to the imaging of cortical blood flow at the level of individual capillary blood vessels in the rat neocortex through openings created in the crania [191]. 2PE fluorescence was excited with a 830 nm laser beam, attaining a depth resolution down to 600 µm. This is several hundred micrometers deeper than can be attained with conventional confocal microscopy. Fig. 14.34a represents a horizontal view in the vicinity of a capillary reconstructed from a set of scans in the x–y direction. These scans were acquired at between 310 and 410 µm depth in 1 µm steps. The inset shows the intensity profile along the cross-section for a scan that passed through the central axis of the selected capillary. The vessel cross-section ('caliber') was estimated from the number of pixels with intensity above the background level. Figure 14.34b gives successive planar images through a small vessel, acquired with a 16 ms interval. The change in position of unstained objects, interpreted as red blood cells, is indicated by the series of arrows. The velocity of the red blood cell was estimated to be 0.11 mm s^{-1}.

14.2.3.6 Future Prospects of MPE Fluorescence Spectroscopy

As can be concluded from the previous examples, MPE fluorescence is useful, but must at the present stage still be considered as a complementary tool to existing, more straightforward bioanalytical techniques. There are still many instances in

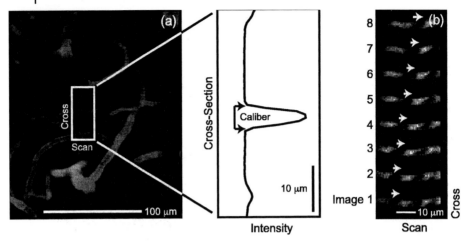

Fig. 14.34 MPE fluorescence images of structures in the rat *neocortex* (reproduces from ref. 191, with permission): (a) Image of single microvessels, whose caliber was determined from the measured cross section in a planar image (a, inset), (b) successive, rapidly acquired planar images of the microvessel, revealing the movement of dark objects (non-fluorescent red blood cells) in the microvessel containing serum spiked with a fluorescent label.

which single-photon excitaton or single photon confocal scanning microscopy yields equivalent or even better results with more economic equipment [181]. Presently, the high cost of Ti:Sapphire lasers presents a substantial barrier to the construction of cheap devices, but novel diode lasers may contribute in the near future to reducing the cost. MPE fluorescence offers presently the greatest advantages in the characterization of biological samples and tissues. Additionally, there are interesting new perspectives in immunoassays based on MPE, because it enables fast single-step, separation-free immunoassays and DNA hybridisation assays in very small volumes [188]. Recently 2PE fluorescence polarization measurements have also been presented in optically dense specimens [192]. As indicated, the method may also be very useful for high throughput screening of drugs. Monitoring of the release of products (e.g. various cytokines, tumor necrosis factor, Fasligand etc.) can be made in a one-step assay without coated tubes and separation procedures. The method is applicable also for whole blood samples because the laser illumination is far beyond the absorption of hemoglobin. Thus, MPE as introduced in this section, has the potential to become a new generic platform for *in vitro* diagnostics when less expensive pulsed laser light sources also become a reality.

14.2.4
Bioluminescence, Chemiluminescence and Electrochemiluminescence

Luminescence phenomena are used in many bioanalytical assays. A first overview of BL and CL reagents and bioanalytical techniques was described in 1978 in a volume of Methods in Enzymology, edited by Marlene DeLuca [193], and hereafter there have been various new developments, both in new reagents and in instrumental design, which have placed these techniques at the forefront of bioanalysis [194, 195]. At present, electrochemically generated chemiluminescence (ECL) is gaining the larger interest for realizing highly sensitive bioassays, particularly immunoassays [196, 197].

A classical bioluminescent system is the luciferase system from the firefly *photinus pyralis*. This system has been widely used for the determination of ATP and analytes that are involved in the conversion of ATP. In the firefly luciferase reaction, ATP and the reduced form of the dye compound luciferin is converted to AMP and an oxidized and decarboxylated form of luciferin in the presence of magnesium and the luciferase enzyme (Fig. 14.35A). This reaction produces light with a quantum efficiency of almost 100%. Classic applications of the luciferine/luciferase reaction are, for instance, the determination of ATP in mitochondria. Enzymatic reactions can be coupled via ATP to other enzymatic reactions (e.g. kinases), which broadens the application potential of the luciferase system to a large variety of analytes. Magnesium activity can also be determined very sensitively, because the reaction depends greatly on magnesium. A second class of bioluminescent reactions comprises the bacterial luciferases, which are part of the electron-transport pathway from reduced substrates to oxygen via flavins. Thus, these bioluminescent reactions can be coupled to enzymes that convert flavins like NADH and $FADH_2$. Most bacterial luciferases convert the reduced form of flavinmononucletide (FMNH2) to its oxidised form (FMN) producing light very efficiently (Fig. 14.35B).

The classical CL reaction is that of luminol with hydrogen peroxide catalyzed by horseradish peroxidase (HRP). Such reactions are frequently coupled to hydrogen peroxide-producing enzymatic reactons, e.g. the oxidase enzymes [198], providing a sensitive alternative to HRP-based absorbance or fluorescence detection (Fig. 14.35C). New CL reagents have recently been developed, including novel luminol and isoluminol derivatives [199, 200], acridinium ester labelling reagents, [201–203] and 1,2-dioxetane derivatives [204, 205]. Recently a regenerable immunosensor has been described by Marquette and Blum for the detection of the herbicide 2,4-dichlorophenoxyacetic acid (2,4-D), based on the luminol/HRP system [206]. A CL-based fiberoptic sensor in a flow injection analysis (FIA) configuration with a competitive immunodetection scheme was used in this system.

In ECL the luminescence is produced as the result of an electrochemical reaction. A reactive species is produced electrochemically at an electrode and diffuses into the bulk solution and reacts with chemicals in the vicinity of the electrode. There are various mechanisms by which ECL can be initiated: (1) by electrochemical initiation of a conventional CL reaction (e.g. of luminol), (2) by electrochemical modification of an analyte molecule into a species which can take part in a CL re-

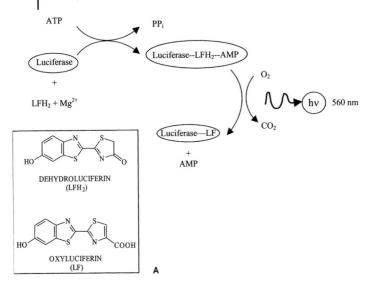
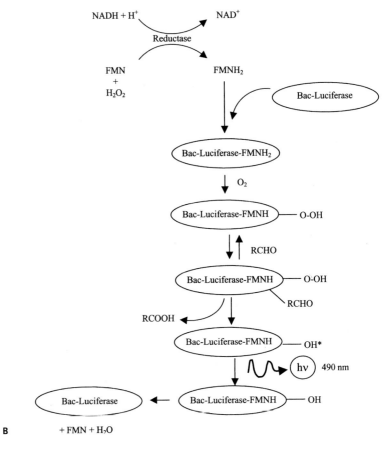

C

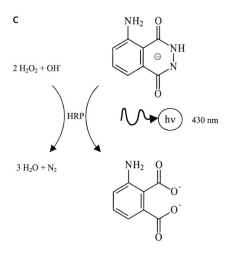

D

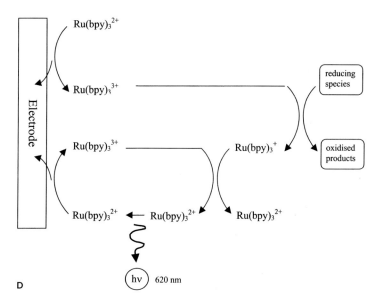

Fig. 14.35 Examples of various luminescent reactions. (A) The firefly Luciferase reaction. (B) The bacterial luciferase reaction. (C) The luminol reaction with horseradish peroxidase catalyst. (D) The electroluminescence of $Ru(bpy)_3$.

action, (3) by high energy transfer of electrons to or from electrochemically generated organic radicals (e.g. radical ion recombinations between polyaromatic hydrocarbons), (4) through high energy electron transfer reactions between inorganic ions (e.g. between transition metal complexes) and (5) through emissions from certain oxide-covered valve metal electrodes (i.e. cathodic luminescence). Electrochemiluminescence has initially relied much on the luminol reaction, but presently the majority of analytical techniques are based on ruthenium trisbipyridyl (Ru(bpy)$_3$) complexes, which has the great advantage that the luminescent reagent can be regenerated at the electrode. This results in an enhanced detection sensitivity (Fig. 14.35D). The advantages of ECL detection over CL and BL are that the luminescence reaction (initiation, rate and course) can be electronically controlled and even modulated, while the light is produced very close to the electrode. This gives major advantages in sensitivity. Additionally, ECL is more compatible with flow injection analysis techniques and additional information can be obtained about the reaction via electrochemistry. As a disadvantage can be mentioned that sensing applications are still difficult to realise due to the need to add reagents to the assay system.

Immunoassays presently form probably the most active area of exploitation of CL, ECL and BL, as evidenced by the extensive commercialization of luminescent immunoassay analyzers and test kits. Among these, acridinium ester-based CL assays for alkaline phosphatase and amplified CL assays using peroxidase labels have become the most widespread [207, 208]. Besides immunosassays CL and BL have been used in nucleic acid assays [209–211] and in cellular studies concerning, for instance, phagocytosis [212]. ECL immunoassays based on ruthenium trisbipyridyl labels are now becoming popular in clinical immunoassay analyzers [213].

Recently the incorporation of bacterial luciferase genes ("Lux genes") into bacteria (e.g., recombinant *E. coli*) has also been successfully exploited. The group of Karp and co-workers in Finland has pioneered this area and recently reported a BL-based biosensor for the specific detection of the tetracycline group of antibiotics using a bioluminescent *Escherichia coli K-12* strain [214]. The *E-coli* contained a 'sensor' plasmid, containing five genes from the bacterial luciferase operon of *Photorhabdus luminescens*, which yielded a tetracycline-dependent light production.

14.2.5
Bioanalytical Applications of NIR Absorption Spectroscopy

Near-infrared absorption spectroscopy is increasingly used in agriculture, food science, medicine, life sciences, pharmaceuticals, textiles, general chemicals, polymers, process monitoring, food quality control and in clinical *in vivo* measurements [215, 216]. The increase in popularity is largely due to the availability of miniaturised NIR-spectrometers by a variety of vendors (e.g. Ocean Optics Inc.). The most promising applications of NIR-absorbance spectroscopy clearly lie in process control, because of the relatively low complexity of the sample in chemical and biochemical processes, e.g. compared to biological tissues. Also in food quality control

NIR-IR has been successfully used, as exemplified by the analysis of fat in raw milk [217].

NIR-spectroscopy has also been prospected for use in *in vivo* monitoring of important tissue and blood metabolites, such as glucose, creatinine, lactate, urea, cholesterol, oxygen and hemoglobin [326]. NIR spectroscopy can be easily applied to the measurement of tissues due to the great transparency of skin for near-IR radiation. The most interesting analyte for in vivo monitoring is, unquestionably, glucose. There is still a huge need for the reliable diagnosis of diabetes mellitus, and self-monitoring of blood glucose in people with diabetes with a non-invasive method would give substantial advantages over the present finger-prick methods. For in vivo glucose monitoring promising systems have been designed, which rely heavily on multivariate calibration methods [218, 219]. Unfortunately, the present NIR absorbance methods, as based on multivariate calibration, are still quite unreliable. This has been recently demonstrated by Arnold and coworkers [220]. Essentially, the unreliability is due to the small signals produced by glucose relative to other signals and the complexity of the tissue matrix. Presently, *in vivo* glucose monitoring is surfacing more rapidly in the patent literature as in the scientific literature.

14.2.6
Bulk Optical Sensing Techniques

Within the rapidly growing field of chemical sensor and biosensors, optical sensing strategies are still very much at the center of interest, as evidenced from the untiring quest for new principles of transduction and the commercialisation of optical sensing instruments [221]. A chemical sensor generally comprises the integration of a chemical recognition element with a particular detection system. In the case of a biosensor, a biomolecule is used as the recognition element. Chemical sensor systems should be capable of providing continuous, specific quantitation without the need for addition of external reagents [222]. A variety of approaches for the generation of an optical signal from a selective biological binding event have been evaluated. Although most of the earlier developed systems relied on absorption or fluorescence detection, research has now shifted much towards the detection of refractive index changes. The next section will be dedicated mainly to the evanescent wave-based techniques, while this section briefly introduces the methods based on bulk absorbance and fluorescence transduction.

The receptors that can be used in optical sensors can be either synthetic or from a biological source, while there are also many approaches to chemically or genetically engineered biological receptors for use in sensors [223]. Presently, combinatorial chemistry and molecular imprinting are at the forefront of receptor research, particularly for the more complicated analytes, such as drugs and chemical warfare agents. Cation receptors were the first prototypes of designed receptors, based on the rules of supramolecular chemistry [224]. Cation receptors based on calixarenes have been used, e.g. for the direct measurement of sodium in blood [225]. The design of anion receptors has also been the subject of intensive research due to the fundamental role that anions play in biochemical processes [226]. Anion sensing, however, is yet a degree more complex than cation sensing and has, as a consequence, been slower in its development. Anions are much larger and their shapes are more variable, while they exist only in a limited pH range. Thus, a potential anion receptor must be designed to satisfy the particular anion's unique characteristics of size, shape and pH dependence.

In ion-selective electrodes a signal is generated by the permselective exchange of ions into a hydrophobic membrane phase. For optical transduction the ion-receptor complexation has to be coupled to a complementary chemical process in an absorbing or fluorescent dye compound. Such sensing matrices are used frequently in conjunction with fiber optics for the construction of sensor systems. A first way to achieve optical transduction is to include a lipophilic acidochromic dye (chromoionophore) into the membrane phase together with the receptor. The complexation of the metal ion by the ligand will expel a proton, which changes the absorbance or fluorescence spectrum. The advantage is that no synthetic modification of the receptor or dye is needed. Originally, lipophilic Nile Blue (phenoxazine) derivatives were much used, but also acridine, fluoresceine and various other synthesised chromoionophores have been described with signal-transducing properties [227–230]. Chromoionophores that are compatible with lipophilic matrices and have absorptions in the NIR region have also been described [80]. Many cation receptors and chromoionophores are commercially available. For instance, the ETH series of chromoionophores is available from Fluka AG, which has also issued a practical guide for making optical sensing matrices.

That these sensors have practical significance is illustrated by the publications of Hisamoto et al. and Wang et al., which have described the measurement of common ions directly in serum [228, 229]. Alternatively, the complexation of the metal ion may influence the partition of the lipophilic dye in the membrane phase, which can affect the fluorescence yield of the dye. Recently this principle has been demonstrated with potential sensitive dyes by Wolfbeis and Mohr [231, 232]. This approach is also applicable to the sensing of neutral species and anions, as exemplified by sensor matrices for 2-phenetylamine [233] and nitrate [234].

In a second approach, the dye is made an integral part of the receptor, such that absorbance and or fluorescence parameters are directly modified by the complexation reaction. This is a more developed form of indicator chemistry and is in supramolecular terminology defined as "semiochemistry" [224, 235]. Receptor molecules modified with photosensitive groups may display very large changes in their photophysical properties upon the binding of analytes, and this can enable very sensitive detection. For instance, Chapoteau et al. reported already in 1992 a colorimetric method for the determination of lithium in blood serum, not requiring any sample pretreatment or solvent-extraction steps [236]. The chromogenic ionophore exhibited exceptionally high selectivity for lithium over sodium (>> 4000:1). Many ion receptors, and also receptors for small organic compounds have been reported in the literature, based on azophenol [236, 237], spirobenzopyran [238], antracene [239] or coumarine [240]. Presently, calixarenes with covalently linked chromogenic or fluorogenic groups are still at the center of interest [241]. Although already in

1992 Shinkai et al. and Diamond et al. had reported on different azophenol-modified calixarenes for quantitation of lithium [242, 243], more complex structures have recently been produced, of which the usefulness in bulk optodes and in realistic sample matrices still needs to be assessed. Modified calixarenes have been produced for sensing of anions [244] and small organic molecules, such as dopamine [245] and 2-phenetylamine [233], while chiral recognition by calixarene derivatives has also been studied [246]. Furthermore, calixarenes have been produced in which fluorescence can be switched on upon binding of the analyte, a principle that may afford extremely sensitive sensors [247]. Recently, a variety of new squarilium dyes with signal transducing properties have been reported in the literature. Akkaya et al. have reported on squarine dyes with phenylboronic acid groups for the detection of carbohydrates [248] and similar dyes with sensitivity for zinc ions [249] and calcium ions [250].

Imprinted polymers have lately shifted into the center of interest, particularly in the quest for sensors for analytes more complex than ions [251]. The principle of molecular imprinting is basically simple: a well-optimised mixture of monomers with various functional groups is polymerised in the presence of the analyte (the 'template'). Hereafter, the analyte is eluted from the polymer matrix, which leaves complementary binding sites in the polymer for the analyte. This polymer can then be used in an analytical system or sensor. Artificial enzymes can also be produced by molecular imprinting, in which a transition state analogue of a chemical reaction is used as the template molecule [252].

There are already some reports of the combination of fluorescence with imprinted polymers [253]. If the functional monomers are fluorescent and are designed to have specific chemical interactions with the analyte it is possible to directly monitor the binding via changes in fluorescence of the polymer [254, 255]. For reasons of background and sensitivity, the activation of fluorescence by the binding of analyte to an imprinted polymer would be the most preferred situation. Turkewitsch et al. described a first example of this novel design for template-selective recognition sites in imprinted polymers with the analyte cyclic AMP [254]. The polymer included a fluorescent dye, a dimethylaminostyryl-pyridinium derivative, as an integral part of the recognition cavity, thus serving as both the receptor and transducing element for the fluorescence detection of cAMP in aqueous media. The imprinted polymer displayed quenching of fluorescence in the presence of cyclic AMP in aqueous solution, whereas almost no effect was observed in the presence of cyclic GMP. The affinity constant of the polymer for cyclic AMP was about 10^5 M^{-1}.

14.2.7
Evanescent Wave Spectroscopy and Sensors

14.2.7.1 Introduction
To perform spectroscopy on biomolecules at a surface, it would be favourable to confine the measurement to a thin layer at the surface, in the region where the molecules are immobilised. Typically, this region would range from 5 nm for

small proteins to 50 nm for large protein assemblies such as ribosomes. Methods based on evanescent waves permit such spectroscopy by guiding light parallel to the surface of the optical substrate upon which the biomolecules are immobilised. According to the Fresnell equations, light is internally reflected at interfaces between high and low refractive index at angles larger than a critical angle. The electrical field, however, penetrates into a thin layer of the medium with a lower refractive index to some degree and with a strong exponential decay at increasing distances form the surface (the *evanescent wave*). Before discussing the various applications the theory of evanescent waves will be briefly reviewed.

14.2.7.2 Theory of Total Internal Reflection

Reflection and refraction

At a dielectric interface, we can have both reflection and transmission, as shown in Fig. 14.36 [256]. The following conditions must be satisfied:

$$\theta_3 = \theta_1$$

$$\frac{\sin\theta_2}{\sin\theta_2} = \frac{n_1}{n_2} = \left(\frac{\varepsilon_1}{\varepsilon_2}\right)^{1/2} \quad (3)$$

Where n is the refractive index, ε the permittivity and μ the permeability of the dielectric media. The incident, reflected and refracted rays are coplanar (located in the x–z plane, the plane of incidence, in Fig. 14.36). Transverse electric (TE), perpendicular ($\perp$) or s-polarised light has its electric vector perpendicular to the plane of incidence (x–z plane) in Fig. 14.36, while transverse magnetic (TM), parallel ($\parallel$) or p-polarised light has its magnetic vector perpendicular to the plane of incidence.

The Fresnel equations

The Fresnel equations describe the reflection and transmission coefficients at the interface of two optical media. The polarisation of the incident light affects the magnitude of these coefficients. It is possible to derive expressions for the intensities of the reflected and refracted rays. These differ for the TE and TM polarisations as follows:

$$R_{TE} = \frac{\sin^2(\theta_2 - \theta_1)}{\sin^2(\theta_2 + \theta_1)} = \frac{\left(n_1 \cos\theta_1 - \sqrt{n_2^2 - n_1^2 \sin^2\theta_1}\right)^2}{\left(n_1 \cos\theta_1 + \sqrt{n_2^2 - n_1^2 \sin^2\theta_1}\right)^2} \quad (4)$$

$$T_{TE} = \frac{\sin 2\theta_1 \sin 2\theta_2}{\sin^2(\theta_2 + \theta_1)} = \frac{4n_1 \cos\theta_1 \sqrt{n_2^2 - n_1^2 \sin^2\theta_1}}{\left(n_1 \cos\theta_1 + \sqrt{n_2^2 - n_1^2 \sin^2\theta_1}\right)^2} \quad (5)$$

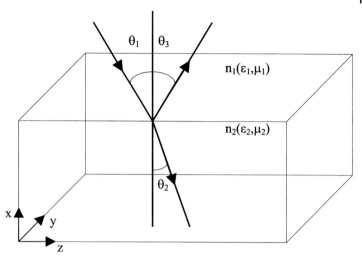

Fig. 14.36 Reflection and refraction of light at the interface between two dielectric media having different refractive indices.

$$R_{TM} = \frac{\tan^2(\theta_1 - \theta_2)}{\tan^2(\theta_1 + \theta_2)} = \frac{\left(n_2^2 \cos\theta_1 - n_1\sqrt{n_2^2 - n_1^2 \sin^2\theta_1}\right)^2}{\left(n_2^2 \cos\theta_1 + n_1\sqrt{n_2^2 - n_1^2 \sin^2\theta_1}\right)^2} \quad (6)$$

$$T_{TM} = \frac{\sin 2\theta_1 \sin 2\theta_2}{\sin^2(\theta_1 + \theta_2)\cos^2(\theta_1 - \theta_2)} = \frac{4n_1 n_2^2 \cos\theta_1 \sqrt{n_2^2 - n_1^2 \sin^2\theta_1}}{\left(n_2^2 \cos\theta_1 + n_1\sqrt{n_2^2 - n_1^2 \sin^2\theta_1}\right)^2} \quad (7)$$

These expressions are only valid for non-normal incidence ($\theta_1 > 0$). For the special case of normal incidence, these expressions become:

$$R = \left(\frac{n_1 - n_2}{n_1 + n_2}\right)^2 \quad (8a)$$

$$T = \frac{4n_1 n_2}{(n_1 + n_2)^2} \quad (8b)$$

Note that the polarisation has no effect at normal incidence, as it is not possible to distinguish a particular plane of incidence.

Total internal reflection (TIR)

If we have the condition $\theta_2 = \pi/2$, then the transmission coefficients for both TE and TM polarised light go to zero, because the term $\sin 2\theta_2$ becomes zero. Thus, the transmission coefficient is zero when $\theta_2 = \pi/2$. In this case,

$$\sin\theta_c = \frac{n_2}{n_1} \tag{9}$$

And the angle of incidence at which this condition is true is called the critical angle. Since the sine function can only produce values from −1 to +1 for real angles, we can see that for this condition to be satisfied $n_2 < n_1$. In other words, total internal reflection can only occur when light travels from a high index to a lower index medium. For angles of incidence greater than or equal to the critical angle, total internal reflection occurs regardless of the polarisation.

Evanescent waves
When the angle of incidence at a high-low refractive index boundary is greater than the critical angle, light is totally internally reflected. It can be shown that, although there is no energy transmitted across the interface, an electromagnetic wave, the evanescent wave, is present on the other side of the boundary. This wave propagates parallel to the interface and decays away exponentially with distance from the boundary. The penetration depth of the evanescent wave is given by:

$$d = \frac{\lambda}{2\pi\sqrt{n_1^2 \sin^2\theta_1 - n_2^2}} \tag{10}$$

Where γ is the vacuum wavelength of the light. The penetration depth is the distance over which the intensity of the evanescent field decays to $1/e$ of its original intensity. This means that, except near the critical angle, the penetration of the evanescent wave into the optically rarer medium is no more than a wavelength. As an example, if we take an SF10 prism ($n=1.732$) in contact with water ($n=1.333$) at a wavelength of 633 nm (HeNe laser), then the critical angle is 50.32 degrees. If our angle of incidence is 60 degrees, the penetration depth is approximately 146 nm.

While the evanescent field does not propagate into the rarer medium, it does propagate parallel to the interface and can interact with the lower index medium. Thus, if the lower index medium is absorbing, fluorescent or scattering, some of the light will be absorbed and the reflectivity will be reduced from unity. In the case of fluorescent or scattering media, fluorescence emission or scattered light from the evanescent field will be observed.

Optical absorption
The attenuation factor a links the imaginary part K of the complex refractive index to the loss per unit length:

$$I_x = I_0 e^{-ax} \quad \text{or} \quad \ln\frac{I_x}{I_0} = -ax \tag{11}$$

Where I_0 is the initial intensity and I_x is the intensity after the light has propagated a distance x through the absorbing medium, and

$$a = \frac{4\pi K}{\lambda} \tag{12}$$

But the Lambert-Beer law states that

$$\log \frac{I_0}{I_x} = \varepsilon c x \tag{13}$$

Where ε is the molar extinction coefficient (m^2 mol^{-1}) and c the concentration (mol m^{-3}). Combining Eq. (11), (12) and (13) yields:

$$K = \frac{\varepsilon c \lambda \ln 10}{4\pi} \tag{14a}$$

Thus, we can calculate the imaginary part of the complex refractive index of a dielectric medium if we know the extinction coefficient and concentration of the absorbing species. If the lower index medium in a total internal reflection configuration has a complex refractive index, that is it is lossy, then the reflection coefficient is reduced from unity, even for angles of incidence above the critical angle.

Attenuated total internal reflection (ATR)

The small decrease in reflectivity observed when light is reflected by TIR at the interface between non-absorbing and absorbing media is termed attenuated total internal reflection. If the light is scanned over a range of wavelengths, then an absorption spectrum can be generated. Figure 14.37 illustrates the effect of the ima-

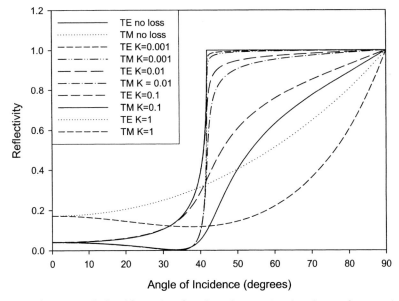

Fig. 14.37 Reflectivities calculated for angles of incidence between 0 and 90 degrees for internal reflection at the interface between dielectric media having refractive index 1.5 and 1.0 + iK.

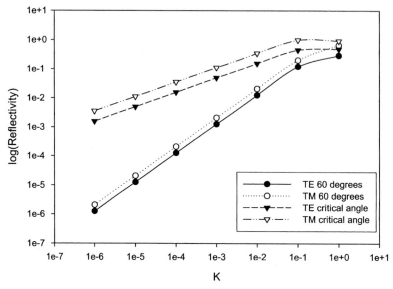

Fig. 14.38 Log(reflectivity) for internal reflection at the interface between dielectric media having refractive index 1.5 and 1.0 + iK for two angles of incidence, 60 degrees and the critical angle (41.8 degrees).

ginary part of the refractive index on the reflectivity in a simple TIR configuration. When the imaginary part of the complex refractive index is quite small (0.001), the reflectivity curves only deviate significantly from the curves for the no-loss situation around the critical angle. At higher loss values, the curves deviate strongly from the no-loss values.

Figure 14.38 shows log(reflectivity) for internal reflection at the interface between dielectric media having refractive index 1.5 and 1.0 + iK for two angles of incidence, 60 degrees and the critical angle (41.8 degrees). It can be seen that the TM polarisation gives slightly higher sensitivity, and that operating at the critical angle for $K \approx 0.1$ also gives higher sensitivity. Above these values of K, the reflectivity is a non-linear function of K.

Effective path length and sensitivity

We can use the preceding calculations to determine an effective path length for a material having a given value of K. This is calculated as a fraction of the path length of the material in a conventional spectrophotometer. Since Eq. (14) can be written as:

$$\varepsilon c = \frac{4\pi K}{\lambda \ln 10} \tag{14b}$$

In addition, we can use the reflectivity of the ATR configuration to calculate an effective absorbance or optical density:

$$A = \log \frac{I_0}{I} = -\log R \qquad (15)$$

The ratio of the sensitivities for the ATR and conventional systems allows us to calculate an effective path length:

$$r_s = \frac{-\log R}{\varepsilon c l} = \frac{-\gamma \ln 10 \log R}{4\pi K l} = \frac{-\gamma \ln R}{4\pi K l} \qquad (16)$$

For example, if we use ATR at the interface between dielectric media having refractive index 1.5 and 1.0 + 0.00001i, we can calculate that εc = 86.24 m^{-1} or 0.8624 cm^{-1}. Operating at the critical angle for such a system, using TM polarisation, we obtain a value for R of 0.97487, which gives a value of 0.011055 for $-\log(R)$. This shows that the sensitivity of the ATR configuration is approximately 1/78 (0.011055/0.8624) of that of a conventional spectrophotometer, or the effective path length is 1/78 cm compared to a conventional 1 cm cuvette.

Although the effective path length is considerably shorter for a simple ATR configuration, the volume of material sensed is very small, since the penetration depth into the sensed material is generally very small. If we assume a circular input beam of radius r for both the conventional and ATR configurations, then the input beam will form an elliptical spot at the interface between the two media whose area will be given by:

$$A = \frac{\pi r^2}{\cos \theta_1} \qquad (17)$$

The volume of material sensed will be given by the product of the area of the spot and the penetration depth d:

$$V = \frac{\pi r^2 \lambda}{2\pi \cos \theta_1 \sqrt{n_1^2 \sin^2 \theta_1 - n_2^2}} \qquad (18)$$

And for the conventional cuvette the volume sensed will be given by:

$$V = \pi r^2 l \qquad (19)$$

Where l is the path length of the cuvette. The ratio of these two volumes is:

$$r_V = \frac{\lambda}{2\pi l \cos \theta_1 \sqrt{n_1^2 \sin^2 \theta_1 - n_2^2}} \qquad (20)$$

One obvious way to increase the sensitivity of ATR systems is to increase the number of reflections. This can be achieved by using a thin transparent slab and coupling the light in and out through the end faces of the slab. Many different configurations have been devised utilising this method, including schemes where the light enters and exits by the same face and single- and double-sided slabs. If the reflectivity for a single total internal reflection is R, then for n reflections the total reflectivity is R^n. ATR slab waveguides in IR-transparent materials such as silicon, germanium, zinc selenide and KRS-5 are commercially available for FTIR instruments.

Waveguides

We can utilise the phenomenon of TIR to construct a waveguide. If we have a slab of transparent dielectric of a higher refractive index than its surroundings, light can be waveguided in the slab. Even a simple microscope slide can act as a slab waveguide if light can be coupled into it in such a way as to exceed the critical angle on reflection at the boundaries of the slab. Figure 14.39 shows a simple ray model of the propagation of light along the waveguide at a characteristic angle of incidence by successive reflections. To minimise losses, TIR is used as the reflection mechanism.

It might appear that θ can adopt any angle, but this is not the case. Consider an observer moving along the z-axis who sees only the transverse (x-direction) motion. For a self-consistent picture, this observer must see the ray having the same phase every time the ray reaches him. If this were not the case, over a large number of reflections the phase shifts would cancel out, giving an intensity of zero. For this not to occur, the total phase shift over a complete cycle from $x = 0$ to $2a$ and back to 0 must again be an integral multiple of 2π. This is termed the transverse resonance condition.

If $2a$ is large compared to the wavelength of the light, then there will be a large number of angles θ that satisfy the transverse resonance condition. As the thick-

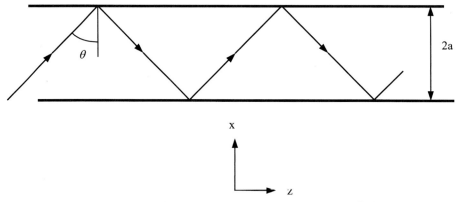

Fig. 14.39 A simple model of a slab waveguide showing the propagation of light in a zig-zag path.

ness of the waveguide is reduced, the number of possible solutions decreases until, in the limit, there is only one angle at which the light can propagate. Each discrete angle of propagation is termed a mode, and below a critical thickness (the cut-off thickness) no waveguide modes can be supported. Waveguides are termed monomodal if they can only support one mode or multimodal if they can support more than one mode. Since a thin, monomodal, waveguide will support the largest number of internal reflections, we find the sensitivity of monomodal waveguides to be higher than multimodal waveguides. Since the thickness of a monomodal waveguide is typically less than a wavelength, these waveguides are too fragile to be used without a thick substrate for support. This reduces the sensitivity, as only one side of the waveguide is exposed to the sample to be analysed.

A guided mode in a waveguide also produces an evanescent field in the lower-index layers outside the waveguide. In essence, some of the light is travelling outside the high-index waveguiding layer. The effective refractive index (N) of the waveguide depends on the bulk refractive indices of the waveguide and cladding, and the relative proportions of the light in the waveguide and the cladding layers. This means that the effective index is always between the waveguide and cladding indices.

The other implication of this is that if the refractive index of the cladding layer changes, the effective index of the waveguide changes, which can be detected by a change in phase. Thus, waveguides are sensitive to changes in refractive index as well as absorption outside the waveguide layer. To maximise this sensitivity, we need to increase the amount of light in the evanescent field. To do this, we require a large difference between the waveguide and cladding refractive indices and the waveguide should be thin enough to support only a single mode. Because of the nature of the equations that determine the supported modes of a waveguide, exact analytical solutions are not possible and numerical solutions are required. Figure 14.40 shows the sensitivity of a planar waveguide to changes in the external bulk refractive index as a function of waveguide thickness, Figure 14.41 shows the sensitivity of the waveguide to changes in thickness of an adsorbed layer on the surface as a function of waveguide thickness and Fig. 14.42 shows the sensitivity of the waveguide to losses in the cover layer.

These graphs clearly show that there is an optimum waveguide thickness that maximises sensitivity to refractive index change for a particular mode, that these optimum thicknesses are different for TE and TM modes, that TM modes are generally more sensitive than TE modes and that the zero order modes are the most sensitive. The general form of all three graphs is the same, as they are effectively measuring the fraction of the light present in the evanescent field that is interacting with the material being sensed.

The calculations in the preceding section have been performed for a simple planar slab waveguide of infinite extent but finite thickness sandwiched between semi-infinite substrate and cover layers. Other waveguide configurations are possible, most notably the cylindrically symmetrical case of a high refractive index rod surrounded by a lower index cladding. This is the fiber-optic configuration used widely in telecommunications.

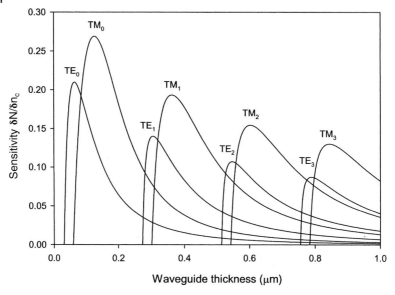

Fig. 14.40 Graph of the sensitivity (ratio of the change of waveguide effective index to change in refractive index of cover layer) as a function of waveguide thickness for the first four TE and TM modes. Substrate index 1.46, waveguide index 2.00 and cover layer index 1.333 calculated for illumination at 660 nm.

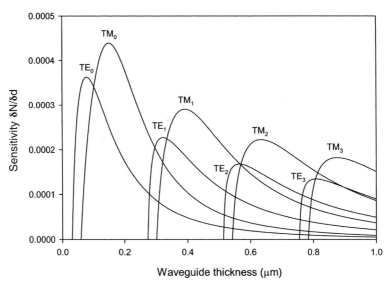

Fig. 14.41 Graph of the sensitivity (ratio of the change of waveguide effective index to change in thickness of a thin adlayer) of the waveguide as a function of waveguide thickness for the first four TE and TM modes. Substrate index 1.46, waveguide index 2.00, adlayer index 1.45 and cover layer index 1.333 calculated for illumination at 660 nm.

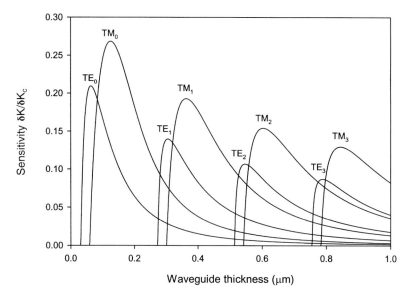

Fig. 14.42 Graph of the sensitivity (ratio of the change of the imaginary part of waveguide effective index to change in the imaginary index of the cover layer) as a function of waveguide thickness for the first four TE and TM modes. Substrate index 1.46, waveguide index 2.00, cover layer index 1.333 calculated for illumination at 660 nm.

14.2.7.2 Measurement Configurations

There are various possible configurations, which utilise evanescent waves for biomolecular sensing and spectroscopy [221, 257]. Waveguides (either planar or in an optical fiber) can be used as prism couplers, grating couplers, mode couplers (interferometers), and in surface plasmon resonance (SPR), reflectometric interference and frustrated total internal reflection (FTR) experiments. Grating couplers use a grating fabricated in either the substrate or waveguide to couple light in and out of the waveguide. As the surface refractive index changes, the coupling angle changes, giving a change in angular position of the out-coupled light on an appropriate detector. In this mode, grating couplers work very much like SPR and resonant mirror (RM) sensors. They can also be used purely as a means of coupling light in and out of the waveguide to permit detection of absorption or fluorescence changes. Interferometers (such as Mach-Zender or Michelson devices) use a separate reference channel that does not undergo the chemical binding stage to act as a phase reference. A periodic change in intensity at the output is observed, caused by interference between the sensing and reference channels as the refractive index on the sensing channel changes. A problem with interferometers is that the output can be ambiguous, as a large change in refractive index can take the output through more than one cycle. These devices are generally very sensitive.

Ellipsometric sensors make use of the phase changes between the TM and TE modes propagating in a waveguide. If equal intensities of TE and TM are excited at the input, the polarisation state of the output will depend on the relative phase shift between TE and TM at the output. Since the sensitivity of the waveguide is different for TE and TM modes, a change in refractive index on the waveguide surface will result in a change in polarisation at the output. Like interferometric sensors, the output is cyclic and can result in ambiguous outputs. The sensitivity is very high.

In frustrated total internal reflection (FTR) experiments a leaky high index waveguide, which permits in- and out-coupling at well-defined coupling angles, is used. The coupling angle depends strongly on the cover layer refractive index. They can also be used to monitor absorption, as the reflectivity is a strong function of the loss in the cover layer.

Fiber optics
Optical fibers are cylindrical waveguides consisting of a high index core surrounded by a low index cladding layer. Because of their low losses, fiber optics are widely used in the telecommunications industry, although the properties of telecommunications fiber optics make them poor sensors. Specially designed fiber optics can be used as biosensors, using a thin or no cladding layer on a monomode fiber core to permit interaction of the evanescent field with the material to be sensed [25]. Conventional fiber optics have also been widely used for sensing applications, but in the majority of cases they have been used as a convenient means of delivering and collecting light for conventional spectroscopies.

End-fire coupling
End-fire coupling, as its name implies, is carried out by focusing a laser beam into a diffraction-limited spot centered on the waveguiding layer. Although end-fire coupling is conceptually very simple, it is not easy to perform in practice. One example where end-fire coupling has been successfully used in biological sensing applications is the difference interferometer [258–259, 260, 261]. If we consider Fig. 14.41, we can see that there are waveguide thicknesses at which the difference in sensitivity between the TE and TM modes is at a maximum. If linearly polarized light is launched at 45° to the plane of the waveguide, equal intensities of the appropriate TE and TM modes will be launched. Since end-fire coupling will generally excite more than one mode, the waveguide thickness is chosen such that only monomode operation will be possible at the excitation wavelength, and that the difference in sensitivity between the TE and TM mode is maximized. As the surface refractive index changes, the retardation of the TE and TM modes will change by different amounts, changing the relative phase of the TE and TM modes, and hence the polarization state of the out-coupled light. When the TE and TM components are in phase, the outcoupled light will be polarized in the same direction as the in-coupled light. As the phase difference between TE and TM increases, the

out-coupled light will become elliptically polarized, reaching circular polarization when the phase difference reaches $\pi/2$. At a phase difference of π, the light will be linearly polarized again, but at 90° to the polarization of the in-coupled light. As the phase difference increases further, the light will again become circularly polarized at $3\pi/2$, but with the opposite handedness, while at 2π the cycle will be complete and the out-coupled light will be polarized identically to the in-coupled light. This cycle can be followed using an appropriate arrangement of beam splitters, polarizers and detectors, returning the TE–TM phase difference after processing the detector outputs. Since the phase difference is a cyclical function of surface refractive index, this device does not give an unambiguous measurement of surface refractive index. It can track slow changes in refractive index, but can only track the index successfully if the index change causes a smaller than $\pm\pi$ phase change between successive samples of the output. Because the path lengths employed in this type of sensor can be quite large (~ 10 mm), the sensitivity is quite high. Effenhauser et al. [262] were able to detect 10^{-11} M concentrations of human IgG using anti-human IgG antibodies with a difference interferometer configuration and a thin titanium dioxide waveguide. In studies with a bidiffractive grating coupler system, Kubitchko et al. have used nanoparticles to enhance the sensitivity for the detection of analytes like the thyroid-stimulating hormone (TSH) [306]. Without amplification the detection limit was 430 pM and with amplification by a latex-conjugated sandwich antibody the detection limit dropped to 0.11 pM, which is the clinically interesting range (0.3–667 pM TSH).

Grating couplers

A grating structure at the substrate-waveguide or waveguide-cover layer interface can be used to couple light into the waveguide [263]. A major advantage of this method is that the waveguide can be fabricated on a simple planar substrate and coupling can take place through the bottom of the substrate. Careful design of the waveguide and grating can permit the use of TM-polarised light at the Brewster angle, which eliminates reflection at the bottom surface of the substrate.

Dubendorfer et al. [264] employed a chirped grating coupler design, where the grating period varies continuously along the width of the coupling area. This means that light will only couple into the waveguide at the location where the grating equation is satisfied. Thus, the position in space of the out-coupled beam is a simple function of the surface refractive index. They were able to detect refractive index changes of 5×10^{-6}, corresponding to changes in surface loading of 5 pg mm^{-2}.

Resonant mirror (RM)

The RM device consists of a high-index substrate (~1 mm thick lead glass, n_d = 1.72825), a thin low-index spacer (about 1000 nm of magnesium fluoride or silica) and a very thin monomode waveguiding layer (about 100 nm of titanium oxide, zirconium oxide, hafnium oxide or silicon nitride). It can be used to monitor refractive index and absorbing or fluorescent species within the evanescent field above the waveguide surface [265]. Light incident above the critical angle on the substrate/spacer interface is coupled into the waveguiding layer via the evanescent field in the spacer layer when the propagation constants in the substrate and waveguide match. For monochromatic light, this occurs over a very narrow range of angles, typically spanning less than 10 arc minutes. The device has been termed the resonant mirror because it contains a resonant cavity (the waveguide) and it acts as a nearly perfect reflector for light incident above the critical angle. Since the waveguiding layer acts as a resonant cavity, the light reflected from the RM device undergoes a full 2π phase change as we scan across the resonance. To detect the resonance position, a phase reference must be provided, which is substantially constant in phase. This could be provided by splitting off part of the input beam and recombining it with the light from the sensor, but this is instrumentally difficult, and means that the object and reference beams travel by widely separated paths. Ideally, the two beams should travel by identical paths, so that phase-shifting effects (such as temperature changes) are common to both beams. This can be accomplished in the RM sensor by using the TE mode as reference for the TM resonance and vice versa. This is only feasible because the resonance positions for TE and TM are widely separated.

The resonance positions may be determined in two ways:

1. In angle, using monochromatic input light covering a range of input angles (angular scan mode) [265].
2. In wavelength, using broadband input light at a fixed input angle (wavelength scan mode) [266].

To determine the resonance angles or wavelengths, at which light couples into the waveguiding layer, linearly polarised light at 45° to the plane of the waveguiding layer is applied to the device, exciting equal intensities in the TE and TM modes. The output light from the device is passed to a crossed analyser, which only passes light that has undergone a π phase change in the sensor device. In the angular measurement mode, the input light is a converging monochromatic wedge beam, covering a sufficiently wide range of input angles to permit the resonance angle (for the fixed wavelength) to be determined for the required range of surface refractive indices. In the wavelength measurement mode, the input light is a well-collimated beam of white light, covering a sufficiently wide range of wavelengths to permit the resonance wavelength (at the fixed input angle) to be determined.

Buckle et al. [267] used the RM to monitor antigen/antibody and enzyme/substrate/inhibitor interactions using several methods for immobilization of the biomolecules at the sensor surface. Sensitivities in the nM range were reported. Watts et al. [268] used the RM sensor to monitor the binding of microbial cells to antibodies immobilized on the surface of the sensor. They detected the binding of *Staphylococcus Aureus* (Cowan-1 strain) to human immunoglobulin G (hIgG), covalently immobilized on the sensor surface via aminopropylsilane, at concentrations between 8×10^6 and 8×10^7 cells cm^{-3}. By employing a sandwich assay using a hIgG–gold particle conjugate, detection limits were reduced by a factor of 1000. It appeared feasible to detect between 4×10^3 and 1.6×10^6 cells cm^{-3} in spiked milk samples.

A range of instruments based on the RM is commercially available (the IASys system of Labsystems/Affinity Sensors), using the angular scan mode to determine the resonance angles. These instruments can only measure refractive index changes.

14.2.7.4 Surface Plasmon Resonance (SPR)

Surface plasmon waves are excited in thin metal films when appropriate coupling conditions are met. The surface plasmon wave is an oscillation of the free electrons in the metal under the influence of the electric field of the light [257]. For sensor applications, the Kretschmann configuration [269–270, 271] is most often used, where a thin metal layer (usually 50 nm gold) is deposited on a glass prism, and light is coupled into the metal film at the coupling angle. This configuration is shown in Fig. 14.43. Only TM polarized light shows this behaviour, where the mode is localised at the metal-cover layer interface. TE polarized light does not show this behaviour. Since the metal film is very lossy (it has a large imaginary component of the complex refractive index), the reflected light is very strongly attenuated at the coupling angle. The coupling angle is a very strong function of the cover layer refractive index. Thus, as material binds to the metal layer there is a change in the angle at which the drop in reflectivity occurs. This is the basic sensing mechanism for SPR biosensors, as shown in Fig. 14.44.

Instrumentation for SPR sensors generally comprises a monochromatic light source (a filtered incandescent lamp, LED or laser) followed by a cylindrical lens to provide illumination over a wide range of angles. A CCD detector is then used to monitor to position of the dip in intensity (angle measurement mode),

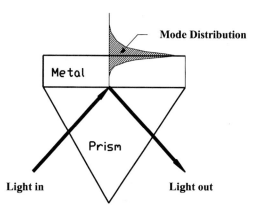

Fig. 14.43 The Kretschman configuration for excitation of SPR in a thin metal layer.

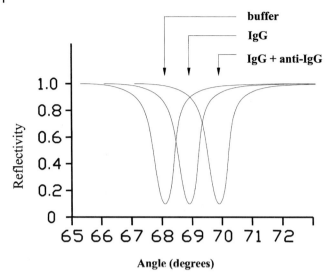

Fig. 14.44 Illustration of the shift in resonance angle at various stages in the adsorption of biomolecules to the metal surface.

which indicates the coupling angle. Alternatively, a collimated beam of white light may be used at a fixed angle of incidence, and the wavelength position of the dip determined (wavelength measurement mode). Systems of this kind have no moving parts, and thus can follow very rapid changes in the angular position of the reflectivity dip. The metals most often used are gold or silver, but copper can also be used. These metals all have the high conductivity needed to give a relatively narrow dip in reflectivity, making the task of following small changes in the position of the dip easier.

Commercial SPR biosensors

Commercial systems using SPR sensing methods have become available from a number of suppliers. These systems range from large multichannel laboratory bench instruments to simple single-channel integrated devices in an integrated-circuit style package. Since its commercialisation SPR has been used for the detection of a large variety of biomolecules. Reviews on the applicability of SPR detection have appeared rather regularly, discussing biomolecular interactions of proteins, nucleic acids, lipids and carbohydrates [272–277]. SPR has also shown some usefulness in the study of conformation changes of adsorbed proteins. For instance, Boussaad et al. have used multiwavelength SPR to study conformational and electronic changes induced by the electron-transfer reaction in cytochrome c [278], and Sota et al. studied the acid-induced denaturation of an immobilized protein [279]. DNA hybridization has also been studied by SPR [280]. Through many studies much new knowledge has been acquired on the kinetics of antigen–antibody reactions at surfaces [281].

The BiaCore system was the first commercialised SPR instrument based on the Kretschmann configuration. The BiaCore is a multi-channel instrument that has been extensively used for biological assay work, and in the monitoring of both thermodynamic (dissociation constants) and kinetic (on and off rates) parameters of a wide range of biomolecular interactions. A full overview of the work done with the BiaCore can be obtained from the manufacturer's website [282]. Recent applications include the determination of streptomycin residues in whole milk [283], bovine insulin-like growth factor (IGF)-binding protein-2 [284], global conformational transitions in human phenylalanine hydroxylase [285], peptides from the GH loop of foot-and-mouth disease virus [286], gentamicin residues [287], penicillin residues in milk [288], HIV-1 protease inhibitor interactions [289] and acylated proteins [290]. A competitive instrument to the BiaCore, the "IBIS" series of biosensors, has also been developed in the Netherlands at the Twente Technical University and is based on the conversion of angle data into a time interval by the use of a vibrating mirror system [291, 292].

The Spreeta™ device manufactured by Texas Instruments is a complete miniaturised SPR sensor containing an infra-red LED emitting at 830 nm, a cast polymer prism with a gold SPR layer, a 256 element linear CMOS image sensor and a 4 Kbit serial EEPROM memory [293]. It is packaged as a single unit of dimensions 41.40 by 28.92 by 13.49 mm with a standard 16 pin 0.3 in dual-in-line connector. Melendez et al. have reported a resolution of 10^{-5} refractive index units using this device [194]. Elkind et al. describe the use of the Spreeta device as a refractive index sensor for a direct assay for human creatine kinase MB (CK-MB, a marker for heart attacks) using adsorbed anti-CK-MB antibodies, obtaining a change in index of $\sim 4 \times 10^{-4}$ for binding of 100 ng ml^{-1} of CK-MB [295]. They were also able to assay trinitrotoluene (TNT) using a competitive assay format by immobilising a trinitrobenzene–bovine serum albumin adduct to the sensor surface, then reacting the TNT sample with anti-TNT antibodies and passing the resultant solution over the sensor surface. If high concentrations of TNT were present in the sample, then most of the antibody binding sites were occupied, resulting in a reduced rate of binding of the antibody to the trinitrobenzene–BSA coated sensor surface. With this assay it was possible to detect 7 ng ml^{-1} of TNT. Sesay and Cullen have used the Spreeta sensor to monitor endocrine disrupting chemicals (EDC) in aqueous samples [296]. Estrone-3-glucuronide (E3G) was used as a model EDC, and a competitive assay format was developed. Ovalbumin-E3G conjugate was immobilised by adsorption onto the sensor surface, while the sample to be assayed was pre-incubated with a known concentration of anti-E3G antibody. The pre-incubated sample was passed over the treated sensor surface for a fixed time, followed by washing with phosphate-buffered saline solution to remove any non-specifically bound material. The SPR peak shift was then determined and used to derive the concentration of E3G in the original sample by reference to a calibration curve. The sensor surface could be regenerated using a strong domestic detergent (Persil biological liquid) to remove all the biological material. The detection range was 10–150 ng ml^{-1}.

Fiber and waveguide SPR

Fiber optic devices are attractive in many applications because they can be physically very small, provide remote operation and electrical isolation from the associated instrumentation. For these reasons, many designs for fiber-optic SPR sensors have been developed. They all rely on phase matching between the optical mode in the fiber and the surface plasmon wave. Since the control over the coupling angle is lost in a fiber, wavelength scanning is often employed. Slavik et al. [297] have shown that a wavelength-scanned fiber-optic SPR sensor can detect very low concentrations of proteins. They used a single-mode optical fiber with a cut-off wavelength of 724 nm which was polished down to remove all but 500 nm of the cladding. The resulting device was coated with 65 nm of gold to support surface plasmons and then 19 nm of tantalum pentoxide to adjust the sensor's operating range to refractive indices between 1.329 and 1.353. This device was capable of detecting 40 ng ml^{-1} of human immunoglobulin G (IgG) using anti-human IgG antibodies immobilised on the sensor surface by cross-linking with glutaraldehyde.

Similar behaviour is observed when a thin metal layer is coated on a planar waveguide. Harris et al. [298] have developed the theory of waveguide-coupled SPR sensors. Brecht and Gauglitz have compared the performance of a waveguide-coupled SPR sensor against a grating coupled waveguide, a channel waveguide interferometer and a thin-film reflectance sensor for the detection of pesticides using a competitive assay format with anti-triazine antibodies [299]. Instead of using a full spectrum scan, the intensity at a fixed wavelength was monitored. The limit of detection for triazine was found to be 0.15 ng cm^{-3}.

Sensitivity and applications of SPR-based biosensor instruments

In a recent review by Homola et al., the analytical characteristics of various SPR instrumentation configurations for SPR have been compared with respect to the type of coupling (prism or grating) and the mode of measurement (angular, wavelength and intensity read-out) [300, 301]. In principle, the prism-coupler (Kretschman) configuration was assessed as being capable of yielding the highest sensitivity (in terms of refractive index change) in comparison with a grating coupler based set-up, particularly when the method of interogation was based on intensity measurements. Lekkala and Sadowski have used the Kretschmann configuration in conjunction with lock-in amplifiers for the highly sensitive detection of intensity changes at near-resonance conditions [302]. The system was used for the optimisation of antibody orientation on various lipid-modified surfaces [303].

Although the intrinsic sensitivity of SPR measurements (without any amplification schemes) had been initially assessed to be of the order of 1 nM or 150 ug l^{-1} for IgG [304], lower detection limits for small analytes have been reported [305, 306]. In most studies equilibrium dissociation constants ranged from nanomolar to micromolar, with a few studies in the picomolar range [307, 308]. The resolution of SPR measurements has to some extent been improved by using well-designed reference surfaces and sophisticated data processing techniques [309]. SPR has been combined with other techniques, such as mass spectrometry [310] and AFM + electrochemistry [311]. Particularly the combination of multichannel SPR with MALDI-TOF can be regarded as a powerful tool for proteomics research (see Section 14.4.6).

In general, SPR-based biosensors have detection limits around 50 pM or 100 pg cm^{-2}. This is not always sufficient for the detection of low molecular weight analytes where concentrations fall below picomolar [312]. In such cases amplification of the signal by sandwich assay formats is presently the only solution. Already in 1988 Mandenius and Mosbach had used quartz particles to amplify ellipsometric measurements of biospecific interactions [313]. Similar labels, which have a large effect on the refractive index near the surface, will be suitable for use with SPR. For instance, Lyon et al. have described the use of colloidal Au nanoparticles to enhance the SPR response, through strong optical coupling between the film and the particle, achieving detection limits for human IgG down to 6.7 pM using a sandwich immunoassay format [314]. Besides a large shift in incident angle, the colloidal gold gives a broadening of the plasmon resonance peak and an increase in minimum reflectance.

14.2.7.5 Reflectometric Interference Spectroscopy (RIf S)

Reflectometric measurements are technically easy to perform and can yield useful information about thin films and their interfaces. The partial reflection of incident light from two interfaces of a thin film with a thickness of about 1 µm can function as a Fabry-Perot interferometer with low reflectance. In the situation depicted in Fig. 14.45a, the reflected light will show an interference pattern as a function of wavelength, according to:

$$R = R_1 + R_2 + 2\sqrt{R_1 R_2} \cos\left(\frac{4\pi n d}{\lambda}\right) \qquad (21)$$

where R is the reflectance, λ the wavelength, n the effective refractive index and d the effective thickness of the optical layer.

An increase in optical thickness of the thin film, caused by e.g. ligand adsorption, will shift the interference spectrum to a higher wavelength and widen the distance between the minima and maxima in the inteference spectra as illustrated in Fig. 14.45b. This is the principle behind reflectometric interference spectroscopy, or RIfS [315]. Due to the high sensitivivity of the detection (ppm levels of phase shifts can be measured [316]), the RIfS device has been successfully used for the study of various biological interactions at surfaces, such as mouse anti-atrazine/atrazine [317] and DNA–ligand interactions [318]. The principle of RIfS also allows the construction of low-cost devices.

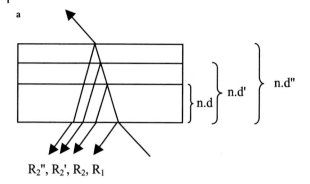

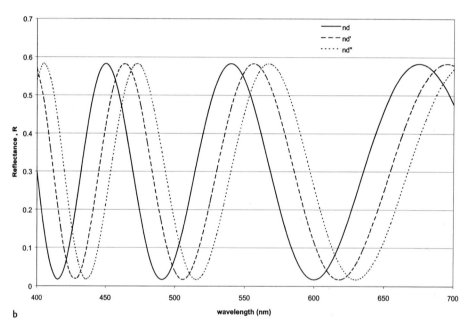

Fig. 14.45 The principle of interference reflectometry at a thin film of increasing optical thickness: (a) The thin optical layer with several reflections from increasing optical thicknesses nd, nd" and nd". (b) Calculated interference curves with: $R_1=0.2$, $R_2=0.1$, $n=1.35$ and nd=1000 nm, nd'=1030 nm and nd"=1050 nm.

14.2.7.6 Total Internal Reflection Fluorescence (TIRF) and Surface Enhanced Fluorescence

In assays based on fluorescence labels there is generally a low degree of discrimination of the signal from molecules in the bulk solution from molecules bound to the surface (e.g. of a sensor). Surface enhanced fluorescence can be used to partly discriminate the signal from surface bound species from those in the bulk solution. This can be done either by using normal evanescent waves or using amplification of evanescent waves through a thin metal layer via SPR. Additionally it is possible to excite plasma oscillations in metal nanoparticles, which causes a strongly enhanced near-field at the particle surface. Molecules located in this near-field feel a higher light field strength, which results in enhanced light-induced molecular responses. The dipole momentum of the excited molecules interacts with the metal. Surface-enhanced fluorescence of molecules adsorbed on silver island films is already a well-known phenomenon, and is also related to other surface enhanced optical phenomena, such as Raman scattering [319, 320]. Time resolved surface enhanced fluorescence of molecules positioned close to submicroscopic metal particles has also been investigated by Aussenegg et al. [321]. Measurement of picosecond time-resolved fluorescence indicated a significant shortening of the lifetime of the excited molecular state.

Recently, Schalkhammer et al. have discussed the relevance of surface enhanced fluorescence in immunosensing applications (surface-enhanced fluoroimmunoassay, SE-FIA) [322]. The enhancement mechanism was explained by an electrodynamic model and the interaction between metal particle and fluorophore for the excitation and emission process was discussed. It was shown that the discrimination power increased with decreasing quantum efficiency of the fluorophore. This suggested that in SE-FIA a low-quantum efficiency fluorophore needs to be used, as was shown by experiments with model compounds.

Due to the surface sensitivity surface enhanced fluorescence has become particularly popular in the characterisation of thin molecular films, such as Langmuir–Blodgett films and self-assembled biomembranes. Two surface enhanced spectroscopic techniques (surface enhanced IR absorption, SEIRA, and surface enhanced fluorescence, SEF) were recently applied to the study of biomembrane systems by the group of Reiner Salzer [323]. With SEIRA, specific fingerprints of biomolecules could be obtained with a tenfold IR intensity enhancement. With SEF signal enhancement factors greater than 100 were obtained. The enhancement factor was very dependent on the properties and structure of the metal clusters used. With the two techniques biomembranes formed from vesicles with embedded nicotinic acetylcholine receptors were spectroscopically characterized.

14.2.8
Infrared and Raman Spectroscopy in Bioanalysis

14.2.8.1 FTIR, FTIR Microscopy and ATR-FTIR

Fourier transform infrared spectrometry has very much revolutionized the investigation of peptides and proteins, and is used in various clinical chemistry applications [324–326]. FTIR is most frequently used in protein chemistry to determine the secondary structure content [327–330], to assess conformational changes in proteins [331–333] and to study protein unfolding [334, 335]. FTIR has been used to characterise basic structural motifs of model peptides, such as α-helices and β-turns [336–338]. Hydration of proteins has also been extensively studied, particularly with relevance to the determination of the structural integrity of lyophilized protein preparations [339]. A relatively new trend is the characterisation of the secondary structures of peptide fragments of proteins in different solvents [340–343]. FTIR studies of proteins and peptides are usually concerned with the amide groups and their deuterium substituted analogues. The carboxyl groups are also commonly analyzed by FTIR studies [344].

Although technically less easy to perform, FTIR spectroscopy is also an ideal tool for studying membranes and membrane-associated proteins, because of the possibility of examining samples in different physical states. Secondary structure and conformational changes have been described for ATPases, photosystem II reaction center proteins and bacteriorhodopsin, as well as proteins in the membrane of gastric carcinoma cells and erythrocytes [345–350]. The orientation in lipid bilayers can be studied with polarized FTIR, which yields insight into the orientation of toxins, ion channels and proteins in cell membranes [351–355]. FTIR has also been extensively applied to the study of protein adsorption to solid surfaces [357]. These studies often involve the use of attenuated total reflectance (ATR) accessories [357]. Thus, studies have been performed on the adsorption of IgG and F(ab')$_2$ to siliceous surfaces [358], serum albumin to organotrichlorosilane-derivatised surfaces [359] and proteins on alkylmercaptan-derivatised surfaces [360]. FTIR microscopes have given new possibilities for the study of proteins in their native environment, such as in serum, whole blood, bone, brain tissue and many other matrices [361–363].

The application of FTIR directly to clinical studies and diagnosis has been very much debated. Methods for in vivo monitoring of glucose have until now been more presented in the patent literature than in the scientific literature. IR spectroscopy methods, however, have been described for the *in vivo* monitoring of glucose, hemoglobin, urea, albumin, phosphocreatine, and nitric oxide [326].

14.2.8.2 Raman Spectroscopy

Raman spectroscopy is increasingly being applied not only to the analysis of proteins and peptides [364–366], but also to the analysis of polymers [367], viruses [368] and skin lesions [369]. Generally higher concentrations of samples are required as with FTIR, although resonance enhancement techniques can alleviate this problem. Although Raman spectroscopy is much used for protein structure elucidation, FTIR and circular dichroism (CD) are still more commonly employed. This is mainly due to better instrument availability and less problems with sample preparation. A recent technical improvement of Raman spectroscopy is the use of near-infrared excitation. Today such spectrometers are all well available. As standard FTIR is concerned with main-chain vibrations, Raman spectroscopy focusses more on side-chain vibrations. Tryptophan and tyrosine are the most frequently investigated side chains, but cysteine and histidine side chain resonances are also much studied [370–376]. Raman spectroscopy is frequently applied to the study of heme vibrations in porphyrin-containing proteins due to the possibility to use resonance enhancement [377].

14.2.8.3 Surface Enhanced Raman Spectroscopy (SERS)

Since its discovery in 1974, surface enhanced Raman spectroscopy, or surface enhanced Raman scattering (SERS), has been an important method for examining surface adsorption of organic molecules [378, 379]. For instance, SERS has been employed in studies of adsorption of small peptides to silver [380]. Recent investigations have focussed on the study of photosynthetic membranes and cytochromes [381, 382]. Rather novel applications include elucidation of virus structure and proteins adsorbed onto vaccine adjuvants [383] and gene diagnostics [384]. SERS has recently also been applied in non-resonant conditions to observe colorless, single biomolecules [385]. SERS has been combined with scanning probe techniques to obtain Raman imaging of DNA-molecules at a 100 nm resolution [386]. The combination of a confocal microscope and careful surface treatments extends the application of SERS to the study of adsorbates on a wider range of surfaces [387], while the applicability of SERS has recently been expanded to the study of interfacial materials by overlayer deposition [388]. In the latter technique, materials that do not display Raman enhancement themselves, are deposited as an ultrathin overlayer on Raman-active surfaces, expanding the application of SERS to metallic, semiconductor and insulator films. A novel spectroelectrochemical cell for surface Raman spectroscopy has also been developed, which has been used for studies on cytochrome P450 [389].

14.2.9
Circular Dichroism

Circular dichroism (CD) is a form of light absorption spectroscopy that measures the difference in absorbance of right- and left-circularly polarized light (rather than the absorbance of isotropic light) by a sample in solution. CD is related to optical rotary dispersion (ORD), which is the variation of optical rotation as a function of wavelength. CD spectra recorded between 180 and 260 nm enable fast routine determination of secondary structural types in proteins, i.e. the α-helix, parallel and antiparallel β-sheet and β-turn content. The use of CD spectroscopy in bioanalysis is expected to grow, due to the expanding markets for chiral compounds [390]. Contemporary CD instrumentation allows levels of accuracy for the determination of 97% for α-helices, 75% for β-sheet, 50% for β-turns and 89% for the other structure types. The most common application of CD is the study of conformation changes (secondary and tertiary structure). As such the technique lends itself well to the study of folding and unfolding induced by temperature, pH and chaotropic ions [391, 392]. Model synthetic peptides have been examined by CD to explore the physical basis of the formation of α-helix and β-sheet formations [933, 394]. CD is increasingly used to examine tertiary structural features of proteins through aromatic amino acid mutant proteins [395, 396]. The use of CD in membrane and surface adsorption studies has been, unfortunately, limited by differential light scattering and absorption flattening [397].

Vibrational circular dichroism (VCD) was developed as a supplementary technique in the 1970s, and is the IR counterpart of CD [398]. VCD measures the differential absorption of left and right circularly polarized IR light by chiral molecules. Since there are more spectral lines in the IR region than in the UV and each can have a chiral response, a more elaborate stereochemical and structural analysis is possible than on the basis of CD. Researchers have started to use VCD for conformational studies of all classes of chiral molecules, especially carbohydrates [399].

14.3
NMR Spectroscopy of Proteins

14.3.1
Introduction

Nuclear magnetic resonance spectroscopy is likely the most versatile method for the study of biomolecules. The power of NMR originates from the fact that practically every atom with a magnetic nucleus gives rise to an individual signal in the NMR spectrum that carries spatial and temporal information of the local chemical environment of that atom. Solution conditions in high resolution NMR experiments mimic the natural biological environment and results relate to functional assays.

Proteins, owing to their many biological functions, display versatile phenomena to investigate. In addition there are spectroscopic reasons in favor of proteins. Proteins, as a class of molecules, give particularly well-dispersed NMR spectra because of rich local diversity in their physical-chemical structures. Proteins can also be obtained, using the methods of molecular biology, in large quantities enriched uniformly or selectively with ^{13}C and ^{15}N as well as deuterated or specifically reprotonated to make use of a large repertoire of heteronuclear NMR experiments. These are the main reasons why NMR spectroscopy of proteins is highly developed.

Progress in spectroscopy, instrumentation, computational methods, data analysis and sample production has resulted in astonishingly versatile and powerful methods well-documented in the original papers, recent reviews (e.g. [400–406]) and textbooks on NMR spectroscopy (e.g. [407, 408]). The most important recent innovations in the field of NMR include the introduction of weakly aligned systems [409] and the transverse relaxation optimized spectroscopy [410]. These discoveries have significantly expanded the realm of high resolution NMR spectroscopy by including new directional and range-independent information and by significantly enlarging the set of molecules amenable to NMR. Many of the NMR methods that were originally developed for proteins have been amended to be suitable for studies of other classes of molecules, in particular nucleic acids and carbohydrates.

14.3.2
Protein Sample

14.3.2.1 Solubility and Stability

Successful NMR spectroscopy begins from successful sample preparation [411]. Usually it is very difficult to compensate for a poorly "behaving" molecule by spectroscopic means. Indeed the preparation of a highly soluble labeled protein sample is often the limiting factor, not the spectroscopy, in the studies of proteins and their complexes by NMR. Problems with samples manifest themselves typically as low signal-to-noise ratio (SNR) or fast transverse relaxation i.e. broad lines or multiple signals. These can be signs of poor solubility, oligomerization, aggregation, conformational isomerism or perhaps paramagnetic ion impurities, all deleterious for carrying out detailed NMR studies. However, low-quality spectra may also be indicative of an interesting phenomenon perhaps worthy of an investigation itself.

The signal-to-noise ratio of NMR has increased many folds over the decades due to the progress in instrumentation and methodology. Nevertheless, NMR remains a comparatively insensitive method, owing to the minute nuclear magnetic moments, compared with many forms of optical spectroscopy. Therefore fairly large amounts of material are required for experiments. Even though compounds on a nanogram scale can be detected using special NMR probes, detailed structural studies, e.g. determination of three-dimensional structures, need at least a few hundred micromolar protein solutions in few hundred microliter volumes. Proteins, that are obtained by overexpression in bacterial or eukaryote hosts and subsequently purified by affinity, ion-exchange or gel-filtration etc., come in large quantities and are usually sufficiently pure for NMR spectroscopy. Small molecule impurities are easy to detect and often indifferent for heteronuclear NMR spectroscopy, unless occurring in much higher concentrations than the solute of interest or unless they happen to possess a particular activity in the system of interest.

The high solubility of a biomolecule required for NMR cannot be taken for granted. The spectral characteristics might be compromised, even without an apparent aggregation, seen as weak signals or broad lines. Often it is not obvious if or how unspecific binding or aggregation or oligomerization, that lead to increase in rotational correlation times, can be avoided. Temperature, pH and ionic strength are among the easiest parameters to vary in search of appropriate solvent conditions. Knowledge of the properties of a protein, such as isoelectric point (pI), acquired during the production and purification may provide valuable clues about the conditions and additives, i.e. co-solvents that may improve the spectral appearance. Some of the commonly used co-solvents include surfactants such as CHAPS, octylglucoside, SDS, alcohol such as trifluoroethanol, zwitter-ionic compounds such as glycine and also sometimes denaturants such as urea and DMSO. Ligands or inhibitors of enzymes or cofactors of proteins may improve the spectral appearance. Chimeric constructs have also been used to improve the behavior [412]. Nevertheless, it is difficult to take into account all aspects that have an influence on the spectral appearance and suitable conditions are often found after trial and error.

The presence of more than one conformation may give rise to multiple signals or broadened lines due to exchange and may significantly reduce SNR, complicating assignment and interpretation of the spectra. The solvent conditioning may alter the equilibrium towards a single conformation. Provided that there is some evidence or even hypothesis of conformational heterogeneity, e.g. proline *cis–trans* isomerism or flexibility in the catalytic site due to the lack of ligand or hinge motions in a multi-domain structure, it may help to understand and eventually to solve the problem of conformational isomerism by protein engineering [413]. Of course, it should be kept in mind not to change the solvent conditions or to engineer the protein to render the original biological question meaningless. Functional assays help to judge the appropriateness of the solvent conditioning or protein engineering. NMR itself, e.g. when used to follow reaction kinetics, can be used to assess functional properties.

A stable sample is a necessity for extensive studies such as the determination of three-dimensional structure, which require several weeks of measurement time. Bacterial growth, due to non-sterile conditions, in the sample tube can be largely eliminated by using sodium azide (NaN_3) at 0.02 % (w/v). Proteins are also subject to degradation by proteolytic enzymes that may be present in the sample in minute amounts after the purification. Application of a "cocktail" of inhibitors against the common proteases largely eliminates the enzymatic breakdown. Exposed cystein residues not paired in disulfide bonds may, with time, form intermolecular disulfide bonds, generating dimers or higher oligomers depending on the number of exposed cysteins. Dithiotreitol (DTT) in a stoichiometric excess to cysteins can be used to keep *all* cysteins in the reduced state. Alternatively the unpaired cysteins could often be mutated to serines to avoid the problem of unnatural pairing. Deamidation of asparagine (Asp) over time is a potential problem in certain sequences but can be avoided by engineering the corresponding sequence.

In practice, it is difficult to take into account or rationalize all aspects that influence the sample solubility and stability. Fortunately, the protein stability and solvent conditions are easy to monitor, in particular by heteronuclear correlation spectroscopy. Amide resonances are sensitive reporters of changes in the protein structure or in the solvent conditions. Mass spectroscopy that requires very little material is a very practical method of monitoring the protein sample. Once optimal conditions are found experiments should be conducted promptly.

14.3.2.2 Isotope Labeling

The purpose of isotope labeling is to produce the molecule of interest enriched with the isotopes of preferable nuclear magnetic properties in order to extract more information or to facilitate the interpretation of the information and, more frequently, to make the study feasible in the first place. In many applications it is advantageous to increase the proportions of ^{13}C and ^{15}N that are both spin-½ nuclei, from their natural abundance 1.1% and 0.4% respectively, as high as possible, to nearly 100%, for the detection. The labels facilitate assignment by multidimensional heteronuclear correlation spectroscopy. In other applications selective labeling provides the means to simplify the interpretation, e.g. to distinguish intermolecular correlations from intramolecular signals or segmental labeling [414, 415] e.g. to reduce the number of signals.

In large systems perdeuteration is the efficient way to slow the transverse relaxation [416–419) and specific reprotonation serves to maintain crucial information [420]. The labels allow measurements of dihedral via scalar couplings [421] and projection angles via cross correlation rates [422] as well as internuclear directions via residual dipolar couplings [421]. Furthermore site-specific spin labels, i.e. paramagnetic electrons, can be used to extract long-range distance information [423]. Today isotope labeling is an integral part of biomolecular NMR spectroscopy [402].

^{13}C and ^{15}N labeling

Uniform ^{13}C and ^{15}N labeling provides improved signal dispersion and editing by multidimensional experiments or filtering by isotope or spin-state as well as various routes for coherence transfer and ways to measure scalar and dipolar couplings. Solvent suppression in heteronuclear correlation experiments is also much easier than in homonuclear correlation experiments.

Uniformly labeled proteins are customarily obtained via overexpression in bacteria grown in a culture medium having enriched metabolites as the sole sources of carbon and nitrogen. For bacterial expression ^{13}C-glucose (or ^{13}C-acetate) and ^{15}N-ammonium sulfate or chloride are commercially available and commonly used. The use of the minimal medium easily leads to reduction in growth and to a reduced yield compared to the use of a rich medium. The choice of a host strain and vector, and at times also the codon usage, have to be optimized to reach a yield that is sufficient, given the current price of ^{13}C-glucose. The ^{15}N-compounds are inexpensive and thus can be used for testing much of the performance of the protein. More recently, rich growth media with isotope labels have become commercially available to facilitate the uniform enrichment. To make proteins with post-translational modifications, expression in eukaryote hosts, primarily in yeast, reaches sufficient expression levels [424].

For studies of protein complexes intermolecular correlations can be distinguished from intramolecular correlations by filtered heteronuclear experiments, provided that one of the components is labeled selectively and differently from the other. This approach is also useful for large protein complexes to acquire simplified spectra to alleviate the assignment. For the same end, segments of a polypeptide chain can selectively be labeled by ^{13}C and ^{15}N using inteins, splicing enzymes [415]. The inteins also provide the way to obtain cyclic peptides with labels that are significantly more inexpensive than labeled amino acids from the solid state peptide synthesis. DNA and RNA molecules are nowadays obtained with labels from cell-free synthesis using uniformly labeled ribonucleotide triphosphates (NTP) [425–428].

Deuteration
Perdeuteration has become an indispensable means for circumventing problems of transverse relaxation (T_2) inherent to large biomolecules and their complexes [419]. Protons, due to their large nuclear magnetic moment, are the main cause of dipolar relaxation of protons themselves as well as the directly bound heteronucleus. The relaxation problem can largely be reduced by uniform deuteration. Perdeuterated proteins are obtained by growing host microorganisms in deuterium oxide (D_2O) often concurrently with ^{15}N and/or ^{13}C labeling. The expression levels in D_2O are usually comparable to that in H_2O but the growth is compromised. When the labeled protein is dissolved in H_2O the amide deuterons exchange for protons which serve as the main source of magnetization for coherence transfer in heteronuclear correlation experiments and also for the direct detection of the signal during the acquisition. The high degree of labeling is advantageous in reducing the transverse relaxation of the aliphatic carbons and amide protons in transverse relaxation optimized spectroscopy (TROSY) at high magnetic fields [410, 429]. Obviously perdeuteration reduces significantly the distance information available via interproton nuclear Overhauser enhancements (NOE). A compromise between the favorable relaxation properties and the extent of NOE data may be found by adjusting the degree of deuteration with the D_2O content in the culture medium. A suitable degree of random fractional deuteration is often above 50%.

Alternatively, an important portion of the conformational restrictive NOEs can be recovered by relabeling for protonated methyl rotors of aliphatic residues [420]. This is accomplished using either selectively labeled amino acids that are more readily incorporated during the expression than de novo synthesized amino acids or more affordably by using selectively protonated non-glucose carbon sources. It depends on the metabolism of a particular metabolite how the protons become incorporated in specific positions in the amino acids. The methyl group of pyruvate is a precursor for the methyls of alanine, valine, leucine and γ-rotor of isoleucine. More recently, protonated methyl rotors of valine, leucine and isoleucine have been derived using α-ketoglutarate. It has been demonstrated that these labeling strategies are sufficient to obtain short-range distance data for determination of three-dimensional structures when the reduced distance information from highly perdeutratated proteins is compensated partly by using the directional restraints and hydrogen bond restrains. More sophisticated labeling strategies are designed to reprotonate C^{α} to be used as the sole source of magnetization for experiments in D_2O.

14.3.2.3 Dilute Liquid Crystals

Perhaps surprisingly, biomolecules as large as proteins tumble in water so stochastic that it is difficult to observe any degree of molecular alignment [430]. It is only proteins with prosthetic groups of high magnetic anisotropic susceptibility, such as globins with heme-groups, that acquire a noticeable degree of alignment in isotropic milieu [431]. In general it is when molecules are first dissolved in a dilute liquid crystal that they assume a minute but measurable degree of molecular alignment [409]. The molecular alignment arises from the interaction between the molecule of interest and the liquid crystal particles. The steric hindrance, i.e. the shape of the molecule, determines the size and direction of the alignment [432]. The surface charges on the biomolecule, the charge texture on the liquid crystalline particle and the ions or co-solvents in the solution may make a contribution to the size and direction of alignment.

The molecular alignment manifests itself as residual dipolar couplings (RDC) and chemical shift anisotropy (CSA) that carry information about directions and molecular shape. The preferred degree of molecular alignment is small, of the order of 10^{-3}, to retain essentially all the characteristics of high-resolution NMR spectra. Otherwise full span dipolar couplings and CSAs would render the spectrum of a complex molecule hopelessly uninterpretable. RDCs reveal the directions of internuclear vectors and frequency shifts the directions of the components of CSA tensors [433]. The information content is highly complementary to the torsion angles available from three-bond scalar couplings and to the interproton distances available from nuclear Overhauser enhancements when determining the three-dimensional structure. Furthermore, the anisotropy data implicitly contain information about the molecular shape via simulations [432]. The directional dependence of dipolar couplings and chemical shifts is easy to interpret and to compare with values computed from known three-dimensional structures. In this way structures and conformational changes and the formation of quaternary structures are easy to examine and to analyze without extensive structure determination.

A number of aqueous liquid crystalline media suitable for inducing a weak alignment of biomolecules are known [405]. It should be understood that the choice for a particular liquid crystalline system depends on the properties of the biomolecule, liquid crystalline particles and temperature, ionic strength and eventual co-solvents. Some of the critical factors affecting the formation of the liquid crystal and the size and direction of the solute alignment may be difficult to anticipate beforehand and preliminary experiments are often required to find a suitable system. Furthermore, the set of aqueous liquid crystals for aligning biomolecules is likely to continue to expand as the search for novel media is currently in progress.

Liquid crystal composed of organic compounds

Certain amphiphatic molecules arrange themselves in discoidal bilayered structures known as bicelles. The bicelles have highly anisotropy magnetic susceptibility and thus the property to align themselves in the magnetic field with the plane normal perpendicular to the field [434]. This results in a uniaxial medium suitable for

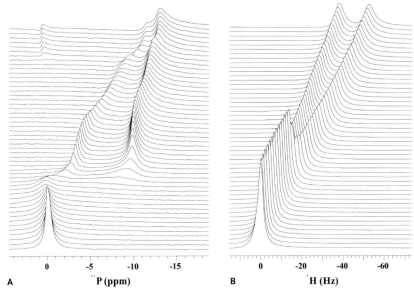

Fig. 14.46 Deuterium (A) and phosphor (B) spectra as a function of temperature of a liquid crystal composed of DHPC:DMPC (1:3) dissolved in water (5 % w/v). The liquid crystalline phase emerges near 30 °C and isotropic phase reappears above 50 °C.

aligning solute molecules. Bicelles can be prepared, for example, from a mixture of (3:1) dry dimyristolphophatidylcholine and dihexanoylphosphatidylcholine dissolved in water to a concentration above 3 % (w/v) [435]. The long-chain lipid is thought to make the bilayer and the short-chain lipid confines and protects the perimeter of the disc. The solubilization is best carried out at low temperature near 0 °C where the viscosity is low and components mix properly. The liquid crystalline phase is stable in the temperature range from approximately 30 to 50 °C depending slightly on the exact composition of the medium. The correctly prepared phase should be uniform and transparent. The presence of the liquid crystalline phase is easy to confirm by measuring the deuterium quadrupolar splitting of D_2O or from the shifted phosphor signals of the phosphatidylcoline head groups (Fig. 14.46).

The size of the alignment can be tuned by varying the concentration of the liquid crystal, however, above the critical concentration required for the liquid crystalline phase to form [435]. The tuning is a convenient way to adjust the range residual dipolar couplings and chemical shift anisotropy optimal for the measurements. Provided that the alignment is induced, not exclusively by the molecular shape, but also by the electric interactions between the surfaces of the solute and liquid crystal particle the ionic strength of the solution will influence both the size and the direction of the alignment. This phenomenon can be exploited by "doping" the bicelles with a small amount of charge. In this way it is possible to create several non-redundant alignments and to collect non-redundant data for the structure determination.

Liquid crystal phases can also be prepared by dissolving various polyethylene glycols, referred to as $C_m E_n$ compounds such as $C_{12}E_5$, in water above 3 % (w/v) [436]. A small amount of short chain alcohol such as hexanol helps to form the phase by stabilizing the exposed edges of the lamellae. The liquid crystalline phase is found in the temperature range from 30 to 40 °C. However, the actual composition of polyethylene glycol or their mixtures or any of their derivatives may noticeably alter the phase diagram and transition temperatures. The phase transitions to and from the liquid crystalline phase, again depending on the composition, show hysteresis as a function of temperature. Another lamellar phase can be prepared from cetylpyridinium bromide/hexanol/sodium bromide [437]. The liquid crystalline phase is stable and robust, tolerating different buffer conditions, temperature ranges and concentrations.

Liquid crystals prepared from biological material
Many particles of biological origin are also suitable for making lyotropic liquid crystals. For example filamentous phages most notably Pf1, fd, and tobacco mosaic virus (TMV) readily align their long axis along the magnetic field [438–440]. The full degree of alignment of Pf1 is obtained already at modest field strengths e.g. at the field of a 300 MHz NMR spectrometer. The liquid crystalline phase is stable over a wide range of temperatures, above 5 °C, and it does not depend on the concentration as the phage particles align individually, unlike bicelles and lamellae where a critical concentration has to be exceeded. It is in the nature of the phages to carry high charge on their surfaces. Consequently electric interactions often play a role in the alignment, in addition to the molecular shape. The presence of electric contributions can often be deduced by varying the salt concentration. The Pf1 particles are negatively charged and therefore suitable for work with DNA which is also negatively charged. On the other hand an attractive Coulombic interaction may lead to an exchange dominated alignment [441]. Likewise, when using particles of biological origin more specific interactions may come into play and express themselves already at very dilute concentrations of liquid crystalline particles as very large dipolar couplings. The low pI ~ 4 of Pf1 implies that particle aggregation becomes a problem in acidic conditions. The bacterial phages can be produced in an infected bacterial culture and harvested after cell lysis and purified by precipitation and finally in a salt-gradient by an ultracentrifuge.

Fragments of purple membranes (PM) of *Halobacterium salinarum* align in the magnetic field [442]. The amount of the alignment can be tuned by varying the concentration of purple membranes and the solute alignment is also affected by salt that screens electric interactions between the negatively charged membranes and solutes. At tens of mM concentrations of monovalent ions, however, the PM membranes undergo transition to a gel where the membrane patches can no longer be oriented with the magnetic field. An order of magnitude lower concentrations of divalent salt are required for a similar phase transition to gel.

Oriented matrices

A cross-linked polyacrylamide gel customarily used in electrophoresis can be compressed or strained to bring about an oriented medium suitable for weakly aligning biomolecules [443]. The oriented matrix is robust, i.e. insensitive to the solvent conditions such as pH, temperature and co-solvents, which makes it a suitable medium for the study of e.g. denatured proteins. Furthermore, the matrix is expected to be insensitive to the surface charge of the biomolecule. The director of the medium is determined by the direction of the strain, not by the magnetic field, which opens possibilities for acquiring data of a sample in different orientations with respect to the magnetic field. The main disadvantage of the polymeric systems is that the solute molecule has to be included in the solution prior to the polymerization or driven to the matrix by electrophoresis or diffusion. Alternatively, the matrix can be dried and made to reswell by solution containing the solute of interest.

Oriented matrices can also be prepared from a liquid crystal composed of membranes and bacterial phages by polymerization to the acrylamide gel. The resulting "frozen" medium is no longer a liquid crystal but an oriented matrix whose director is independent of the magnetic field.

14.3.3
Proton NMR Experiments

The large magnetic moment of a proton, a spin-½ particle, and its high natural abundance are the reasons for the historical predominance of ^{1}H NMR spectroscopy. There are numerous homonuclear ^{1}H experiments, however, certain of them have proven to be most applicable to biomolecular NMR spectroscopy and are mentioned here. The emphasis is on the principles of the experiments, their processing and information content. The reader is encouraged to consult textbooks [407, 408] for a proper description of the experiments using the appropriate nomenclature, which is not possible to introduce here. Furthermore, the operation of an NMR spectrometer and procedures for setting experiments, including calibrations, are usually found in text books or spectrometer manuals and are not discussed here.

In general all the present-day pulse sequences employ pulse field gradients (PFG) for selection of desired coherences and/or for removal of artifacts. Furthermore, phase sensitive experiments are preferred over their magnitude mode counterparts due to the superior resolution and the possibilities for analyzing the multiplet patterns for coupling constants. For the quadrature detection, the states-TPPI protocol [444] is often the most advantageous to displace artifacts to the edges of spectra. For small proteins and peptides and for many other biomolecules the basic proton correlation experiments are the most practical and require no isotope labeling. Many of the elements of the basic proton correlation experiments are also embedded in many heteronuclear experiments.

14.3.3.1 One-dimensional NMR Experiment

Customarily studies of a biomolecule by NMR spectroscopy will begin by recording a one-dimensional (1D) proton spectrum. The 1D spectrum allows consideration of the signal-to-noise ratio (SNR) for further studies. Signals must be easily visible above the noise from a spectrum acquired by averaging 32 scans or preferably less for two-dimensional correlation experiments to be feasible (Fig. 14.47). For equimolar concentrations larger proteins have lower apparent SNR because they have larger line widths than smaller proteins and peptides.

1D spectra of peptides and small proteins usually permit one to judge from the line widths whether aggregation or exchange processes are likely to compromise the performance. For larger proteins in particular it may be difficult to find sufficiently isolated signals in the 1D spectrum to measure line widths. Furthermore, the proton line widths of a large protein may not be adequate to evaluate the feasibility of further experiments. The feasibility of NMR studies for larger proteins is better judged on the basis of heteronuclear experiments and at least a ^{15}N-labeled sample is required.

The overall dispersion of the signals seen in 1D is an important attribute of the structural characteristics of a protein. Well-structured proteins display signals over large spectral regions whereas resonances of unstructured or unfolded proteins lump together and have values similar to that found in short (random coil) pep-

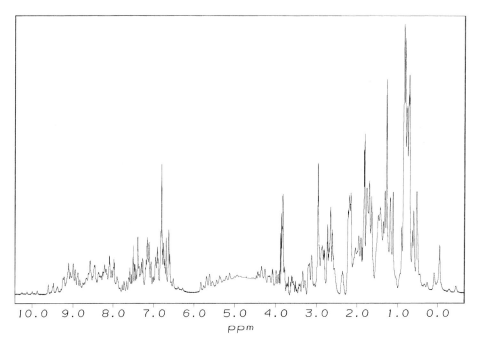

Fig. 14.47 One-dimensional spectrum of a 3-mM human SH3 domain of Tec protein dissolved in H$_2$O/D$_2$O (90/10 v/v) recorded at 25 °C and pH 6 with an 500 MHz NMR spectrometer. Water signal was filtered post acquisition. Courtesy of Hanna Avikainen.

tides. In particular, the chemical shift dispersion in amide, alpha and methyl proton resonance regions is indicative of the structure. 1D can be used to monitor the effects of variables such as pH, temperature, ionic strength or co-solvents on the SNR and the dispersion of signals. However, when using non-deuterated co-solvents the solvent signal suppression may become problematic.

Owing to the high concentration of water it is necessary to record the 1D proton spectrum using solvent suppression. The same applies to other non-deuterated solvents. There are various ways to reduce the solvent lines to a level comparable to the signals of the solute in order to utilize the full dynamic range of detection of the NMR spectrometer. Saturation of the water line by selective radiation during the recycle delay is a simple, effective and robust method. Therefore the presaturation method is commonly used. The main disadvantage of the method is that the solute signals near the solvent signals are also easily saturated, thus complicating the assignment. Even a partial saturation will compromise the quantitative spectroscopy. The exchangeable protons, such as amide protons, are also subject to the saturation transfer that diminishes their signal intensity.

More selective schemes for the solvent suppression employ selective pulses, spin-lock purge pulses and field-gradient pulses. Some of these methods aim to saturate the water signal in milliseconds rather than during the recycle delay, while others are designed to maintain the water magnetization in equilibrium. To destroy the water magnetization selective pulses are used to generate orthogonal components for the solvent and solute magnetizations and subsequently to destroy the solvent signal by RF-inhomogeneity or to dephase the solvent coherence by pulsed-field gradients. To place the water magnetization along the z-axis during the acquisition, selective pulses or binomial sequences are employed for selective excitation, and pulsed-field gradients are used to destroy any residual transverse magnetization caused by pulse imperfections [445]. In this way radiation damping is largely avoided and also noticeable gains in the signal intensity of the exchangeable protons are obtained.

Post-acquisition methods are used to improve the *appearance* of the spectrum. Filtering the low-frequency components by convolution is one of the most widely used methods to suppress the residual solvent signal [446]. The filtered data set is constructed by subtracting the data "smeared" by the filter function from the original FID (Fig. 14.47). If the solvent signal is away from the carrier frequency, i.e. high-frequency components are to be removed, then the original data are initially digitally frequency shifted to zero by multiplication by an exponential function and subsequently filtered and finally restored to the original position by multiplication by the complex conjugate of the exponential function.

A flat baseline is obviously an important merit of a spectrum of any dimensionality. There are various reasons for a poor baseline. The baseline will have an offset and curvature if the signal phase at the beginning of the acquisition period or the indirect dimension is not a multiple of 90° and if the sampling delay is not adjusted to zero, i.e. $t_o = 1$, or to the inverse of twice of the spectral width, i.e. $t_o = 1/(2sw)$. The Hahn echo can be used to adjust the initial sampling delay for the acquisition dimension. For the indirect dimensions in two- or higher dimensional spectra the phase evolution during e.g. off-resonance pulses and mixing periods must be taken into account to set the initial sampling delay correctly. The baseline distortions may also result from corrupted data points at the beginning of the acquisition period. These can be corrected by reconstruction of the initial points by a backward linear prediction. A receiver gain that is too high compared with the dynamic range of the analog–digital converter will result in signals with "wiggles". Also a finite rise time of audio filters may cause a baseline roll. In modern spectrometers this is circumvented by oversampling, i.e. sampling the signal with the maximal frequency of the analog-to-digital converter (ADC), and reconstructing the signal mathematically afterwards for processing.

14.3.3.2 Correlation Experiments

The principle purpose of correlation experiments is to establish a one-to-one mapping from the signal to its source i.e. to the particular atomic nucleus in the molecule. This assignment task involves identification of the members in the coupling network, referred to as the spin system. In addition, correlation experiments, as such or with modifications, are suitable for measurements of scalar and dipolar couplings. Correlation in the two dimensions is the most natural dimensionality because the spin–spin interactions are pair wise. Three-dimensional or experiments of higher dimensionality are constructed from concatenated two-dimensional experiments. Homonuclear three-dimensional experiments, such as TOCSY-NOESY, are not considered here because in many cases the multidimensional heteronuclear experiments are superior.

For all homonuclear experiments there are certain common guidelines to follow. When preparing a correlation experiment it is the expected line shapes and widths that determine the lengths of acquisition and indirect dimension and the processing. Narrow in-phase lines allow a considerable degree of freedom for the setting experiments whereas broad antiphase lines are the most difficult. The repetition rate of the acquisition for signal averaging should be adjusted properly to reach optimal SNR or to obtain quantitative spectra.

COSY-type experiments

Correlated spectroscopy (COSY) was among the first two-dimensional (2D) NMR experiment realized [447, 448] and it is still among the most useful NMR experiments. COSY generates cross peaks in the 2D spectrum at the intersection of resonances of coupled spins (Fig. 14.48). In proteins cross peaks are observed for geminal, i.e. over two bonds, and vicinal, i.e. over three bonds, protons and in small peptides also couplings over four bonds may be detected. Thus the COSY spectrum allows the identification of spin systems for the assignment. However, apart from peptides, the overlap and degeneracy in chemical shifts is likely to prevent one from obtaining entire spin systems exclusively from the COSY spectrum; additional experiments are required.

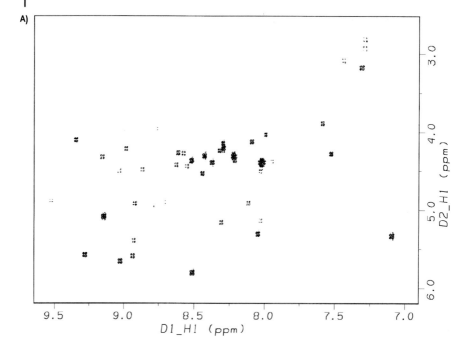
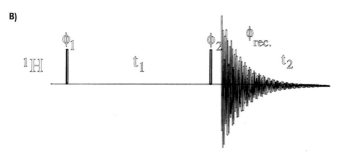

Fig. 14.48 COSY spectrum of a 1-mM SH3 sample in H$_2$O/D$_2$O (90/10 v/v) recorded at 30 °C with a 500 MHz NMR spectrometer using the phase sensitive COSY experiment (A) spectrum (B) pulse sequence.

The COSY spectrum also provides information about the coupling constants in the fine structures of cross peaks. The value of the three-bond scalar coupling is related to the torsion via the Karplus equation [449] and knowledge of the couplings together with NOEs provides stereospecific assignment. The coupling can be determined from the frequency separation of the antiphase absorptive doublet. For two coupled spins i.e. without passive coupling partner such as the $^3J_{HNH\alpha}$ of residues other than glycines (that have two H$_a$) the lobes of the cross-peak are separated by the coupling constant along the directly observed direction. However, as the linewidth increases comparable to the coupling constant, or even larger, the positive and negative lobes of the antiphase line began to cancel each other more and more, thereby increasing the apparent separation of the multiplet components. This aberration from the true coupling value has to be taken into account when measuring coupling constants. Analyses of the antiphase lines for extracting the true separation of the multiplet components require that the spectrum has been processed to maintain Lorentzian line shapes. In practice when the line width is twice as large as the coupling constant or more it is difficult to obtain an accurate value. Therefore, scalar coupling constants are preferably measured by heteronuclear experiments. Due to the self-cancellation of the broad antiphase lines the sensitivity of the experiment also decreases with increasing molecular weight. Otherwise for peptides and small proteins COSY is among the most sensitive of experiments.

The COSY pulse sequence is very simple, consisting of the recycle delay, the first 90° pulse succeeded by the incremental delay t_1, the other 90° pulse followed by the acquisition period t_2 (Fig. 14.48B). The phases of the pulses and the receiver are cycled according to CYCLOPS and axial peaks are suppressed by inverting the phases of the first pulse and receiver. The residual axial peaks are moved to edges of the spectrum by the States-TPPI protocol. A product operator analysis shows that the cross-peak has an antiphase lineshape and that the diagonal peak differs by 90° and thus cannot be phased simultaneously for absorption. The nature of the antiphase lines requires that the data in the indirect dimension t_1 are collected sufficiently long, at least longer than $1/(4J)$, otherwise the signal intensity is reduced due to self-cancellation of the antiphase lines. For the measurement of couplings the acquisition dimension t_2 can well be longer than $1/(2J)$ as the recycle delay can be shortened correspondingly. The COSY spectra are customarily processed for resolution using sine bell weight functions. The analysis of coupling constants requires that the F_2 slices have been adequately zero-filled, e.g. to about 0.5 Hz per point.

There are a number of variant COSY experiments and extensions to COSY. One of the main goals of the descendants has been to remove the dispersive tails of the diagonal peaks that may obscure the near-diagonal cross-peaks. The double-quantum filtered (2QF) COSY experiment [450] yields a spectrum with absorptive antiphase lines for the diagonal peaks of the coupled spins and the attenuated diagonal resonance of the uncoupled spins. The drawback of 2QF-COSY over COSY is the two-fold decrease in sensitivity. Also the total acquisition time for a concentrated sample may become long because of the longer phase cycle. The P-COSY experiment is designed for the same purpose as 2QF-COSY but offers greater sensitivity.

E-COSY is an abbreviation for methods to generate coupling patterns with, exclusively, the active components. The fine structure of the E-COSY cross-peak is a superposition of two-quantum and three-quantum COSY peaks so that the two components that stem from the passive couplings cancel each other to leave a simplified coupling pattern. The active coupling can be measured directly from the antiphase separation on either of the multiplets as usual and the passive couplings can be measured from the displacements of the two antiphase multiplets. Importantly, the method will not be subject to the errors due to finite linewidths. Furthermore, the E-COSY experiment serves to determine the relative sign of the passive couplings. The same principle is used in heteronuclear experiments. The pre-TOCSY-COSY experiment aims to recover resonances attenuated by the solvent presaturation. The mixing sequence prior to the first 90° pulse serves to transfer magnetization to H_α that is near to the irradiated water line.

Relayed COSY [451] is a straightforward extension to COSY. An additional coherence transfer delay is inserted in the pulse sequence before the acquisition period in order to relay coherence from the second spin to a third spin. This will result in a spectrum with the cross peak between the first and the third spin even if they are not directly coupled but only via the second spin. The chemical shifts are monitored before the coherence transfer steps, i.e. t_1, and after, i.e. t_2. The mixing time for the second coherence transfer depends, of course, on the couplings, for proteins the delay from 20 to 40 ms is usually a good compromise between transfer efficiency and relaxation losses. There is no new information compared to COSY, however, in the case of degenerate shifts, R-COSY provides means of assessing whether the spins belong to the same spin system. The acquisition and processing of R-COSY is similar to COSY.

Multiple quantum spectroscopy
Multiple quantum spectroscopy offers complementary information to COSY to elucidate scalar coupling networks [452]. The multiple quantum transitions are observed indirectly during t_1. The two-quantum (2Q) experiment is commonly used to circumvent problems in COSY due to diagonal peaks, self-cancellation of signals and solvent suppression.

The idea is to generate the 2Q coherence that evolves with the frequency sum of the chemical shifts $\Omega_1 + \Omega_2$ of two coupled spins I_1 and I_2. The active coupling J_{12} will not evolve during t_1. During the acquisition cross-peaks appear along F_2 at frequencies of Ω_1 and Ω_2 with the absorptive antiphase splittings corresponding to J_{12} and along F_1 at a frequency of $\Omega_1 + \Omega_2$ with absorptive in-phase lineshape and a dispersive antiphase component that add constructively. These cross-peaks are referred to as direct peaks. In addition three-spin coherences that originate from the J_{12} and J_{13} couplings and from the J_{13} and J_{23} couplings become observable and result in cross-peaks along F_2 at frequencies of Ω_3 and along F_1 at frequencies of $\Omega_1 + \Omega_2$. These cross-peaks with double antiphase dispersive nature along F_2 and with dispersive antiphase and absorptive in-phase components along F_1 are referred to as remote peaks. In general the direct and remote peaks can be distinguished from each other by the opposite line shape along F_2. Considering the lineshapes, the acquisition and processing parameters along t_2 are similar to those used in COSY, whereas along the indirect dimension t_1 it is not necessary to collect data extensively and cosine apodization is appropriate.

Total correlation spectroscopy

Total correlation spectroscopy (TOCSY) also known as homonuclear Hartmann–Hahn (HOHAHA) experiment provides all relayed connectivities within a spin system [453, 454]. The primary intention in TOCSY as in other relayed experiments is to establish connectivities in less crowded spectral regions to facilitate the assignment of spin systems.

The key idea in TOCSY is to transfer in-phase magnetization from spin to spin throughout the spin system by an isotropic mixing sequence rather than to rely on transferring antiphase coherence during free precession periods. During the mixing sequence the system is governed by the strong coupling Hamiltonian. The intensity of a cross-peak will depend on the topology of the spin system, i.e. the ways the spins are coupled, the various coupling constants and the properties of the mixing sequence. There are several mixing sequences. The DIPSI-2 sequence [455] is better than WALTZ-16 [456] or MLEV-17 when relaxation is neglected (Fig. 14.49). The actual performance of a mixing sequence over the others depends on the implementation and the accuracy of the calibrations.

Dipolar couplings during the mixing period may give rise to the rotating-frame Overhauser effect (ROE). Attenuation of the in-phase TOCSY peaks due to the ROE peaks of opposite sign can be particularly important for larger proteins but can be largely eliminated by interrupting the mixing train so that laboratory-frame nuclear Overhauser enhancements with opposite sign to ROEs can develop to cancel each other. These sequences are referred to as clean (TOCSY). Furthermore, adiabatic mixing sequences (WURST-8) [456] have low sensitivity to RF field inhomogeneity and miscalibration of the field strength. The sensitivity enhanced TOCSY pulse sequence retains two orthogonal components of magnetization for post-acquisition reconstruction of the spectrum to achieve a gain of $\sqrt{2}$ in the SNR.

The evolution during the indirect dimension and acquisition are not constrained by the scalar couplings because the peaks in TOCSY are mostly in-phase and only resolution should be considered when setting the acquisition and indirect dimensions and weight functions in processing. It is seldom worthwhile to collect t_1 longer than $1.5 T_2$ where T_2 is the transverse relaxation time. The length of the isotropic mixing sequence is subject to two opposing conditions. On the one hand a long mixing sequence is preferred to obtain a correlation between the spins at the opposite ends of the network and on the other hand all correlations are preferably observed. These constraints are difficult to meet simultaneously and therefore usually two or more TOCSY spectra with mixing times from 30 to 120 ms, depending on the relaxation, are collected. The interpretation of TOCSY is subject primarily to two concerns. Since the experiment provides correlations along the network of spins it may remain uncertain which nucleus in the spin system gives rise to which signal in particular when chemical shifts are not unambiguous. The other concern is that cross-peaks may not always be observed due to the intensity variation of the signal as a function of the mixing time. Furthermore, if TOCSY is used to estimate coupling constants it must be kept in mind that the signal intensity depends on *all* couplings in the spin system. Short mixing times transfer mostly via a single scalar coupling and are more reliable for coupling constant measurements.

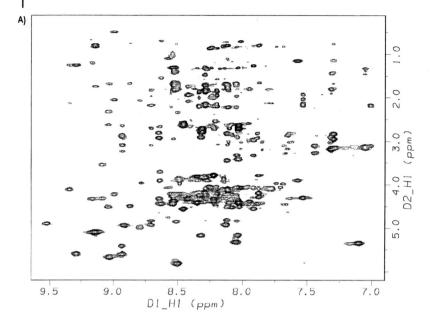
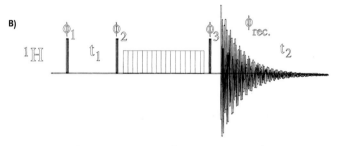

Fig. 14.49 (A) TOCSY spectrum of a 1-mM SH3 sample in H$_2$O/D$_2$O (90/10 v/v) recorded at 30 °C with a 500 MHz NMR spectrometer using the DIPSI-2 mixing sequence (B).

14.3.3.3 Cross-relaxation Experiments

The elucidation of the scalar coupling network by the correlation experiments is, apart from small molecules, not sufficient for the unambiguous, sequential and stereo-specific assignment. The complementary information of spatially adjacent protons is obtained via cross-relaxation experiments, the laboratory-frame nuclear Overhauser enhancement spectroscopy (NOESY) and the rotating-frame nuclear Overhauser effect spectroscopy (ROESY). These experiments provide also the distance restraints for the structure determination and help to recognize exchange processes.

NOESY

The NOESY experiment is composed of the frequency labeling part, i.e. t_1 flanked by two 90° pulses, ensued by the magnetization transfer via dipolar couplings during the mixing period τ_m before the read pulse for the acquisition t_2 (Fig. 14.50). During the mixing period relaxation processes govern the longitudinal magnetization. The autorelaxation results in the diagonal peak and the cross-relaxation results in the cross-peaks in the NOESY spectrum. Both peaks are absorptive in-phase in both dimensions. The phase cycling will remove higher coherences than zero. Consequently, in addition to the longitudinal magnetization the cross peak may have an undesirable zero-quantum contribution. The zero-quantum peak is antiphase in both dimensions, just as a COSY peak, but dispersive with respect to the NOE peaks. The intensity of the zero-quantum peak depends on the chemical shift difference of the spins in question and the mixing time. Furthermore, the transverse zero-quantum term relaxes during the mixing period faster than the desirable longitudinal magnetization. Therefore, the zero-quantum artifacts are usually observed when the mixing time is short but can further be suppressed by varying the length of the mixing time, a procedure known as z-filtration. However, for short mixing times it may be difficult to vary the mixing time suficiently without introducing problems in the interpretation due to the averaging.

When setting the NOESY experiment it is worthwhile to make sure that the baseline is flat and the total measuring time will yield adequate SNR. The recycle delay should be $3T_1$, where T_1 is the longitudinal relaxation time, otherwise the intensity of the cross-peaks can be affected by the steady-state conditions. The length of the mixing time is subject to contradictory demands. A long mixing time, of the order of T_1, will give high intensities for the cross-peaks and have small zero-quantum artifacts. On the other hand multiple magnetization transfers, referred to as spin-diffusion, will affect the intensity and obscure the interpretation. Therefore, a compromise depending on the size of the protein has to be made or a series of NOE spectra must be collected for more precise calibration of intensities to distances. For large proteins the dipolar relaxation is effective and short mixing times are appropriate whereas for small proteins or peptides longer mixing times are required. Furthermore, the cross-relaxation rate constant depends on the rotational correlation time and changes its sign at short correlation times, thereby making the observation of NOEs for small molecules difficult unless there is a freedom to change temperature so as to affect the rotational correlation time. For proper integration of the cross-peaks during processing, window functions for the resolution enhancement are to be applied with caution because signals differing in their linewidths will then not retain relative intensities. Otherwise the in-phase absorptive peaks allow considerable freedom for choosing lengths acquisition and indirect dimensions and their apodization.

Chemical exchange, not too slow compared with the mixing time, can result in cross-peaks in the NOESY spectrum. The chemical exchange peaks are hard to distinguish from the NOE peaks but they have a different sign from the ROE peaks. Furthermore, coupled exchange and spin-diffusion can result in complicated spectra.

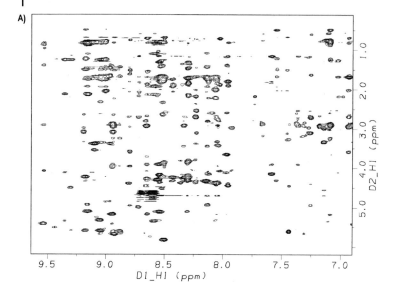

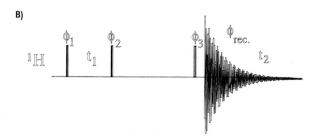

Fig. 14.50 (A) NOESY spectrum of a 1-mM SH3 sample in H$_2$O/D$_2$O (90/10 v/v) recorded at 30 °C with a 500 MHz NMR spectrometer using the NOESY sequence (B).

ROESY

The ROESY experiment [454, 458] follows the cross-relaxation in the rotating-frame that is established by locking the spins with a train of RF-pulses. During the spin-lock any components of magnetization orthogonal to the RF-field are dephased by the inhomogeneity of the RF-field. The longitudinal magnetization is governed by rotating-frame relaxation rates. The rotating-frame cross-relaxation rate constant is positive for all correlation times, which makes ROESY particularly suitable for studies of small peptides. The ROESY cross-peaks depend, in addition, on the resonance offset from the carrier. The offset dependence appears as a phase error along F_1 and F_2 which can be removed during processing. However, the intensity of cross-peaks for the off-resonance spins will have an opposing NOE contribution in addition to ROE. The offset dependence has to be taken into account in quantitative analysis of the cross-peak intensities. The ROE mixing time is usually kept shorter than the corresponding NOE mixing time because the ROE build-up is twice as fast as the NOE build-up. The in-phase absorptive cross-peaks and diagonal peaks but of opposing sign to each other allow considerable freedom in setting the parameters for the acquisition and indirectly detected dimensions and their processing.

The ROE experiment is potentially subject to artifacts rising from isotropic mixing. The TOCSY transfer will reduce the intensity of the ROE cross-peak and obscure the quantification. In particular consequent TOCSY and ROE transfers have the same sign as the authentic ROE peaks and may easily be misinterpreted. Therefore, as long as the long spin-lock is not sufficiently strong to match the Hartmann–Hahn condition there is only need to watch for artifacts for those spins that have closely similar chemical shifts or shifts that are placed on opposite sides with respect to the carrier. Eventual artifact related problems can be noticed by taking spectra with different carrier positions and varying the strength of the spin-lock.

In contrast to the NOE experiment the cross-peaks of the ROE experiment are of opposite sign with respect to the chemical-exchange peaks. This is a particularly valuable property in studies of complex formation where peaks arise both from cross-relaxation and chemical exchange. Furthermore, the assignment of stereospecific restraints to the methylene protons that are often subject to spin-diffusion can be facilitated by the ROESY measurement. The ROE cross-peaks that stem from the relayed transfer have opposite sign with respect to the direct transfer and thus the methylene ROE signals do not tend to equal values.

14.3.4
Heteronuclear NMR Experiments

Heteronuclear NMR experiments have many virtues over the homonuclear ^{1}H experiments when labeled samples are available for the measurements. It is for the high molecular weight proteins and protein complexes that the heteronuclear experiments are imperative. Heteronuclear methods compared with homonuclear experiments provide improved resolution via increased dimensionality and editing and filtering possibilities and various ways for coherence transfer and for measurement of scalar and dipolar couplings to increase the information content. It is in the nature of amino acids in folded proteins to contain distinguishable nuclei that are coupled to each other with various strengths of couplings that lay the foundation for the rich repertoire of heteronuclear experiments. Furthermore, relaxation measurements of ^{15}N and also ^{13}C are much more amenable to interpretation than those of protons. Today heteronuclear experiments are indispensable for biomolecular NMR spectroscopy.

14.3.4.1 Basic Heteronuclear Correlation Experiments
Many heteronuclear multidimensional correlation experiments appear superficially complicated but actually comprise simple building blocks for coherence transfer and for recording chemical shift and/or coupling evolution. These basic elements

are simple and robust. The usual protocol begins by transferring proton magnetization to the heteronucleus i.e. ^{15}N or ^{13}C, which is subject to further actions and finally the coherence is returned to the proton for detection. This indirect detection scheme is superior in sensitivity to direct methods because of the proton's large gyromagnetic ratio and favorable relaxation properties. The most frequently used experiments are the heteronuclear multiple quantum coherence (HMQC) [459] and heteronuclear single quantum coherence (HSQC) experiments [460]. More recently it was discovered that significant gains in SNR for large proteins at high magnetic fields are obtained by selecting particular multiple components for the detection to result in transverse relaxation optimized spectroscopy (TROSY) [410].

HMQC

The HMQC experiment begins by a coherence transfer from proton to the heteronucleus during a delay matched to $2\tau = 1/(2J)$, $J_{HN} = 94$ Hz and $J_{HC\alpha} = 140$ Hz, to result in a multiple-quantum (MQ) coherence that evolves during t_1. The heteronuclear coupling is not active under MQ during t_1 but the ^{1}H chemical shift evolution must be refocused by a 180° proton pulse. The MQ coherence is transferred to the proton for detection. Before the acquisition antiphase dispersive components can be purged by a 90° proton pulse. During the acquisition t_2 heteronuclear decoupling is employed. Homonuclear scalar couplings during the coherence transfer periods will cause a phase error along F_1. The size of the error is fairly small but it depends on the size of the coupling and cannot be simply compensated. The HMQC experiment is very simple with only a few pulses and it is therefore a favored building block in more complicated experiments.

HSQC

The HSQC experiment (Fig.14.51) is based on the INEPT (insensitive nuclei enhanced by polarization transfer) sequence which converts proton magnetization to the antiphase single-quantum (SQ) coherence that evolves during t_1 [460]. The scalar coupling evolution is refocused by a 180° proton pulse or decoupled. Homonuclear scalar couplings do not contribute to the lineshape along F_1. The reverse transfer from the heteronucleus to proton is analogous. Homonuclear scalar couplings during the coherence transfer periods will affect the amplitude of the peak and cause an antiphase dispersive ontribution along F_1 which can be purged by a 90° proton pulse just before the acquisition. During the acquisition the t_2 heteronucleus is decoupled.

The decoupled HSQC experiment employs the refocused INEPT sequence to generate an in-phase SQ coherence that is desirable e.g. for measurements of heteronuclear relaxation rates. The transfer function of the refocused INEPT depends on the proton multiplicity. i.e. the number of protons bound to the heteronucleus. This can be used to edit the spectrum or, when an overall maximal transfer efficiency is required, $2\tau = 1/(3J)$ makes a good compromise. The constant time

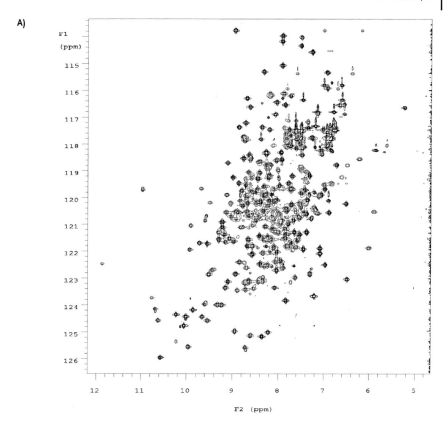

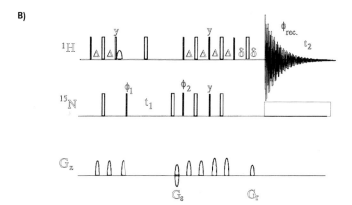

Fig. 14.51 (A) HSQC spectrum of a 1-mM cellulase sample in H$_2$O/D$_2$O (90/10 v/v) recorded at 40 °C with an 800 MHz NMR spectrometer using the gradient selected sensitivity enhanced HSQC sequence (B). Courtesy of Outi Salminen.

(CT) HSQC has an evolution period $t_1 = T$ fixed in length. This brings in two advantages. The homonuclear couplings do not modulate the signal during t_1 and consequently there is no homonuclear multiplet structure along F_1. This is particularly useful for fully ^{13}C-enriched samples to remove the ^{13}C–^{13}C couplings of aliphatic carbons that cannot be selectively decoupled. The other advantage is the narrow lines. The linewidth along the indirect dimension does not depend on the transverse relaxation rate but is mainly determined by the window function used for processing. The transverse relaxation rate appears only as a constant factor that scales the intensity. The optimum trade-off between intensity and resolution depends on each particular case.

TROSY

At high magnetic field strengths the chemical shift anisotropy (CSA) interaction of ^{15}N is a comparable source of relaxation as the dipole–dipole (DD) interaction. The same is true for the amide proton CSA and DD. Therefore the four multiplet components of the amide spin pair have different relaxation rates, depending on whether the DD and CSA mechanism interfere constructively or destructively. When other sources of relaxation such as those due to remote protons are negligible the differential relaxation effect on the multiplet components is large and also, in practice, attainable at high magnetic fields for large perdeuterated proteins. It is the most slowly relaxing multiplet component that is of particular interest and it is possible to filter away the other multiplet components for spectral simplification. This is the essence of the transverse relaxation optimized spectroscopy (TROSY) [410, 462]. The resulting spectrum is superior to an HSQC experiment under the aforementioned conditions.

The pulse sequence resembles superficially that of the sensitivity enhanced HSQC [462]. Obviously neither proton during t_1 nor ^{15}N during the acquisition t_2 should be decoupled to maintain the multiplet components. It is the latter part of the pulse sequence used for the reverse transfer from ^{15}N to ^{1}H including the gradient selection that chooses the most slowly relaxing multiplet component for the detection (Fig. 14.52).

The intensity of the TROSY component can be further boosted by sensitivity enhancement and more importantly by taking advantage of the ^{15}N steady state magnetization [462]. The TROSY principle has been implemented in many heteronuclear correlation experiments such as those used for the main chain assignment and measurement of scalar and dipolar couplings that exploit coherence transfer between the amide nuclei before the acquisition. The TROSY principle works also for the CH pairs in aromatic rings and side chain NH_2 groups [463]). TROSY has significantly increased the molecular size limit of proteins and their complexes feasible for NMR studies [464, 465].

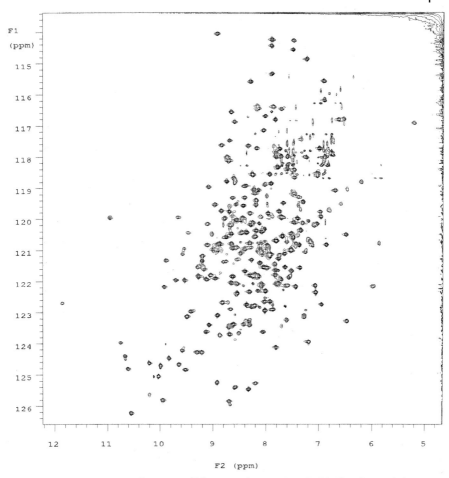

Fig. 14.52 TROSY spectrum of a 1-mM cellulase sample in H_2O/D_2O (90/10 v/v) recorded at 40 °C with an 800 MHz NMR spectrometer using the gradient selected TROSY sequence.

14.3.4.2 Edited and Filtered Experiments

The basic heteronuclear experiments are easy to combine with the two-dimensional homonuclear experiments to produce three- or four-dimensional edited spectra. In this terminology editing means selection of the protons that are attached to the heteronucleus. The main purpose of these experiments is to reduce the signal overlap of the homonuclear two-dimensional experiments.

Combinations of NOESY with HSQC (or HMQC) and TOCSY with HSQC are among the most useful edited three-dimensional experiments (D. Marion, P. C. Driscoll, L. E. Kay, P. T. Wingfield, A. Bax, A. M. Gronenborn, G. M. Clore. Biochemistry 28, 6150–6156 (1989). E. R. P. Zuiderweg, S. W. Fesik Biochemistry 28, 2387–2391 (1989). For example the NOESY-HSQC (Fig. 14.53) begins with the NOESY sequence and is followed directly by the HSQC sequence. It is important

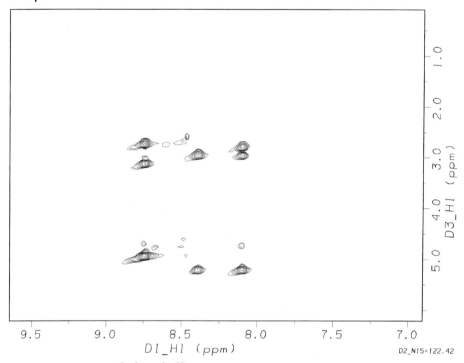

Fig. 14.53 Plane ^{1}H-^{1}H of ^{1}H-^{15}N-NOESY-HSQC spectrum of a 3-mM SH3 sample in H$_2$O/D$_2$O (90/10 v/v) (A) and a plane ^{1}H-^{1}H of ^{1}H-^{15}N-TOCSY-HSQC (A) and ^{1}H-^{15}N-NOESY-HSQC (B) recorded at 25 °C with a 500 MHz NMR.

to acquire enough data points (typically 128 and 2048) along the proton dimensions (t_1) and (t_3) whereas fewer increments (typically 32) along the ^{15}N editing dimension (t_2) will suffice. The F1–F3 plane at each F2 displays the homonuclear correlations, just as in the two-dimensional spectrum but edited with respect to the heteronucleus. It will take about three days to acquire a spectrum with good SNR.

Clearly the homonuclear and the heteronuclear experiments could be combined in the reverse order i.e. HSQC-NOESY and HSQC-TOCSY. The main advantage in these schemes relates to ^{15}N-edited experiments in which the narrower amide proton spectral width is sampled during t_1 and the full proton spectral width is collected during t_3. On the other hand water suppression is more effective when HSQC follows NOESY. Also the sensitivity enhancement can be incorporated into the NOESY-HSQC experiment. For the TOCSY-HSQC or HSQC-TOCSY it does not matter because both the TOCSY and HSQC sequences can be implemented with the sensitivity enhancement. It should be mentioned that the TOCSY type of transfer is more effective between ^{13}C nuclei than between protons and therefore the HCCH-TOCSY experiment is preferred when a doubly labeled sample is available.

Even if the heteronuclear editing provides a big advantage over the two-dimensional homonuclear experiments, this is not sufficient when two protons with a mutual correlation have degenerate chemical shifts. This can be of importance, in particular, in NOESY that contains a large number of cross-peaks. The four-dimensional HMQC-NOESY-HMQC experiment provides editing in two dimensions to alleviate the overlap problem. The NOESY part is flanked by two HMQC sequences. Obviously the two HMQC parts can be chosen to edit with respect to the same or different heteronucleus i.e. ^{15}N or ^{13}C. The total acquisition times tend to be long, several days to achieve sufficient resolution along the indirect dimensions.

For studies of intermolecular interactions, filtered experiments are valuable. The essence of filtered experiments is to distinguish protons on the basis of the heteronucleus to which they are bound. In the terminology, filtering means rejection of the protons that are bound to the heteronucleus. Therefore, provided that the heteronuclei in the subunits that make the quaternary complex are labeled differently i.e. either the target protein or the ligand is labeled, it is possible to employ filtering experiments to distinguish e.g. intermolecular proton NOEs from intramolecular NOEs. While for the heteronuclear edited experiments it is sufficient to have only a modest degree of labeling, it is mandatory for the filtered experiments to have a very high degree of labeling, i.e. $\geq 98\%$. The most robust filters are the double spin-echo type. However, it should be kept in mind that none of the filter sequences will work properly if the labeling degree is not close to 100%. Leakage of signal will lead to misinterpretation.

14.3.4.3 Triple Resonance Experiments

The assignment of signals is, of course, a prerequisite for any detailed interpretation of NMR spectra. The advent of triple resonance experiments [466] for double labeled and, more recently, for perdeuterated proteins has been very important in resolving the assignment problem. The fundamental principle of all triple resonance experiments is a directional relay of the magnetization from nucleus to nucleus via the scalar coupling network concurrent with a multi-dimensional detection of the resonance frequencies. In this way it is possible to correlate several nuclei with enough dispersion to avoid degeneracy, even for large proteins. Today many of the triple resonance experiments are well established (see e.g. [400, 404 and 467]). The principles of how to use them are described below whereas a more thorough discussion can be found in the reviews. In addition the triple resonance experiments are the skeletons for experiments designed to measure scalar and dipolar couplings.

The assignment of resonances to the chemical structure commences from the main chain atoms HN, N, CA, CO and CB. The experiments, i.e. the pulse sequences for this purpose, appear superficially complex but are in fact built from concatenated parts of heteronuclear polarization transfers. The magnetization is often derived from the amide proton and also from the amide nitrogen when using transverse relaxation optimized spectroscopy. Subsequently the polarization is relayed to amide nitrogen and further to carbons. Three-dimensional spectra are produced by acquiring the frequencies of the amide proton directly and recording

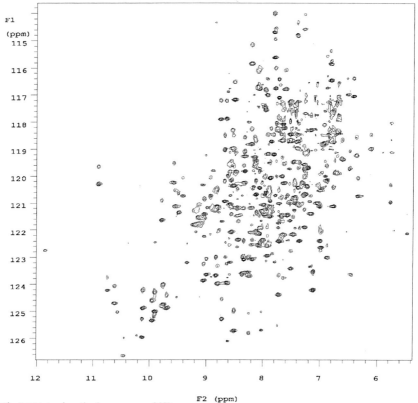

Fig. 14.54 A (continuing on page 121).

indirectly the frequencies of amide nitrogen and either aliphatic carbon or carboxyl carbon (or both in four- dimensional experiments).

It is the coupling network that dictates the possibilities to select correlations to be observed. The key idea is to produce a mutually complementary pair of spectra, e.g. HNCA and HN(CO)CA. The former displays both intra- and inter-residue correlations and the latter exclusively the inter-correlations. Alternatively, an exclusive inter-HNCA can be recorded. The pair of spectra displaying for each amide nuclei resonances both intra- and inter-carbon resonances allow one to trace the polypeptide backbone from one residue to another. The sequential walk is interrupted at prolines and at times potentially due to exchange broadening.

The other commonly used pair of experiments involves also resonances of CB i.e. HNCACB and HN(CO)CACB. Of course, these experiments are less sensitive since the total intensity is divided among CA and CB, but the knowledge of CB shifts will ascertain the type of residue with more confidence than with CA only. The third pair of experiments is the HNCO and HN(CA)CO. The principle is the same as above but now the carbonyl carbons are used instead. The former experiment is among the most sensitive triple resonance experiment whereas the latter is among the least sensitive, making this pair unbalanced. Certain modifica-

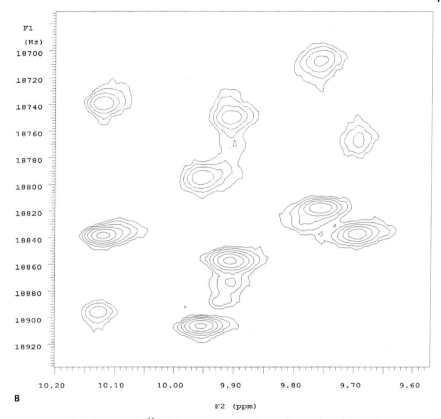

Fig. 14.54 (A) Spin-state edit ^{15}N-1H correlation spectra of a 1-mM cellulase sample in H$_2$O/D$_2$O (90/10 v/v) recorded at 40 °C in a dilute liquid crystal composed of filamentous phages with an 600 MHz NMR spectrometer using the gradient selected TROSY sequence modified to select various spin states. The displacement of the multiplets contain in addition to the 94 Hz scalar coupling residual dipolar contributions (B).

tions to the latter experiment improve the performance, however, at the cost of non-uniform performance for the various amino acid residues.

Today there are many variants of the basic sequences. The most important modifications are the transverse relaxation optimized experiments suitable for perdeuterated proteins, the sensitivity enhanced versions and the constant time (CT) implementations for improved resolution. Furthermore there are many experiments based on the basic sequences designed to measure scalar and residual dipolar couplings, often also implemented with spin-state selection as exemplified in Fig. 14.54.

The assignment of side chain resonances is a more laborious undertaking. The proton detected versions of the aforementioned experiments such as HN(CA)HA and HN(CACB)HAHB reveal the protons at the stem of the side chain. Customarily carbons and aliphatic protons are obtained by CC(CO)NH, HC(C–CO)NH, HCCH-COSY and -TOCSY experiments. Furthermore there are special experiments to correlate nuclei in the aromatic and polar side chains.

14.4
Bioanalytical Mass Spectroscopy

14.4.1
Introduction

Mass spectroscopy (MS) has flourished in various bioanalytical applications for the last 20 years mainly due to the increased demand for fast and sensitive analysis of proteins and nucleic acids separated on gels (2DE) and in capillary electrophoresis (CE). Mass spectrometric analysis requires no fluorescence or radioactive labeling and the mass spectra offer a far more complete analysis. This section will give a summarized impression of the field as it is developing today. Since there are a multitude of modifications and expansions to MS, there is still not very much clarity as to which technique will become 'standard' in proteomics and genomics research. At the moment electrospray (ESI) MS and matrix-assisted laser desorption (MALDI) MS are competing techniques, while various combined forms of MS and tandem MS (MS/MS) have also recently been explored to further increase the resolution of MS spectra of biomolecules and enable multiplexing, higher throughput etc.

Although ESI and MALDI overlap in their scope of application, MALDI-TOF mass spectrometry has been used by most research groups due to the high duty cycle of the TOF spectrometer (acquisition of a spectrum within one second), which has made it more amenable to high-throughput analysis. The MALDI technique also allows the analysis of the highest mass-to-charge ratios (m/z up to 300 kD). Arguably MALDI-TOF, as it is useful for the analysis of solid phases and bioarrays, will remain the technique of choice for bioarrays. ESI-MS, however, is also gaining momentum, because of the better compatibility of ESI with microchips, which is a major innovation force in proteomics today [468]. Various successful demonstration experiments with microchips have recently been reported for performing on-chip sample preparations, such as solid phase extraction, tryptic digestion, pre-separation and pre-concentration. The combination of microchips with ESI-MS seems to be one of the most promising developments today for proteomics research.

14.4.2
MALDI-TOF

Matrix-assisted laser desortion ionization (MALDI) was first described by Karas and Hillenkamp in 1988 [469]. At that time it was a revolutionary method for the ionisation and analysis of large biomolecules. Now many more MS methods have been devised, but the analysis of complete protein masses is still only possible by MALDI [470]. With MALDI, a 'matrix' (crystals of small organic molecules) with a small amount of analyte is ionised by a short laser pulse at a wavelength close to the adsorption band of the matrix molecules. This produces predominantly singly charged molecular ions, which are detected by the TOF spectrometer. The analyte ions are generally produced by a gas-phase proton-transfer from the matrix to the analyte. The advantages of the method are that it is a fast process (ionisation in milliseconds) and that it yields an absolute intrinsic m/z ratio of the biomolecule. MALDI thus also enables the analysis of complex mixtures. As matrices, generally carboxylic acid or hydroxyl-containing aromatic compounds are used, which can form stable carbanions in the gas phase. For proteins, matrices like nicotinic acid, cinnamic acid or 2,5-dihydroxybenzoic acid have been much used. For nucleic acid analysis 3-hydroxypicolic acid has been favoured most, but also 4-nitrophenol and 8-hydroxyquinoline have been recently proposed as efficient matrices [520]. The effectiveness of mass spectrometry is evident in its capability to identify a variety of biomolecules up to 300 kDa. Presently, MALDI-TOF MS is mostly used for the analysis of solid phases, due to the solid matrix. Due to the laser excitation source, however, the area interrogated can be very small. The method thus has a strong potential not only for the reading of DNA-arrays, but also for peptide mapping from microbeads [471].

14.4.3
Electrospray Methods (ESI-MS)

Of the mild ionization techniques electrospray ionization (ESI) has been one of the first to be used for organic compounds, but was introduced for the analysis of biomolecules only recently by Fenn et al. [472]. With ESI-MS, liquid is sprayed into a mass spectrometer with the aid of a very high electric field from a needle-type injector (e.g. a gold-coated capillary). This is a soft ionization technique suitable for the characterisation of biomolecules ranging from several hundred to several thousand in molecular weight. For proteomics research, however, the staggering number of proteins to be analysed from a single organism still places large demands on the sample preparation, and this is the main driving force behind the combination of ESI-MS with microfluidic chips [468, 473]. The first demonstration system of a microchip-ESI interface was described by Karger et al. and consisted of a microfabricated multiple-channel glass chip fabricated by standard photolithographic, wet chemical etching, and thermal bonding procedures (Fig. 14.55) [474]. With this device, separations of model proteins were investigated (myoglobin, recombinant human growth hormone, ubiquitin and endorphin). A high voltage was applied individually from each buffer reservoir to spray the samples sequentially from each channel into the mass spectrometer. To maintain a stable electrospray a liquid flow of 100–200 nL min^{-1} was used. The detection limit of the microchip MS experiment for myoglobin was below 6×10^{-8} M. Samples in 75% methanol and aqueous samples could be successfully analyzed with good sensitivity.

Although the interface of a TOF spectrometer with ESI has been traditionally difficult, orthogonal injection has recently proven to be a good method to enable the ESI method to be coupled with a TOF spectrometer [475]. This also gives possibilities of studying biomolecular complexes. ESI has also recently been coupled to ion mobility spectrometry, which enables a higher resolution in comparison with

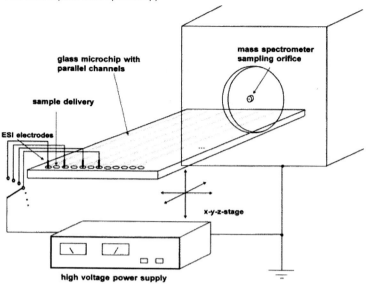

Fig. 14.55 A microchip interface to ESI-MS. (Karger et al 1997, Ref. 474).

normal ESI-MS, as exemplified with the tryptic digest of hemoglobin [476]. Smith et al. have also elaborated on the use of ESI with a Fourier transform ion cyclotron resonance (FTICR) detector [477]. The combination of the multiple charging phenomena of ESI with the superior mass resolution of FTICR gives a superior performance for the analysis of small peptides.

14.4.4
Tandem-MS

Various forms of tandem mass spectroscopy (MS/MS) have also been used in the analysis of biomolecules. Such instruments consist of an ionisation source (ESI or MALDI or other) attached to a first mass analyser followed by a gas-phase collision cell. This collison cell further fragments the selected ions and feeds these ions to a second mass detector. The final mass spectrum represents a 'ladder' of fragment ions. In the case of peptides the collision cell usually cleaves the peptides at the amide bond. The ladder of resulting peptides reveals the sequence directly [496]. Thus, tandem MS instruments, such as the triple quadrupole and ion-trap instruments have been routinely applied in LC-MS/MS or ESI-MS/MS for peptide sequencing and protein identification via database searching. New configurations, which have been moving into this area include the hybrid Q-TOF [498], the MALDI-TOF-TOF [499] and the Fourier transform ion cyclotron resonance instruments [500].

14.4.5
TOF-SIMS

Secondary ion mass spectroscopy (SIMS) is a surface analysis technique based on the bombardment of the surface with a highly focused ion beam. The ions are usually Cs^+ or Ga^+. The bombardment disrupts the uppermost layer of the substrate, giving rise to the emission of surface elements and intact molecules characteristic of the composition in the uppermost layer of the surface. SIMS is routinely used for depth profiling of elements in various inorganic materials and is usually a rather destructive method [478]. However, when limiting the ion dose per surface area below a certain limit more intact molecules of a larger mass can be emitted from the surface ("static SIMS"). SIMS is generally used in conjunction with a time-of-flight (TOF) mass selector, which gives, in principle, possibilities for the detection of very large molecular masses, as with MALDI-TOF.

Static SIMS has usually been applied to the analysis of small organic molecules on surfaces. One of the first investigations comprised the self-assembly of biotinylated compounds on gold and the variation of the density of biotin for optimal binding to avidin [479]. Some other examples are self-assembled layers of alkylmercaptans on silver and gold [480, 481] and π-extended viologens and alkylmercaptans on gold [482]. Metal surfaces generally give the best yield of intact organic ions. For instance, in the self-assembly of alkylmercaptans and other compounds on gold, various aspects of the chemisorption process have been elucidated with SIMS [480, 481]. Additionally, SIMS has been used in the study of protein adsorption. For instance Davies et al. studied the adsorption of proteins to steel, glass, polypropylene, and silicone surfaces [483] and Pradier studied protein adsorption to stainless steel [484]. Protein immobilization onto self-assembled films on gold has also recently been studied [485].

Although inorganic surfaces have been traditionally studied most by SIMS, presently the technique is also applied to polymeric substrates. For instance, Volooj et al. have studied the adsorption of cationic detergents on keratin fibers [486]. Keller et al. reported the TOF-SIMS analysis of small molecules, such as peptides, Nile Blue and cholesterol, on surfaces using Nafion 117 as a matrix for controlled formation of molecular ions [487, 488]. This already points to new possibilities for ionization with SIMS for the analysis of larger fragments. To date, however, only masses up to 1760 MU (for insulin) have been reported [489].

An interesting capability of TOF-SIMS, however, is to image the presence of molecules at micrometer spatial resolution via characteristic fragments [491]. For instance, Galla et al. have imaged the pulmonary surfactant protein C in mono-, bi- and multilayers of lipids with TOF-SIMS [492]. In a recent report Belu et al. described TOF-SIMS imaging of (recombinant) ^{15}N-labelled streptavidin micro-patterned by light activation on PET substrates [493]. SIMS spectra of normal streptavidin and ^{15}N-labelled streptavidin were compared. The positive ion SIMS-spectra yielded characteristic peaks for streptavidin at m/z 70 (from proline and arginine) and m/z 130 (from tryptophan), while ^{15}N-labelled streptavidin gave corresponding peaks at m/z 71 and m/z 131. The negative SIMS spectra gave peaks at m/z 26

(CN⁻) and m/z 42 (CNO⁻) for streptavidin and m/z 27 and m/z 43 for the ^{15}N-labelled streptavidin. The anion at m/z 27 ($C^{15}N^-$) enabled the most unambiguous imaging of the streptavidin on the protein microarray.

14.4.6
MS in Protein Analysis

The task of a proteome analysis is quite formidable, as it may be necessary to identify thousands of proteins from a single organism. Besides the application of mild ionisation techniques, the use of new computational methods for correlating mass spectrometric data with information in protein sequence databases has led to the capability for 'mass fingerprinting', in which large numbers of proteins can be rapidly identified [493–496]. This type of analysis is generally referred to as 'descriptive proteomics'. As opposed to the genome, the proteome is not a static system. Various clinical studies need to assess the protein expression of one organism with that of another to establish the mechanism by which genetic mutations, infections or drugs modify the functions of the organism. Therefore, more and more emphasis is placed on 'dynamic proteomics', the capability to study the proteome's time-dependent changes, as well as 'quantitative proteomics' instead of mere descriptive proteomics.

The classic technique for separation of proteins, before identification by MS, has been (and still is in many laboratories) 2-dimensional electrophoresis (2DE), a technique that has been known since 1975 [497]. In 2DE proteins are mapped in one direction by net charge (using isoelectric focussing, IEF) and in a second, perpendicular direction by size (using sodium dodecyl sulfate polyacrylamide gel electrophoresis, SDS-PAGE). 2DE allows analytical scale separation of up to 10,000 individual proteins as well as micropreparative scale separation of proteins for further analysis. The identification of the proteins separated on the gel generally starts with the isolation of the protein spots from the gel slab and their enzymatic digestion to smaller fragments. Hereafter MS can be used in two ways for the final identification: either directly by ESI-MS or MALDI-TOF-MS, or indirectly by chromatographic separation of the fragments followed by tandem mass spectroscopy (MS/MS). At the Swiss Institute of Bioinformatics highly automated methods for the mapping of proteins on gels have been used, in which the proteins are first digested in the gel and then transferred to a PVDF membrane. This membrane is then directly scanned by MALDI-TOF-MS [493]. The identification is hereafter performed with specialised software, to create a 2-dimensional map that contains the fingerprint data as well as the identification results. These results are accessible via the ExPASy server (See Table 14.1).

There have been recent reports of new MS configurations for proteomics research, particularly new MS/MS systems. One such development is the combination of MALDI with a Q-TOF mass spectrometer, which enables, besides an enhanced resolution and sensitivity, interesting opportunities for automation [498]. MALDI TOF-TOF has been described by Medzihradski et al. and has the same benefits as MALDI-Q-TOF but is also capable of collisional-induced dissociation (CID) and very fast scan rates [499]. The most recent addition to the repertoire is the

Fourier transform ion cyclotron resonance spectrometer, with a mass accuracy better than 1 ppm. Such instruments have been used in the characterisation of hundreds of intact molecules in a single analysis of proteins from *E. coli* [500].

Despite the development of new MS methods and instrumentation, the scanning and interpretation of proteins of 2DE gels has many limitations in the study of the proteome. Proteins of low abundance, high lipophilicity and low solubility (e.g. membrane proteins), and proteins that have an extremely low or high isoelectric point or have extremely low or high masses are not easily assessed with 2DE. Also, protein complexes cannot be studied with 2DE. In all cases the sample presentation to the spectrometer is the bottleneck in obtaining high-quality spectroscopic data, sometimes necessitating multi-step procedures. For instance, researchers have used multidimensional chromatographic sample preparation systems, such as two-dimensional HPLC [501] and solid-phase microextraction, multistep elution, CE and MS/MS [502] for protein mapping. Methods have been developed to either replace the SDS-PAGE dimension or the IEF dimension by a liquid chromatographic (LC) method prior to MS analysis [503, 504]. Aebersold and coworkers described a completely chromatographic system coupled to tandem MS spectroscopy (LC/LC-MS/MS) for the analysis of low abundance proteins in yeast [505], while Link et al. described a similar LC/LC-MS/MS system in which protein mixtures from the sample were first digested and the resulting peptides were separated by 2-dimensional LC (cation exchange LC in one dimension and reverse phase LC in the other dimension) [506]

In addition to scanning gels and sampling proteins from HPLC and CE runs, the possibility of sampling proteins separated on microfluidic cartridges has been explored. Multistep procedures can be performed very efficiently on microchips and thus combinations of microchips, particularly with ESI-MS, are in the center of interest. A first example is offered by the work of Marko-Varga and coworkers, who used three microfabricated components in an automated set-up for proteomics analysis [507]: a microchip digestion unit, a piezoelectric dispenser and a silicon microvial array (Fig. 14.56.). The microchip digestion unit was anisotropically etched in 110-silicon and contained 30 parallel trenches, 250 µm deep, 50 µm wide and 1.2 mm long, in which trypsin or chymotrypsin was immobilized [508]. The unit enabled on-line protein digestion. The microdispenser was processed from silicon by a pn-etch stop process and furnished with a piezoceramic actuator and moreover functioned as a normal ink-jet printer head. This actuator provided a convenient technique to spot the protein digests in nanovials on silicon substrate plates, which could then be analysed by MALDI-TOF. To obtain highly homogeneous sample-matrix layers in the nanovials, a 'seed-layer' method was used [509].

Researchers have recently explored the possibility of combining the BiaCore instrument with MS, in which analytes immobilized on the sensor chip were analyzed with MALDI-TOF directly from the area of the flow cells (Fig. 14.57a) [310, 510, 511]. Detection limits in the low-femtomol range could be realised from studies with a model sandwich assay system, and the presence of antibody/antigen species retained during the interaction analysis could be confirmed (human myoglo-

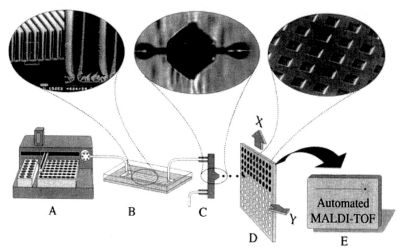

Fig. 14.56 The integrated microanalytical workstation, with the inserts showing micromachined parts of the system. The system comprises: (A) automated sample pretreatment and injection, (B) ■-chip IMER (photo insert shows a SEM picture of the lamella structure with the porous layer), (C) a microdispenser used to deposit sample into nanovials (D) shallow nanovials (300 × 300 × 20 m) on the MALDI target plate; and (E) automated MALDI-TOF MS analysis.
(reproduced from ref. 507 with permission)

bin at m/z = 17,200 Da and anti-human myoglobin at m/z = 144500 Da, Fig. 14.57b). Although there are as yet no reports of coupling the BiaCore with ESI-MS, the general compatibility of the flow regimes in the BiaCore and that of ESI-MS could be a reason to attempt combination of these techniques into an on-line instrument in the near future.

Apart from proteome analysis there are various other types of analysis on proteins that have been performed with MS. These include analysis of binding constants and epitope analysis. ESI-MS has been used for the analysis of binding constants for various complexes formed in solution, ranging from small organic host–guest complexes to large biological complexes [512]. Kempen and Brodbelt recently described an improvement in the methodology by the use of an added second host or guest with known affinity K [513]. Closely related to the determination of binding constants is the investigation of enzyme kinetics with MALDI-TOF MS, which has recently been explored by Houston et al. [514]. They assessed the kinetics of a tyrosin phosphatase by using the formation of a covalent phosphoenzyme intermediate and following the appearance of this complex in real-time. Another field of work, which has also appeared rather recently, is the identification of epitopes directly by MALDI-MS [515]. Epitopes could be identified directly by comparing the MS spectrum of the complex directly with that of an unreacted control sample.

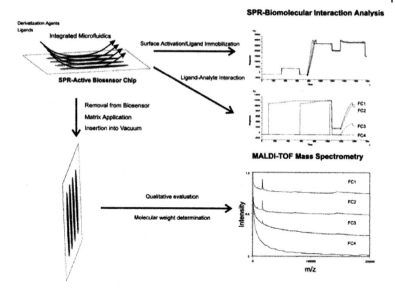

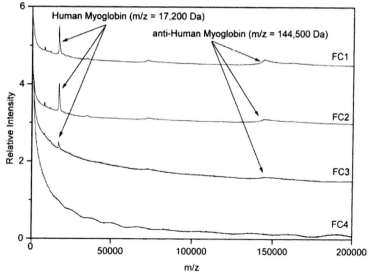

Fig. 14.57 General scheme of the chip-based SPR-BIA/MS as reported by Nelson et al. [510]. (a) The microfluidics of the BiaCore are capable of delivering nanoliters of solution through flow cells either serially or in parallel (each flow cell has the dimension of 0.5 mm × 2.0 mm). The real-time monitoring capabilities of SPR is used for determination of the kinetic parameters of the interaction, while MALDI-TOF analysis of the flow cells directly from the surface of the sensor chip was used to evaluate analytes retained during the SPR experiments. (b) MALDI-TOF mass spectra of analytes retained during the SPR analysis of the myoglobin/anti-myoglobin system. (reproduced from Anal. Chem. with permission)

14.4.7
MS in Nucleic Acid Analysis

In nucleic acid analysis considerable progress has recently been reported for methods based on MS. Although ESI-MS can also be used in nucleic acid analysis, MALDI-TOF-MS has been used by most research groups. Various reviews on the use of MALDI-TOF for nucleic acid analysis have appeared, which focus on DNA characterization [516, 517], DNA sequencing [518] and DNA sizing [519].

The genomics and proteomics tasks for MS are mainly sizing and sequencing of DNA and the identification of genetic changes and polymorphisms and the applications are now also moving fast into the clinical diagnostics arena (e.g. early cancer detection) [520]. In determining the size of DNA strains MALDI-TOF has already replaced conventional gel electrophoresis, because with MALDI-TOF a mass accuracy of better than 1% can be obtained for DNA up to 2200 bases in length. Such an accuracy can only be surpassed by full sequencing. The sensitivity of MALDI-TOF is also superior to gel electrophoresis: below femtomol sensitivities are easily attained. MALDI-TOF MS is now also rapidly replacing the conventional Sanger DNA sequencing methods. The relevance of MALDI-TOF MS in single nucleotide polymorphism studies (SNP genotyping) is also increasing rapidly [521]. Enzymatic DNA sequencing coupled with MALDI-TOF MS has been shown to be effective for discovering previously unknown types of SNPs, either via DNA (mini)sequencing, direct mass analysis or via peptide nucleic acid hybridization probes (PNSs). Single-stranded DNA (ssDNA) has been sequenced by MALDI-TOF MS by using either digestion of the nucleotides or 'in-source' fragmentation. Only a little work has been reported on the sequencing of double-stranded DNA (dsDNA). The direct sequencing of DNA has recently been accelerated by the use of exonuclease III from *E. coli* and a much improved sample clean-up procedure [522].

14.5
Conclusions

From the numerous developments in the areas described above it will be clear that many types of conventional analysis with large, expensive laboratory instruments will still be needed for elucidation of the complex structure and specific functions of biomolecules (IR, NMR and MS). Besides elaborate spectroscopic measurements for structure and sequence determination, the need for dynamic studies in real samples by in situ, in vivo and real-time measurements is also expanding. Although many new techniques will make the more conventional fluorescent labels obsolete, it can be expected that specialised fluorescence probes, particularly those with signal generation capability, will be needed in the future for fast diagnostics products. In this arena novel dyes with an increased Stokes' shift and an emission in the near-IR remain in the center of interest, as well as novel dyes for two-photon excitation. Novel fluorescent particles are also increasingly used in fast diagnostics tests.

Spectroscopy has, unfortunately, also shown some of its limitations. For instance, the near-IR absorbance spectroscopy for the detection of glucose *in vivo*, has not significantly advanced. Also the commercialisation of biosensors in the (bio)chemical process industry has been slowing down due to difficulties with long-term stability and calibration. In clinical diagnostics, only a few optical biosensor devices have been able to meet the demands on sensitivity and selectivity and labelling of reagents will still be required to obtain reliable measurements below nanomolar concentrations in real samples. Some devices, however, are based on labels that can be confined within the sensor, and these may operate reversibly or can be regenerated for continuous use. Otherwise disposable devices are still likely to be used for a long time in the diagnostics area. For instance, electrophoretic separations can presently also be performed on injection-moulded plastic substrates [523].

As seen in some of the examples given above, spectroscopic methods are very much combined with other techniques to provide more information more quickly. Optical sensors using injection-molded microfluidics systems have already been on the market for 10 years (the BiaCore), but presently the combination of spectrometers or optical sensors with microfluidics, capillary electrophoresis and DNA arrays is still very much in the center of interest. For instance, an optical biosensor can be combined with mass spectroscopy, allowing very rapid sequential determination of the binding constants, charge, size and sequence of a target protein [511].

Micromechanical systems will probably also enter the arena of bioanalysis within a few years ("BioMEMS"), which opens up unique possibilities for sample processing (and also detection) at the micro-scale (μ-TAS) [524, 525]. New advances in *silicon processing* have already resulted in novel, proven concepts for miniaturised sample handling and readout in diagnostics, genomics and proteomics applications [526, 527]. The efforts towards miniaturisation are mainly due to the favourable effect of miniaturisation on the sample throughput [528]. Processing in silicon has the great advantage that the electronics can be placed with the analysis system on the same chip. The throughput of analysis systems is further enhanced through sensor arrays and by including more sophisticated fluid handling systems, which may be particularly useful in proteomics and genomics and in drug discovery [529].

References

1. M. J. Dutt, K. H. Lee, *Curr. Opin. Biotechnol.* **2000**, *11*, 176.
2. V. Wasinger, S. Cordwell, A. Cerpa-Poljak et al., *Electrophoresis*, **1995**, *16*, 1090.
3. L. Koutny, D. Schmalzing, O. Salas-Solano et al., *Anal. Chem.* **2000**, *72*, 3388.
4. I. Kheterpal, R. A. Mathies, *Anal. Chem.*, **1999**, *71*, 31A.
5. *Proteome Research: New Frontiers in Functional Genomics*, eds. M. R. Wilkins, K. L. Williams, R. D. Appel, D. F. Hochstrasser, Springer Verlag, Berlin 1997.
6. *Proteome and Protein Analysis*, eds. R. M. Kamp, D. Kyriakidis, T. Choli-Papadopoulou, Springer Verlag, Berlin 1999.
7. *Topics in Fluorescence Spectroscopy, 4: Probe Design and Chemical Sensing*, ed. J. R. Lakowicz, Plenum Press, New York 1994.
8. J. S. Andersen, M. Mann, *FEBS Lett.*, **2000**, *480*, 25.
9. R. E. Banks, M. J. Dunn, D. F. Hochstrasser et al., *The Lancet*, **2000**, *356*, 1749.
10. P. Hensly, D. G. Myszka, *Curr. Opin. Biotechnol.*, **2000**, *11*, 9.
11. *Biosensors. Fundamentals and Applications*, ed. A. P. F. Turner, I. Karube, G. Wilson, Oxford University Press, Oxford 1987.
12. *Handbook of Biosensors and Electronic Noses*, ed. E. Kress-Rogers, CRC Press, Boca Raton, FL 1996.
13. F. W. Scheller, U. Wollenberger, A. Warsinke et al., *Curr. Opin. Biotechnol.*, **2001**, *12*, 35.
14. H. H. Rashidi and L. K. Buehler, *Bioinformatics Basics, Applications in Biological Science and Medicine*, CRC Press, Boca Raton, FL 1999.
15. *Bioinformatics: Sequence, Structure, and Databanks: A Practical Approach*, ed. D. Higgins, W. Taylor Oxford University Press, Oxford 2000.
16. J. Galbán, Y. Andreu, J. F. Sierra et al., *Luminescence*, **2001**, *16*, 199–210.
17. S. D'Auria, J. R. Lakowicz, *Curr. Opin. Biotechnol.*, **2001**, *12*, 99.
18. *Applied Fluorescence in Chemistry, Biology and Medicine*, ed. W. Rettig, B. Strehmel, S. Schrader et al., Springer Verlag, Berlin 1998.
19. ACS Symp. Ser., **1993**, *538*.
20. *Fluorescent and Luminescent Probes for Biological Activity*, ed. W. T. Mason, Academic Press, New York 1999.
21. R. P. Haugland, *Handbook of Fluorescent Probes and Research Chemicals*, 7th edition, Molecular Probes Inc., Eugene, OR 1999.
22. M. M. Martin, L Lindqvist, *J Lumin.*, **1975**, *10*, 381; B. Khodorov, O. Valkina, V. Turovetsky, *FEBS Lett.*, **1994**, *341*, 125; J. A. Thomas, R. N. Buchsbaum, A. Zimniak et al. *Biochemistry*, **1979**, *18*, 2210–2218; A. M. Paradiso, R. Y. Tsien, T. E. Machen, *Proc. Natl. Acad. Sci.*, **1984**, *81*, 7436; W-C. Sun et al., *J Org. Chem.*, **1997**, *62*, 6469; M. Nedergaard, S. Desai, W. Pulsinelli, *Anal. Biochem.* **1990**, *187*, 109; C. S. Huang, S. J. Kopacz, C. P. Lee, *Biochim. Biophys. Acta*, **1983**, *722*, 107–115; C. C. Overly, K. D. Lee, E. Berthiaume et al., *Proc. Natl. Acad. Sci. USA*, **1995**, *92*, 3156.

23 A. Minta, R. Y. Tsien, *J. Biol. Chem.*, **1989**, *264*, 19449–19457; K. Meuwis, N. Boens, F. C. De Schryver et al., *Biophys. J.*, **1995**, *68*, 2469–2473; H. Szmacinski, J. R. Lakowicz, M. L. Johnson, *Methods Enzymol.*, **1994**, *240*, 723–748; S. Jayaraman, Y. Song, L.Vetrivel et al., *J. Clin. Invest.*, **2001**, *107*, 317–324.

24 *Calcium Signaling Protocols*, ed. D. Lambert, Humana Press, Totowa, NJ, USA 1999; *A Practical Guide to the Study of Calcium in Living Cells*, ed. R. Nuccitelli, Academic Press, New York 1994, 342 pp.; D. A. Williams, K. E. Fogarty, R. Y. Tsien et al., *Nature*, **1985**, *318*, 558–561; H. Iatridou, E. Foukaraki, M. A. Kuhn et al., *Cell Calcium*, **1994**, *15*, 190–198; S. Girard, A. Luckhoff, J. Lechleiter et al., *Biophys. J.*, **1992**, *61*, 509–517; A. Takahashi, P. Camacho, J. D. Lechleiter et al., *Physiol. Rev.*, **1999**, *79*, 1089–1125; R. Y. Tsien, *Biochemistry*, **1980**, *19*, 2396–2404; Y. Ohmiya, T. Hirano, *Chem. Biol.*, **1996**, *3*, 337–347; E. Murphy, C. C. Freudenrich, L. A. Levy et al., *Proc. Natl. Acad. Sci. USA*, **1989**, *86*, 2981–2984; J. P. van der Wolk, M. Klose, J. G. de Wit JG et al., *J. Biol.Chem.*, **1995**, *270*, 18975–18982.

25 L. M. Canzoniero, D. M. Turetsky, D. W. Choi, *J. Neurosci.*, **1999**,*19*, RC31; C. J. Frederickson, E. J. Kasarskis, D. Ringo et al., *J. Neurosci. Methods*, **1987**, *20*, 91–103; M. A. Kuhn et al., *Proc. SPIE-Int. Soc. Opt. Eng.*, **1995**, *2388*, 238.

26 O. S. Wolfbeis, E. Urbano, *J. Heterocycl. Chem.*, **1982**, *19*, 841; A. S.Verkman, M. C. Sellers, A. C. Chao et al., *Anal. Biochem.*, **1989**, *178*, 355–361; J. Biwersi, A. S. Verkman, *Biochemistry*, **1991**, *30*, 7879–7883; J. Biwersi, B.Tulk, A. S. Verkman, *Anal. Biochem.*, **1994**, *219*, 139–143; H. B. Li, F. Chen, *Fresenius' J. Anal. Chem.*, **2000**, *368*, 501–504.

27 A. Sano, M. Takezawa, S. Takitani, *Anal. Sci.*, **1986**, *2*, 491.

28 H. Deakin, M. G. Ord, L. A. Stocken, *Biochem. J.*, **1963**, *89*, 296; N. S. Kosower, E. M. Kosower, G. L. Newton et al., *Proc. Natl. Acad. Sci. USA*, **1979**, *76*, 3382–3386.

29 R. H. Upson, R. P. Haugland, M. N. Malekzadeh et al., *Anal. Biochem.*, **1996**, *243*, 41–45.

30 H. Kojima et al. *Angew. Chem. Int. Ed. Engl.*, **1999**, *38(21)*, 3209; T. P. Misko, R. J. Schilling, D. Salvemini et al., *Anal. Biochem.*, **1993**, *214*, 11–16; A. Buldt, U. Karst, *Anal. Chem.*, **1999**, *71*, 3003–3007.

31 M. Krieg, *Biochim. Biophys. Acta*, **1992**, *1105*, 333–335; G. Bottiroli, A. C. Croce, P. Balzarini et al., *Photochem. Photobiol.*, **1997**, *66*, 374–383; D. C. Neckers, O. M. Valdes-Aguilera, *Adv. Photochem.*, **1993**, *18*, 315; S. Miranda, C. Opazo, L. F. Larrondo et al., *Prog. Neurobiol.*, **2000**, *62*, 633–648.

32 R. Y. Tsien, *Nature Biotechnol.*, **1999**, *17*, 956; G. S. Baird, D. A. Zacharias, R. Y. Tsien, *Proc. Natl. Acad. Sci. USA*, **2000**, *97*, 11984.

33 (a) D. A. Armbruster, R. H. Schwarzhoff, E. C. Hubster et al., *Clin. Chem.*, **1993**, *39(10)*, 2137–46; (b) J. Szollosi, S. Damjanovich, L. Matyus, *Cytometry*, **1998**, *34(4)*, 159–179; (c) Y. Z. Zhang, C. Kemper, A. Bakke et al., *Cytometry*, **1998**, *33*, 244–248; (d) D. L. Luchtel, J. C. Boykin, S. L. Bernard et al., *Biotechnol. Histochem.*, **1998**, *73*, 291–309 ; J. M. Brinkley, R. P. Haugland, V. Singer, *1994 US Pat.*, 5,326,692.

34 Z. Diwu, D. H. Klaubert, R. P. Haugland, *Proc. SPIE-Int. Soc. Opt. Eng.*, **1999**, *3602*, 265; K. D. Larison, R. BreMiller, K. S. Wells et al., *J. Histochem. Cytochem.*, **1995**, *43*, 77–83.

35 J. F. Breininger, D. G. Baskin, *J. Histochem. Cytochem.*, **2000**, *48*, 1593–1599; M. N. Bobrow, T. D. Harris, K. J. Shaughnessy et al., *J. Immunol. Methods*, **1989**, *125*, 279–285; R. P. van Gijlswijk, H. J. Zijlmans, J. Wiegant et al., *J. Histochem. Cytochem.*, **1997**, *45*, 375–382.

36 O. H. Lowry et al., *J. Biol. Chem.*, **1951**, *193*, 265; S Fazekas de St Groth, R. Webster, A. Datyner, *Biochim. Biophys. Acta*, **1963**, *71*, 377; M. M. Bradford, *Anal. Biochem.*, **1976**, *72*, 248–254; S. Udenfriend, S. Stein, P. Bohlen et al.,

Science, **1972**, *178*, 871–872; C. C. Goodno, H. E. Swaisgood, G. L. Catignani, Anal. Biochem., **1981**, *115*, 203–211; P. K. Smith, R. I. Krohn, G. T. Hermanson et al., *Anal. Biochem.*, **1985**, *150*, 76–85; L. J. Jones et al., *FASEB J.*, **1996**, *10*, A1512, abstract #2954; Y. Liu, R. S. Foote, S. C. Jacobson et al., *Anal. Chem.*, **2000**, *72*, 4608–4613; W. F. Patton, *Electrophoresis*, **2000**, *21*, 1123–1144; W. F. Patton, *Biotechniques*, **2000**, *28*, 944–948; J. X. Yan, R. A. Harry, C. Spibey et al., *Electrophoresis*, **2000**, *21*, 3657–3665.

37 H. Rinderknecht, *Experientia*, **1960**, *16*, 430; L. C. Felton, C. R. McMillion, *Anal. Biochem.*, **1961**, *2*, 178; G. Weber, *Biochem. J.*, **1952**, *51*, 155; N. Seiler, *Methods Biochem. Anal.*, **1970**, *18*, 259–337; J. K. Lin, J. Y. Chang, *Anal. Chem.*, **1975**, *47*, 1634–1638; K. Muramoto, H. Kamiya, H. Kawauchi, *Anal. Biochem.*, **1984**, *141*, 446–450; S. W. Jin, G. X. Chen, Z. Palacz et al., *FEBS Lett.*, **1986**, *198*, 150–154; L. L. Maggiora, C. W. Smith, Z. Y. Zhang, *J. Med. Chem.*, **1992**, *35*, 3727–3730; H. Maeda, H. Kawauchi, *Biochem. Biophys. Res. Commun.*, **1968**, *31*, 188–192.

38 D. L. Taylor, Y. L. Wang, *Proc. Natl. Acad. Sci. USA*, **1978**, *75*, 857–861; K. Barber, R. R. Mala, M. P. Lambert et al., *Neurosci. Lett.*, **1996**, *207*, 17–20; D. Szczesna, S. S. Lehrer, *J. Muscle Res. Cell Motil.*, **1993**, *14*, 594–597.

39 J. B. Le Pecq, M. Le Bret, J. Barbet et al., *Proc. Natl. Acad. Sci. USA*, **1975**, *72*, 2915–2919; A. R. Morgan, J. S. Lee, D. E. Pulleyblank et al., *Nucleic Acids Res.*, **1979**, *7*, 547–569; G. K. McMaster, G. G. Carmichael, *Proc. Natl. Acad. Sci. USA*, 74, 4835-4838 (1977); W. Schnedl, A. V. Mikelsaar, M. Breitenbach et al., *Hum. Genet.*, **1977**, *36*, 167–172; B. Weisblum, E. Haenssler, *Chromosoma*, **1974**, *46*, 255–260; H. A. Crissman, Z. Darzynkiewicz, J. A. Steinkamp et al., *Methods Cell Biol.*, **1990**, *33*, 305–314; I. Saiki, C. D. Bucana, J. Y. Tsao et al., *J. Natl. Cancer Inst.*, **1986**, *77*, 1235–1240; W. Muller, D. M. Crothers, *J. Mol. Biol.*, **1968**, *35*, 251–290; B. Festy, M. Daune, *Biochemistry*, **1973**, *12*, 4827–4834; T. L. Netzel et al., *J. Phys. Chem.*, **1995**, *99*, 17936; M. Poot, L. L. Gibson, V. L. Singer, *Cytometry*, **1997**, *27*, 358–364; L. M. Popa, S. Winter, G. Lober, *Biochem. Mol. Biol. Int.*, **1994**, *34*, 1189–1196; W. Beisker, E. M. Weller-Mewe, M. Nusse, *Cytometry*, **1999**, *37*, 221–229; E. S. Mansfield, J. M. Worley, S. E. McKenzie et al., *Mol. Cell Probes*, **1995**, *9*, 145–156; S. J. Ahn, J. Costa, J. R. Emanuel, *Nucleic Acids Res.*, **1996**, *24*, 2623–2625; L. Reyderman, S. Stavchansky, *J. Chromatogr. A*, **1996**, *755*, 271–280; L. J. Jones, S. T. Yue, C. Y. Cheung et al., *Anal. Biochem.*, **1998**, 265, 368–374; M. Fujita, S. Tomita, Y. Ueda et al., *Mol. Pathol.*, **1998**, *51*, 342; F. Karlsen, H. B. Steen, J. M. Nesland, *J. Virol. Methods*, **1995**, *55*, 153–156; D. M. Schmidt, J. D. Ernst, *Anal. Biochem.*, **1995**, *232*, 144–146; Y. Hamaguchi et al., *Environ. Health*, **1997**, *60*, 14.

40 E. Schrock, S. du Manoir, T. Veldman et al., *Science*, **1996**, *273*, 494–497; B. Forghani, G. J. Yu, J. W. Hurst, *J. Clin. Microbiol.*, **1991**, *29*, 583–591.

41 C. Schneeberger, P. Speiser, F. Kury et al., *PCR Methods Appl.*, **1995**, *4*, 234–238; J, Huang, F. J. DeGraves, D. Gao et al., *Biotechniques*, **2001**, *30*, 150–157.

42 J. C. Wiegant, R. P. van Gijlswijk, R. J. Heetebrij et al., *Cytogenet. Cell Genet.*, **1999**, *87*, 47–52; H. J. Tanke, J. Wiegant, R. P. van Gijlswijk, *Eur. J. Hum. Genet.*, **1999**, *7*, 2–11; J. C. Alers, J. Rochat, P. J. Krijtenburg et al., *Genes Chromosomes Cancer*, **1999**, *25*, 301–305.

43 V. B. Paragas, Y. Z. Zhang, R. P. Haugland et al., *J. Histochem. Cytochem.*, **1997**, *45*, 345–357.

44 E. J. Speel, A. H. Hopman, P. J. Komminoth, *Histochem. Cytochem.*, **1999**, *47*, 281–288.

45 R. D. Kaiser, E. London, *Biochim. Biophys. Acta*, **1998**, *1375*, 13–22; J. A. Monti, S. T. Christian, W. A. Shaw et al., *Life Sci.*, **1977**, *21*, 345–355; A. Chattopadhyay, E. London, *Biochim. Biophys. Acta*, **1988**, *938*, 24–34; H. J. Galla, E. Sackmann, *J. Am. Chem. Soc.*, **1975**, *97*, 4114–4120; Y. Barenholz, T. Cohen, E. Haas et al., *J. Biol. Chem.*, **1996**, *271*, 3085–3090; A. S. Waggoner,

L. Stryer, *Proc. Natl. Acad. Sci. USA*, **1970**, *67*, 579–589; P. Luan, M. Glaser, *Biochemistry*, **1994**, *33*, 4483–4489; L. A. Sklar, B. S. Hudson, R. D. Simoni, *Proc. Natl. Acad. Sci. USA*, **1975**, *72*, 1649–1653.

46 P. J. Sims, A. S. Waggoner, C. H. Wang et al., *Biochemistry*, **1974**, *13*, 3315–3330; J. A. London, D. Zecevic, L. B. Cohen, *J. Neurosci.*, **1987**, *7*, 649–661; E. Fluhler, V. G. Burnham, L. M. Loew. *Biochemistry*,**1985**, *24*, 5749–5755; S > T. Smiley, M. Reers, C.Mottola-Hartshorn et al., *Proc. Natl. Acad. Sci. USA*, **1991**, *88*, 3671–3675; L. M. Loew, *Methods Cell Biol.*, **1993**, *38*, 195–209; P. Schaffer, H. Ahammer, W. Muller et al., *Pflugers Arch.*, **1994**, *426*, 548–551; Y. Tsau, P. Wenner, M. J. O'Donovan et al., *J. Neurosci. Methods*, 1996, 70, 121–129; A. Grinvald, R. D. Frostig, E. Lieke et al., *Physiol. Rev.*, 1988, 68, 1285–1366; M. Zochowski, M. Wachowiak, C. X. Falk et al., *Biol. Bull.*, 2000, 198, 1–21.

47 V. A. Rakhmanova, R. C. MacDonald, *Anal. Biochem.*, **1998**, *257*, 234–237; D. E. Jones, Jr, D. M. Cui, D. M. Miller, *Oncogene*, **1995**, *10*, 2323–2330; A. G. Rao, P. Flynn, *Biotechniques*, **1990**, *8*, 37; D. C. Young, S. D. Kingsley, K. A. Ryan et al., *Anal. Biochem.*, **1993**, *215*, 24–30; M. Zhou, C. Zhang, R. H. Upson et al., *Anal. Biochem.*, **1998**, *260*, 257–259; A. Tronsmo, G. E. Harman, *Anal. Biochem.*, **1993**, *208*, 74–79.

48 R. Lottenberg, U. Christensen, C. M. Jackson et al., *Methods Enzymol.C*, **1981**, *80*, 341–361; B. G. Rosser, S. P. Powers, G. J. Gores, *J. Biol. Chem.*, **1993**, *268*, 23593–23600; S. P. Leytus, L. L. Melhado, W. F. Mangel, *Biochem. J.*, **1983**, *209*, 299–307; S. S. Twining, *Anal. Biochem.*, **1984**, *143*, 30–34.

49 B. Rotman, J. A. Zderic, M. Edelstein, *Proc. Natl. Acad. Sci. USA*, **1963**, *50*, 1; F. Leiraa, J. M. Vieitesa, M. R. Vieytesb et al., *Toxicon*, **2000**, *38*, 1833–1844; S. Avrameas, *J. Immunol. Methods*, **1992**, *150*, 23–32; K. R. Gee, W. C. Sun, M. K. Bhalgat et al., *Anal. Biochem.*, **1999**, *273*, 41–48; R. D. Petty, L. A. Sutherland, E. M. Hunter, I. A. Cree, *J. Biolumin. Chemilumin.*, **1995**, *10*, 29–34.

50 M. Zhou, Z. Diwu, N. Panchuk-Voloshina, *Anal.Biochem.*, **1997**, *253*, 162–168; D. M. Amundson, M. Zhou, *J. Biochem. Biophys. Methods*, **1999**, *38*, 43–52; M. Zhou, N. Panchuk-Voloshina, *Anal. Biochem.*, **1997**, *253*, 169–174; K. E. McElroy, P. J. Bouchard, M. R. Harpel et al., *Anal. Biochem.*, **2000**, *284*, 382–387; M. Zhou, C. Zhang, R. P. Haugland, *Proc. SPIE-Int. Soc. Opt.Eng.*, **2000**, *3926*, 166.

51 D. W. Rosenberg, H. Roque, A. Kappas, *Anal. Biochem.*, **1990**, *191*, 354–358; E. Solito, C. Raguenes-Nicol, C. de Coupade et al., *Br. J. Pharmacol.*, **1998**, *124*, 1675–1683; H. S. Hendrickson, E. K. Hendrickson, I. D. Johnson et al., *Anal. Biochem.*, **1999**, *276*, 27–35; H. S. Hendrickson, P. N. Rauk. *Anal. Biochem.*, **1981**, *116*, 553–558; H. S. She, D. E. Garsetti, M. R. Steiner, *Biochem. J.*, **1994**, *298*, 23–29; M. Zhou, C. Zhang, R. P. Haugland, *Proc. SPIE-Int. Soc. Opt. Eng.*, **2000**, *3926*, 166; H. S. Hendrickson, K. J. Kotz, E. K. Hendrickson, *Anal. Biochem.*, **1990**, *185*, 80–83; R. C. Trievel, F. Y. Li, R. Marmorstein, *Anal. Biochem.*, **2000**, *287*, 319–328; D. E. Hruby, E. M. Wilson, *Methods Enzymol.*, **1992**, *216*, 369–376; C. K. Lefvre, V. L. Singer, H. C. Kang et al., *J. Histochem. Cytochem.*, **1999**, *47*, 545–550.

52 R. M. Guasch, C. Guerri, J. E. O'Connor, *Exp. Cell Res.*, **1993**, *207*, 136–141; R. Renthal, A. Steinemann, L. Stryer, *Exp. Eye Res.*, **1973**, *17*, 511–515; M. W. Miller, J. A. Hanover, *J. Biol. Chem.*, **1994**, *269*, 9289–9297; N. Ito, S. Imai, S. Haga et al., *Histochem. Cell Biol.*, **1996**, *106*, 331–339; I. Virtanen, *Histochemistry*, **1990**, *94*, 397–401; A. K. Kenworthy, N. Petranova, M. Edidin, *Mol. Biol.Cell*, **2000**,*11*, 1645–1655; G. Sahagun, S. A. Moore, Z. Fabry, *Am. J. Pathol.*, **1989**, *134*, 1227–1232.

53 L. B. Chen, *Methods Cell Biol.*, **1989**, *29*, 103–123; J. Habicht, K. Brune, *Exp. Cell Res.*, **1980**, *125*, 514–518; J. Bereiter-Hahn, *J. Int. Rev. Cytol.*, **1990**, *122*, 1–63; M. Septinus, W. Seiffert, H. W. Zimmermann, *Histochemistry*, **1983**, *79*, 443–456; M. Septinus, T. Berthold, A. Naujok et al., *Histochemistry*, **1985**,

82, 51–66; S. J. Rembish, M. A.Trush, *Free Radical Biol. Med.*, **1994**,*17*, 117–126; M. Poot, Y. Z. Zhang, J. A. Kramer et al., *J. Histochem. Cytochem.*, **1996**, *44*, 1363–1372; A. J. Pereira, B. Dalby, R. J. Stewart et al., *J. Cell Biol.*, **1997**, *136*, 1081–1090; J. Sakanoue, K. Ichikawa, Y. Nomura et al., *J. Biochem. (Tokyo)*, **1997**, *121*, 29–37.

54 R. E. Pagano, O. C. Martin, *Cell Biology: A Laboratory Handbook*, 2nd edition, academic press, New York, ed. J. E. Celis, 1998, Vol. 2, 507–512; M. Terasaki, J. Song, J. R. Wong et al., *Cell*, **1984**, *38*, 101–108; L. Cole, D. Davies, G. J. Hyde et al., *J. Microsc.*, **2000**, *197*, 239–249; M. Terasaki, L. A. Jaffe, *J. Cell Biol.*, **1991**, *114*, 929–940; L. Cole, D. Davies, G. J. Hyde et al., *Fungal Genet. Biol.*, **2000**, *29*, 95–106.

55 A. H. Guse, C. P. da Silva, I. Berg et al., *Nature*, **1999**, *398*, 70–73; Y. Cui, A. Galione, D. A. Terrar, *Biochem. J.*, **1999**, *342*, 269–273; L. Wong, R. Aarhus, H. C. Lee et al., *Biochim. Biophys. Acta*, **1999**, *1472*, 555–564; X. Zhang, J. Wen, K. R. Bidasee et al., *Biochem. J.*, **1999**, *340*, 519–527; B. Abrenica, J. S.Gilchrist, *Cell Calcium*, **2000**, *28*, 127–136; D. A. Malencik, T. S. Huang, S. R. Anderson, *Biochem. Biophys. Res.Commun.*, **1982**, *108*, 266–272; J. Baudier, E. Bergeret, N. Bertacchi et al., *Biochemistry*, **1995**, *34*, 7834–7846; J. A. Putkey, M. N. Waxham, *J. Biol. Chem.*, **1996**, *271*, 29619–29623; K. Torok, D. J. Cowley, B. D. Brandmeier et al., *Biochemistry*, **1998**, *37*, 6188–6198.

56 A. Ogamo, K. Matsuzaki, H. Uchiyama et al., *Carbohydr. Res.*, **1982**, *105*, 69–85; M. Sobel, D. F. Soler, J. C. Kermode et al., *J. Biol. Chem.*, **1992**, *267*, 8857–8862; L. Berry, A. Stafford, J. Fredenburgh et al., *J. Biol. Chem.*, **1998**, *273*, 34730–34736; K. L. Bentley, R. J. Klebe, R. E. Hurst et al., *J. Biol. Chem.*, **1985**, *260*, 7250–7256.

57 C. S. Chen, M. Poenie, *J. Biol. Chem.*, **1993**, *268*, 15812–15822; B. A. Newton, *J. Gen. Microbiol.*, **1955**, *12*, 226; I. Takahashi, S. Nakanishi, E. Kobayashi, *Biochem. Biophys. Res. Commun.*, **1989**, *165*, 1207–1212; T. Hiratsuka, *J. Biol. Chem.*, **1982**, *257*, 13354–13358.

58 M. J. Anderson, M. W. Cohen, *J. Physiol.*, **1974**, *237*, 385–400; N. H. Salzman, F. R. Maxfield, *Subcellular Biochemistry, Vol. 19: Endocytic Components: Identification and Characterization*, 1993, Kluwer Academic Publ, Dordrecht 95–123; Y. Wang, Q. Gu, F. Mao et al., *J. Neurosci.*, **1994**, *14*, 4147–4158; J. L. Reid, J. Vincent, *Cardiology*, **1986**, *73*, 164–174; H. Heithier, D. Hallmann, F. Boege et al., *Biochemistry*, **1994**, *33*, 9126–9134; H. Wang, K. M. Standifer, D. M. Sherry, *Vis. Neurosci.*, **2000**, *17*, 11–21; M. R. Tota, S. Daniel, A. Sirotina et al., *Biochemistry*, **1994**, *33*, 13079–13086; B. Walker, J. Gray, D. M. Burns et al., *J. Peptides*, **1995**, *16*, 255–261; H. Haller, C. Lindschau, B. Erdmann et al., *Circ. Res.*, **1996**, *79*, 765–772; V. M. Kolb, A. Koman, L. Terenius, *Life Sci.*, **1983**, 33 Suppl. 1, 423–426; O. T. Jones, D. L. Kunze, K. J. Angelides, *Science*, **1989**, *244*, 1189–1193; D. Schild, H. Geiling, J. Bischofberger, *J. Neurosci.Methods*, **1995**, *59*, 183–190; T. R. Kleyman, E. J. Cragoe, Jr., *J. Membr. Biol.*, **1988**, *105*, 1–21; P. A. Fortes, *Biochemistry*, **1977**, *16*, 531–540; B. Lohrke, M. Derno, B. Kruger et al., *Pflugers Arch.*, **1997**, *434*, 712–720; G. Fricker, H. Gutmann, A. Droulle et al., *Pharm. Res.*, **1999**, *16*, 1570–1575.

59 D. A. Bass, J. W. Parce, L. R. Dechatelet et al., *J. Immunol.*, **1983**, *130*, 1910–1917; T. C. Ryan, G. J. Weil, P. E. Newburger et al., *J. Immunol. Methods*, **1990**, *130*, 223–233.

60 J. P. Aubry, A. Blaecke, S. Lecoanet-Henchoz et al., *Cytometry*, **1999**, *37*, 197–204; N. A. Thornberry, Y. Lazebnik, *Science*, **1998**, *281*, 1312–1316.

61 M. Matsuoka, Synthetic Design of Infrared Absorbing Dyes, in *Infrared Absorbing Dyes*, eds. A. R. Katritzky and G. J. Sabongi, Plenum Press, New York 1990.

62 R. Zahradnik, *Fortschr. Chem. Forsch.*, **1968**, *10*, 158.

63 R. J. Beunker, S. D. Peyerimhoff, *Theoret. Chim. Acta*, **1974**, *35*, 33.

64 *Density-functional Theory of Atoms and Molecules*, eds. R. G. Parr and W. Yang, Oxford University Press, New York 1989.
65 F. J. Green, *The SigmaAldrich Handbook of Stains, Dyes and Indicators*, Aldrich Chemical Company Inc., Milwaukee, WI, 1990, p. 374.
66 J. L. Riggs, R. J. Seoward, J. H. Burckhalter et al., *Am. J. Pathol.*, **1958**, *34*, 1081.
67 E. Terpetschnig, O. S. Wolfbeis, Luminescent Probes for NIR Sensing Applications, in *Near Infrared Dyes for High Technology Applications*, eds. S. Daehne, U. Resch-Genger, O. S. Wolfbeis, NATO ASI Series, Kluwer, Dordrecht, The Netherlands 1998, 161–182.
68 D. S. Wilbur, P. M. Pathare, D. K. Hamlin et al., *Bioconjugate Chem.*, **2000**, *11*, 584.
69 M. Thompson, N. W. Woodbury, *Biochemistry*, **2000**, *39*, 4327.
70 D. Staerk, A. A. Hamed, E. B. Pederson et al., *Bioconjugate Chem.*, **1997**, *8*, 869.
71 J. B. Randolph, A. S. Waggoner, *Nucleic Acids Res.*, **1997**, *14*, 2923.
72 M. Lipowska, G. Patonay, L. Strekowski, *Synth. Commun.*, **1993**, *23*, 3087.
73 C. V. Owens, Y. Y. Davidson, S. Kar et al., *Anal. Chem.*, **1997**, *69*, 1256.
74 J. H. Flanagan, C. V. Owens, S. E. Romero et al., *Anal. Chem.*, **1998**, *79*, 2676.
75 J. Ren, J. B. Chaires, *J. Am. Chem. Soc.*, **2000**, *122*, 424.
76 L. Strekowski, unpublished results
77 L. Strekowski, M. Lipowska, G. Patonay *J. Org. Chem.*, **1992**, *57*, 4578.
78 D. Keil, H. Hartmann, C. Reichardt. *Liebigs Ann. Chem.*, **1993**, 935.
79 H. Meier, R. Petermann *Tetrahedron Lett.*, **2000**, *41*, 5475.
80 D. Citterio, L. Jenny, S. Rásonyi. *Sens. Actuators B*, **1997**, *202*, 38–39.
81 E. Terpetschnig, H. Szmacinski, A. Ozinskas et al., *Anal. Biochem.*, **1994**, *217*, 197.
82 B. Oswald, L. Patsenker, J. Duschl et al., *Bioconjugate Chem.*, **1999**, *10*, 925.
83 B. Oswald, F. Lehmann, L. Simon et al., *Anal. Biochem.*, **2000**, *280*, 272.
84 H. Ali, J. E. van Lier, *Chem. Rev.*, **1999**, *99*, 2379.
85 N. B. McKeown, Phthalocyanine Synthesis, in *Phthalocyanine Materials: Synthesis, Structure and Function*, eds. B. Dunn, J. W. Goody, A. R. West, Cambridge University Press, Cambridge, 1998, p. 331.
86 N. Brasseur, T. L. Nguyen, R. Langlois et al., *J. Med. Chem.*, **1994**, *37*, 415.
87 N. Brasseur, R. Ouellet, K. Lewis et al., *Photochem. Photobiol.*, **1995**, *62*, 1058.
88 B. L. Wheeler, G. Nagasubramanian, A. J. Bard et al., *J. Am. Chem. Soc.*, **1984**, *106*, 7404.
89 W. E. Ford, M. A. J. Rodgers, L. A. Schechtman et al., *Inorg. Chem.*, **1992**, *31*, 3371.
90 F. Sanger, S. Nicklen, A. R. Coulson, *Proc. Natl. Acad. Sci. USA*, **1977**, *74*, 5463.
91 National Research Council USA, *Sequencing the Human Genome*, National Academy Press, Washington, DC, 1998.
92 US Department of Energy, Office of Energy Research, Office of Health and Environmental Research, *Human Genome 199192 Program Report*, DOE/ER0544P, US Government Printing Office, Washington, DC, 1998.
93 N. J. Dovichi, Capillary Gel Electrophoresis for Large Scale DNA Sequencing: Separation and Detection, in *Handbook of Capillary Electrophoresis*, ed. J. P. Landers, CRC press, Boca Raton, FL, 1997, p. 545.
94 K. Murray, *J. Mass Spectrom.*, **1996**, *31*, 1203.
95 H. Swerdlow, J. Z. Zhang, D. Y. Chen et al., *Anal. Chem.*, **1991**, *63*, 2835.
96 R. A. Mathies, X. C. Huang, *Nature*, **1992**, *359*, 167.
97 X. C. Huang, M. A. Quesada, R. A. Mathies, *Anal. Chem.*, **1992**, *64*, 2149.
98 P. C. Simpson D. Roach, A. T. Wooley et al. Proc. Natl. Acad. Sci., **1998**, *95*, 2256.
99 H. R. Starke, J. Y. Yan, J. Z. Zhang et al., *Nucleic Acids Res.*, **1994**, *22*, 3997.
100 H. Lu, E. Arriaga, D. Y. Chen et al., *J. Chromatogr. A.*, **1994**, *680*, 497.
101 M. L. Smith, J. Z. Sanders, R. J. Kaiser et al., *Nature*, **1986**, *321*, 674.

102 J. M. Prober, G. L. Trainor, R. J. Dam et al., *Science*, **1987**, *238*, 336.
103 W. Ansorge, B. S. Sproat, J. Stegemann et al., *J. Biochem. Biophys. Methods*, **1986**, *13*, 315.
104 S. Carson, A. S. Cohen, A. Belenkii et al., *Anal. Chem.*, **1993**, *65*, 3219.
105 L. M. Smith, *Nature*, **1991**, *349*, 812.
106 A. S. Cohen, D. R. Najarian, B. L. Karger, *J. Chromatogr.*, **1990**, *516*, 46.
107 A. E. Karger, J. J. Harris, R. F. Gesteland, *Nucleic Acids Res.*, **1991**, *19*, 4955.
108 H. Swerdlow, R. Gesteland, *Nuclic Acids Res.*, **1990**, *18*, 1415.
109 Y. Wang, J. Ju, B. A. Carpenter et al., *Anal. Chem.*, **1995**, *67*, 1197.
110 Y. Wang, J. M. Wallin, J. Ju et al., *Electrophoresis*, **1996**, *17*, 1485.
111 S. C. Hung, R. A. Mathies, A. N. Glazer, *Anal. Biochem.*, **1997**, *252*, 78.
112 S. C. Hung, J. Ju, R. A. Mathies et al., *Anal. Biochem*, **1996**, *243*, 1527.
113 L. MiddendoRF, J. Amen, R. Bruce et al., Near-Infrared Fluorescence Instrumentation for DNA anaylsis, in *Near Infrared Dyes for High Technology Applications*, eds. S. Daehne, U. Resch-Genger, O.S. Wolfbeis, NATO ASI Series, Kluwer, Dordrecht, 1998, p. 141.
114 D. B. Shealy, M. Lipowska, J. Lipowski et al., *Anal. Chem.*, **1995**, *67*, 247.
115 D. C. Williams, S. A. Soper, *Anal. Chem.*, **1995**, *67*, 3427.
116 T. Imasska, H. Nakagawa, T. Okazaki et al., *Anal. Chem.*, **1990**, *62*, 2404.
117 A. E. Boyer, M. Lipowska, J. Zen et al., *Anal. Lett.*, **1992**, *25*, 415.
118 R. J. Williams, N. Narayanan, G. A. Casey et al., *Anal. Chem.*, **1994**, *66*, 3102.
119 K. Sauda, T. Imasaka, N. Ishibashi, *Anal. Chem.*, **1986**, *58*, 2649.
120 K. Sauda, T. Imasaka, N. Ishibashi, *Anal. Chim. Acta*, **1986**, *187*, 353.
121 A. J. G. Mank, H. Lingeman, C. Gooijer, *Trends Anal. Chem.*, **1992**, *11*, 210.
122 R. J. Williams, M. Lipowska, G. Patonay et al., *Anal. Chem.*, **1993**, *65*, 601.
123 D. Andrews-WilbeRForce, G. Patonay, *Spectrochim. Acta, Part A*, **1990**, *46*, 1153.
124 T. Fuchigami, T. Imasaka, M. Shiga, *Anal. Chim. Acta*, **1993**, *282*, 209.
125 T. Higashijima, T. Fuchigami, T. Imasaka et al., *Anal. Chem.*, **1992**, *64*, 711.
126 J. H. Flanagan, B. L. Legendre, Jr., R. P. Hammer et al., *Anal. Chem.*, **1995**, *67*, 341.
127 S. McWhorter, S. A. Soper, *Electrophoresis*, **2000**, *21*, 1267.
128 S. P. McGlynn, T. Azumi, M. Kinoshita. *Molecular Spectroscopy of the Triplet State*, Prentice Hall, Englewood Cliffs, NJ 1969.
129 G. Kavarnos, G. Cole, P. Scribe et al., *J. Am. Chem. Soc.*, **1971**, *93*, 1032.
130 S. R. Davidson, R. Bonneau, J. Joussot-Dubien et al., *Chem. Phys. Lett.*, **1980**, *74*, 318.
131 S. A. Berson. R. S. Yalow, *J. Clin Invest.*, **1959**, *38*, 1996.
132 V. C. W. Tsang, K. Hancock, M. Wilson. Enzyme-linked Immunotransfer Blot Technique for Human T-lymphotropic Virus Type III/Lymphoadenopathy Associated Virus (HTLVIII/LAV) Antibodies, in *Immunology Series no. 15. Procedural Guide*, Centers for Disease Control, Atlanta, GA, 1986.
133 V. C. W. Tsang, J. A. Brand, A. E. Boyer, *J. Infect. Diseases.*, **1989**, *159*, 50.
134 V. C. W. Tsang, K. Hancock, A. L. Beatty et al., *J. Immunol.*, **1984**, *132*, 2607.
135 K. Hancock, V. C. W. Tsang, *J. Immunol. Methods*, **1986**, *92*, 167.
136 A. R. Swamy, Ph.D. Thesis, Georgia StateUniversity, Atlanta, GA, 1999.
137 W. C. Sun, K. R. Gee, D. H. Klaubert et al., *J. Org. Chem.* **1997**, *62*, 6469.
138 J. Karolin et al., *J. Am. Chem. Soc.*, **1994**, *116*, 7801.
139 N. Panchuk-Voloshina, R. P. Haugland, J. Bishop-Stewart et al., *J. Histochem. Cytochem.*, **1999**, *47*, 1179–1188.
140 A. R. Swamy, J. C. Mason, H. Lee et al., Near-Infrared Absorption/Luminescence Measurements, in *Encyclopedia of Analytical Chemistry*, Wiley, London 2000.
141 M. J. Baars, G. Patonay, *Anal. Chem.*, **1999**, *71*, 667.
142 G. A. Casay, D. B. Sheally, G. Patonay, *Top. Fluoresc. Spectrosc.*, **1994**, *4*, 183.
143 N. Narayanan, G. Patonay, *J. Org. Chem.*, **1995**, *60*, 2391.

144 N. Narayanan, L. Strekowski, M. Lipowska et al., *J. Org. Chem.*, **1997**, *62*, 9387.
145 M. I. Danesvar, G. A. Casay, G. Patonay et al., *J. Fluoresc.*, **1996**, *6*, 69.
146 M. I. Danesvar, J. M. Peralta, G. A. Casay et al., *J. Immunol. Methods*, **1999**, *226*, 119.
147 I. Wengatz, F. Szurdoki, A. R. Swamy et al., *Proc. SPIE-Int. Soc. Opt. Eng.*, **1995**, *2388*, 408.
148 R. A. Vijayendran, D. E. Leckband., *Anal. Chem.*, **2001**, *73*, 471
149 C. A. Rowe, S. B. Scruggs, M. J. Feldstein et al., *Anal. Chem.*, **1999**, *71*, 433.
150 S. Pilevar, C. C. Davis, F. Protugal. *Anal. Chem.*, **1998**, *70*, 2031.
151 B. M. Cullum, T. Vo-Dinh, *TIBTECH*, **2000**, *18*, 388.
152 T. Ha, T. A. Laurence, D. S. Chemla et al., *J. Phys Chem. B*, **1999**, *103*, 6839.
153 J. N. Forkey, M. E. Quinlan, Y. E. Goldman. *Prog. Biophys. Mol. Biol.* **2000**, *74*, 1.
154 E. Terpetschnig, H. Szmacinski, J. R. Lakowicz, *Anal. Biochem.*, **1995**, *227*, 140.
155 E. Terpetschnig, H. Szmacinski, J. R. Lakowicz. *Methds Enzymol.*, **1997**, *278*, 295.
156 L. Ye, X. Chris Le, J. Z. Xing et al., *J. Chromatogr B.*, **1998**, *714*, 59.
157 K. Pettersson, H. Siitari, I. Hemmila et al., *Clin. Chem.*, **1983**, *29*, 60.
158 I. Hemmilä, Time-resolved Fluorometry: Advantages and Potentials, in *High Throughput Screening*, ed. J. P. Devlin, Marcel Dekker, New York 1997.
159 I. Hemmilä, V.-M. Mukkala, H. Takalo. *J. Alloys Compd.*, **1997**, *249*, 158.
160 H. Karsilayan, I. Hemmilä, H. Takalo et al., *Bioconjugate Chem.*, **1997**, *8*, 71.
161 J. Yuan, G. Wang, H. Kimura et al., *Anal. Biochem.*, **1997**, *154*, 283.
162 J. Yuan, K. Matsumoto, H. Kimura. *Anal. Chem.*, **1998**, *70*, 596
163 E. Zuber, L. Rosso, B. Darbouret et al., *J. Immunoassay*, **1997**, *18*, 21.
164 T. Lövgren, P. Heinonen, P. Lehtinen et al., *Clin. Chem.*, **1997**, *43*, 1937.
165 W.-B. Chang, B.-L. Zhang, Y.-Z. Li et al., *Microchem. J.*, **1997**, *55*, 287.
166 Q. Qin, M. Christiansen, T. Lövgren et al., *Immunol. Methods*, **1997**, *205*, 169.
167 L. Stryer, R. P. Haugland, *Proc. Natl. Acad. Sci. USA*, **1967**, *58*, 719.
168 E. F. Ullman, M. Schwartzberg, K. E. Rubenstein, *J. Biol. Chem.*, **1976**, *251*, 4172.
169 A.-P. Wei, J. N. Herron, D. A. Christensen, *ACS Symp. Ser.*, **1992**, *511*, 105.
170 J. M. Brinkley and R. P. Haugland, US Pat., 5 326 692, 1994.
171 V.T. Oi, A.N. Glazer, L. Stryer. *J. Cell Biol.*, **1982**, *93*, 981.
172 R. MacColl, *J. Fluoresc.*, **1991**, *1*, 135.
173 G. Sohn, C. Sautter, *J. Histochem. Cytochem.*, **1991**, *39*, 921.
174 N. Nakamura, T. Matsunaga, *Anal Lett.*, **1991**, *24*, 1075.
175 N. Baumgarth, M. Roederer, *J. Immunol.Methods*, **2000**, *243*, 77–97.
176 *Green Fluorescent Proteins*, eds. K. F. Sullivan and S. A. Kay, Academic Press, New York 1999.
177 M. V. Matz, A. F. Fradkov, Y. A. Labas et al., *Nature Biotechnol.*, **1999**, *17*, 969.
178 D. Yarbrough, R. M. Wachter, K. Kallio et al., *Proc. Natl. Acad. Sci. USA*, **2001**, *98*, 462.
179 L. A. Gross, G. S. Baird, R. C. Hoffman et al., *Proc. Natl. Acad. Sci. USA*, **2000**, *97*, 11990.
180 C. Xu, W. Zipfel, J. B. Shear, R. M. Williams et al., *Proc. Natl. Acad. Sci. USA*, **1996**, *93(20)*, 10763–10768.
181 J. B. Shear, *Anal. Chem.*, **1999**, *71*, 598A.
182 A. A. Rehms, P. R. Callis, *Chem. Phys. Lett.*, **1993**, *208*, 276.
183 B. Kierdaszuk, I. Gryczynski, A. Modrak-Wojcik et al., *Photochem. Photobiol.*, **1995**, *61*, 319.
184 S. Maiti, J. B. Shear, R. M. Williams et al., *Science*, **1997**, *275*, 530.
185 M. L. Gostkowski, J. B. McDoniel et al., *J. Am. Chem. Soc.*, **1998**, *120*, 18.
186 C. Xu, W. W. Webb, *J. Opt. Chem. Am. B*, **1996**, *13*, 481.
187 M. Albota, D. Beljonne, J.-L. Bredas et al., *Science*, **1998**, *281*, 1653.
188 P. E. Hänninen, A. Soini, N. J. Meltola et al., *Nature Biotechnol.*, **2000**,*18*, 548.
189 J. M. Song, T. Inoue, H. Kawazumi et al., *J. Chromatogr. A*, **1997**, *765*, 315.
190 S. Maiti, J. B. Shear, W. W. Webb, *Biophys. J.*, **1996**, *70*, A210.

191 D. Kleinfeld, P. P. Mitra, F. Helmchen et al., *Proc. Natl. Acad. Sci. USA*, **1998**, *95*, 15741.

192 G. A. Baker, S. Pandey, F. V. Bright, *Anal. Chem.*, **2000**, *72*, 5748.

193 M. A. DeLuca (ed.) *Methods Enzymol.*, **1978**, *57*.

194 L. J. Kricka, *Anal. Chem.* **1999**, *71*, 293R.

195 *Bioluminescence and Chemiluminescence: Molecular Reporting with Photons*, eds. W. Hastings, L. J. Kricka, P. E. Stanley, Wiley, Chichester 1997.

196 A. W. Knight, G. M. Greenway, *Analyst*, **1994**, *119*, 879.

197 A. W. Knight, *Trends Anal. Chem.*, **1999**, *18*, 47.

198 W. R. Seitz, in ref. 193, Ch. 38, p. 445.

199 H.-S. Zhuang, Q.-E. Wang, F. Zhang et al., *Chin. J. Chem.*, **1997**, *15*, 123–129.

200 H. Yoshida, K. Todoroki, K. Zaitsu et al. *Daigaku Chuo Bunseki Senta Hokoku*, **1996**, *14*, 20–26; *Chem. Abstr.*, **1997**, *12 6*, 327211.

201 O. Suzuki, G. Masuda, N. Shiohata et al., *Eur. Pat.*, 812823, 1997; *Chem. Abs.*, **1998**, *128*, 72655.

202 U. Piran, J. J. Quinn, *PCT Int. Appl.*,WO 9834109, 1998; *Chem. Abs.*, **1998**, *129*, 158856.

203 H. Akhavan-Tafti, Z. Arghavani, R. Desilva, *PCT Int. Appl.*, WO 9726245, 1997; *Chem. Abs.*, **1997**, *127*, 176354

204 M. Matsumoto, N. Watanabe, H. Kobayashi et al., *Jpn. Pat.*, 09157271, 1997; *Chem. Abs.*, **1997**, *127*, 50626.

205 I. Bronstein, B. Edwards, A. Sparks et al., *PCT Int. Appl.*, WO 9714954, 1997; *Chem. Abs.*, **1997**, *127*, 25442

206 C. A. Marquette, L. J. Blum, *Talanta*, **2000**, *51*, 395.

207 P.Luppa, C. Bruckner, I. Schwab et al.,*Clin. Chem.*, **1997**, *43*, 2345.

208 R. John, R. Henley, N. Oversby, *Ann. Clin. Biochem.*, **1997**, *34*, 396.

209 M. Verhaegen, T. K. Christopoulos, *Anal. Chem.*, **1998**, *70*, 4120.

210 I. Alexandre, N. Zammatteo, P. Moris et al., *J. Virol. Methods*, **1997**, *66*, 113.

211 M. Ronaghi, M. Uhlen, P. Nyren, *Science*, **1998**, *281*, 363.

212 D. Crocnan, R. Olinescu, G. R. Turcu, *Rom. J. Biophys.*, **1997**, *7*, 87.

213 K.Erler, *Wien. Klin. Wochenschr.*, **1998**, *110*(Suppl. 3), 5–10.

214 M. T. Korpela, J. S. Kurittu, J. T. Karvinen et al., *Anal. Chem.*, **1998**, *70*, 4457.

215 I. Murray, I. A. Cowe, *Making Light Work: Advances in Near Infrared Spectroscopy*, Wiley-VCH, Weinheim 1992.

216 *Near-Infrared Applications in Biotechnology (Practical Spectroscopy, Vol. 25)*, ed. R. Raghavachari, Marcel Dekker, New York 2000.

217 S. Sasic, Y. Ozaki, *Anal. Chem.*, **2001**, *73*, 64.

218 C. Fischbacher, K. U. Jagemann, K. Danzer et al., *Fresenius' J. Anal. Chem.*, **1997**, *359*, 78.

219 H. J. Kim, Y. A. Woo, S. H. Chang et al., *Anal. Sci. Technol.*, **1998**, *11*, 47.

220 M. A. Arnold, J. J. Burmeister, G. W. Small. *Anal. Chem.*, **1998**, *70*, 1773.

221 G. Gauglitz, *Sens. Update*, **1996**, *1*, 1.

222 D. R. Thévenot, K. Toth, R. A. Durst et al., *Pure Appl. Chem.*, **1999**, *71*, 2333–2348.

223 F. W. Scheller, U. Wollenberger, A. Warsinke et al., *Curr. Opin. Biotechnol.*, **2001**, *12*, 35.

224 J.-M. Lehn, *Supramolecular Chemistry*, VCH, Weinheim 1995.

225 R. Forster, D. Diamond, *Anal. Chem.*, **1992**, *64*, 1721.

226 F. P. Schmidtchen, M. Berger, *Chem. Rev.*, **1997**, *97*, 1609.

227 S. O'Neill, S. Conway, J. Twellmeyer et al., *Anal. Chim. Acta*, **1999**, *398*, 1.

228 E. Wang, L. Zhu, L. Ma et al., *Anal. Chim. Acta*, **1997**, *357*, 85.

229 H. Hisamoto, N. Miyashita, K. Watanabe et al., *Sens. Actuators, B*, **1995**, *29*, 378.

230 K. Suzuki, H. Ohzora, K. Tohda et al., *Anal. Chim. Acta*, **1990**, *237*, 155.

231 O. S. Wolfbeis, *Sens. Actuators, B*, **1995**, *29*, 140

232 G. J. Mohr, I. Murkovic, F. Lehmannet al., *Sens. Actuators, B*, **1997**, *39*, 239.

233 Y. Kawabata, T. Yamashiro, Y. Kitazaki et al., *Sens. Actuators, B*, **1995**, *29*, 135.

234 G. J. Mohr, O. S. Wolfbeis, *Anal. Chim. Acta*, **1995**, *316*, 239.

235 A. W. Czarnik (ed.), *ACS Symp. Ser.*, **1993**, *538*.

236 E. Chapoteau, B. P. Czech, W. Zazulak et al., *Clin. Chem.*, **1992**, *38*, 1654.

237 K. R. A. L. Sandanayake, I. O. Sutherland, *Tetrahedron Lett.*, **1993**, *34*, 3165.

238 K. Kimura, T. Yamashita, M. Yokoyama, *J. Chem. Soc., Perkin Trans.*, **1992**, *92*, 613.
239 F. Fages, J.-P. Desvergne, K. Kampke et al., *J. Am. Chem. Soc.*, **1993**, *115*, 3658.
240 J. Bourson, J. Pouget, B. Valeur, *J. Phys Chem.*, **1993**, *87*, 4552.
241 D. Diamond, K. Nolan. *Anal. Chem.*, **2001**, *73*, 22A.
242 K. Iwamoto, K. Araki, H. Fujishima et al., *J. Chem. Soc., Perkin Trans. 1*, **1992**, *92*, 1885.
243 S. J. Harris, G. Barrett, M. A. McKervey et al., *J. Chem. Soc., Chem. Commun.*, **1992**, *92*, 1287.
244 P. A. Gale, Z. Chen, M. G. B. Drew et al., *Polyhedron* **1998**, *17*, 405.
245 K. Odashima, K. Yagi, K. Tohda et al., *Bioorg. Med. Chem. Lett.*, **1999**, *9*, 2375.
246 T. Grady, S. J. Harris, M. R. Smyth et al., *Anal. Chem.*, **1996**, *68*, 3775.
247 I. Aoki, T. Sakaki, S. Tsutsui et al., *Tetrahedron Lett.*, **1992**, *33*, 89.
248 B. Kukrer, E. U. Akkaya, *Tetrahedron Lett.*, **1999**, *40*, 9125.
249 G. Dilek, E. U. Akkaya. *Tetrahedron Lett.*, **2000**, *41*, 3721.
250 E. U. Akkaya, S. Turkyilmaz. *Tetrahedron Lett.*, **1997**, *38*, 4513.
251 G. Wulff, *Angew. Chem. Int. Ed. Engl.*, **1995**, *34*, 1812–1832.
252 J.-M. Kim, K.-D. Ahn, G. Wulff, *Macromol. Chem. Phys.*, **2001**, 1105–1108.
253 O. S. Wolfbeis, E. Terpetschnig, S. Piletsky et al., in *Applied Fluorescence in Chemistry, Biology and Medicine*, ed. W. Rettig, Springer Verlag, Heidelberg 1999, 277–295
254 P. Turkewitsch, B. Wandelt, G. D. Darling et al., *Anal. Chem.*, **1998**, *70*, 2025–2030.
255 D. L. Rathbone, D. Su, Y. Wang et al., *Tetrahedron Lett.*, **2000**, *41*, 123–126.
256 E. Hecht, *Optics*, 3rd edition, Addison Wesley, London 1997.
257 T. Vo-Dinh, L. Allain, in *Biomedical Photonics*, ad. T. Va-Dinh, CRC Press, Boca Raton, **2003**, 20-1–20-40.
258 C. Fattinger, H. Koller, D. Schlatter et al., *Biosens. Bioelectron.*, **1993**, *8*, 99.
259 W. Huber, R. Barner, C. Fattinger et al., *Sens. Actuator, B.*, **1992**, *6*, 122.
260 C. Stamm, W. Lukosz, *Sens. Actuator, B*, **1993**, *11*, 177.
261 C. Stamm, W. Lukosz, *Sens. Actuator, B*, **1994**, *18*, 183.
262 D. Schlatter, R. Barner, C. Fattinger et al., *Biosens. Bioelectron.*, **1993**, *8*, 109.
263 D. Clerc, W. Lukosz, *Sens. Actuator, B*, **1994**, *19*, 581.
264 J. Dubendorfer, R.E. Kunz, *Appl. Opt.*, **1998**, *37*, 1890.
265 R. Cush, J.M. Cronin, W.J. Stewart et al., *Biosens. Bioelectron.*, **1993**, *8*, 347
266 N. J. Goddard, D. Pollard-Knight, C.H. Maule, *Analyst*, **1994**, *119*, 583
267 P.E. Buckle, R.J. Davies, T. Kinning et al., *Biosens. Bioelectron.*, **1993**, *8*, 355.
268 H.J. Watts, C.R. Lowe, D.V. Pollard-Knight, *Anal. Chem.*, **1994**, *66*, 2465
269 E. Kretschmann, H. Raether, *Z. Naturforsch. A*, **1968**, *23*, 2135
270 A. Otto, *Z. Phys.*, **1968**, *216*, 398.
271 B. Liedberg, C. Nylander, I. Lundström, *Sens. Actuators*, **1993**, *4*, 299.
272 A. Szabo, L. Stolz, R. Granzow, *Curr. Opin. Struct. Biol.*, **1995**, *5*, 699.
273 M. Raghavan, P. J. Björkman, *Structure*, **1995**, *3*, 331.
274 P. A. van der Merwe, A. N. Barclay, *Curr. Opin. Immunol.*, **1996**, *8*, 257.
275 D. G. Myszka, *Curr Opin. Biotechnol.*, **1997**, *8*, 50.
276 P. Schuck, *Annu. Rev. Biophys. Biomol. Struct.*, **1997**, *26*, 541.
277 M Fivash, E. M. Towler, R. J. Fisher, *Curr. Opin. Biotechnol.*, **1998**, *9*, 97.
278 S. Boussaad, J. Pean, N. J. Tao, *J. Anal. Chem.*, **2000**, *72*, 222.
279 H. Sota, Y. Hasegawa, *Anal. Chem.*, **1998**, *70*, 2019.
280 E. Kai, S. Sawata, K. Ikebukuro et al., *Anal. Chem.*, **1999**, *71*, 796–800.
281 D. G: Myszka, X. He, M. Dembo et al., *Biophys. J.*, **1998**, *75*, 583.
282 http://www.biacore.com/customer/references.phtml
283 G.A. Baxter, J.P. Ferguson, M.C. O'Connor et al., *J. Agric. Food Chem.*, **2001**, *49*, 3204.
284 F.E. Carrick, B.E. Forbes, J.C. Wallace, *J. Biol. Chem.*, **2001**, *276*, 27120
285 T. Flatmark, A.J. Stokka, S.V. Berge, *Anal. Biochem.*, **2001**, *294*, 95
286 P. Gomes, E. Giralt, D. Andreu, *Mol. Immunol.*, **2000**, *37*, 975
287 W. Haasnoot, R. Verheijen, *Food Agric. Immunol.*, **2001**, *13*, 131.

288 V. Gaudin, J. Fontaine, P. Maris, *Anal. Chim. Acta*, **2001**, *436*, 191.
289 P.O. Markgren, M.T. Lindgren, K. Gertow et al., *Anal. Biochem.*, **2001**, *291*, 207.
290 M.O. Roy, M. Pugniere, M. Jullien et al., *J. Mol. Recognit.*, **2001**, *14*, 72.
291 R. P. H. Kooyman, A. T. M. Lenferink et al., *Anal. Chem.*, **1991**, *63*, 83.
292 http://www.ibis-spr.nl/homeframe.htm
293 http://www.ti.com/spreeta
294 J. Méléndez, R. Carr, D. Bartholew et al., *Sens. Actuators, B*, **1997**, *38(1-3)*, 375.
295 J.L. Elkind, D.I. Stimpson, A. A. Strong et al., *Sens. Actuators, B*, **1999**, *54*, 182.
296 A.M. Sesay, D.C. Cullen, *Environ. Monitoring Assess.*, **2001**, *70*, 83.
297 R. Slavik, J. Homola, J. Ctyroky et al., *Sens. Actuators, B*, **2001**, *74(1-3)*, 106.
298 R.D. Harris, J.S. Wilkinson, *Sens. Actuators, B*, **1995**, *29(1-3)*, 261.
299 A. Brecht, G. Gauglitz, *Anal. Chim. Acta*, **1997**, *347(1-2)*, 219.
300 J. Homola, S. S. Yee, G. Gauglitz, *Sens. Actuators, B*, **1999**, *54*, 3.
301 J. Homola, I. Koudela, S. S. Yee, *Sens. Actuators, B*, **1999**, *54*, 16.
302 J. O. Lekkala, J. W. Sadowski, Surface Plasmon Immunosensors, in *Chemical Sensor Technology*, ed. M. Aizawa, Kodansha Ltd, Tokyo 1994, Vol. 5, p. 199.
303 I. Vikholm, T. Viitala, W. M. Albers et al., *Biochim. Biophys. Acta*, **1999**, *1421*, 39.
304 M. J. Eddowes, *Biosensors*, **1987/1988**, *3*, 1.
305 T. A. Morton, D. G. Myszka, *Methods Enzymol.*, **1998**, *295*, 268.
306 Kubitchko, J. Spinke, T. Bruckner et al., *Anal. Biochem.*, **1997**, *253*, 112.
307 D. G. Myszka, *J. Mol. Recognit.*, **2000**, *12*, 390.
308 J. E. Pearson, J. W. Kane, I. Petraki-Kalloti et al., *J. Immunol. Methods*, **1998**, *221*, 87.
309 D. G. Myszka, *J. Mol Recognit.*, **1999**, *12*, 279.
310 C. Williams, T. A. Addona, *TIBTECH*, **2000**, *18*, 45.
311 A. Badia, S. Arnold, V. Scheumann et al., *Sens. Actuators, B*, **1999**, *54*, 145.
312 W. M. Albers, I. Vikholm, T. Viitala et al., in *Handbook of Surfaces and Interfaces of Materials*, ed. H. S. Nalwa Academic Press, New York 2001, Vol. 5, Ch. 1.
313 C. F. Mandenius, K. Mosbach, *Anal. Biochem.*, **1988**, *170*, 68.
314 L. A. Lyon, M. D. Musick, M. J. Natan, *Anal. Chem.*, **1998**, *70*, 5177.
315 G. Gauglitz, W. Nahm, *Fresenius' J. Anal. Chem.*, **1991**, *341*, 279.
316 A. Brecht, G. Gauglitz, *Fresenius' J. Anal. Chem.*, **1994**, *349*, 360.
317 A. Brecht, J. Piehler, G. Lang et al., *Anal. Chim. Acta*, **1995**, *311(3)*, 289–299.
318 J. Piehler A. Brecht, G. Gauglitz et al., *Anal. Biochem.*, **1997**, *249*, 94.
319 D. A. Weitz, S. Garoff, J. I. Gersten et al., *J. Electron Spectrosc. Relat. Phenom.*, **1983**, *29*, 363.
320 S. A. Maskevich, G. A Gachko, A. A Maskevich et al., *Proc. SPIE-Int. Soc. Opt. Eng.*, **1995**, *2370*, 131.
321 F. R. Aussenegg, A. Leitner, M. E. Lippitsch. *Rev. Roum. Phys.*, **1988**, *33*, 349.
322 T. Schalkhammer, F. R. Aussenegg, A. Leitner et al., *Proc. SPIE-Int. Soc. Opt. Eng.*, **1997**, *2976*, 129.
323 G. Steiner, C. Kuhne, B. Leupolt et al., *Proc. SPIE-Int. Soc. Opt. Eng.*, **1998**, *3256*, 106.
324 B. C. Smith, *Fundamentals of Fourier Infrared Spectroscopy*, CRC Press, Boca Raton 1996.
325 M. L. McKelvy, T. R. Britt, B. L. Davis et al., *Anal. Chem.*, **1998**, *70*, 119R.
326 L. M. Ng, R. Simmons, *Anal. Chem.*, **1999**, *71*, 343R.
327 L.-J.Jiang, W.-Y. Sun, M.-H. Shu et al., *Spectrosc. Lett.*, **1998**, *31*, 347.
328 H. R. Constantino, J. D. Andya, S. J. Shire et al., *Pharm. Sci.*, **1997**, *3*, 121.
329 A. Stevens, D. Michael, T. J. Schleich, *Mol. Biol. Biophys.*, **1998**, *1*, 221.
330 A. A. Tulub, *Biofizika* **1997**, *42*, 1208
331 M. Trinquier-Dinet, M.-T. Boisdon, J. Perie et al., *Spectrochim. Acta, Part A*, **1998**, *54*, 367.

332 Y.-Y. Sun,. Y.-C. Liu,. M.-H. Ma et al., *Gaodeng Xuexiao Huaxue Xuebao,* **1998,** *19,* 135.
333 B. Caughey, G. J. Raymond, R. A. Bessen,. *J. Biol. Chem.,* **1998,** *273,* 32230.
334 H. Zhang, Y. Ishikawa, Y. Yamamoto et al., *FEBS Lett.,* **1998,** *426,* 347.
335 N. B From, B. E. Bowler, *Biochemistry,* **1998,** *37,* 1623.
336 E. Vass, E. Lang, J. Samu et al., *J. Mol. Struct.,* **1998,** *440,* 59.
337 K. Taga, M. G. Sowa, J. Wang et al., *Vib. Spectrosc.,* **1997,** *14,* 143.
338 D. K.Graff, B. Pastrana-Rios, S. Y. Venyaminov et al., *J. Am. Chem. Soc.,* **1997,** *119,* 11282.
339 J. F. Carpenter, S. J. Prestrelski, A. Dong, *Eur. J. Pharm. Biopharm.,* **1998,** *45,* 231.
340 R. A.Shaw, G. W. Buchko, G. Wang et al., *Biochemistry,* **1997,** *36,* 14531.
341 W. C.Wigley, S. Vijayakumar, J. D. Jones et al., *Biochemistry,* **1998,** *37,* 844.
342 J. Corbin, N. Méthot, H. H. Wang et al., *J. Biol. Chem.,* **1998,** *273,* 771.
343 S. Nishimura, H. Kandori, A. Maeda, *Biochemistry,* **1998,** *37,* 15816.
344 S. Kim, B. A. Barry, *Biophys. J.,* **1998,** *74,* 2588.
345 F. Tanfani, G. Lapathitis, E. Bertoli et al., *Biochim. Biophys. Acta,* **1998,** *1369,* 109.
346 J. De Las Rivas, J. Barber, *Biochemistry,* **1997,** *36,* 8897.
347 H. Shi, L. Xiong, K.-Y. Yang et al. *J. Mol. Struct.,* **1998,** *446,* 137.
348 H. Kandori, N. Kinoshita, Y. Shichida, *J. Phys. Chem. B,* **1998,** *102,* 7899.
349 J.-J. Wang, C.-W. Chi, S.-Y. Lin et al., *Anticancer Res.,* **1997,** *17,* 3473.
350 Q. Zhou, S. Sun, S.Zhang et al., *Guangpuxue Yu Guangpu Fenxi,* **1998,** *18,* 162.
351 A. Menikh, M. T. Salch, J. Gariépy et al., *Biochemistry,* **1997,** *36,* 15865.
352 H. Kandori, *J. Am. Chem. Soc.,* **1998,** *120,* 4546.
353 J. Le Coutre, H. R. Kaback, C. K. N. Patel et al., *Proc. Natl. Acad. Sci. USA,* **1998,** *95,* 6114.
354 M. K. Bahng, N. J. Cho, J. S. Park et al., *Langmuir,* **1998,** *14,* 463.
355 H. S. Kim, J. D. Hartgerink, M. R. Ghadiri, *J. Am. Chem. Soc.,* **1998,** *120,* 4417.
356 P. Tengvall, I. Lundstrom, B. Liedberg, *Biomaterials,* **1998,** *19,* 407.
357 K. Oberg, A. L. Fink, *Anal. Biochem.,* **1998,** *256,* 92.
358 J. Buijs, W. Norde, J. W. Th. Lichtenbelt, *Langmuir,* **1996,** *12,* 1605.
359 S. Ge, K. Kojio, A. Takahara et al., *J. Biomater. Sci. Polym. Ed.,* **1998,** *9,* 131.
360 M. Lestelius, B. Liedberg, P. Tengvall, *Langmuir,* **1997,** *13,* 5900.
361 Y. Gotshal, R. Simhi, B.-A. Sela et al., *Sens. Actuators, B,* **1997,** *42,* 157.
362 S. M. Levine, D. L. Wetzel, *Free Radical Biol. Med.,* **1998,** *25,* 33.
363 D. L. Wetzel, D. N. Slatkin, S. M. Levine, *Cell. Mol. Biol.,* **1998,** *44,* 15.
364 Y. Yoshimura, *Kagaku to Kogyo,* **1997,** *71,* 548.
365 L. A. Lyon, C. D. Keating, A. P. Fox, A. P. et al., *Anal. Chem.,* **1998,** *70,* 341R.
366 G. Xue, *Prog. Polym. Sci.,* **1997,** *22,* 313.
367 R. Callender, H. Deng, R. Gilmanshin, *J. Raman Spectrosc.,* **1998,** *29,* 15.
368 S. A.Overman, G. J. Thomas, *J. Raman Spectrosc.,* **1998,** *29,* 23.
369 S. Fendel, B. Schrader, *Fresenius' J. Anal. Chem.,* **1998,** *360,* 609.
370 N. Okishio, R. Fukuda, M. Nagai et al., *J. Raman Spectrosc.,* **1998,** *29,* 31.
371 S. Hashimoto, M. Sasaki, H. Takeuchi, *J. Am. Chem. Soc.,* **1998,** *120,* 443.
372 T. Miura, T. Satoh, A. Hori-I, et al., *J. Raman Spectrosc.,* **1998,** *29,* 41.
373 M. Unno, J. F. Christian, J. S. Olson et al., *J. Am. Chem. Soc.,* **1998,** *120,* 2670.
374 X. Zhao, D. Wang, T. G. Spiro, *J. Am. Chem. Soc.,* **1998,** *120,* 8517.
375 X. Zhao, D. Wang, T. G. Spiro, *Inorg. Chem.,* **1998,** *37,* 5414.
376 T. Miura, T. Satoh, A. Hori-I et al., *J. Raman Spectrosc.,* **1998,** *29,* 41.
377 J.-S. Wang,. *Huaxue,* **1997,** *55,* 65.
378 M. Fleischmann, P. J. Hendra, A. J. McQuilan, *Chem. Phys Lett.,* **1974,** *26,* 163.
379 D. J. Jeanmaire and R. P. Van Duyne, *J. Electronanal. Chem.,* **1977,** 84.1
380 T. M. Herne, A. M. Ahern, R. A. Garrell, *J. Am. Chem. Soc.,* **1991,** *113,* 846.

381 E. Y. Kryukov, A. V. Feofanov, A. A. Moskalenko et al., *Asian J. Spectrosc.*, **1997**, *1*, 65.

382 S. Lecomte, H. Wackerbarth, P. Hildebrandt et al., *J. Raman Spectrosc.*, **1998**, *29*, 687.

383 M. Jang, I. Cho, P. Callahan, *J. Biochem. Mol. Biol.*, **1997**, *30*, 352.

384 T. Vo-Dinh, L. Allain, D. L. Stokes, *J. Raman. Spectrosc.*, **2002**, *33*, 511.

385 K. Kneipp, H. Kneipp, G. Deinum et al., *Appl. Spectrosc.*, **1998**, *52*, 175.

386 V. Decker, D. Zeisel, R. Zenobi et al., *Anal. Chem.*, **1998**, *70*, 2646.

387 Z. Q. Tian, J. S. Gao, X. Q. Li et al., *J. Raman Spectrosc.*, **1998**, *29*, 703.

388 M. J. Weaver, S. Zou, H. Y. H. Chan, *Anal. Chem.*, **2000**, *72*, 38A.

389 G. Niaura, A. K. Gaigalast, V. L. Vilker, *J. Raman Spectrosc.*, **1997**, *28*, 1009.

390 E Zubritsky, *Anal. Chem.*, **1999**, *71*, 545A.

391 Y.-W.Yang, C.-C. Teng, *Int. J. Biol. Macromol.*, **1998**, *22*, 81.

392 D. N.Georgieva, S. Stoeva, S.A. Ali et al., *Spectrochim. Acta, Part A*, **1998**, *54*, 765

393 S. Padmanabhan, M. A. Jiménez, D. V. Laurents et al., *Biochemistry*, **1998**, *37*, 17318.

394 S. E.Blondelle, B. Forood, R. A. Houghten et al., *Biochemistry*, **1997**, *36*, 8393.

395 M. M.Juban, M. M. Javadpour, M. D. Barkley, *Methods Mol. Biol.*, **1997**, *78*, 73.

396 S. M. Kelly, N. C. Price, *Biochim. Biophys. Acta*, **1997**, *1338*, 161.

397 H. Hermel, in *Particle Surface Characterization Methods*, eds. R. H. Mueller, W. Mehnert, G. E. Hildebrand, Medpharm Scientific Publishers, Stuttgart 1997, p. 159.

398 L. A. Nafie, T. A. Keiderling, P. J. Stephens, *J. Am. Chem. Soc.*, **1976**, *98(10)*, 2715.

399 P. K. Bose, P. L. Polavarapu, *J. Am. Chem. Soc.*, **1999**, *121*, 6094.

400 G. M. Clore, A. M.Gronenborn, *Curr. Opin. Chem. Biol.*, **1998**, *2(5)*,564.

401 A. E. Ferentz, G. Wagner, *Q. Rev. Biophys.*, **2000**, *33(1)*, 29.

402 K. H. Gardner, L. E. Kay, *Annu. Rev. Biophys. Biomol. Struct.*, **1998**, *27*, 357.

403 L. E. Kay, *Biochem Cell Biol.*, **1998**, *76(2-3)*,145.

404 K. Pervushin, *Q. Rev. Biophys.*, **2000**, *33(2)*,161.

405 J. H. Prestegard, H. M.al-Hashimi, J. R. Tolman, *Q. Rev. Biophys.*, **2000**, *33*, 371.

406 G. Wider, *Biotechniques*, **2000**, *29*, 1278 and 1292.

407 J. Cavanagh, W. J. Fairbrother, A. G. Palmer et al., *Protein NMR Spectroscopy: Principles and Practice*, Academic Press, New York 1996.

408 F. J. M. Van de Ven, *Multidimensional NMR in Liquids: Basic Principles and Experimental Methods*, VCH, Weinheim 1995.

409 N. Tjandra, A. Bax, *Science*, **1997**, *278*,1111.

410 K. Pervushin, R. Riek, G. Wider, K. Wüthrich *Proc. Natl. Acad.Sci. USA*, **1997**, *94*, 12366.

411 S. Bagby, K. I. Tong, M. Ikura, *Methods Enzymol.*, **2001**, *339*, 20.

412 P. Zhou, A. A. Lugovskoy, G. Wagner, *J. Biomol. NMR*, **2001**, *20*, 11.

413 S. Bagby, K. I. Tong, D. Liu et al., *J. Biomol. NMR*, **1997**,*10*, 279.

414 T. Yamazaki, T. Otomo, N. Oda et al., *J. Am. Chem. Soc.*, **1998**, *120*, 5591.

415 T. Otomo, N. Ito, Y. Kyogoku et al., *Biochemistry*, **1999**, *38*, 16040.

416 D. M. LeMaster, F. M. Richards, *Biochemistry*, **1988**, *27*, 142.

417 D. A. Torchia, S. W. Sparks, A. Bax, *J. Am. Chem. Soc.*, **1988**, *110*, 2320.

418 M. Sattler, S. W. Fesik, *Structure*, **1996**, *4*, 245.

419 L. E. Kay, K. H. Gardner, *Curr. Opin. Struct. Biol.*, **1997**, *7*, 722.

420 N. K. Goto, K. H. Gardner, G. A. Mueller et al., *J. Biomol. NMR*, **1999**, *13*, 369.

421 A. Bax, G. Kontaxis, N. Tjandra, *Methods Enzymol.*, **2001**, *339*, 127.

422 H. Schwalbe, T. Carlomagno, M. Hennig, *Methods Enzymol.*, **2001**, *338*, 35.

423 J. L. Battiste, G. Wagner, *Biochemistry*, **2000**, *39*, 5355.

424 H. Denton, M. Smith, H. Husi et al., *Protein Expr. Purif.*, **1998**,*14*, 97.

425 D. P. Zimmer, D. M. Crothers, *Proc. Natl. Acad. Sci. USA*, **1995**, *92*, 3091.

426 R. T. Batey, N. Cloutier, H. Mao, Nucleic Acids Res., **1996**, *24*, 4836.
427 E. P. Nikonowicz, Methods Enzymol., **2001**, *338*, 320.
428 M. H. Werner, V. Gupta, L. J. Lambert et al., Methods Enzymol., **2001**, *338*, 283.
429 M. Salzmann, K. Pervushin, G. Wider et al., Proc. Natl. Acad. Sci. USA, **1998**, *95*, 13585.
430 N. Tjandra, S. Grzesiek, A. Bax, J. Am. Chem. Soc., **1996**, *118*, 6264.
431 J. R. Tolman, J. M. Flanagan, M. A. Kennedy et al., Proc. Natl. Acad. Sci. USA, **1995**, *92*, 9279.
432 M. Zweckstetter, A. Bax, J. Am. Chem. Soc., **2000**, *122*, 3791.
433 J. A. Losonczi, M. Andrec, M. W. Fischer MW et al., J. Magn. Reson., **1999** *138*, 334.
434 C. R. Sanders, J. H. Prestegard, Biophys. J., **1990**, *58*, 447.
435 A. Bax, N. Tjandra, J. Biomol. NMR, **1997**, *10*, 289–92.
436 M. Rückert, G. Otting, J. Am. Chem. Soc., **2000**, *122*, 7793.
437 L. G. Barrientos, C. Dolan, A. M. Gronenborn, J. Biomol. NMR, **2000**, *16*, 329.
438 M. R. Hansen, L. Mueller, A. Pardi, Nat. Struct. Biol., **1998**, *5*, 1065.
439 M. R. Hansen, P. Hanson, A. Pardi, Methods Enzymol., **2000**, *317*, 220.
440 G. M. Clore, M. R. Starich, A. M. Gronenborn, J. Am. Chem. Soc., **1998**, *120*, 10571.
441 D. D. Ojennus, R. M. Mitton-Fry, D. S Wuttke, J. Biomol NMR 14, 175–179, **(1999)**.
442 J. Sass, F. Cordier, A. Hoffmann et al., J. Am. Chem. Soc., **1999**, *121*, 2047.
443 R. Tycko, F. J. Blanco, Y. Ishii, J. Am. Chem. Soc., **2000**, *122*, 9340.
444 D. Marion, M. Ikura, R. Tschudin et al. J. Magn. Reson., **1989**, *85*, 393.
445 M. Piotto, V. Saudek, V. Sklenar, J. Biomol. NMR, **1992**, *2*, 661.
446 D. Marion, M. Ikura, A. Bax, J. Magn.Reson., **1989**, *84*, 425.
447 J. Jeener, Lecture, Ampère Summer School, Basko Polje, Yugoslavia, 1971.
448 W. P. Aue, E. Bartholdi, R. R. Ernst, J. Chem.Phys., **1976**, *64*, 2229.
449 M. Karplus, J. Phys. Chem., **1959**, *30*, 11.
450 U. Piantini, O. W. Sorensen, R. R.Ernst, J. Am. Chem.Soc., **1982**, *104*, 6800.
451 G. Eich, G. Bodenhausen, R. R. Ernst, J. Am. Chem. Soc., **1982**, *104*, 3731.
452 C. Dalvit, M. Rance, P. E. Wright, J. Magn. Reson., **1986**, *69*, 356.
453 L. Braunschweiler, R. R. Ernst, J. Magn. Reson., **1983**, *53*, 521.
454 A. Bax, D. G. Davis, J. Magn. Reson., **1985**, *63*, 207.
455 M. Rance, J. Magn. Reson., **1987**, *74*, 557.
456 A. J. Shaka, J. Keeler, R. Freeman, J. Magn. Reson., *53*, 313–340 **(1983)**.
457 E. Kupce, P. Schmidt, M.Rance et al., J. Magn. Reson., **1998**, *135*, 361.
458 A. A. Bothner-By, R. L. Stephens, J.-M. Lee et al., J. Am. Chem. Soc., **1984**, *106*, 811.
459 L. Müller, J. Am. Chem. Soc., **1979**, *101*, 4481.
460 G. Bodenhausen, D. J. Ruben, Chem. Phys. Lett., **1980**, *69*, 185.
461 M. Salzmann, G. Wider, K. Pervushin et al., J. Biomol. NMR, **1999**, *15*, 181.
462 J. Cavanagh, M. Rance Annu. Rep. NMR Spectrosc., *27*, 1–58, **(1993)**.
463 K. Pervushin, D. Braun, C. Fernandez et al., J. Biomol. NMR, **2000**, *17*, 195.
464 R. Riek, G. Wider, K. Pervushin et al., Proc. Natl. Acad. Sci. USA, **1999**, *96*, 4918.
465 R. Riek, K. Pervushin, K. Wuthrich, Trends Biochem. Sci., **2000**, *25*, 462.
466 M. Ikura, L. E. Kay, A. Bax, Biochemistry, **1990**, *29*, 4659.
467 G. M. Clore, A. M.Gronenborn, Methods Enzymol., **1994**, *239*, 349; G. M. Clore, A. M.Gronenborn, Trends Biotechnol., **1998**,*16*, 22.
468 R. D. Oleschuk, D. J. Harrison, Trends Anal. Chem., **2000**, *19*, 379.
469 M. Karas, F. Hillenkamp, Anal. Chem., **1988**, *60*, 2299–2301.
470 F. Hillenkamp, M. Karas, Int. J. Mass Spectrom., **2000**, *200*, 71.
471 A. Douchette, D. Craft, L. Li, Anal. Chem., **2000**, *72*, 3355.
472 J. B. Fenn, M. Mann, C. K. Meng et al., Science, **1989**, *246*, 64–71.
473 C. Henry, Anal. Chem., **1997**, *69*, 359A.

474 Q. Xue, F. Foret, Y. M. Dunayevskiy et al., *Anal. Chem.*, **1997**, *69*, 426.

475 I. V. Chernushevich, W. Ens, K. G. Standing, *Anal. Chem.*, **1999**, *71*, 452A.

476 R. Guevremont, D. A. Barnett, R. W. Purves et al., *Anal. Chem.*, **2000**, *72*, 4577.

477 R. D. Smith, *Int. J. Mass Spectrom.*, **2000**, *200*, 509.

478 R. Falcone, D. Mello, A. Passaro et al., *Surf. Interface Anal.*, **2000**, *30*, 251.

479 B. Hagenhoff, M. Deimel, A. Benninghoven, *Proceedings of the 9th International Conference on SIMS*, Wiley, Chichester 1994, 792–795.

480 M. J. Tarlov, J. G. Newman, *Langmuir*, **1992**, *8*, 1398–1405.

481 B. Hagenhoff, A. Benninghoven, J. Spinke et al., *Langmuir*, **1993**, *9*, 1622–1624.

482 W. M. Albers, J. Likonen, J. Peltonen et al., *Thin Solid Films*, **1998**, *330*, 114.

483 J. Davies, C. S. Nunnerley, A. J. Paul, *Colloids Surf. B*, **1996**, *6*, 181.

484 C. M. Pradier, P. Bertrand, M. N. Bellon-Fontaine et al., *Surf. Interface Anal.*, **2000**, *30*, 45.

485 N. Patel, M. C. Davies, C. Martyn et al., *Langmuir*, **1997**, *13*, 6485.

486 S. Volooj, C. M. Carr, R. Mitchell and J. C. Vickerman, *Surf. Interface Anal.*, **2000**, *29*, 422.

487 B. A. Keller, P. Hug, *Proceedings of the 12th International Conference on Secondary Ion Mass Spectroscopy*, Brussels, 1999, p. 885. (Publ. in Sims XII, A. Benninghoven, P. Bertrand, H. N Migeon, M. W. Werner (Eds), Elsevier Science, Amsterdam, 2000).

488 B. A. Keller, P. Hug, *Proceedings of the 12th International Conference on Secondary Ion Mass Spectroscopy*, Brussels, 1999, p. 749. (Publ. in Sims XII, A. Benning Noven, P. Bertrand, H. N Migeon, M. W. Werner (Eds), Elsevier Science, Amsterdam, 2000).

489 K. J. Wu, R. W. Odom, *Anal. Chem.*, **1996**, *68*, 873–882.

490 G. Gillen, J. Bennet, M. J. Tarlov et al., *Anal. Chem.*, **1994**, *66*, 2170–2174.

491 H.-J. Galla, N. Bourdos, A. von Nahmen et al., *Thin Solid Films*, **1998**, *632*, 327–329.

492 A. M Belu, Z. Yang, R. Aslami et al., *Anal. Chem.*, **2001**, *73*, 143–150.

493 P.-A. Binz, M. Müller, D. Walther et al., *Anal. Chem.*, **1999**, *71*, 4981.

494 V. Egelhofer, K. Bussov, C. Luebbert et al., *Anal. Chem.*, **2000**, *72*, 2741.

495 M. J. Chalmers, S. J. Gaskell, *Curr. Opin. Biotechnol.*, **2000**, *11*, 384.

496 S. P. Gygi, R. Aebersold, *Curr. Opin. Chem Biol.*, **2000**, *4*, 489.

497 P. O'Farrell, *J. Biol. Chem.*, **1975**, *250*, 4007–4021.

498 A. Shevchenko, A. Loboda, A. Shevchenko et al., *Anal. Chem.*, **2000**, *72*, 2132.

499 K. F. Medzihradsky et al., *Anal. Chem.*, **2000**, *72*, 552.

500 L. Pesa-Tolic et al., *J. Am. Chem. Soc.*, **1999**, *121*, 7949.

501 G. I. Opiteck, S. M. Ramirez, J. W. Jorgenson et al., *Anal. Biochem.*, **1998**, *258*, 349.

502 W. Tong, A. Link, J. K. Eng et al., *Anal. Chem.* **1999**, *71*, 2270.

503 J. A. Loo et al., *Electrophoresis*, **1999**, *20*, 743.

504 Y. Oda et al. *Proc. Natl. Acad. Sci. USA*, **1999**, *96*, 6591.

505 S. P. Gygi, G. L. Corthals, Y. Zhang et al., *Proc. Natl. Acad. Sci. USA*, **2000**, *97*, 9390.

506 A. J. Link, J. Eng, D. M. Schieltz et al., *Nature Biotechnol.*, **1999**, *17*, 676.

507 T. Laurell, J. Nilsson, G. Marko-Varga, *J. Chromatogr. B*, **2001**, *752*, 217.

508 J. Drott, L. Rosengren, K. Lindström et al., *Mikrochim. Acta*, **1999**, *131*, 115.

509 P. ÖnneRFjord, S. Ekström, J. Bergqvist, J. Nilsson, T. Laurell and G. Marko-Varga, *Rapid Commun. Mass Spectrom.*, **1999**, *13*, 315.

510 R. W. Nelson, J. R. Krone, O. Jansson, *Anal. Chem.*, **1997**, *69*, 4363.

511 R. W. Nelson, D. Nedelkov, K. A. Tubbs, *Anal. Chem.*, **2000**, *72*, 404A.

512 M. Przbylski, M. O. Glocker, *Angew. Chem. Int. Ed. Engl.*, **1996**, *35*, 806.

513 E. C. Kempen, J. S. Brodbelt, *Anal. Chem.*, **2000**, *72*, 5411.

514 C. T. Houston, W. P. Taylor, T. S. Widlansi et al., *Anal. Chem.*, **2000**, *72*, 3311.

515 J. G. Kiselar, K. M. Downard, *Anal. Chem.*, **1999**, *71*, 1792.

516 P. William, C. Chou, D. M. Schielta, in *Automation Technologies for Genome Characterization*, ed. T. J. Beugelsdijk, Wiley, New York 1997, p. 227.

517 P. Crain, J. A. McCloskey, *Curr. Opin. Biotechnol.*, **1998**, *9*, 25.

518 C. R. Cantor, K. Tang, J. H. Graber et al., *Nucleosides Nucleotides*, **1997**, *20*, 591.

519 D. M. Lubman, J. Bai, Y. Liu et al., in B. S Larsen and C. N McEwen (Eds), *Mass Spectrometry of Biological Materials*, eds. B.S. Larsen, C. N. McEwen, 2nd edition, Marcel Dekker, New York 1998, p. 405.

520 B. Guo, *Anal. Chem.*, **1999**, *71*, 333R.

521 T. J. Griffin, L. M. Smith, *TIBTECH*, **2000**, *18*, 77.

522 U. Puapaiboon, J. Jai-nhuknan, J. A. Cowan, *Anal. Chem.*, **2000**, *72*, 3338.

523 R. M. McCormick, R. J. Nelson, M. G. Alonso-Amigo et al., *Anal. Chem.*, **1997**, *69*, 2626.

524 D. Figeys, D. Pinto, *Anal. Chem.*, **2000**, *72*, 330A.

525 J. P. Kutter, *Trends Anal. Chem.*, **2000**, *19*, 352.

526 A. G. Hadd, D. E. Raymond, J. W. Halliwell et al., *Anal. Chem.*, **1997**, *69*, 3407.

527 L. Martynova, L. E. Locascio, M. Gaitan et al., *Anal. Chem.*, **1997**, *69*, 4783.

528 G. H. W. Sanders, A. Manz, *Trends Anal. Chem.*, **2000**, *19*, 364.

529 D. G. Myszka, R. L. Rich, *Pharmaceutical Science & Technology Today*, **2000**, *3*, 310.

Section VIII
Applications 2: Environmental Analysis

Introduction

Damia Barcelo

The two chapters that were selected for this topic one on GC-ion trap mass spectrometry, by Sablier and Fujii and the other by Schröder on LC-MS in environmental analysis give an excellent contribution to the application of GC-MS and LC-MS to environmental analysis. Both chapters include many practical aspects and examples in the environmental field and also cover the historical perspective of the techniques and show the perspective on ionisation and scanning modes. Advances achieved in GC-ion trap by the use of external ion sources and GC/MS/MS possibilities are discussed. The LC-MS chapter provides an overview of the first applications of LC/MS interfacing systems, such as moving belt, direct liquid introduction (DLI) and particle beam (PB), and then on the more recent soft ionisation techniques, like thermospray and atmospheric pressure ionisation interfacing systems.

Perhaps the most interesting aspect is the number of applications reported. There are many applications in the environemtal field, especiallly of LC-MS, such as the analysis of dyes, explosives, polycyclic aromatic hydrocarbons (PAH), surfactants, toxins, pesticides (including herbicides, fungicides etc.), quaternary amines, toluidines and thiocyanate compounds, carbamates, organophosphorus compounds, phenoxycarboxylic acids, phenylureas, thioureas and sulfonylureas, triazines, estrogenic compounds, haloacetic acids and disinfection byproducts, organoarsenic compounds, sulfonic acids and antifouling pesticides. The advantage of using a particular ionisation method, for instance atmospheric pressure ionisation versus electrospray, are discusses for each class of compunds. In this respect we think that these two chapters will give a useful picture of the application of mass spectrometry to the environmental chemistry field. Finally, I would like to thank the authors of these two chapters for their time and effort in preparing their contributions Without their engagement the application of mass spectrometry to environmental analysis would certainly have not been possible.

15
LC-MS in Environmental Analysis

H. Fr. Schröder

15.1
Introduction

15.1.1
Historical Survey of the Development of LC-MS

The history of the development of liquid chromatography-mass spectrometry (LC-MS) from its beginnings in the early 1970s till the 1990s has been briefly outlined by Niessen [1]. The main reason for its rapid progress was the lack of a substance-specific detector system for the analysis of non-volatile polar, thermolabile compounds; these not being amenable to gas chromatography coupled with mass spectrometric detection (GC-MS). This had been a considerable disadvantage in any analytical research that had involved handling complex mixtures composed of either nonpolar and polar or just polar compounds.

The initial step, the introduction of liquids and liquid mixtures through a narrow capillary into a high vacuum system, was first undertaken by Talroze et al. [2] about 30 years ago. This step developed into an approach that combined a high resolution liquid chromatographic technique and the universal detector mass spectrometer ((HP)LC-MS). The following three different interfacing strategies resulted from this first analytical research approach for handling complex mixtures of polar, non-volatile compounds [3–5]:

(1) Atmospheric-pressure chemical ionisation was favoured by Horning et al. [3], whereas (2) Scott et al. [4] applied a moving-wire system which became transformed and finally resulted in the moving-belt interface. (3) The research of Arpino and his co-workers [5] led further in the direction initiated by Talroze [2], which after all had brought about the direct liquid introduction interface.

The first two commercially available LC-MS interfaces were the moving-belt interface and the direct liquid introduction interface. These hyphenated techniques promoted pharmacological research at several stages of drug development. The polar pharmaceutical compounds that were under research in pharmacological experiments, their polar by-products from chemical synthesis or even the very polar metabolites of pharmaceuticals could be substance-specifically analysed in underivatized forms. Since compounds are degraded under such pharmacological experimental conditions, that is, in body fluids, renal excretions or faeces of test animals and test subjects, such compounds for research purposes are often more polar than the precursor compounds. GC separation prior to determination necessitated the derivatization of polar compounds in order to identify the volatile derivatives by MS or other detector systems. This procedure may however discriminate against compounds which do not react with the derivatization agents. LC-MS as an alternative separation and detection technique to GC-MS allows analysis without any discriminating derivatization step prior to ionisation and determination. So identification and peak purity assessment that had been impossible with unspecific detector systems applied previously, such as ultraviolet (UV), refractive index (RI) or fluorescence detection, then became realizable with the on-line combination of LC and MS.

"The history of LC-MS is characterised by many attempts to solve the difficult problems faced; only a few of these attempts resulted in successful LC-MS interfaces, which also became commercially available, [1]. Niessen sketched the development of LC-MS from the late 1970s till the early 1990s by scanning the number of papers published annually according to the LC interfaces: moving belt (MBI), direct liquid introduction (DLI), particle beam (PBI), fast atom bombardment (FAB) or continuous flow FAB (CF-FAB), thermospray (TSP), electrospray (ESI), and atmospheric-pressure chemical ionisation (APCI). A tremendous increase in the overall LC-MS publications was observed from the mid-1990s. At the same time, atmospheric pressure ionisation interfaces (API) such as APCI or ESI came much into favor. After 30 years of investigation and development of the on-line combination of liquid chromatography and mass spectrometry (LC-MS), both these techniques, APCI and ESI, became routine laboratory methods.

15.1.2
First applications of LC-MS

Today, attention focuses on three application areas: LC-MS is mainly applied in the pharmaceutical industry but also plays an important role in environmental analysis and natural-product analyses, i.e. in biochemical and biotechnological research. In addition, there is large-scale use of LC-MS in industrial research, where the analysis and elucidation of compound mixtures of personal care products, detergents and cleaning agents, etc., i.e. unknown mixtures of polar and nonpolar compounds, produced by competitors are of great interest. In all these applications in which polar compounds are involved, LC-MS has become the method of choice.

In environmental analysis, the LC-MS technique proved a very convenient and robust analytical technique for determining and identifying polar pollutants, predominantly in aqueous matrices, such as groundwater, all types of surface waters, wastewaters, leachates and eluates of soil samples. Aqueous eluates and organic extracts of fruits and vegetables containing either polar precursor compounds or their metabolites were examined in order to detect, identify and quantify the polar compounds. Now, the comprehensive literature covering general aspects of LC-MS in environmental chemistry [6–11] or different aspects of the environmental application of LC-MS [12–25] is available in article-, chapter-, and book-form.

Over the last few years several general reviews in the field of environmental analysis have been published describing the use of mass spectrometry in general as a basic analytical technique for the determination of contaminants in environmental matrices [7, 8, 26, 27]. Recently, reviews have dealt with the topics of compound classes or interfacing techniques. Polar organic pollutants such as pesticides and herbicides were the most extensively reviewed environmental compounds. Brief descriptions of the state of the art of various mass spectral techniques are given and their various applications are outlined. LC-MS application reviews in environmental analysis have been reported by several authors. The papers of, e.g., Barceló [20], Clench et al. [28], Moder et al. [27], Ferrer et al. [29], Slobodnik et al. [22], Stan et al. [30] deal with the identification and quantification of pesticides in the environment. Polar pesticides and their metabolites were presented in an overview published by Slobodnik et al. [31] and the overview of Barceló et al. [32] reported sample handling strategies and analysis of pesticides in environmental waters. Besides overviews of triazine herbicides and their metabolites [33, 34] the determinations of phenoxyacetic acid [35] and quaternary ammonium [36] herbicides by LC-MS were reported. Even an interlaboratory study for the validation of liquid chromatography-mass spectrometry methods in the pesticide analysis of chlorinated herbicides, carbamates and benzidines by LC-MS with different MS instruments and interfaces, such as PBI and TSP, was published [37]. TSP, PBI, APCI and ESI applications were discussed for the analyses of pesticides, surfactants, dyestuffs and PAHs in environmental matrices [28].

The reviews of Berger et al. [38], Schröder [23] and Schröder et al. [21] dealt with structure elucidation and quantification by LC-MS and MS^n, while Kiewiet et al. [39], DiCorcia [40] and Marcomini et al. [41] reported how surfactants and their degradation products in an aquatic environment were tracked down. Dyes were the topics of reviews published by Hites [42] and Riu et al. [43]. Reemtsma reported about the application of API techniques in water analysis [44, 45], Clench et al. [28] described applications of LC-MS for a minor part of environmental contaminants, whereas the overview of Barceló [24] succeeded in covering the topic for the whole spectrum of contaminants.

The objective of this contribution is to present a comprehensive, up-to-date overview of the increasingly widespread use of LC-MS in environmental analysis to determine polar contaminants such as aromatic sulfonates, complexing agents, drugs and diagnostic agents, dyes, explosives, haloacetic acids, PAHs, pesticides, phenols, organoarsenic compounds, surfactants, toxins and xenoestrogens. LC-MS analytical methods for determining the anthropogenic precursor compounds as well as their biodegradation or physicochemical degradation products in the environment will be reported. Published results are compiled within tables and will therefore provide a convenient overview of the interfaces used and the compounds analysed.

This survey will include reports about applications with interfaces which have been used during the last decade when the role of LC-MS analysis extended so that LC-MS has now become, besides GC-MS, the most important analytical technique in environmental organic analytical chemistry. The results which are reviewed here were generated with the following interfaces: direct liquid introduction (DLI), particle beam (PBI), fast atom bombardment (FAB) and continuous flow FAB (CF-FAB), thermospray (TSP) and the most used actual atmospheric pressure interfaces (API), the electrospray, also designated as "ion spray" (ESI), and the atmospheric-pressure chemical ionisation (APCI) interface.

Besides the history of LC-MS that Niessen [1] rendered he gave an excellent description of these types of interfaces and their different principles of operation. Moreover, he extensively discussed LC-MS interfacing strategies as combined with these different types of interfaces.

At the outset of the LC-MS evolution, the main problem was how to determine the small quantities of analytes dissolved in large quantities of eluents. In order to determine the small quantities of analytes contained in the column effluents, the analytes must be brought into a system operating under high vacuum conditions. In addition, chromatographic integrity and mass spectrometric sensitivity must be maintained:

Therefore three principal strategies for handling the effluents of the LC-columns were under research: 1. Removal of solvent by vaporization and subsequent ionisation of the analytes first led to the moving-belt interface which, later, was followed by the development of particle beam ionisation. 2. Direct ionisation was the basic principle of the continuous flow-FAB interface, whereas 3. nebulization of the column effluent was the basic principle of DLI, TSP, APCI or ESI ionisation [1].

All these interfaces were applied for the determination of medium to strong polar environmental contaminants in water and aqueous eluates or suspensions. Their relevance for the examination of environmental samples varied. During the last years of the previous decade, ESI and APCI have seen a spectacular rise, while TSP, that had been employed as the "work-horse" under routine conditions in the late 1980s and early 1990s was used less and less, and DLI ionisation was almost completely abandoned.

Particle beam ionisation, however, for a short time, became the interface of preference, since its spectra are similar to those electron impact (EI) spectra listed in the NIST-library, and so sustain any identification of unknown compounds. Its sensitivity, though, was quite unsatisfactory.

15.2
Applications of LC-MS Interfaces in Environmental Analyses

With a certain delay, various types of interfaces that had been developed and applied in pharmacological and pharmaceutical research during the past three decades came to be used in environmental analytical applications. The following survey of "LC-MS in environmental analysis" will start with a description of the moving belt interface (MBI), followed by other interface types - DLI, PBI, FAB, TSP, and will finish with the applications and results obtained with the API interfaces APCI and ESI.

MBI was already phased out more than 10 years ago, so that applications with the help of MBI, reported on during the late 1980s, will be described here just for the sake of completeness. DLI and FAB or CF-FAB, which were there from the very beginning of the LC-MS evolution, and were applied on a large scale in the early 1990s, are hardly used nowadays [1], as the literature discloses.

15.2.1.
Moving Belt Interface (MBI)

As mentioned before, the first steps in substance-specific environmental LC-MS analysis were undertaken by interfacing LC and MS with the help of the moving wire system which was later modified into the moving belt technique. For a couple of years, the MBI technique, which, different from the interfaces used later and now, provided electron impact-like mass spectra, was used. First, this LC-MS interface type that had been used in pharmaceutical and pharmacological research concerned with the analysis of drugs and their metabolites was also successfully applied for the analysis of all kinds of anthropogenic chemicals or natural products, as well as for all kinds of pollutants, low volatile nonpolar pollutants and polar pollutants (e.g. benzidines, nitrosamines, anilines, nitroaromatics, dinitroaromatics, hydrazines, amides, phenylenediamines, organophosphites, acrylates, pyridines, phthalates, nitrophenols, pesticides, halogenated pyridines and alkyl tins [46]) present in the environment.

So different types of pesticides, i.e., carbamates [47, 48], chloro-phenoxyacetic acids [49], phenyl- and sulfonylureas [50–54], halogenated triazines [46], as well as non-ionic surfactants [55], polycyclic aromatic hydrocarbons (PAH) [56–58] and polar pharmaceutical compounds [59] were determined. This technique first used a steel wire which later was substituted by a Kapton® ribbon. However, it was soon replaced by the particle-beam interface, because the complex mechanical device led to considerable difficulties with the endless, continuously moving belt.

15.2.2
Direct Liquid Introduction (DLI)

During the short time of their application in the early 1980s, DLI interfaces were often applied for substance-specific analysis [60] of various types of pesticides and herbicides (triazines, carbamates, organophosphorus compounds) [61–67], chlorinated phenoxyacetic acids, phenylureas, analides (alachlor, propachlor and aldicarb) [63].

Because of its unsatisfactory sensitivity, a result of the low flow rate and the clogging of the diaphragms separating the eluting analytes from the high vacuum of the ion source, the application of the DLI interface technique was successively reduced. Because of all the disadvantages observed with its application, the DLI approach was soon replaced by the application of the more robust TSP interface, permitting high flow rates without any split into the mass spectrometer. Such high flow rates, combined with TSP ionisation (DLI:TSP flow rates = 0.1 : 2.0 mL min^{-1}), improved sensitivity as a result of the high concentrations of analytes contained in the column effluents.

As a result of new interfacing techniques that arose during the mid-80s, techniques more advantegeous in LC-MS analyses, the number of contributions which presented DLI data decreased tremendously in the 1990s and today these data are no longer being cited.

15.2.3
Particle Beam Interface (PBI)

Particle beam interfacing of LC-MS was designed for the analysis of less volatile compounds using the advantage of the ability to record positive (PCI) and negative (NCI) chemical ionisation and electron impact (EI) mass spectral data on the analytes examined [60]. This ability to collect spectra which can be used for library matching or structural elucidation of unknown compounds with the help of the EI fragmentation pattern available from EI-GC-MS analysis was examined to produce library-searchable EI spectra which were then compared with NIST-library data [68]. Several pesticides were used for optimization [68, 69] resulting in spectra that could be compared favourably with EI spectra from pure samples. But calibration curves for quantification purposes were found to be generally non-linear under the conditions applied [68].

A commercially available PBI interface was the product of results Willoughby and Browner obtained from monodisperse aerosol generation interface, also known by its acronym MAGIC-LC-MS [70]. But the impact on LC-MS analyses was reduced because of unsatisfactory detection limits associated with the use of a broad spectrum of analytes. This drawback could not be compensated by the advantage that the production of library-searchable EI spectra represented.

The considerable and increasing number of applications where this interface operated in parallel to the TSP interface was the beginning of a fruitful development in LC-MS analysis. The method in general was reviewed in several papers and was also partly compared to results obtained by other interface types [6, 29, 32, 71]. In the field of environmental analysis, that is, predominantly in the detection, identification and quantification as well as in the confirmation after UV-DAD [72] of pesticides, herbicides and their biochemical or physicochemical degradation products, PBI-MS was applied. These results can be found in the literature together with a few results on surfactants and dyes.

The spectrum of pesticides examined contained all different types of pesticides. So, besides the biogenic pesticide rotenone in water after SPE [73] acidic pesticides in water [74, 75] or soil [76] as well as non-acidic compounds were under research. Cappiello et al. identified and quantified [77] 13 acidic and 32 basic–neutral pesticides in water samples. With the application of large-volume injections for the analyses of these compounds [75, 78] an improvement in sensitivity was observed which made trace analysis possible. The variation of the ionisation mode, EI,

PICI or NICI, could also improve the determination efficiency. Application of NICI improved the detection efficiency of chlorinated compounds in off-line [79] or on-line determinations [80] and of triazines, anilides and organo-phosphorus pesticides [81].

One of the most intensively examined type of pesticides handled under PBI conditions were the chlorinated phenoxy acids and their esters which were determined in water [82–86] and soil samples [83, 86]. Even results of an interlaboratory comparison study of 10 chlorinated phenoxy acids using PBI or TSP ionisation were published by Jones et al. [37] [87]. Statistically significant differences were observed between the interfaces and under these conditions PBI was found to have a better precision than TSP [87]. Betowski et al. [88] observed thermal degradation induced by residence time in the ion source and the influence of ion source temperature in the ionisation of the chlorinated phenoxy acid derivatives 2,4-D and MCPA. Several authors successfully examined a large number of different carbamate pesticides [67, 89, 90] and their transformation products by PBI-LC-MS [89, 90], by PBI-FIA-MS (flow injection analysis) [67] or by supercritical fluid chromatography (SFC) PBI-interfaced to MS [91].

Application of PBI to phenylurea pesticides [74, 92–94] and their chlorinated compounds such as diuron, linuron and monuron [95] in environmental samples allowed their determination in surface and drinking waters in underivatized form. An improvement in sensitivity for the determination of phenylurea and isocyanates was obtained with a post-column alcohol addition [96].

For triazines and their polar hydroxy metabolites PBI-MS was also very effective. These pesticides could be determined in drinking and surface water [97] and in humic soil extracts after SFE [98] with good results. Soil samples were also under research for acidic and non-acidic herbicides [76] which also were determined by PBI-MS, e.g., imazamethabenz-methyl in three different soil extracts [99]. Diquat and paraquat as quaternary ammonium pesticides were observed in drinking water after SPE (Dowex resin) with a recovery of >75% by PBI-MS [100].

For the determination of several different groups of polar pesticides [101, 102] the application of LC-MS in combination with PBI-MS was very useful. The application of this technique because of its EI-like fragmentation pattern was helpful in the identification of biochemical, physicochemical and chemical degradation products, as confirmed with the analysis of the biodegradation of 3-chloro-p-toluidine-HCl in soils [103] or with the photodegradation products of alachlor, aldicarb and methiocarb. 4 alachlor-, 8 aldicarb- and 4 methiocarb degradation products were observed [104]. Petrier et al. applied PBI-MS to monitor the degradation of pentachlorphenol and atrazine under ultrasonification [105] and the results of degradation of bromacil in a water matrix by UV photolysis were published [106] while the degradation products of bromacil ozonation were reported by Acher et al. and Hapeman et al. [106, 107].

The non-atmospheric pressure ionisation interfaces (non-API) PBI and TSP besides the API interfaces ESI and APCI were applied for the analysis of the N-methylcarbamate pesticides (methomyl, aldicarb, aldicarb sulfoxide, aldicarb sulfone, carbaryl, methiocarb, carbofuran, and 3-hydroxycarbofuran). The PBI interface provided the worst results [108] of all other interfaces under research.

Simultaneously with the use of PBI for the analysis of pesticides and agrochemicals, both dispersed in large quantities in the environment [109], this interface type was also applied to perform the determination of a broad spectrum of pollutants generated by degradation processes, mobilized from waste disposals and contained in the leachates [110] and finally found in the aquatic environment. The analysis of 500 L samples of drinking water made the pollution of these waters with alkylphenol ethoxylates (APEOs) and alkylphenol carboxylates (APECs) obvious [111]. As polar constituents of wastewater samples non-ionic surfactants of NPEO type and their acidic metabolites, plasticizers, and plastic additives could be confirmed by the application of PBI-LC-MS [112].

The precursors of the potential carcinogenic aromatic amines, azo and diazo dyes were also under research by PBI. So 14 commercially available azo and diazo dyes were characterised and analysed by PBI, TSP and ESI-LC-MS. While PBI gives molar mass information and various fragments, TSP and ESI result only in molecular ions with little fragmentation and therefore little structural information [113].

The eight triphenylmethane dyes, malachite green, leucomalachite green, gentian violet, leucogentian violet, brilliant green, pentamethyl gentian violet, N',N'-tetramethyl gentian violet, and $N',N,$-tetramethyl gentian violet, were characterised by PBI-LC-MS while in parallel the six cationic dyes were reduced in the ion source to form the corresponding leuco compounds [114]. Maguire published a study of the Yamaska River in Canada from the period 1985–1987, that proved the occurrence of 15 dyes in water, suspended solids, and sediment downstream of textile mills [115]. PBI-LC-MS was used by Voyksner et al. [116] to identify reduction products of azo and diazo dyes.

PBI-LC-MS could also be applied for the simultaneous detection and quantification of the nonpolar, lipophilic PAHs and their more polar metabolites. So PBI-LC-MS was used to identify and quantify PAHs [28, 117, 118] and nitro-PAHs [119]. Flow-injection detection limits at the pg level could be obtained under NICI conditions. A broad range of organic pollutants besides PAHs, such as aromatic sulfonic acids and chlorinated phenoxy acid herbicides in soil and water were determined quantitatively and and could be confirmed by PBI LC-MS [86, 117]. The method was specific for PAHs with molar masses > 170 Da [86] (> 178 Da [117]). PAHs were ionised as molecular ions, as base peak and doubly-charged molecular ions. The capacity of PBI for providing library-searchable EI spectra was confirmed. PBI besides moving-belt and APCI (operated as heated pneumatic nebulizer interface). LC-MS interfaces were compared for qualitative and quantitative analyses of PAHs. While PBI ionisation resulted in the detection of 7 PAHs, the APCI interface showed good results and made it possible to detect all 16 target PAHs plus coronene in coal tar. MBI was inefficient for more volatile compounds [120]. Pace et al. [121] applied PBI in order to study PAHs with molar masses within the range 300–450 Da in soils. Spectra varied with the ion distribution ratio of the single- to the doubly-charged molecular ion, dependent on the molecular weight, source temperature and concentration. Doerge et al. [118] reported PBI-LC-MS work to detect and quantify PAHs and oxygenated metabolites in sediment and water samples from the the Exxon Valdez oil spill in Alaska with ng detection limits. The PAH spectra were identical with EI library spectra and oxygenated PAH spectra contained molar mass information and diagnostic fragment ions.

Biological degradation of PAHs was brought about in water samples and precursor compounds and metabolites were analysed by PBI. Naphthalene, 1-methyl- and 2-methylnaphthalene, acenaphthene and acenaphthylene were studied [122]. The limitations of the data in the GC-MS library for comparison with the PB LC-MS results were emphasized and the disadvantages of the PBI method were discussed.

PBI was evaluated for packed-column SFC-MS [91]. Fundamental studies on the effects of various operational parameters were reported and the sensitivity of the system was evaluated.

After these first tentative attempts at combining LC with MS, some respectable results could be achieved, however, technical problems, e.g., the service life of the belts (which were exposed to considerable differences of temperature and pressure with the MBI interface) or the quite poor sensitivity of the DLI or PBI interfaces, led to substitution of these interfaces. In thir place, MBI, DLI or PBI have gained firm acceptance among the various approaches to LC-MS. But even the advantage of the PBI interface, its capability to produce library searchable spectra, could not prevent its disappearance. The disadvantages of MBI, DLI or PBI, which are quite evident in comparison with those types of interfaces that will be presented in the following, were simply too grave to let them survive in routine application under the conditions of ultra-trace analysis.

15.2.4
Fast Atom Bombardment (FAB) and Continuous Flow FAB (CF-FAB)

While the removal of solvent by vaporization and its subsequent ionisation was the principle of MBI that came to be followed up later by PBI in the FAB process, a quite different ionisation technique, direct ionisation of compounds from a target, is the principle of the FAB technique.

Besides analysis for pharmaceutical, biochemical and biological research operating with FAB or CF-FAB, environmental applications were reported between the late 1980s and early 1990s. Their application reached its peak in about 1992 [1].

Some reviews were published dealing with this type of interface and its application in environmental analysis [24, 42, 123]. Qualitative and quantitative analysis of polar pollutants by FAB or CF-FAB was performed with extracts of aqueous matrices, such as wastewater, surface water, seawater, raw and drinking water [124–129], for all types of surfactants (non-ionics, anionics, cationics and amphoterics) in urban wastewaters, receiving waters (rivers and costal receiving areas), and groundwater [124–148], for metabolites of surfactants [130, 149–153], and brominated surfactants [137, 154].

Some of these results are worth mentioning because of their importance to FAB studies, particularly of surfactants. Results of flow injection analysis (FIA) by CF-FAB and by ESI-MS and MS/MS were compared for MS-quantification of cationic dodecyl-, tetradecyl- and hexadecyltrimethyl-ammonium surfactants and their deuterated analogues dissolved in water [155]. FIA-ESI, however, was the method with greater specificity. Linear alkylbenzenesulfonates (LAS) in filtered river water and sewage samples could be determined qualitatively and quantitatively by CF-FAB-MS and MS/MS by Borgerding et al. with a detection limit of 1 ng abs [135]. Simultaneously the differentiation between linear and branched AS homologues was enabled by scanning the parent ions of m/z 183 or m/z 197. LC-MS with a frit-FAB interface was applied in order to analyse non-ionic surfactants. The mass spectra contained $[M+H]^+$ ions together with a series of fragment ions of $\Delta\ m/z$ 44, which are characteristic of ethoxylated alcohols. This interpretation was supported by MS/MS measurements [132]. Unlike in this work, significant fragmentation will occur if mixtures of polyethoxylated and polypropoxylated surfactants are analysed by SFC MS [156].

With FAB, TOF, MALDI, ESI-MS, field desorption (FD) MS and GC-MS technique the analytical capabilities for non-ionic gemini surfactants were compared [157]. Parees et al. reported on the analysis of a series of oligomeric ethoxylated surfactants of this type which showed an improved surface activity. Even an antibacterial lipopeptide biosurfactant, lichenysin A, cultured and isolated, was analysed by FAB-MS and FAB-MS/MS, ESI-MS and various other methods [158]. The compound was characterised and the lipid moiety contained a mixture of 14 linear and branched β-hydroxy fatty acids from C_{12} to C_{17}.

Beside these bipolar, surface active compounds a broad spectrum of polar compounds, sulfonated azo dyes [159, 160], sulfonates [161], additives in commercial dyes [145], benzo[a]pyrene conjugates [162, 163], DNA adducts [164], ozonization products of surfactants [165, 166], explosives [167] and so forth applying FAB or CF-FAB were qualitatively determined or even quantified with success. Grange et al. [159] used accurate mass measurements for identification and confirmation of e.g. sulfonated azo dyes [159].

However, FAB ionisation was most frequently applied for the detection of pesticides due to both the broad dispersion of these substances throughout the environment combined with a need for information about these compounds of concern and the improved detection limits of this method. So the whole spectrum of polar pesticides belongs to this group of compounds which again were often over-proportional under comprehensive research [138]. CF-FAB-MS was applied to the detection of glyphosate [168], and CF-FAB-MS and electrospray ionisation were compared by recording herbicide spectra on the same magnetic sector instrument and the same LC system [169]. Both techniques provide accurate high-resolution measurements. Non-volatile and thermally labile components, e.g., pesticides in natural or purified drinking water, were investigated by Bruchet et al. [170] who applied FAB-MS/MS and TSP-LC-MS/MS. A series of sulfonyl urea herbices, with phenyl, norbornene, pyridine or pyrimidine ring substituents [54, 171], were characterised by CF-FAB-MS. These methods here permitted characterization of the molecular weights in spite of their high thermolability and their low volatility [172]. In the same manner, simultaneous determination of phenylurea and carbamate herbicides diuron, linuron, siduron, methyldymron, chlorpropham, and swep in water was performed by Okura et al. [173], and Tondeur et al. studied quaternary ammonium pesticides, e.g., paraquat, diquat and dibenzoquat [174].

In a LC-MS study [175] applying FAB besides TSP, APCI, ESI, plasma desorption (PD) ionisation a series of N- and P-containing pesticides were studied. Collision-induced dissociation (CID) spectra were recorded. Pesticide residues could readily be identified, confirmed and quantified. The results for several different polar pesticides were compared with APCI and ESI and were presented [176].

The metabolism of compounds predominantly results in more polar compounds than the precursor molecules, so for the follow-up of pesticides in the environment in the late 1980s and early 1990s FAB and CF-FAB-MS at this stage of MS development met the need. CF-FAB-MS was applied by Reiser et al. [171] to unknown sulfonylurea herbicide metabolites using packed capillary columns. New and unusual heterocyclic ring-opened metabolites and hydrolysis products were identified and metabolic pathways proposed. The metabolism of cycloate, a thiocarbamate herbicide, was investigated [177].

The analysis of dyes as constituents in environmental samples was reported by different groups. The fragmentation of carbocyanine dyes was studied by FAB-MS/MS and ESI-MS by Melnyk et al. [178], whereas Soubayrol et al. [179] discussed the use of FAB-MS and ESI-MS in the analysis of alizarin dyes from ancient materials. The degradation of four sulfonated azo dyes by lignin peroxidase labeled with ^{18}O was studied and the sulfonated products were analysed by FAB-MS and ESI-MS [180]. Wastewaters were under research for dyes by FAB, so trace level dyestuffs components in mixtures of process sewage containing chlorinated diphenyl sulfides and dyestuffs were identified or confirmed by accurate mass measurements [159]. Food and cosmetic dyes (Acid Blue 9, Acid Violet 17, Quinoline Yellow, Acid Red 51, Acid Red 87 and Acid Red 92) were determined qualitatively and quantitatively by CF-FAB-MS in colored municipal wastewater samples. Results of balances in the sewage treatment process indicating a resistance to degradation and sorption were presented [181]. Evaluation of various separation techniques by Brumley et al. led to the application of micellar electrokinetic chromatography (MEKC) and capillary LC in the analyses of environmental matrices for seven selected synthetic dyes [182]. Recovery data for spiked water and soil matrices were obtained. Capillary LC detection was performed using CF-FAB-MS, which allowed confirmation.

The polar metabolites of PAHs were important analytes and were studied with FAB in MS and MS/MS mode [24, 183, 184]. Quantitative determinations by FAB-MS required that the sample ions were completely separated from the background ions of glycerol [185]. A CF-FAB-MS method was developed by Teffera [186] for analysis of conjugates of benzo[a]pyrene (BAP)In the negative ion mode. The PAH–DNA acetylaminofluorene adduct could be determined with detection limits improved by up to 3 orders of magnitude in the low pmol and low fmol ranges [187]. Tsuruda et al. [188] analysed all the major peaks from isolated perfused liver metabolism of ^{14}C-labelled naphthalene by FAB-MS/MS. The results were compared with in vivo urinary acidic metabolites. CF-FAB-MS and -MS/MS were used to investigate the detection and structural characterization of amino-PAH–DNA nucleoside adducts in ng and pg concentrations [189] and the adducts of deoxyguanosine with a series of PAHs and amino-PAHs [190]. The complementary nature of positive and negative product ions was discussed [189].

Explosives were also examined by FAB-MS and MS/MS. The spectra of cyclic nitramine and nitroguanidine were compared with results obtained by ESI-MS and MS/MS. While FAB resulted in fragmentation (loss of H_2O, NO_2 and NO) ESI CID spectra were simple [167].

Despite an improvement in recording conditions by the use of the array detector to collect time-resolved data, which resulted in spectra with significantly enhanced sensitivity if compared with normal FAB [191], both techniques, FAB and CF-FAB-MS, however, suffered from the background generated by the essential FAB matrix. Though application of the continous flow technique in combination with FAB improved the signal-to-noise (S/N) ratio, the improvement of detection limits was not sufficiently satisfactory. What is quite advantageous and more marked than in other interface types is, because of soft ionisation, the possibility of determining molecular mass by FAB and the identification after fragmentation by MS/MS.

15.3
LC-MS Interfaces Applied in Environmental Analysis During the Last Decade

15.3.1
Achievements and Obstacles

Right from the outset of the 1990s, a selection of those interfaces that could be adapted to a routine LC-MS analysis was observable. This trend had been initiated by pharmacological and pharmaceutical research, although it had the TSP interface at its disposal, which was a well-adapted and reliable type of interface that had shown its full capacity in manifold appliances. The sample material, being available only in very limited quantities for such research, and improved separation techniques, as, for example, capillary electrophoresis (CE) or capillary zone electrophoresis (CZE) necessitated different types of interfaces that could be operated with considerably smaller amounts of sample than the TSP interface, which reached its optimized sensitivity with flow rates of about 2 mL min^{-1}. Such a desirably lower sample demand is guaranteed by atmospheric pressure ionisation (API) interfaces, atmospheric pressure chemical ionisation (APCI) and electrospray ionisation (ESI) interface.

The limitations in the flows here were determined by the LC or CE-technique applied for on-line separation prior to MS detection.

As the abbreviation "API" implies, ionisation takes places under atmospheric pressure conditions. Therefore the API techniques APCI and ESI can handle samples eluting from an LC column or by-passing the analytical column when FIA is applied, and in this feature are quite different from the TSP ionisation technique as the name implies and as reported on in the literature [1, 22]. Nevertheless the qualitative results obtained by means of TSP, APCI or ESI interfaces, which were used predominantly in environmental analysis during the last decade, may be identical or may be quite different. Three different results form the worst case.

Fig. 15.1a–c exemplify a mode of presentation for mass spectra of a mixture of non-ionic polyethylene glycol (PEG) surfactant homologues obtained by TSP, APCI and ESI in the positive FIA-MS mode. The equidistant ions in the spectra that are equally spaced by $\Delta m/z$ 44 u ($-CH_2-CH_2-O-$) are caused by different polyethyleneglycol chain lengths of the alkylpolyethyleneglycol ether homologues (alkylethoxylates; AEO). A comparison between the signal patterns of these FIA spectra showed that for all three spectra the patterns were similar, i.e., the results of the ionisation were comparable.

The analysis of another non-ionic surfactant mixture of alkylpolypropyleneglycol ether type (alkylpropoxylates; APO) was performed in the same manner, FIA-MS(+), and resulted in the FIA-MS spectra shown in Fig. 15.2a–c. Differing in their ethoxy chain links, now polypropylene glycol (PPG) chain links instead of PEG units, the signals were also equidistant in the FIA-MS spectra yet they now differed by $\Delta m/z$ 58 u equivalent to ($-CH(CH_3)-CH_2-O-$). The differences in the results obtained with the different interfaces were tremendous. TSP (a) provided a distribution of homologues in the form of ammonium adduct ions ($[M+NH_4]^+$) and molecular ions ($[M+H]^+$), APCI (b) ionisation showed a predominance of ammonium adduct ions, whereas under ESI ionisation the sensitivity was low and resulted in background ions without relevance to the expected compounds.

The third quite impressive example was obtained from a study with alkylethersulfates recorded as TSP, APCI and ESI-LC-MS total ion currents or in the form of select mass traces. The correspondences or the differences in the ionisation results observed or pointed out before for the analyses of alkylpolyether surfactants, respectively, were recorded under positive ionisation, the adequate ionisation mode

Fig. 15.1, 15.2, 15.3 on pages 165–167

Fig. 15.1. a–c: Overview spectra of non-ionic polyethylene glycol (PEG) surfactant mixture (AEOs with general formula: $C_nH_{2n+1}O-(CH_2-CH_2-O)_x-H$ ($n = 13$; $x = 1–17$)) recorded in flow injection mode (FIA) bypassing the analytical column and ionised in the positive mode by means of the different interface types: (a) TSP-FIA-MS(+) ($[M+NH_4]^+$), (b) APCI-FIA-MS(+) ($[M+NH_4]^+$ and $[M+H]^+$) and (c) ESI-FIA-MS(+) ($[M+NH_4]^+$ and $[M+H]^+$).

Fig. 15.2. a–c: Overview spectra of non-ionic polypropylene glycol (PPG) surfactant mixture (APOs with general formula: $C_nH_{2n+1}O-(CH(CH_3)-CH_2-O)_x-H$ ($n= 7$; $x = 1–10$) recorded in flow injection mode (FIA) bypassing the analytical column and ionised in the positive mode by means of the different interface types: (a) TSP-FIA-MS(+) ($[M+NH_4]^+$), (b) APCI-FIA-MS(+) ($[M+NH_4]^+$ and $[M+H]^+$) and (c) ESI-FIA-MS(+) (no ions observable).

Fig. 15.3. a–j: LC-MS total ion current traces (RIC) and selected mass traces of $[C_{12}-O(EO)_3-SO_3]^-$ ions or $[C_{14}-O(EO)_3-SO_3]^-$ ions of alkylpolyethersulfate blend recorded by means of the different interface types and ionisation modes: TSP(-) (a,b), APCI(+) (c,d), APCI(-) (e,f), ESI(+) (g,h) and ESI(-) (i,j). Gradient elution separated on RP-C_{18} column.

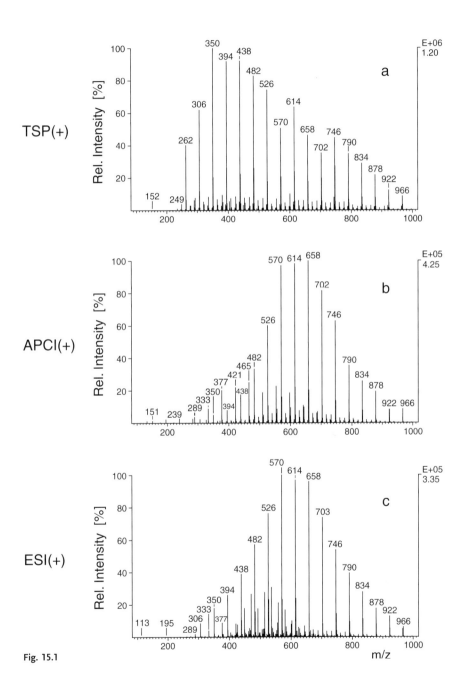

Fig. 15.1

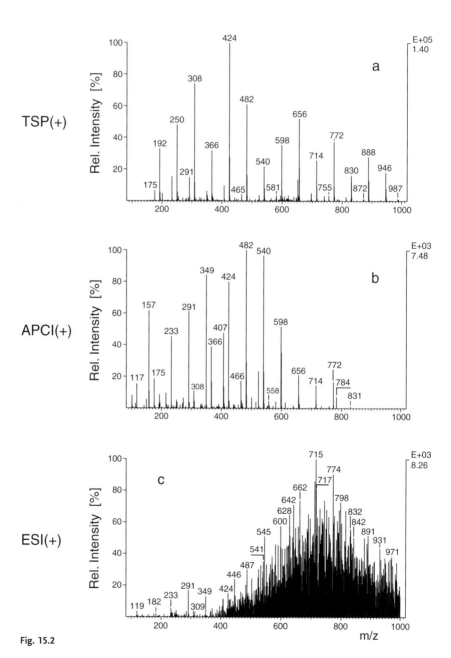

Fig. 15.2

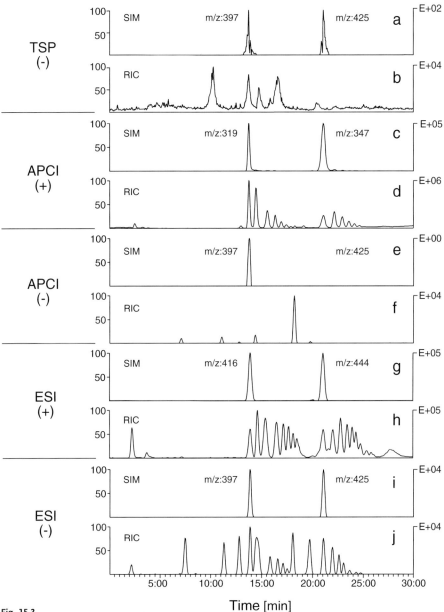

Fig. 15.3

for this type of compounds. In order to illustrate the differences observed in the ionisation process when using TSP, APCI or ESI interfaces and recording either in positive or negative mode, a mixture of C_{12}- and C_{14}- homologues of alkylethersulfates (C_nH_{2n+1}–O– $(CH_2$–CH_2–$O)_x$–SO_3H; n = 12, 14) was separated by LC, ionised by TSP(-) (Fig. 15.3a,b) [23], APCI(+) (Fig. 15.3c,d), APCI(-) (Fig. 15.3e,f), ESI(+) (Fig. 15.3g,h) and ESI(-) (Fig. 15.3i,j) [21] and detected by MS. Again the results demonstrate a considerable variation in the ionisation efficiencies of different interface types operated either in positive or negative mode. Within a comparison of the results of all interfaces, only the LC-ESI-MS(-) total current trace (Fig. 15.3j) presented the entire spectrum of the separated AES homologues on the grounds of their alkyl chain lengths and in accordance with PEG chain lengths. The signals in the mass traces m/z 397 (C_{12}-AES) or 425 (C_{14}-AES) (Fig. 15.3i) belong to the $[C_{12}$–O– $(EO)_3$–$SO_3]^-$ ion or $[C_{14}$–O– $(EO)_3$–$SO_3]^-$ ion, respectively.

The intensities of ions obtained under TSP-conditions (Fig. 15.3a,b) were weak [23] whereas APCI(+)-TIC (Fig. 15.3c,d) looked fine but ionisation had resulted in $[M$–$SO_3]^+$ ions ($[C_nH_{2n+1}$–O– $(CH_2$–CH_2–$O)_x$–$H]^+$). APCI(-) ionisation was impossible (Fig. 15.3e,f) and ESI(+) (Fig. 15.3g,h) ionised the AES homologues C_nH_{2n+1}–O– $(CH_2$–CH_2–$O)_x$–SO_3H with $x \geq 3$.

On the one hand, these results presented the correspondence in an impressive manner; on the other hand, they showed the differences in the analytical results obtained by TSP, APCI or ESI interfaces. TSP as the work-horse for routine analysis could be used to examine medium-polar and strongly polar compounds, while APCI ionisation led to successful ionisation of medium-polar or even lipophilic pollutants such as PAHs. The ESI interface (Fig. 15.2c), though inappropriate for ionisation of polypropylene glycolethers, enabled researchers to cope with the strongly polar ionic AES compounds in an excellent manner.

15.3.2
Soft Ionisation Interfaces (TSP, APCI and ESI)

One of the most serious drawbacks that has been observed in the ionisation process with TSP, APCI, ESI interfaces, and also with FAB, is the soft ionisation of the analytes which mostly leads to molecular ions or molecular adduct ions. Though molecular mass information is provided, there is little or no structural information at all observable with PBI or electron impact (EI) MS. This soft ionisation is clearly disadvantageous for any identification of environmental contaminants, since it generates either considerably less or no fragments at all, and hence is unable to confirm the presence of such compounds of environmental concern. With the commercial availability of tandem devices, tandem mass spectrometry (MS/MS) helped to overcome these identification obstacles via collision-induced dissociation (CID) in MS/MS mode or via ion trap in MS^n mode. Today, even bench-top machines provide the possibility of MS^n. However, when TSP began to become the method of choice in environmental analysis and became commercially available, MS/MS technology was still quite expensive. Users of TSP ionisation with spectrometers not amenable for MS/MS had the possibility to record spectra in discharge-on- or, alternatively, in filament-on-mode to induce or support the generation of fragment ions. The fragmentation observed under these conditions, however, was quite divergent and less extensive than in the EI mode and often identification could not be achieved because the fragmentation was not reproducible. API ionisation methods installed on conventional mass spectrometers and applied for analysis in the early 1990s allowed one to perform fragmentation recording mass spectra under so-called in-source-CID (also termed pre-analyser CID or cone voltage fragmentation (CVF)) conditions by applying an appropriate voltage difference between two regions of the API source. The method was followed up to solve the identification problem, since for small molecules no structure elucidation could be achieved [192–195]. Many analyses could be performed in this way, though the results were not at all satisfactory. This method was sometimes denoted as "poor-man's MS/MS" [196].

Although in-source-CID is a useful technique, the method has distinct limitations, which should be pointed out here. To demonstrate the capabilities of in-source-CID as applied to environmental samples, a separation and fragmentation of a mixture of nonylphenolethoxylate homologues (NPEO) contained in SPE extracts was performed and the results were compared to MS/MS product ion spectra recorded by triple quad MS. A SPE extract was submitted to APCI-MS(+) under in-source-CID-conditions or to APCI-MS/MS(+) after normal-phase-C_{18}-LC. While under in-source-CID conditions all NPEO homologue ions were fragmented, under MS/MS CID conditions the $[M+NH_4]^+$ ion at m/z 634 was selected to pass into the collision cell to be fragmented. The separation in both cases resulted in a single peak in the ion current trace (TIC) containing the whole mixture of NPEO homologues. Product ion spectra were recorded by adjusting different capillary voltages or tube lens voltages and by variation of the collision energy, respectively.

Figures 15.4a–c present in-source-CID spectra. While under "normal" LC-APCI-MS conditions no fragmentation could be observed (Fig. 15.4a), the complex mixture of precursor and product ions is observable (Fig. 15.4b), since with in-source-CID all ions that are present in the ion source are subjected to CID. The choice of more elevated potentials, however, led to a complete fragmentation with less expressiveness for identification (cf. Fig. 15.4c). CID, however, performed in a collision cell of an MS/MS instrument or in an ion trap, first leads to a selection of a particular precursor ion with improved signal-to-noise (S/N) ratios and fragmentation takes place in a more controlled way, as shown in Fig. 15.5a–c. Due to the collision energy imposed, abundant variations of precursor ions or product ions can be observed yet the characteristic fragment ions can be recognized.

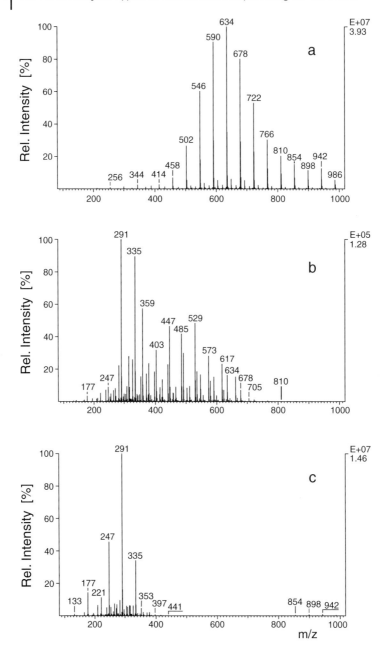

Fig. 15.4. a–c: LC-MS/MS spectra of nonylphenolpolyglycolether blend (NPEO) recorded in "in-source CID mode". Spectra were recorded for comparison of in-source CID results and results obtained with conventional MS/MS (cf. Fig. 15.5). An alkylphenolpolyglycolether blend (APEO) was submitted to "in-source CID" under variation of capillary: tube lens voltages. (a) capillary : tube lens: 20 V / 60 V, (b) 110 V/ 150 V and (c) 190 V/ 190 V.

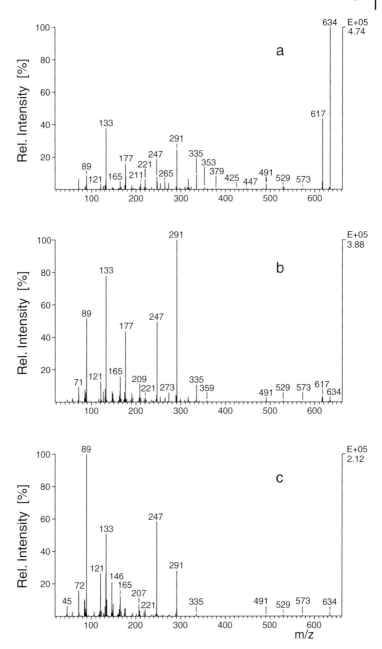

Fig. 15.5. a–c: LC-MS/MS(+) of nonylphenol-polyglycolether blend (NPEO) recorded by CID of selected [M+NH$_4$]$^+$ parent ion m/z 634 in collision chamber of TSQ in MS/MS mode. Collision gas: argon. Due to the collision energy imposed: (a) -25 eV, (b) -35 eV and (c) -50 eV, the precursor ions and product ions varied in abundance.

15.3.3
The Applications of Soft Ionising Interfaces

15.3.3.1 Applications Using Thermospray Ionization Interface (TSP)

During the second half of the 1980s, the TSP interface, which had first been introduced in 1983 [197], became the most widely-used technique for coupling LC and MS. The main disadvantage of DLI, its low sensitivity, had soon led to its replacement by the TSP interface, before, in the mid-1990s, the commercial breakthrough of API technology took place.

The option of high flow rates, combined with TSP ionisation, helped to improve the sensitivity because of the quantitative transfer of analytes in the column effluents into the mass spectrometer. High flow rates under reversed-phase conditions (RP) as well as normal-phase separations (NP) were amenable to this interface type. Thermospray ionisation takes place by means of a solvent-mediated chemical ionisation (CI) process, where a filament or discharge electrode is employed, or by an ionisation process which is enabled, and supported by a volatile buffer such as, for example, ammonium acetate, that is added to the eluent to improve positive ionisation.

While the number of reviews and papers reporting on TSP ionisation in environmental application had increased in the 1980s, the 1990s, with the development and commercial availability of interfaces using atmospheric pressure ionisation techniques, brought about a decrease in TSP publications. During this period TSP reviews covered applications in environmental analysis [6] in general or in the whole spectrum of polar pollutants [6, 198], while others reported on specific types of compounds or compound classes such as pesticides [22, 32, 199]. In addition strategies for the rapid characterization of organic pollutants in solid wastes and contaminated soils using LC-MS with TSP and other types of interfaces (PBI, ESI, and APCI) in combination with flow injection analysis and MS and MS/MS detection [71] were presented. Riu et al. [43] reported, in an overview, on current methodologies for determining sulfonated azo dyes in environmental waters elaborated by TSP, APCI, and ESI (ion spray) using LC-MS. TSP analyses of azo dyes were reviewed by Yinon et al. [200]. The unequivocal identification of isomers, oligomers and homologues of surfactants and their biodegradation intermediates in environmental samples at trace levels by LC-MS were also reviewed, with particular attention to thermospray (TSP) [23, 40] and ESI [40]. But the most comprehensive research work was done on pesticides, herbicides and fungicides, determined by TSP-FIA or LC-MS and MS/MS [22, 32, 199]. In accordance with the widespread use of these compounds in agriculture, papers dealing with the application of TSP ionisation were found in the literature.

Table 15.1 reflects the applications of TSP used as the interface to couple MS with different liquid chromatographic separation techniques in environmental analytical applications. A list from the literature is presented for different groups of compounds.

Table 15.1 Environmental applications in LC-MS performed by thermospray interface (TSP).

Topic	References
Reviews	
General reviews	6, 198
Compound class specific reviews	
– Dyes	43, 200
– Pesticides	22, 32, 199
– Surfactants	23, 40
Compound classes	
– Dyes	28, 43, 113, 200–202
– Explosives	198, 203–206
– Polycyclic aromatic hydrocarbons	28, 207–209
– Surfactants	23, 28, 40, 210–234
– Toxins	235
– Pesticides (including herbicides, fungicides etc.)	
– Anilides, quaternary amines, toluidines and thiocyanate compounds	175, 239–245, 248, 264, 501
– Carbamates	32, 108, 175, 239, 242, 244–257
– Organophosphorus compounds	166, 175, 176, 239, 242, 244, 247, 256, 259, 261–263, 287, 523
– Phenoxycarboxylic acids	176, 244, 264–269, 501
– Phenylureas, thioureas and sulfonylureas	175, 176, 239, 240, 242, 244, 246–248, 254–256, 270–272
– Triazines	94, 175, 176, 239, 240, 242, 244–247, 256, 266, 273–284
– Miscellaneous	266, 287–290
– Comparison of TSP interfaces with different types of interfaces	32, 40, 71, 94, 108, 113, 166, 175, 176, 199, 200, 205, 240, 257, 271, 275, 276, 311, 501, 503, 509, 523

Dyes

Dyes, and especially azo dyes, compounds of environmental concern because of their carcinogenic potential, were studied with TSP-LC-MS [28, 201]. An overview presented current methodologies for determining sulfonated azo dyes in environmental waters using TSP [200]. Results obtained by LC-MS and capillary zone electrophoresis (CZE) MS, coupled with TSP-, APCI-, and ESI interface were discussed [43].

A standard commercial TSP interface was modified to increase the sensitivity to sulfonated azo dyes and to permit their analysis. The sensitivity could be increased in order to determine sub-µg amounts of the dyes. A by-product of AZO 4 (2,2'-dihydroxy-4-sulfonyl-6-nitro-1,1'-azobisnaphthalene) which could not previously be identified by LC-MS, was then confirmed to be a structural isomer of AZO 4 [202].

Fourteen commercial azo and diazo dyes were characterised and analysed by three different LC-MS interfaces: the TSP-, PBI- and ESI interfaces. TSP-LC-MS mainly resulted in [M+H]$^+$ ions with little fragmentation [111]. Aqueous solutions of the monosulfonated azo dyes with concentrations between 50 and 200 ppm were analysed by LC-MS coupled with TSP-, ESI- and APCI interfaces. TSP was not as successful as ESI or APCI which gave more structural information. Loss of [Na]$^+$ (Δ m/z 23 u) in the fragmentation proved to be common in all dyes.

Explosives

TSP-LC-MS with explosives was one of the topics Arpino [198] presented when he reviewed the applications of TSP interfacing. General operating principles, optimization strategies and possible ionisation mechanisms were presented.

TSP-LC-MS/MS was applied to qualitative analyses of explosives such as ethyleneglycol dinitrate, β-cyclotetramethylene tetranitramine (β-HMX), cyclotrimethylene trinitramine (RDX), diethyleneglycol dinitrate, glycerol trinitrate, 2,6-dinitrotoluene (2,6-DNT), 2,4,6-trinitrotoluene (TNT), 2,4-dinitrotoluene (2,4-DNT), 3,4-dinitrotoluene (3,4-DNT) and pentaerythritol tetranitrate (PETN). Detection limits with excellent selectivity were in the pg range, discharge-on mode improved the sensitivity by a factor of 25 for dinitrotoluenes [203]. Nitroaromatic and nitramine explosives in water were quantitatively examined by HPLC with photodiode array detection (UV-DAD) and compared with TSP-LC-MS and LC-MS/MS (SIM and selected reaction monitoring (SRM)) in negative ion mode. MS quantitation limits were 10–100 ng L^{-1}, presenting underestimated concentrations in SIM and SRM mode [204].

TSP-LC-MS in the negative mode was used to identify and quantify the explosives TNT, RDX and hexyl, as well as their degradation products and other pollutants, in groundwater samples of an ammunition hazardous waste site after SPE applying LiChrolut® EN. 31 compounds could be identified, such as nitramines and their by-products, TNT and partially nitrated toluenes, 1,3,5-trinitrobenzene and partially nitrated benzenes, aminonitrotoluenes, nitroanilines, hexyl and nitrophenols [205].

TSP-LC-MS and LC/NMR analyses were applied to characterise the phototransformation products of 2,4,6-trinitrotoluene generated by sunlight. Combined information from both analytical techniques allowed structural characterization of several acidic nitroaromatic compounds and some not commercially available phototransformation products of TNT [206].

Polycyclic aromatic hydrocarbons (PAH)

While PAHs with more than 3 rings are strong lipophilic compounds amenable to normal phase LC, their physicochemical or chemical degradation products are more or less polar compounds which enable the application of RP-LC in combination with TSP interfacing. Clench et al. [28] gave brief descriptions of the state of the art of TSP and some other mass spectral techniques in the analysis of PAHs, in addition to other pollutants of environmental concern. Benzo[a]pyrene (BaP) and its metabolites in air, 1,6-, 3,6- and 6,12-dione derivatives, were studied by TSP or APCI-LC-MS, after they had been extracted from particles. TSP ionisation in the negative mode yielded predominantly [M]$^-$, [M-H$_2$O]$^-$ or [M+CH$_3$COO]$^-$. TSP calibration graphs for BAP were linear at 10–1000, with detection limits of 1–20 ng

[207]. Negative ion spectra provided the richest information when spectra of hydroxylated metabolites (1–4 OH groups) include [M–H]- ions, and fragments attributed to progressive dehydration were recorded [208]. The fragmentation of conjugated metabolites was found to be greater than predicted.

Primary and secondary photolysis products of pyrene, 1-hydroxypyrene or 1,6- and 1,8-pyrenequinones, respectively, in water and in Brij 35 micellar media were analysed and quantified by TSP-LC-MS [209].

Surfactants

LC-MS methodologies applying TSP interfacing for unequivocal identification of isomers, oligomers and homologues of surfactants and their biodegradation intermediates in environmental samples at trace levels have been reviewed [28] very extensively [23, 40].

Trace amounts of alcohol ethoxylates (AEO) in diluted aqueous environmental samples were determined by TSP-LC-MS in sewage treatment plant (STP) effluents and river water which were spiked with a mixture of AEOs. $[M+H]^+$ and $[M+NH_4]^+$ ions were observed and highly branched AEOs could be distinguished from isomeric linear AEOs. The method could be validated [210]. Sum parameter analyses for the determination of different types of surfactants and their primary degradation products, methylene blue active (MBAS) and bismuth active substances (BiAS), were compared with substance-specific methods applying TSP combined with FIA, LC-MS and MS/MS. With MS methods, surfactants and metabolites could be determined quantitatively, while MBAS and BiAS methods failed [211]. Alkyl polyglucosides were followed in the biological wastewater treatment process (WWTP) and could be recognized as easily degradable [212], whereas fluorine-containing surfactants, non-ionics and anionics, were found to be very stable against biochemical and physicochemical treatment [213, 214]. Non-ionic fluorinated compounds with a hydrophilic moiety consisting of PEG chains could be biodegraded to shortened PEG chains carboxylated in the terminal position [213].

TSP methods were used for the determination of polar organic pollutants concentrated by SPE or liquid–liquid extraction [215]. TSP-FIA and LC-MS and -MS/MS methods were described for the qualitative and quantitative analysis of hardly eliminable or non-biodegradable polar compounds, such as surfactants in inflows and effluents from municipal biological wastewater treatment plants [215–224]. Non-ionic surfactants and their metabolites (primary degradation products) as well as anionic surfactants of linear alkylbenzenesulfonate type dominated the effluents. Parent and product ion scans or neutral loss scans in the TSP-MS/MS mode were applied to identify surfactants from WWTP [225] and in surface water samples from the river Elbe [226] and its tributaries [226–228]. Surfactants were observed even in drinking water after a soil filtration process and activated carbon treatment. Pattern recognition by TSP-FIA-MS and MS/MS was applied for confirmation [224, 229–231].

By LC-MS, retention time shifts which occurred because of surface-active compound adsorption on the analytical column could be confirmed. TSP was used to monitor polar pollutants, predominantly several types of surfactants, in textile [232] and municipal wastewaters [233]. Results of biochemical [232, 233] and physicochemical [232] treatment for elimination were reported and precursor compounds and degradation products generated under these treatment steps were identified by CID. The treatment of recalcitrant wastewater constituents of the surfactant type by ozone (O_3) or O_3 combined with UV (O_3/UV) [231, 234] or by hydrogen peroxide/UV [234] was examined and compounds arising were studied by MS/MS for identification [231, 234] and by biotoxicity testing (Vibrio fisheri and Daphnia magna) [231].

Toxins
Only soft ionizing interfaces such as TSP [235], FAB [236–238] and API interfacing techniques (cf. 15.3.3.2 ESI, toxins) were able to handle thermolabile, polar seafood toxins that were classified, e.g., as amnesic shellfish-poisoning toxin, paralytic shellfish-poisoning toxin and as diarrhetic shellfish-poisoning toxin, according to the toxic results observed after ingestion.

Pesticides (including herbicides, fungicides etc.)
Of all the organic pollutants measured by LC-MS in aqueous environmental matrices, e.g., drinking water, surface and groundwater, pesticides with the application of TSP ionisation became the most studied compound group. With the phasing-out of most chlorine-containing pesticides because of their persistence and their bio-accumulation potential the new generation of pesticides were designed to be more polar and therefore more biodegradable.

Anilides, quaternary amines, toluidines and thiocyanate compounds
Anilide-, N-substituted amine-, quaternary amine-, toluidine- and thiocyanate pesticides were not so often used and therefore papers reporting on these pollutants in the environment are quite rare.

A comprehensive paper by Volmer et al. [175] reported on these compounds and other types of pesticides. Residues of 19 amine- and anilide-pesticide derivatives (alachlor, allidochlor, bentazone, butachlor, carboxin, dimethachlor, oxicarboxin, metalaxyl, metazachlor, monalide, pendimetalin, pentanochlor, prochloraz, propachlor, propanil, tebutam, trifluralin) were readily identified, confirmed and quantified by TSP LC-MS and different ionisation techniques (APCI, ESI, FAB and PD) at < 100 ng L^{-1} after C_{18}-SPE [175]. For propioanilide, C_{18} Empore® extraction from river water and spiked seawater was applied prior to detection [239]. In real environmental water samples from estuaries, metolachlor and the anilide alachlor could be confirmed by both LC-MS techniques, ESI and TSP [240]. The chemical and photochemical stability of metolachlor (2-chlor-6'-ethyl-N-(2-methoxy-1-methylethyl)acet-o-toluidide) in organic-free water and lake water was determined by TSP-LC-MS. Sunlight degradation was found to be faster than chemical degradation, however, near-surface half-lives were determined with 22 d in summer and 205 d in winter. Photoproducts and four chemical dechlorination products were observed [241].

The effects of various additives on the sensitivity and selectivity of TSP-LC-MS of thiocyanates and anilines were studied to optimize ionisation conditions. Trialkylammonium formates were found to increase the selectivity and sensitivity of the TSP process [242]. TSP-LC-MS was also used for the characterization of the quaternary amine pesticides paraquat, difenzoquat, diquat, mepiquat and chlormequat from water and soil samples. Base peaks were $[M+H]^+$ and $[M-CH_3+H]^+$ [243]. Difenzoquat, a difficult-to-determine quaternary ammonium pesticide was analysed using a post-column ion-pair extraction system [244].

Carbamates

Carbamates applied as insecticides were often observed in environmental samples because of their widespread use. They could be identified, confirmed and quantified after sample concentration by C_{18} SPE by SIM-TSP at < 100 ng L^{-1} according to the CEC drinking water limit. The interfacing techniques APCI, ESI, FAB and PD supported the elucidation of fragmentation mechanisms in TSP mass spectra [175]. TSP ionisation was evaluated for the LC-MS determination of 128 pesticides (containing 22 carbamates and 2 thiocarbamates besides 16 triazines-, 4 pyrimidine-, 3 triazole-, 2 pyridine-, 3 morpholine-, 1 N-substituted amide-, 12 aniline- and 1 di-nitroaniline, 24 organophosphorus, 15 phenylurea and 1 thiourea, 5 carboxylic, 8 phenoxy acid, 5 carboxylic ester and 4 quaternary ammonium pesticides) having a wide range of polarities. Detection limit, linearity and reproducibility data are given. Possibilities for the confirmatory analysis of carbamates were reported [245].

Besides those of other polar pesticides, mass spectra of carbamates were obtained from several surface and drinking water samples after on-line concentration on a precolumn prior to TSP-LC-MS [246]. $[M+H]^+$ or $[M+NH_4]^+$ of the pesticides were observed as base peaks [244] [247]. In negative mode, fragments such as [M–CONHCH$_3$]$^-$ were observed for carbamates [244]. A multi-residue TSP-LC-MS method was described for carbamates and thiocarbamate pesticides in water samples after C_{18}-SPE [248]. After concentration on C_{18} Empore® disks carbamates from river water and spiked seawater samples at concentration levels of 0.25, 25 and 1000 µg L^{-1} were determined by LC-UV and confirmed by TSP-LC-MS (detection limits: 2–20 µg L^{-1}) [239]. The results obtained for the analysis of the N-methylcarbamate pesticides methomyl, aldicarb, aldicarb sulfoxide, aldicarb sulfone, carbaryl, methiocarb, carbofuran, and 3-hydroxycarbofuran by LC-MS in combination with the non-API interfaces TSP and PBI were compared with the results of API interfacing (ESI and APCI) [108]. The TSP interface performed well, and TSP in the SIM mode resulted in detection limits of a few ng, offering a viable method for confirmation at the regulatory level of about 0.1 ppm. The main problem with TSP interfacing was signal stability [108]. With emphasis on N-methylcarbamates the performances of on-line MS detection coupled to SPE and LC were reviewed when TSP interfaces besides PBI or ESI ionisation were applied [32]. Automated on-line SPE followed by LC techniques for monitoring carbamates and their transformation products in traces in well waters were investigated TSP-LC-MS with time-scheduled SIM could be used for confirmation [249]. Pollution by carbamate pesticides and metabolites was sporadic and exceeded the limit of 0.5 µg L^{-1} for total pesticides allowed by the EEC Drinking Water Directive. The main toxic metabolites of carbamates, 3-hydroxycarbofuran and methiocarb sulfone [249], should be included in future monitoring programs. Comparing the results of TSP and ESI ionisation in SIM modes for the quantitative determination of 10 carbamate pesticides, quite different in polarity, made obvious that ESI was 10–150 times better than using TSP ionisation [250].

The effects of coeluting carbamate pesticides carbofuran, propoxur and pirimicarb combined with ion formation suppression under TSP-LC-MS conditions were discussed applying multivariate curve resolution and ranking analysis [251].

The use of different eluents in LC-RP separations was examined to optimize LC separation and determination of carbamates besides other pesticides by TSP-LC-MS [244]. The effects of additives on LC separation and of the vaporizer temperature on ion formation in TSP-FIA-MS analysis were studied for N-methylcarbamate [252], carbamates [242] and (thio)carbamates [242] pesticides. A strong reduction in abundance of the characteristic ions [M+H–CH$_3$NCO]$^+$ and [M–H–CH$_3$NCO]$^-$ for methiocarb and its sulone were found because of thermal degradation at 90 °C which made quantitation difficult [252]. The addition of trialkylammonium formates increased selectivity and sensitivity, detection limits being < 20 ng in full scan mode [242].

An interlaboratory examination quantifying 16 carbamate pesticides from four classes was performed applying three commercially available TSP interface types. In contrast to several literature reports very similar spectra were obtained, however, thermally labile carbamates gave varying results [253]. In an interlaboratory study with nine participating laboratories a TSP LC-MS method was evaluated to determine three N-methylcarbamates, three N-methylcarbamoyloximes, two substituted urea pesticides and one carbamic acid ester. Data confirmed that interlaboratory variation was greater than that within labs [254]. The multiresidue method for determining 19 thermally labile and non-volatile carbamate pesticides in fruits and vegetables published by Liu et al. [255] involved a single extraction step followed by TSP-LC-MS(+/-) (SIM). Limits of detection in apples, beans, lettuce, peppers, potatoes and tomatoes were observed with 0.025–1 ppm and recoveries ranged from 69–110% [255].

The carbamate pesticides aldicarb, carbaryl and their degradation products, all under research in photochemical degradation studies, were characterised by TSP-LC-MS and MS/MS. A tentative photodegradation pathway for the different pesticides in water was postulated [256].

The degradation of the carbamates carbofuran and methiocarb in estuarine waters was compared with that of aqueous samples exposed to UV light in the laboratory. TSP- and ESI-LC-MS were used for product monitoring and identification of degradation products. The predominant degradation pathways observed here, however, were hydrolytic and microbial degradation [257]. Tap water was analysed for thiobencarb degradation products as a result of chlorination. Several compounds were determined [258].

Organophosphorus compounds
Besides TSP-spectra from organophosphorus and different groups of polar pesticides, APCI, ESI, FAB and PBI spectra were presented. CID allowed the identification of pesticide residues and the confirmation and quantification of these compounds by TSP at concentrations < 100 ng L^{-1} [175]. Barceló et al. [239] extracted river water and spiked seawater samples by C_{18} Empore® disks to concentrate organophosphorus pesticides prior to identification and quantification while Bagheri et al. [247] for the same purpose used on-line SPE for phosphorus pesticides and TSP-LC-MS.

After aircraft spraying, levels of temephos and its degradation products, i.e., temephos sulfoxide and isomers, were determined by TSP-LC-MS in extracts of water samples with detection limits of 1–2 ng and mean recoveries of 80%. Temephos sulfoxide and its isomers were measured [259]. To improve the determination of TSP-LC-MS for polar organophosphorus pesticides, trialkylammonium formates were found to increase selectivity and sensitivity with detection limits of <20 ng even in full scan mode [242]. The use of different eluents in RP separations was examined for an optimized separation and determination of organophosphorus pesticides and their biodegradation products by TSP-LC-MS [244]. Different adduct ions were formed ($[M+NH_4]^+$ or $[M]^-$ and $[M-R]^-$ (R = Me or Et)) depending on the eluent used. Under normal phase (NP) conditions eluents enhanced the response in some cases and provided more structural information [244]. Ion chromatography (IC) coupled with TSP-MS/MS was compared with an ESI interfaced IC for the determination of ionic compounds in agricultural chemicals. The solid-phase suppressor applied allowed methyl phosphate and methyl sulfate determination [260]. Lacorte et al. described a TSP-LC-MS method for quantitative determination of 10 organophosphorus pesticides and their photolytic transformation products in river waters [261] and in estuarine waters [262]. For quantification $[M+H]^+$ and $[M+NH_4]^+$ or $[M+CH_3CN]^+$ were used in the positive mode, while $[M-H]^-$ and $[M+HCOO]^-$ were used under negative ionisation. The limit of detection was found to be 0.01–0.1 µg L^{-1}. TSP-LC-MS was used also for the characterization of the organophosphorus pesticides fenitrothion and parathion-methyl and their degradation products after photolysis [256]. TSP-LC-MS in the positive mode was also applied to study the anaerobic aquatic metabolism of the ^{14}C-pyrimidinyl-tebupirimphos. S-ethyl isomers as metabolites were identified [263].

Phenoxycarboxylic acids
Isolation and trace enrichment in off-line and on-line modes were described together with TSP-LC-MS determination of the spiked phenoxy acetic acid herbicides 2,4-dichlorophenoxy acetic acid (2,4-D), 4-chloro-2-methylphenoxy acetic acid

(MCPA), and 2-(4-chloro-2-methylphenoxy)propionic acid (MCPP) in estuarine waters [264]. Chlorinated phenoxy acids were identified in Rhine water after ion-pair LC on-line with TSP-LC-MS detection [265]. Liquid–liquid extraction was applied to concentrate chlorinated phenoxy acids, triazines and amphoteric pesticides from water samples prior to TSP-LC-MS [266]. FIA-TSP-MS/MS was used as a fully automated screening method for phenoxy acid herbicides in water samples at the low µg L^{-1} levels without sample concentration [267]. The optimization of RP-separation combined with TSP-LC-MS for phenoxy acid analysis in negative mode was studied [244]. Samples with eight chlorphenoxy carboxylic acid herbicides were analysed by Geerdink et al. [268]. Selectivity of measurement was studied using 2,4,5-trichlorophenoxyacetic acid (2,4,5-T), and triclopyr, which differs in just one mass unit. Screening of surface water samples spiked at the 1 µg L^{-1} level was easily achieved without analyte concentration and a minimum of sample clean-up [268]. An interlaboratory comparison of TSP and PBI-LC-MS for the analysis of 10 chlorinated phenoxy acid herbicides was evaluated in seven laboratories by TSP and PBI. Sample extracts were analysed for 2,4-D, 2,4-DP, MCPA, MCPP, 2,4,5-T, dinoseb, dalapon, 2,4,5-TP, dicamba, 2,4-DB and 2,4,5-TP. Statistically significant differences were found between interfaces. PBI generally had better precision than TSP, however, for the quantitative analysis of low levels, TSP in the negative ion mode should be preferred [87]. The anaerobic degradation of the herbicide picloram, a 4-amino-3,5,6-trichloropyridine-2-carboxylic acid, in anoxic freshwater sediments was examined by Ramanand et al. An intermediate and its recalcitrant metabolite were separated and analysed by NMR and TSP-LC-MS [269].

Phenylureas, thioureas and sulfonylureas
Volmer et al. studied phenylureas and sulfonylureas by TSP-LC-MS after sample concentration by C$_{18}$ SPE [175]. 15 phenylurea- and 1 thiourea pesticides besides 112 polar pesticides from other pesticide classes were examined by TSP ionisation, detection limits and TSP mass spectra of these polar compounds were presented [245]. Besides other polar pesticides, the phenylurea pesticides isoproturon and diuron were on-line concentrated on a precolumn from several surface and drinking water samples and then determined by TSP-LC-MS [247]. A multi-residue TSP-LC-MS method was published by Moore et al. for the determination of the urea pesticides chlortoluron, diuron, isoproturon, and linuron in water samples after C$_{18}$-SPE [248]. C$_{18}$ Empore® disks were applied to concentrate phenylureas from river water and spiked seawater samples prior to TSP-LC-MS. Detection limits of 2–20 µg L^{-1} and recoveries between 80 and 125% were observed [239]. TSP-LC-MS (SIM) in the positive mode allowed determination of the urea pesticides chlorbromuron, diuron, linuron, metobromuron, monuron, neburon in apples, beans, lettuce, peppers, potatoes and tomatoes with detection limits of 0.025–1 ppm [255]. 20 other polar pesticides, linuron, which was under suspicion of being a dietary oncogenic risk (US Natl. Res. Council) was determined by TSP-LC-MS a single rapid procedure in vegetables with detection limits of 0.05–0.10 ppm [270]. TSP-LC-MS and ESI were used in a multi-residue method for determination of the sul-

fonylurea herbicides chlorsulfuron and the methyl esters of sulfometuron, metsulfuron, tribenuron, bensulfuron, chlorimuron and primisulfuron. Both TSP and the ESI interface provide mass spectra with three structure-significant ions necessary for unambiguous identification in environmental samples [271]. The validation of the enzyme immunoassay for the herbicide triasulfuron in soil and water samples was performed by means of TSP-LC-MS. Immunoassay results compare favourably with LC-MS ($r=0.88$, in soil) but have a lower detection limit of 0.05 ppb (water) and 0.10 ppb (soil) [272]. Phenylurea pesticides were examined by TSP-LC-MS after on-line SPE [246] while in parallel the use of different eluents [244] or the effects of additives to LC separation [242] were studied. Compounds could be recorded as base peak in positive ion mode as $[M+NH_4]^+$ adduct ions [244]. When TSP-LC-MS was applied in order to characterise the phenylureas (chlortoluron, isoproturon, diuron, linuron and diflubenzuron) in estuarine waters $[M+H]^+$ was generally the base peak [240]. Compared with a parallel applied ESI ionisation TSP results were not satisfactory.

In an interlaboratory study with nine participating laboratories a TSP-LC-MS method was evaluated in order to determine diuron and linuron. Depending on the compounds, the intralaboratory precisions range from 6.5–33 with a relative standard deviation (RSD) of 1% [254]. After linuron had been submitted to photochemical degradation, TSP-LC-MS and MS/MS was used to characterise the degradation products. A tentative photodegradation pathway for the different pesticides in water was postulated by Durand et al. [256].

Triazines
Triazine derivatives and their de-alkylated or hydroxylated degradation products were studied by LC-UV or TSP-FIA-MS/MS after isolation and trace enrichment by liquid–liquid extraction [266], SPE cartridge extraction [246] [273] or by C_{18} Empore® extraction disks [239]. Optimization of SPE procedures influenced the quality of the FIA MS/MS spectra [273]. Recoveries for the atrazine metabolites were found to be only 3–17% if Empore disks were applied [239]. Comprehensive research was published by Volmer et al. who determined 13 [175] or 16 [245] triazines and N-heterocyclic pesticides, which were readily identified, confirmed and quantified by TSP according to the CEC drinking water limit at < 100 ng L^{-1} by time-scheduled SIM. Several surface and drinking water samples were analysed in the SIM mode and low levels of simazine and atrazine with detection limits of 2–90 ng L^{-1} in the course of an on-line low-level screening of polar pesticides in drinking and surface waters [247]. Triazines and other pollutants in river water samples were also determined at low- to sub-µg L^{-1} levels in a full automated SPE-LC-MS using TSP or PBI ionisation [94]. The analysis of 12 triazines and 11 of their biodegradation products was achieved by multiple reaction monitoring in the FIA-MS/MS mode [274]. Detection limits for triazines and hydroxylated degradation products were 0.5–0.15 µg L^{-1} and 0.2–0.45 µg L^{-1} water, respectively [274]. The occurrence of hydroxylated degradation products of atrazine in the stream water from Goodwater Creek watershed was determined by TSP-LC-MS and ESI-

LC-MS/MS. The proportions of the different degradation products were observed with 100% for hydroxy-atrazine, 25% for deethylhydroxy-atrazine and 6% for deisopropylhydroxy-atrazine, with concentration ranges of µg L^{-1} [275]. A SPE method was described for the preconcentration of atrazine and its metabolites from water samples by RP-C$_{18}$ or on a benzenesulfonic acid cation exchange support prior to TSP-LC-MS [276]. The qualitative and quantitative analysis of triazines and their metabolites in environmental samples by selecting different eluents in LC-RP separations [244] or various operating modes in TSP-LC-MS/MS was examined [277]. Studies of degradation products in polluted soil samples and after aquatic photodegradation were also carried out [277]. Chlorotriazines and their hydroxylated metabolites recorded in TSP-MS(+) mode resulted in [M+H]$^+$ base peaks despite the presence of ammonium acetate. The same results were observed when TSP-LC-MS was applied to the characterization of the triazines atrazine, simazine, ametryne, cyanazine, deethylatrazine and deisopropylatrazine and triazine metabolites in estuarine waters after TSP ionisation [240]. The effects of various mobile phase additives on the sensitivity and selectivity of TSP-LC-MS for the determination of 55 polar pesticides were studied to optimize conditions [242]. With the addition of trialkylammonium formates selectivity and sensitivity could be increased [242].

TSP was also used for the characterization of triazines and their degradation products after physicochemical and chemical degradation experiments. Cyanazine was monitored after photolysis [256] whereas atrazine and simazine and their degradation products were determined by TSP-MS and -MS/MS after ozonolysis [278]. The main degradation pathways observed were dealkylation, deamination, dehalogenation and hydroxylation [256, 278]. The photodegradation pathways of atrazine by dealkylation reactions were also observed by Schmitt et al. [279] applying capillary electrophoresis in combination with TSP-MS and MS/MS. LC-MS/MS was applied to screen and characterise chlorotriazines and hydroxytriazines as degradation products after ozonolysis in aquous solutions with the help of product and parent ion scans [280]. The kinetics and mechanism of the chemical oxidation of prometryn and prometryne, terbutryne, ametryne and desmetryne by NaOCl [281], or HClO and ClO$_2$ [282], respectively, were studied. Intermediates formed after NaOCl treatment in the sequence R–S–CH$_3$ (R = prometryn), R–SO–CH$_3$, R–SO$_2$–CH$_3$, R–O–SO$_2$–CH$_3$, were identified by TSP-LC-MS after SPE and confirmed by synthesis [281]. All triazines reacted in the same way, reactions with HClO occurring much faster than with ClO$_2$ giving the sulfoxide, sulfone and the sulfone hydrolysis product. Reactions with ClO$_2$ gave only the sulfoxide. A general pathway for the oxidation of these triazines was proposed [282]. TSP-LC-MS was used to confirm three highly polar metabolites of the triazine herbicide hexazinone in SPE-extracts of soil and vegetation, using SIM of the protonated molecular ions [283]. A SFE method using CO$_2$ containing different modifiers was described for the extraction of atrazine and its polar metabolites from sediments, followed by identification and quantification by TSP-LC-MS [284]. Extraction mixtures which contained strong nucleophiles (MeOH; H$_2$O; triethylamine) caused significant or total degradation of atrazine.

Miscellaneous

TSP ionisation was also applied for the analyses of fungicides. After the isolation and trace enrichment of captan, captafol, carbendazim, chlorothalonil, ethirimol, folpet, metalaxyl and vinclozolin on C_{18} Empore® disks from drinking, river and estuarine water samples TSP-LC-MS analysis was performed but only carbendazim, ethirimol and metalaxyl could be analysed by LC-MS [285].

Surface water was analysed for about 300 pesticides and organic compounds [286]. When LC-MS/MS was applied the fungicide dichloran and bacterial transformation products from a chemostat system could be observed. Some more non-common pesticides were examined under application of TSP-LC-MS. So 2 C_{18} and 2 XAD resin cartridges were used to examine the recoveries of thermally labile pesticides from water. Compounds under research were benazolin, bromofenoxim, ethofumesate, fenamiphos and phenmedipham. Recoveries were highest (>85%) with C_{18} with detection limits of 1–10 ng (TSP full scan) and 60–800 pg (SIM) [287]. TSP-LC-MS was applied for determination and identification of the insecticide imidacloprid which was present as residue in vegetables at 0.01–0.60 mg kg concentrations [288]. TSP-LC-MS was used to confirm positive LC-UV results (>0.05 ppm) of the insecticide hydramethylnon from extracts of pasture grass samples [289].

After a comprehensive extraction procedure the triazolopyrimidine herbicide metosulam could be determined in soil samples by LC-UV and TSP-LC-MS with excellent agreement between the methods [290].

15.3.3.2 Atmospheric Pressure Ionization Interfaces (API)

Applications using atmospheric pressure chemical ionization interface (APCI)

In parallel with the routine application of TSP the improvement in the handling of the atmospheric pressure ionisation (API) techniques for coupling LC with MS in environmental analysis proceeded. During the last decade of the last century, the dissemination of this interfacing technique grew tremendously with the up-coming demand for low detection limits in the analysis of polar organic pollutants of concern present in the environment in very low concentration ranges.

Two different types of API interfaces came into routine use for coupling LC and MS, the atmospheric pressure chemical ionisation (APCI) interface and the electrospray (ESI) or ion-spray interface. Both interfaces ionise outside the mass spectrometer at ambient pressure before the ions enter the high-vacuum mass analyser region. To enter into the high vacuum region of the MS the ions generated at ambient pressure have to pass a very small orifice. This process and the different techniques applied from the different commercially available interfaces were extensively described by Slobodnik et al. [22] when they applied this interfacing technique to the analysis of polar pesticides. This ionisation process is supported by nebulisation which is performed pneumatically in the APCI process or, for ESI, by means of a strong electrical field, while ion-spray uses both nebulisation techniques. Ion spray ionisation can be termed a combination of both, pneumatic nebulisation and in parallel nebulisation by means of an electrical field. The formation of the spray and the ionisation of the compounds were described extensively [22].

The main distinctions between both interface types were the flow rates and the molecular weight ranges that could be handled. In the start of API, it was possible to have high flow rates (2 mL min^{-1}) compared to the small flows (20 µL min^{-1}) amenable to ESI interfacing. The application of heat and pneumatic nebulisation for APCI ionisation allowed the handling of the 0.1–2.0 mL LC-eluents of RP separations. With the invention of ion spray the disadvantage of low flow rates combined with ESI disappeared because flow rates could be increased. The most important advantage of ESI, however, is its capacity for handling very polar organics with molecular weights of 100 up to > 100,000 Da, with which this interface also grew to be the work-horse of analytical protein chemists.

The application of MS/MS or in-source-CID for identification of pollutants brought problems, since no product ion libraries were available. Product ion spectra either had to be interpreted for compound identification or standard comparison had to be performed provided that standards were available. To overcome this disadvantage, analysts prepared their own libraries suitable for the instrument generated on or for instruments of the same type. Attempts at generating mass spectral libraries for polar compounds determined by API methods were reported and the results obtained with their application in real environmental samples were discussed [291, 292]. The generation of IT-MS^n product ion spectra and their use for identification provided the most promising results so far. The preparation of a universally applicable product ion library for the identification of polar compounds, i.e. of a library which could work with various equipments for identification, still remains wishful thinking.

A series of surveys and reviews [28, 44, 45, 293] dealt with the simultaneous determination of a broad range of polar compounds in environmental samples by API interfaces. Possibilities and limitations of structure elucidation by LC-ion trap multiple mass spectrometry (LC-ITMS^n) were the topic overview [38]. As shown later, pesticide residue analysis was the most frequent application of LC-MS in water sample analysis, as the number of review articles on the subject of pesticide analysis and their degradation products demonstrates [20, 22, 29, 30, 32, 199, 294]. The analysis of dyes by means of API interfacing techniques was reviewed by three groups [43, 161, 200], while the LC-MS analysis of surfactants, as compounds of environmental concern, was comprehensively reviewed [21].

In Tab. 15.2 the applications of APCI used as interface to couple MS in environmental analytical applications are listed, specified for different groups of compounds.

Drugs and diagnostic agents
The use of APCI for the analysis of drugs or their metabolites in environmental samples is not yet as common as for ESI applications (cf. 15.3.3.2 ESI, drugs). Predominantly aqueous samples, e.g., effluents of STPs or wastewaters from pharmaceutical industry were studied. Surface and groundwater samples were also under

Table 15.2 Environmental applications in LC-MS performed by atmospheric pressure chemical ionization interface (APCI).

Topic	References
Reviews	
General reviews	28, 44, 45, 293
Compound class specific reviews	
– Dyes	43, 161
– Pesticides	20, 22, 29, 30, 32, 199
– Surfactants	21, 24, 39–41
Compound class applications	
– Drugs and diagnostic agents	292, 295, 296
– Dyes	28, 43, 161, 201, 297
– Estrogenic compounds	298–301
– Explosives	205, 291, 302
– Haloacetic acids	303
– Polycyclic aromatic hydrocarbons	28, 207, 304–314, 394
– Phenols	315–320, 323
– Sulfonic acids	297
– Surfactants	226, 326–328, 330–346, 479
– Pesticides (including herbicides, fungicides etc.)	
– Anilides, quaternary amines, toluidines and thiocyanate compounds	175, 260, 320, 323, 347–352, 385
– Carbamates	108, 175, 320, 325, 353–360
– Organophosphorus compounds	175, 320–322, 324, 353, 354, 360–368, 370–372, 385
– Phenoxycarboxylic acids	176, 320, 325, 348, 361
– Phenylureas, thioureas and sulfonylureas	175, 320, 322, 325, 350, 351, 353, 354, 358, 361, 373–385
– Triazines	175, 260, 320, 323–325, 350, 351, 353, 354, 361, 362, 368, 374–377, 382, 385–387, 389, 390
– Phenolic compounds	320–325
– Antifouling pesticides	324, 368, 377, 378
– Miscellaneous	324, 358, 384, 391–394
– Comparison of ESI interfaces with different types of interfaces	6, 20–22, 28, 29, 32, 43, 71, 108, 155, 161, 175, 176, 185, 199, 205, 291, 297–300, 311, 317, 320, 322, 325–327, 334, 340, 346, 347, 351, 353, 355, 357, 358, 361, 373, 374, 381, 385, 389, 390, 394

research, because drugs and their degradation behavior in the environment have attracted the attention of environmental chemists and the public. With the exception of wastewaters from pharmaceutical industry the water samples contained low concentrations of compounds of concern. Pre-concentration was necessary prior to RP-LC. Methods for the determination of drug residues in water samples by means of APCI-LC-MS or -CE-MS were elaborated and applied to river water samples. Pharmaceuticals such as naproxen, bezafibrate, diclofenac, iboprufen or their degradation product clofibric acid were determined (cf. 15.3.3.2 ESI, drugs) [295]. As drugs were observed in municipal wastewaters or in industrial effluents, APCI and ESI-FIA- and LC-MS and MS/MS besides GC-MS were applied to follow polar and nonpolar organic pollutants (e.g. drugs, their precursor compounds from synthesis and biochemical degradation products generated during wastewater treatment process). An extraction and concentration scheme, MS and MS/MS spectra of pharmaceuticals were presented [296]. Baumann et al. [292] also elaborated a library of APCI and ESI product ion mass spectra of steroids, morphine and drugs of abuse based on wideband excitation in an ion trap mass spectrometer. The spectra were applied to compare results with results obtained from urine, forensic and blood samples.

Dyes
The number of applications dealing with the analysis of dyes for APCI-interfaced MS analysis is small despite azo dyes being compounds of environmental concern because of the carcenogenic potential of their degradation products [297]. ESI was used predominantly [161] because an increased tendency of fragmentation under APCI was observed [201] and therefore a more widespread analytical dissemination could be observed by applying ESI for the determination of these very polar compounds (cf. 15.3.3.2 ESI, dyes). An overview covering the extraction and pre-concentration of dyes from water using SPE was presented together with the current LC-MS methodologies which were coupled with APCI, ESI and TSP to the MS for determining sulfonated azo dyes in environmental waters. Capillary zone electrophoresis (CZE) coupled with MS was also discussed [43]. The application of APCI, ESI, PBI or TSP interfaced to LC-MS was discussed when determinations of dyes in environmental matrices were reported [28] and results from the analysis of sulfonic compounds and sulfonated azo dyes using APCI and ESI-LC-MS were reported. Non-volatile TBA ion pairing agents which were removed on-line prior to MS analysis improved separation [297].

Estrogenic compounds
With increasing sensitivity and decreasing detection limits estrogenic compounds, present in very low concentrations in environmental samples, have become an emerging area of concern. For the determination of metabolites of natural and synthetic estrogens LC-MS is the method of choice, whereas for the analyses of the precursors GC-MS without derivatization or after silylation of the analytes should be favoured. The analysis of 17β-estradiol in the aquatic environment was examined extensively using APCI or ESI-LC-MS (cf. 15.3.3.2 ESI, estrogens) [298]. The estrogens estrone, 17β-estradiol, estriol and 17α-ethinylestradiol were analysed in raw and treated wastewaters after C_{18} SPE [299, 300] and the number of compounds analysed was extended to a total of 10 estrogens and progesterons [301]. APCI(+) ionisation and ESI were applied after RP separation but MRM (multiple reaction monitoring) detection was essential to improve selectivity.

Explosives

Contaminated sites from ammunition production during World War II are often heavily polluted by explosives and their metabolites. Both, precursor nitro compounds and metabolite amino compounds are mobile in the ground and may reach drinking and groundwaters. The API interfaces APCI and ESI were used to compare and to confirm TSP-LC-MS(-) results for the identification of pollutants in ammunition hazardous waste sites (cf. 15.3.3.2 ESI, explosives). 31 compounds could be identified in water samples after SPE using LiChrolut® EN. The precursor compound TNT and partially nitrated toluenes, 1,3,5-trinitrobenzene, nitramines and their by-products were found (cf. 15.3.3.1 TSP, explosives) [205]. 1,3,5-trinitro-1,3,5-triazacyclohexane (RDX) and the nitroso-RDX metabolites were determined by APCI-LC-MS in groundwater samples after SPE (Sep-Pak Porapak RDX). The advantage of APCI was to provide a 20-fold greater signal for nitroaromatics than ESI, whilst ESI produces a 5-fold increase in response for nitramines [302]. As the determination and identification of pollutants emitted from ammunition production and disposal became more and more important a list of compounds examined and available as EI-MS or API-MS (ESI and APCI) spectra, e.g. explosives and pesticide residues, in environmental samples was published by Schreiber et al. [291]. The limits of the application of these libraries for identification were elucidated and discussed.

Haloacetic acids

In the disinfection process of drinking water, besides other chlorinated compounds, halogenated carboxylic acids arise. For quality control of drinking water the determination of these contaminants was established using LC-MS with a preference for ESI (cf. 15.3.3.2 ESI, haloacetic acids). An application of APCI-MS for haloacetic acid analysis was reported after CE separation in a non-aqueous medium [303].

Polycyclic aromatic hydrocarbons (PAH)

PAHs, contaminants of public concern because of the carcinogenic potential of benzo[a]pyrene and some of their homologues preferentially, were determined by LC with fluorescence detection. The chemical or biochemical degradation products, however, being more polar and hardly detectable because of missing knowledge about their characteristic, highly specific excitation and emission wavelengths were determined by APCI-LC-MS and MS/MS.

Basic research work was performed to establish a method for the determination of PAHs and their hydroxylated degradation products, standard mixtures of hydroxy PAHs were determined qualitatively and quantitatively by APCI and ESI-LC-MS. Under S/N = 3:1 conditions detection limits of 0.3–50 µg mL^{-1} were obtained in APCI mode whereas ESI ionisation was less sensitive (cf. 15.3.3.2 ESI, PAH) [304]. The whole spectrum of PAHs determined according to EPA protocol were analysed in solid waste by APCI-LC-MS(+) and HPLC with fluorescence detection after liquid–liquid extraction, however, determination was disturbed by methylated PAH [305]. A capillary column SFC was interfaced by APCI to an

MS detector and PAH analysis was performed with a heated pneumatic nebulizer. CO_2 was used as the mobile phase to determine benz[a]anthracene, constituents of a pond sediment contaminated with coke oven residues and of a standard mixture of PAHs. Detection limits observed were 40 pg for the $[M+H]^+$ ion of chrysene (S/N=2) [306]. APCI-LC-MS/MS was performed to identify and to differentiate isomeric PAHs in coal tar [28, 307]. Fragment ion spectra were presented and peak area ratios were found to be a reliable indicator for isomeric differentiation [307]. In a short overview the state of the art of various mass spectrometric techniques applied to environmental analysis of PAHs in natural matrices was given and the application of different types of interfaces, APCI, ESI, PBI and TSP, was compared and described [28]. Thirty one oxidized PAH derivatives containing up to five condensed aromatic rings and carrying different functional groups (e.g. carboxyl, dicarboxylic anhydride, lactone, hydroxyl and carbonyl) were characterised by APCI-LC-MS/MS in the positive and negative mode applying in-source CID with four different fragmentor voltages. APCI-LC-MS allowed the investigation of PAHs at trace levels [308]. After ozone treatment of 29 PAHs the compounds and their degradation products, benzo[a]pyrene-4,5-dione and 4-oxa-benzo[d,e,f]-chrysen-5-one were characterised from retention by comparison to reference standards and APCI-LC-MS data [309]. APCI-LC-MS and GC-MS were applied to determine and to identify polar products from the ozonolysis of benzo[a]pyrene. Atmospheric degradation by ozone was simulated and BAP ozonolysis products determined as mainly quinones and carboxylic acids resulting from oxidative ring opening reactions [310]. Benzo[a]pyrene and its degradation products (metabolites or physicochemical) in air were studied [207] by APCI- and TSP-LC-MS. APCI ionisation yielded mostly $[M]^-$ and $[M-H]^-$ ions (TSP: $[M]^-$, $[M-H_2O]^-$ or $[M+CH_3COO]^-$). The 1,6-, 3,6- and 6,12-dione derivatives from oxidative BAP degradation were detected adsorbed on air particulates [311]. APCI-LC-MS was also applied to characterise PAHs with high-molecular masses (> 300 Da) in air filters and in biological material (zebra mussels) after extraction with dichloromethane and MeOH. Mass spectra and ion chromatograms showed similar profiles for the different samples. Results suggest that these PAHs contained in coal tar and air particulates were accumulated in tissues of zebra mussel [312]. The determination of PAHs by ELISA kits for the determination of PAHs among the organic analytes in various industrial effluents were examined and confirmed by APCI-LC-MS. Unequivocal identification of ELISA positive target analytes was obtained. The advantages and limitations of the three RAPID-magnetic particle-based ELISA kits were reported [313]. Heteroatoms (N, S, O, P) substituted were analysed in samples of a pond sediment contaminated by coke-oven residues by APCI-LC-MS/MS. The S-containing heterocyclic compounds were identified by SRM and quantified [314].

Phenols

Many phenolic compounds were applied as pesticides because of their toxicity, especially against insects. For the follow-up of these compounds in waters GC-MS was the method of choice, but some LC-MS methods were also established for analysis [315–319]. Molecular ions were observed under negative APCI or ESI ionisation after RP-LC and SPE. APCI was found to be more effective than ESI [317]. The endocrine disruptor compounds, the Bisphenol A, nonyl- and octylphenols, as industrial chemicals with high production rates or as metabolites of non-ionic surfactants had also been analysed by LC-MS, however, detection limits were poor compared to GC-MS. In a study with automated on-line SPE-APCI-MS was observed to be more sensitive for nitrophenolic pesticides (e.g. dinoseb, 4-nitrophenol and dinoterb) than the application of PBI. Limits of detection were reported [320]. The degradation of pentachlorophenol (PCP) in natural waters was studied by LC-DAD and confirmed by APCI-LC-MS both after Lichrolut EN SPE. A half-life of PCP in ground water, in estuarine and river waters of < 2 h was reported [321]. A test mixture of 17 pesticides also containing the nitrophenol pesticide trifluralin was used to develop a quantitative on-line SPE-LC-MS and MS/MS method using APCI or ESI interfaces [322]. While trifluralin was not detected by ESI, detection limits (S/N=3) with APCI were observed with 3.0 or 0.2 µg L^{-1} in full scan or SIM mode, respectively. In a monitoring study of priority pesticides and other organic pollutants in river waters from Portugal using GC-MS and APCI-LC-MS [323, 324] besides other pesticides the phenolic compounds 2,4,6-trichlorophenol, pentachlorophenol, 2,4-dichlorophenol and the 2-, 3- or 4-mono chlorophenols were determined in the negative ionisation mode [324]. Baglio et al. [325], however, observed that APCI was not as effective in the ionisation of 2-methyl-4,6-dinitrophenol as pneumatically assisted ESI (ion spray) when they studied standard solutions of the pesticide mixtures.

Surfactants

From the application of surfactants in household, handicraft and industry a large quantity of these compounds have been discharged with wastewaters and despite biological treatment considerable quantities of them have reached the environment. Therefore surfactants continue to be an environmental concern. Knowledge of the endocrine disrupter potential of some surfactant metabolites had heightened public interest about the fate of these pollutants.

Only the application of LC-MS analysis can be conceived of as reliable surfactant analysis. To elaborate determination methods for the analysis of the anionic surfactant mixture of alkyl ethoxysulfates (AES) APCI and ESI-MS(+/-) studies combined with in-source MS/MS examinations were performed and results were compared (cf. Fig. 15.4 and 15.5 and 15.3.3.2 ESI, surfactants). APCI fragment ion spectra revealed the alkyl chain length and the number of ethoxylate moieties therefore APCI was found to be the method of choice [326]. To confirm determination methods applied in surfactant analyses an inter-laboratory comparison study of LC-MS techniques and enzyme-linked immunosorbent assay for the determination of surfactants in wastewaters was performed in seven laboratories. The quantitative determination of the non-ionic NPEOs, AEOs, coconut diethanol amides (CDEAs) and the anionic LAS, NPEO-sulfates and the secondary alkane sulfonates (SAS) was performed under APCI or ESI-interfacing conditions in positive or negative mode [327] using LC and/or FIA-MS or MS/MS. The stability of SPE concentrated analytes spiked into wastewater effluents and groundwater samples was studied using the non-ionic surfactant mixtures of NPEOs, AEOs and CDEAs and the anionic mixture of LAS. C_{18} SPE cartridges were stored at 4 °C or –20 °C, water samples were kept cold (4 °C) before the compounds were analysed. Stability in a water matrix, even when preservation agents had been used, was very poor, samples preconcentrated on SPE cartridges immediately after sampling were stable for up to 1 month at 4 °C, and for longer at –20 °C (cf. 15.3.3.2 ESI, surfactants) [328].

Qualitative determination methods for so-called gemini surfactants (2 anionic and 1 cationic), a new type of compound with highly improved surface activity, were reported. Results of APCI or ESI-FIA-MS or LC-MS and MS/MS with mass and fragment ion spectra were presented [329] and fragmentation pathways were proposed. Synthetically produced surfactant mixtures may contain undesirable by-products, hardly degradable in the environment and relevant in the drinking water works. Billian et al. described the detection and identification of the synthetic by-product n-decyl-isomaltoside contained in a technical surfactant mixture of APGs. APCI-LC-MS and MS/MS in negative mode in parallel with the use of NMR spectroscopy allowed identification [330].

The analyses of environmental samples confirmed the ubiquitous presence of surfactants in surface and sea water as a result of the surfactants discharged with STP effluents. Analysis of River Elbe (Germany) water samples by GC-MS and APCI-LC-MS and MS/MS confirmed qualitatively the presence of nonpolar and polar organic pollutants of AEO, NPEO, CDEA and aromatic sulfonic acid type, respectively [226]. After C_{18} and/or SAX SPE anionic and non-ionic surfactants were qualitatively and quantitatively analysed in surface water samples by APCI-LC-MS in the negative or positive mode, respectively. Alkylphenol ethoxylates (APEOs) could be confirmed in river water at levels of 5.6 µg L^{-1} [331].

A concentration and detection method for the non-ionic surfactants of APEO type dissolved in river water samples using XAD-16 and APCI-LC-MS and MS/MS, respectively, was developed. Octylphenol ethoxylates (OPEOs) oligomers containing 5–14 ethoxylate groups were confirmed in the Meguro river (Japan) and treated wastewater from a discharge canal [332]. NPEOs and LAS were observed and quantified in sea water and sediment samples from the German Bight of the North Sea or Waddensea marinas and estuaries applying APCI or ESI-FIA and LC-MS and MS/MS in the positive or negative mode, respectively [333].

AEOs spiked into raw wastewaters were applied to elaborate an APCI or ESI-LC-MS method to determine non-ionic surfactants after SPE. Ionisation efficiencies of both interface types were compared and the more effective APCI technique then was applied for quantification [334]. Recoveries observed with standard determination methods for surfactants and MS detection techniques for different types of surfactants (e.g. alkylether carboxylates, sulfosuccinates, fatty acid polyglycol amines, quaternary carboxoalkyl ammonium compounds, modified AEOs, EO/PO compounds, APGs, alkyl polyglucamides, betaine and sulfobetaine) in spiked wastewater samples were compared by applying APCI and/or ESI(+/-).Poor recoveries were obtained by standard methods but good results by MS [335]. APCI and

ESI-FIA-MS/MS and LC-MS/MS analyses were performed to characterise polar pollutants in the wastewaters of Thessakloniki (Greece) STP after C_{18} or LiChrolut EN-SPE. LC-MS(+), MS/MS spectra of sequentially eluted non-ionics as dominaing pollutants in inflow and effluent, (AEOs and NPEOs) and their degradation products (PEGs) and their chromatograms were presented [336]. In municipal wastewaters organic pollutants were analysed by APCI and ESI in FIA and LC-MS(+/-) mode confirming the predominance of surfactants [337, 338]. Results of quantification were reported. The application of FIA-MS screening vs. time consuming LC-MS techniques was discussed. Complex wastewater samples were analysed by Preiss et al. [339] by means of APCI and ESI-MS coupled on-line with ^{1}H-NMR. The identification of compounds was reported and advantages and disadvantages of the techniques used were discussed. A mixture of homologs of NPEOs in the inflow and effluent of a STP were analysed by a novel rapid screening method, combined precursor ion scanning and multiple reaction monitoring, using APCI MS/MS and ESI MS/MS (cf. 15.3.3.2 ESI, surfactants). NPEO concentrations as low as 50 ppt could be detected [340]. Industrial wastewaters (e.g. petrochemical, textile, pulp mill and ammunition plant effluents) were monitored by GC-MS and APCI-LC-MS. Polyethoxylated aliphatic and aromatic non-ionic surfactant compounds and their degradation products were observed by APCI-MS and characteristic ions were listed [341]. The biodegradation of AEOs treated by wastewater biocoenoses of different STPs resulted in multiple different biodegradation pathways monitored and elucidated by APCI-FIA-MS and MS/MS. NPEOs and NPEO-sulfates were degraded aerobically and anaerobically and the metabolites were determined and identification was performed by triple quad and ion trap MS/MS [342]. To improve the elimination efficiencies in a conventional and 3 bio-membrane, assisted wastewater treatment plants operated in parallel were studied by APCI- and ESI-FIA and LC-MS(+/-) and MS/MS, respectively. Qualitative, semi-quantitative and quantitative results on AEOs, NPEOs and LASs were reported. Diagnostic scans were applied for confirmation [343]. The elimination for surfactants and drugs spiked into a conventional and into a biomembrane-assisted STP was monitored by APCI-FIA and LC-MS and MS/MS. NPEOs and OPEOs were observed and followed by a semiquantitative pattern recognition approach [344].

Halogenated APEOs already observed in a FAB study by Rivera et al. [154] in drinking water samples were quantitatively analysed and identified in drinking, surface and wastewater as well as in river sediments and STP sludges by APCI-LC-MS(+/-) (cf. 15.3.3.2 ESI, surfactants) under optimized conditions [345]. Despite the polarity and therefore improved solubility of surfactants in water the content of surfactants in STP sludges cannot be neglected. Methanol/dichloromethane extraction by means of ultrasonification was used to extract anionic and nonionic surfactants and their degradation products. APCI or ESI-LC-MS was applied but for determination of the less polar AEO, NPEO and CDEA APCI-LC-MS(+) was used [346].

Pesticides (including herbicides, fungicides etc.)
Of all the organic pollutants measured in environmental matrices, pesticides still continue to be the most studied. They are the focus of drinking water, surface and groundwater. With every new generation of pesticides coming on the market, compounds become more polar and, in parallel, less persistent in the environment and during separation and detection in the laboratory. LC-MS and MS/MS as method of choice for the analysis of polars provided molecular weight and structural information not always sufficient for identification. The new trend toward the application of an ion trap with the possibility of MS^n offer significant improvement over "conventional" MS/MS.

Anilides, quaternary amines, toluidines and thiocyanate compounds.
APCI- and ESI-LC-MS in the ion-pair LC-mode was used to characterised the quaternary ammonium herbicides diquat, paraquat, difenzoquat, mepiquat and chlormequat in spiked tap water samples after SPE (Sep-Pak silica). Detection limits down to 0.1–4 µg L^{-1} were obtained and reproducibility results were reported [347].

A simultaneous MS method was developed for the determination of acidic and neutral rice herbicides and their degradation products in trace quantities in estuarine waters (Ebro delta). Positive APCI was applied for the determination of molinate and 3,4-dichloroaniline, the major degradation products of propanil whereas negative ionisation was applied to acid compounds. 8-Hydroxybentazone and 4-chloro-2-methylphenoxyacetic acid were successively observed in the samples [348]. After LC-UV-DAD APCI-LC-MS in the more sensitive negative mode was used to confirm propanil and its major degradation product, 3,4-dichloroaniline, in surface water and soil samples after pesticide application to dry rice fields. Propanil was found to be rapidly degradable to 3,4-dichloroaniline [349]. APCI-LC-MS was used to analyse surface water samples for organonitrogen pesticides e.g. metolachlor and triazines. Detection limits after dichloromethane extraction or after SPE (Carbopack B), were < 3 ng L^{-1} applying APCI-LC-MS and < 4 ng L^{-1} if GC/NPD was used [260].

An APCI-LC-MS method was developed and described for the quantitative determination of the anilide pesticides alachlor and metazachlor in ground water samples. After optimization of instrumental conditions detection limits of 0.001–0.005 µg L^{-1} (50–300 pg injected) could be obtained. Recovery, precision and linearity data were reported [350]. The pesticide bentazone was one of the most frequently found compounds determined by APCI-LC-MS in shallow groundwater samples from twosandy and two clay catchment areas [351].

Stability studies of SPE-adsorbed anilides and N-substituted amines (bentazone, molinate and metolachlor) were performed by means of APCI-LC-MS. From river water samples containing the pesticides and their degradation products, the pesticides had been extracted prior to APCI-LC-MS [320]. The SPE adsorbed compounds were stored on SPE cartridges (styrene-divinylbenzene) for up to 3 months at ambient temperature, +4 °C and –20 °C. After 3 month storage at –20 °C on the polymeric cartridges recoveries were > 90% [320].

When metolachlor was submitted to a degradation process by *Cunninghamella elegans* [352] APCI-LC-MS was used to track the metabolic process over a period of 96 h. After incubation had stopped six metabolites were concentrated and separated prior to identification. Predominantly an O-demethylation of the N-alkyl side chain of metolachlor and benzylic hydroxylation of the arylalkyl side chain was observed [352].

Carbamates.
For trace analysis of carbamate pesticides and other pesticide compounds Hogenboom et al. [353] developed a substance specific APCI-LC-MS/MS method using short columns. Detection limits of 0.03–5 µg L^{-1} in full-scan and 2–750 ng L^{-1} in SIM mode were reported. The pesticides could be successfully identified from a search against a pesticide MS/MS library. In a comprehensive TSP study with a series of N- and P-containing pesticides (amines, anilides, carbamates, phosphonates, phenylureas, sulfonylureas and triazines) APCI spectra and CID spectra were obtained and APCI was useful for solving fragmentation mechanisms observed in TSP mass spectra [175]. Doerge et al. [354] examined different classes of pesticides (triazines, carbamates, phenylureas, organophosphates) quantitatively by APCI-MS. With this method they observed that sensitivity was less affected by differences in analyte structure than using TSP or PB-LC-MS. In the analysis of N-methylcarbamate pesticides (methomyl, aldicarb, aldicarb sulfoxide, aldicarb sulfone, carbaryl, methiocarb, carbofuran, and 3-hydroxycarbofuran) [108] all the API interfaces performed better than the TSP and PBI [320] interfaces. Comparing APCI with ESI (ion spray) applications, APCI resulted in a better sensitivity, linearity and covered more carbamates. Molecular weight information and abundant fragment ions were provided while ion spray gave comparable performance but mainly protonated molecular ions and less fragmentation. Comparable results were obtained by Fernández et al. with carbamate residues in fruits and vegetables. Detection limits observed were 10 to 100 times lower than EU maximum residue levels (cf. 15.3.3.2 ESI, carbamates) [355]. An automated on-line extraction and preconcentration based on high-performance immunoaffinity chromatography combined with RP-HPLC and APCI-MS detection were described for the determination of the fungicide carbendazim. For a quick determination an ELISA method was applied with the result that both analytical methods, APCI-LC-MS and ELISA, correlated well [356].

River water samples were under research by APCI and ESI-LC-MS to analyse seven N-methylcarbamate pesticides quantitatively. The effects of varying APCI and ESI conditions were investigated, confirming that APCI resulted in less effective sensitivity [357].

The contamination of fruits and vegetables with pesticides became a problem with the increased application of pesticides because of an intensified agriculture. So the comparison of APCI and ESI-LC-MS for the determination of 10 pesticides of carbamate type (pirimicarb, carbofuran, 3-hydroxycarbofuran, aldicarb, and its metabolites, the sulfoxide and the sulfone), besides others in fruits, met the need and proved that the application of APCI was more efficient than ESI. A positive-negative mode switch in the MS run allowed the determination of eight pesticides in matrix-matched standards [358]. APCI-LC-MS was applied to validate an ELISA method using anti-rabbit immunoglobulin G antibodies for the determination of traces of carbaryl in vegetable and fruit extracts. An illustrative mass chromatogram presenting the degradation of carbaryl to 1-naphthol [359] confirmed that APCI interfacing allowed the detection of pesticides at very low levels. The increased application of pesticides in agriculture led to an accumulation of these pollutants in the food chain with the result that pesticides were observed in body fluids. Itoh et al. [360] elaborated a method for the determination of carbamates by APCI in biological fluids (e.g. serum or urine). They described this method as a very rapid method which requires only an extremely simple pretreatment process. While ESI-LC-MS was used to analyse acidic polar pesticides in water APCI-LC-MS was applied to determine carbamates. Under these conditions the specific fragment ions $[M+H-CONCH_3]^+$ [325] could be observed.

Organophosphorus compounds.
For trace analysis of organophosphorus pesticides Hogenboom et al. [353] developed a substance specific APCI-LC-MS/MS method using short columns. An on-line SPE-APCI-LC-MS(+/-) and MS/MS method was described [322] for the analysis of a test mixture of 17 pesticides containing the organophosphorus compounds dimethoate, fenamiphos, coumaphos, fenchlorphos, chlorpyriphos and bromophos-ethyl at low ng L^{-1} levels applying samples of 100 mL for concentration. Under CID conditions APCI and ESI-MS/MS resulted in similar product-ion spectra obtained from protonated molecular ions. A pesticide MS/MS library could be applied successfully for identification [322, 353]. Automated on-line SPE followed by LC-MS interfaced by APCI or PBI was applied to fenitrothion, malathion, parathion-ethyl and vamidothion. The study demonstrated the higher sensitivity of APCI compared to PBI [320, 361]. A comprehensive study by Volmer et al. of pesticides by applying different types of interfaces (TSP, ESI and APCI) confirmed that the application of APCI was very useful for solving fragmentation mechanisms observed in TSP-MS spectra obtained from the organophosphorus pesticides butonate, dichlorovos and trichlorfon [175]. A modified method was described for quantitative monitoring of various phosphorus pesticides (parathion-methyl, fenitrothion, diazinon and chlorpyrifos) in river water [362], (coumaphos, azinophos-ethyl, azinophos-methyl, triazophos, parathion, fenthion malathion fenitrothion, parathion-methyl, disulfoton, dimethoate, ometoate, mevinphos, trichlorfon) and in ground water [324]. For (E)- and (Z)-mevinphos, dichlorvos, azinphos-methyl, azinphos-ethyl, parathion-methyl, parathion-ethyl, malathion, fenitrothion, chlorfenvinphos, fenthion and diazinon [363] fully automated on-line SPE and LC-UV-DAD or APCI-LC-MS(+/-) were performed. The effects of temperature and extraction voltage on the mass spectra obtained in an APCI-LC-MS examination analyzing the organophosphorus pesticides acephate, azinphos-ethyl, fenitrothion, fensulfothion, fenthion, metamidofos, paraoxon-methyl, parathion-methyl, trichlor-

fon, vamidothion and vamidothion sulfoxide in groundwater samples were studied. Calibration and detection limit data were provided but recoveries were poor [364]. A robust method was developed for the determination of 56 different insecticides and fungicides of predominantly organophosphorus type in groundwater samples for a pilot survey study in Almeria (Spain). Pesticides were analysed using different GC techniques besides APCI-LC-MS. A subsequent interlaboratory study using all methodologies showed good agreement between the techniques [365]. The analyses of fenamiphos and diazinon as potential ground water contaminants by APCI-MS proved that under APCI ionisation sensitivity influenced by differences in analyte structure was less affected than observed with TSP or PBI-LC-MS [354].

Fenthion, temephos and their degradation products were determined by HPLC-DAD and APCI-LC-MS in rice field water samples after SPE (Empore C_{18}). Four fenthion transformation products could be identified in the positive or negative modes whereas the degradation of temephos resulted in six products. For both pesticides the oxo-analogue products predominate [366]. After rice fields had been treated with fenitrothion, crop waters were sampled and analysed by ELISA, LC-DAD and MS coupled by APCI interface. Fenitrothion, fenitrooxon, 3-methyl-4-nitrophenol and the s-Me isomer of fenitrothion were identified by LC-MS [367].

Biodegradation of fenitrothion, ethyl-parathion, methyl-parathion in natural waters was observed and examined by APCI-LC-MS. Various degradation products could be identified unequivocally. All transformation products were observed to be more stable than the parent compound except in ground water [321]. Physicochemical and biochemical degradation was observed after SPE adsorption if storage conditions were not optimised. First attempts to study the stability of different pesticides of organophosphorus type and concentrated on the SPE materials Hysphere-1, IST Envirolut and LiChrolut were reported. APCI-LC-MS analysis was used for confirmation. Complete recoveries were obtained after storage at $-20\,°C$ for 1 month for water spiked at 10 µg L^{-1}. Degradation occurred after storage temperatures of $4\,°C$ and at room temperature [368]. The results were compared to stabilities in acidified and non-acidified ground water [369]. A monitoring procedure for 21 organophosphorus pesticides in biological fluids using APCI-LC-MS [370] was elaborated. Propaphos, isoxathion, iprofenfos, malathion, fenitrothion and chlorpyrifos were determined by APCI in serum or urine. Itoh et al. [360] described a method amenable to these matrices which required only an extremely simple pretreatment process despite complex matrices.

The organophosphorus herbicide butamifos and its 4-nitrophenyl derivative were submitted to photodegradation. Products were analysed by APCI-LC-MS. MS peaks obtained from the precursors and their degradation products were tabulated and a mechanism of degradation was proposed [371]. The ozone degradation pathway of pirimiphos methyl in industrial water was examined [372] by ion-trap EI or PICI in GC-MS, or in APCI and ESI-LC-MS mode. APCI and ESP mass chromatograms of the pirimiphos methyl oxidation products and fragmentation data were reported and an ozonolysis degradation pathway was proposed (cf. 15.3.3.2 ESI, organophosphones compound).

Phenoxy carboxylic acids. The application of APCI in the determination of the polar acidic phenoxycarboxylic acids led to unsatisfactory results. Nevertheless this method was applied by Santos et al. [348] for the determination of the 2,4-D, MCPA and MCPP herbicides and their degradation products in trace quantities in estuarine waters of drainage of the Ebro delta. Response for the very polar acidic compounds preferentially ionised by ESI in the negative mode with excellent sensitivity [176, 325] in the APCI mode, however, was reduced (cf. 15.3.3.2 ESI, organophosphones compounds) [348]. Results obtained by APCI or PBI, however, proved an increased sensitivity of APCI compared to PBI [320].

Phenylureas, thioureas and sulfonylureas.
The capabilities of a modern API source as a universal LC-MS coupling tool were checked by performing pesticide analysis of the polar urea pesticides metoxuron, monuron, monolinuron, chlortoluron, metobromuron, metabenzthiazuron, isoproturon, diuron, linuron in the low pmol range (limits: 10–100 pg) [373].

An on-line SPE-APCI-LC-MS and MS/MS method was described by Slobodnik et al. [322] for the analysis of a test mixture of 17 pesticides containing monoron, diuron and neburon at low ng L^{-1} levels applying samples of 100 mL for concentration.

APCI-LC-MS [325] was applied to determine phenylurea herbicides in water samples with detection limits all in the low pg range, whereas ESI-LC-MS was used to analyse the acidic polar pesticides. Under APCI conditions the specific fragment ion of phenylureas m/z 72 corresponding to $[O=C=N^+(CH_3)_2]$ could be observed [325]. Fragmentation pathways were presented. In a TSP-LC-MS study covering 15 phenylurea and thiourea pesticides APCI-LC-MS/MS was used to elucidate fragmentation behavior observed under TSP conditions [175].

The urea pesticides diuron, fluormeturon, neburon and linuron cited as potential groundwater contaminants from US EPA in the National Pesticide Survey were quantitatively determined by APCI-LC-MS(+) [354]. APCI-LC-MS was also used to test for 46 pesticide compounds in shallow groundwater samples from two sandy and two clay catchment areas. Of the neutral polars observed, isoproturon belonged to the most frequently found compound [351]. Spliid et al. described an APCI-LC-MS method for the determination of isoproturon and different types of pesticides and their degradation products in ground water samples. Detection limits, recovery, precision and linearity data were reported [350].

Surface water samples from Southeastern regions of France and from the St. Lawrence River in Canada were monitored by APCI-LC-MS and MS/MS in the positive mode. Diuron and isoproturon were confirmed by MS/MS. Results obtained by LC-MS and ion trap LC-MS/MS were found to be comparable [374]. A substance-specific robust APCI-LC-MS/MS method using short columns for trace analysis of phenylureas was elaborated and validated. Despite the low quantity of sample applied (15 mL) detection limits of < 5 µg L^{-1} in full-scan and < 750 ng L^{-1} in SIM mode, respectively, could be achieved. Product ion spectra were obtained from $[M+H]^+$ parent ions and by means of a pesticide MS/MS library the majority of the pesticides were successfully identified [353]. When natural waters [375, 376] were analysed by a rapid target analytical technique for quantitative determination of microcontaminants in water by applying on-line single-short-column separation coupled with ion trap APCI-LC-MS and -MS/MS phenylurea herbicides and triazine derivatives were under research. Results obtained from concentration of 4 mL water samples containing eight phenylureas were poor compared to results obtained from triazine determination [376]. Isoproturon and fluormeturon were monitored in environmental waters. Besides these urea pesticides a series of organo-phosphorus, triazine, chlorophenoxy acid, phenolic and thiocarbamates pesticides were also analysed from 200 mL water samples by on-line SPE-APCI-LC-MS and PBI-MS [361]. The analytical capabilities of the different ionisation techniques with the analyses of urea pesticides were discussed and the behavior under different ionisation conditions was documented [320].

Four selected antifouling booster biocides, to which diuron belongs, were determined by APCI-LC-MS after enrichment by C_{18}-SPE from marine waters. Detection limits of $0.01-0.18$ µg L^{-1} [377] or < 5 ng L^{-1} were obtained with recoveries $> 91\%$ [378]. The other antifouling biocides also quantified were Kathon 5287, TCMTB and TCMS pyridine. To validate pesticide analysis methods phenylurea pesticides were determined quantitatively by APCI-LC-MS in an interlaboratory testing performed with aqueous real-life samples (drinking water, surface water and groundwater) containing pesticides ranging from $0.02-0.8$ µg L^{-1} [379]. A coupled column system (LC-LC) was used and detection was performed either with UV or by APCI-MS which allowed direct quantification down to 0.01 µg L^{-1} and in parallel confirmation of the results.

Yarita et al. developed an APCI-LC-MS(+/-) method to analyse diuron and linuron and their respective degradation products, 1-(3,4-dichlorophenyl)urea and 1-(3,4-dichlorophenyl)-3-methylurea before the method was then applied to follow biodegradation of diuron in sewage sludge over a period of 28 d [380].

The determination of sulfonylurea degradation products of chlorsulfuron, metsulfuron-methyl, thifensulfuron-methyl and tribenuron-methyl in soil by LC-UV detection was studied with standards and real environmental samples. Compounds observed were confirmed by APCI- and ESI-LC-MS/MS. SRM was applied to identify five degradation products by LC-MS/MS(+/-). Calibration graphs were linear from $0.05-1$ µg mL^{-1} with detection limits of $10-50$ µg kg^{-1} [381]. Attempts were made to couple an immunoaffinity concentration step with APCI-LC-MS determination for the analysis of phenylureas and triazines in water and sediment samples. Calibration graphs were linear at $0.01-0.2$ µg L^{-1} groundwater, with typical detection limits of $1-5$ ng L^{-1} [382]. An anti-isoproturon immunosorbent immobilized on 500 mg silica was used to concentrate and to purify phenylurea herbicides (chlortoluron, isoproturon, diuron, linuron and diflubenzuron) from aqueous samples prior to quantitative LC-UV-DAD determination and APCI-LC-MS confirmation. With the application of 50 mL of water detection limits < 1 µg L^{-1} in SIM mode were obtained [383]. The comparison of APCI and ESI-LC-MS for the determination of the fluorine-containing phenylurea pesticide diflubenzuron and 10 pesticides of carbamate type in fruits [358] was reported. This compound was also determined in plums, strawberries and blackcurrant-based fruit drinks [384] with recoveries of 76%.

Phenolic compounds. In Portuguese rivers tri-, di and mono-chlorophenols were determined by APCI-LC-MS in negative mode in parallel with other pesticides. SPE was performed with Oasis cartridges [323].

Triazines. The optimization of sensitivity in the quantitative determination of polar pesticides was the purpose of comparing the atmospheric pressure interfaces APCI and ESI in the field of pesticide analysis. Optimized detection limits were observed to be dependent on eluent flow-rates used in the analysis. Triazines were less sensitive to variations in flow-rates under both ionisation techniques (cf. 15.3.3.2 ESI, triazines) [385]. APCI was used for solving fragmentation mechanisms observed by TSP-LC-MS/MS of triazines and other pesticides [175]. APCI-LC-MS was applied to determine triazines whereas ESI-LC-MS was used to analyse the acidic and more polar pesticides in water. MS^n results from triazine derivatives resulting in a cleavage of lateral chains followed by ring opening were presented [325]. To improve specificity and selectivity a substance-specific APCI-LC-MS/MS method for trace analysis of the triazine derivatives propazine and ter-butylazine was developed. Product ion spectra obtained by CID from $[M+H]^+$ ions from the majority of the pesticides could be successfully identified from a search against a pesticide MS/MS library compiled in-house [353]. APCI-LC-MS methods were elaborated and described for the determination of triazine pesticides and their degradation products in ground water [350, 354]. Detection limits were obtained with 0.001–0.005 µg L^{-1}, corresponding to 50–300 pg injected. Recovery, precision and linearity data were reported [350]. Natural waters were analysed for a series of triazine derivatives by ion trap APCI-LC-MS and MS/MS. A rapid target analytical technique for quantitative determination of microcontaminants in water by on-line single-short-column separation was developed and applied and detection limits of 0.1–1 µg L^{-1} could be obtained [376]. Typical MRM chromatograms were reported when herbicides were examined applying the same technique [386]. In Portuguese rivers priority pesticides of triazine type besides phenols were determined by APCI-LC-MS. SPE was performed with Oasis cartridges. "Hot spots" were located [323]. The results obtained by APCI-LC-MS and GC-NPD were compared for the quantification of seven different triazines and their degradation products in surface water samples. Detection limits after dichloromethane extraction or after SPE (Carbopack B), were 0.6–3 ng L^{-1} applying APCI-LC-MS and 0.4–4 ng L^{-1} by GC/NPD, respectively, with in parallel good recoveries [260].

The problems of stability of SPE concentrated pesticides were studied also with triazines as targets [320, 368, 369]. Polymeric SPE materials Hysphere-1, IST Envirolut and LiChrolut were examined using APCI-LC-MS analysis for confirmation. Complete recoveries could be confirmed by APCI-LC-MS analysis after storage at −20 °C for 1 month for water samples spiked at 10 µg L^{-1} [368, 369]. APCI-LC-MS was used to confirm LC-UV-DAD monitoring results of pesticides in water from the Ebro delta. In parallel stability of pesticides which had been adsorbed and stored on SPE cartridges (styrene-divinylbenzene) for up to 3 months was judged [320]. To handle large sample quantities with the increasing number of analyses automated SPE off and on-line coupled to LC systems came into use. Triazines besides other pesticides were concentrated by automated on-line SPE prior to determination by APCI or PBI-LC-MS [320]. Fully automated on-line SPE combined with LC and coupled with UV-DAD or APCI-MS was used in pesticide analysis. A modified SAMOS method was developed and described for quantitative monitoring of the triazine pesticides desethylatrazine, atrazine, terbuthylazine, simazine and propazine besides phosphorus pesticides in river waters [362]. To establish robust determination methods, the triazine derivatives atrazine, simazine and their degradation products desethylatrazin, hydroxyatrazine and hydroxysimazine were spiked into tap and surface water samples. These samples were used to confirm suitability for FIA coupled with APCI to fulfil criteria recommended in the Netherlands for GC-MS analysis of pesticides. The APCI-FIA-MS/MS results obtained were quite promising because the polar desethylatrazine could be confirmed by FIA-MS but was not observed by GC-MS [387]. Samples from 95 Missouri streams and 46 Midwestern state streams were analysed for atrazine and its biodegradation products using APCI-LC-MS after cation exchange and SPE [388]. Several polar degradation products were quantified and confirmed by APCI LC-MS/MS. Maximum concentrations for the streams were reported, proving that ca. 60% of the atrazine load was atrazine metabolites. Detection limits were 0.04–0.1 µg L^{-1} [388]. In Europe and Canada these compounds were observed in environmental water samples when 48 target pesticides belonging to 8 different classes along with their degradation products were monitored in surface waters. API methods were applied for analysis of samples from Southeastern regions of France and from the St. Lawrence River in Canada. Triazine derivatives, as all the other neutral compounds, were determined by APCI-LC-MS in the positive mode, whereas ESI in negative mode was applied to the acidic compounds and sulfonylureas [374]. Priority pesticides (e.g. atrazine, simazine, terbutylazine, Irgarol, deisopropylatrazine and deethylatrazine and other types of pesticides) and other organic pollutants were monitored by APCI-LC-MS in 43 river water samples from Portugal and concentrations were reported [324]. For confirmation of results obtained from analyses of triazines and phenylureas in water and sediment samples operated in coupling with an immunoaffinity column, LC-MS-APCI was used [382].

Since triazines are very mobile in the aquatic environment, the presence of these pesticide types in groundwater samples could be confirmed. Pollution was observed when neutral pesticides contained in shallow groundwater samples from two sandy and two clay catchment areas were analysed in APCI mode [351]. The most frequently (> 300 samples) determined compound was atrazine with its degradation products.

For the biodegradation of atrazine and ^{14}C-labeled atrazine by *Rhizobium sp.* strain APCI-LC-MS was used for qualitative follow-up, for quantification, however, ESI with simazine as an internal standard was applied. Metabolisation but no mineralization could be observed [389]. APCI and ESI-MS were applied to determine humic substances and dissolved organic matter (DOM) together with the fungicide anilazine (2,4-dichlor-6-(2-chloranilino)-1,3,5-triazine) which was bound to DOM. With anilazine bound residues, a high release was found of the main metabolite, the dihydroxy-anilazine [390].

Antifouling pesticides.
Different antifouling pesticides of triazine type were determined in Portuguese rivers by APCI-LC-MS(+) after SPE for concentration, which was performed with Oasis cartridges [323].

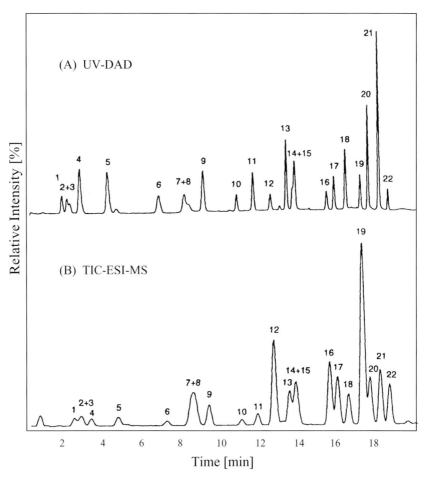

Fig. 15.6 Electropherogram of 22 aromatic sulfonate compounds (standard) with (A) UV-DAD detection and (B) ESI-MS(-) detection. Reproduced with permission from [395] © Elsevier, 1990.

Miscellaneous.
From the large group of pesticides, the triazole herbicide amitrole could be analysed in water samples via an automated SPE-APCI-LC-MS/MS procedure after a precolumn derivatization using 9-fluorenylmethoxy-carbonyl chloride. Recoveries in drinking water and surface water of >95% or 75%, respectively, could be achieved [391]. Degradation experiments in an aqueous medium combined with APCI-LC-MS proved that the total photodegradation of imidacloprid in water under natural sunlight in a pilot plant could be achieved in the presence of TiO_2 as catalyst. Levels of imidacloprid and the degradation product 6-chloronicotinic acid were monitored, five compounds were volatile and were detected by GC/MS [392].

APCI and ESI-LC-MS were compared to determine, besides other substances, acaricidic clofentezine and fungiciddic thiabendazole in fruits. Detection limits observed were equivalent to 0.002–0.033 mg per kg of crop [358]. APCI ionisation was found to be more efficient than ESI. When clofentezine was also determined in plums, strawberries and blackcurrant-based fruit drinks [384] mean overall recoveries of 70% from spiked extracts were observed. An efficient APCI-LC-MS/MS method was developed for the quantitative determination and identification of intact daminozide in apples and apple leaves [393]. APCI mass and MS/MS spectra of daminozide were presented. Recoveries for daminozide in apples and apple leaves were 98–102% and 112–116% and detection limits were observed with 0.008 and 0.02 mg kg^{-1} for apples and leaves, respectively.

The tin-containing pesticide fenbutatin oxide [394] in tomatoes, cucumbers and bananas was determined by APCI-LC-MS. Mean recoveries in SIM mode, recording seven ions of the isotopic cluster at m/z 515.2–521.2, were observed with 88% (tomatoes) and 80% (cucumbers and bananas). Detection limits were 0.06–0.12 ng μL^{-1}.

Applications using electrospray ionization interface (ESI)

As demonstrated (cf. Fig. 15.1 and 15.2) with the results of ionisation observed in the spectra of the non-ionic surfactant mixtures of AEOs or APOs or for ionisation of the anionic surfactant mixture of AES (cf. Fig. 15.3), if APCI or ESI interfaces were applied, both API interface types presented considerable differences in the ionisation processes. These differences were in both the type of ions and the efficiency of ionisation, i.e., either high molecular or low molecular compounds were favoured in ionisation and no ionisation takes places with the one interface whereas the other interface type ionises the compounds with high sensitivity. Obviously ESI is the interface which handles the very polar, partly charged compounds with low as well as high molecular weights in the best way, while the APCI interface can be used successfully for the more lipophilic compounds contained in water samples, e.g. phenol compounds. With the improved flexibility of ESI handling low and high flow rates of eluents ESI-CEz-MS became a powerful tool to separate complex mixtures with an improved separation efficiency never previously observable with any kind of LC (see Fig. 15.6) [395].

Table 15.3 Environmental applications in LC-MS performed by elektrospray interface (ESI).

Topic	References
Reviews	
General reviews	28, 44, 45
Compound class specific reviews	
– Dyes	43, 161, 200
– Pesticides	22, 29, 30, 32, 199
– Surfactants	21, 24, 39–41
– Sulfonates	161
– Toxines	396
Compound classes	
– Complexing agents	397–399
– Drugs and diagnostic agents	292, 295, 296, 400–416
– Dyes	28, 43, 297, 416, 417, 421–429
– Estrogenic compound	298–301, 430
– Explosives	205, 291, 302, 431, 432
– Haloacetic acids and desinfection byproducts	433–439
– Organoarsenic compounds	428
– Polycyclic aromatic hydrocarbons	28, 304, 442, 443
– Phenols	315–319, 346, 444
– Sulfonic acids	161, 297, 445–449
– Surfactants	21, 28, 40, 155, 212, 226, 326–329, 333, 334, 336, 339, 340, 343, 345, 346, 422, 450–482
– Toxins	396, 410, 483–488
– **Pesticides (including herbicides, fungicides etc.)**	
– Anilides, quaternary amines, toluidines and thiocyanate compounds	175, 240, 347, 385, 489–506
– Carbamates	108, 175, 257, 355, 357, 358, 424, 500, 502, 503, 506–515, 519, 537
– Organophosphorus compounds	155, 175, 322, 353, 385, 397, 500, 510, 515–523, 537
– Phenoxycarboxylic acids	325, 351, 374, 500–502, 524–532
– Phenylureas, thioureas and sulfonylureas	175, 240, 271, 325, 358, 373, 374, 381, 385, 500, 502, 503, 506, 508, 511, 512, 525, 527, 533–537, 539–554
– Triazines	175, 240, 275, 276, 325, 385, 389, 390, 424, 500, 502, 506, 510, 511, 513, 519, 536, 537. 546, 553, 555–562
– Antifouling pesticides	563
– Phenolic pesticides	317, 322, 325, 351, 527
– Miscellaneous	358, 502, 540, 564–576
– **Comparison of ESI interfaces with different types of interfaces**	21, 28, 29, 32, 40, 43, 108, 113, 155, 161, 175, 176, 185, 199, 200, 205, 240, 257, 271, 275, 276, 291, 297–300, 317, 322, 325–327, 334, 340, 347, 351, 353, 355, 357, 358, 373, 374, 381, 385, 389, 390, 501, 503, 509, 523, 533

These effects observable while comparing APCI and ESI led the users to apply preferentially ESI rather than APCI ionisation, with the result that more papers dealing with ESI applications were published.

The overviews and reviews dealing with the applications of the ESI interface in environmental matrices in general have been mentioned with the applications of the APCI interface because of the overlap which exists between both interface types [28, 44, 45]. Pesticide residue analysis again was the most frequent application of ESI-LC-MS in the analysis of water samples [22, 29, 30, 32, 199] with the result that there was a tremendous increase in papers published within the last few years. Several reviews reporting ESI-MS results obtained with dyes [43, 161, 200], surfactants [21, 24, 39–41], sulfonates [161] or toxins [396] were published in the literature.

Table 15.3 reflects the applications of ESI used as the interface to couple MS with manifold liquid chromatographic separation techniques (LC, SFC, IC, CE, CZE, FAI (high-field asymmetric wave form ion mobility spectrometry)) in environmental analytical applications. The literature for different groups of compounds is presented.

Complexing agents

Many complexing agents are hardly degradable and therefore can be observed in environmental waters. Since the complexing agents such as ethylenediamino tetra-acetic acid (EDTA), nitrilotriacetic acid or aminophosphonic acids are extremely polar compounds, they were determined preferentially by ESI-LC-MS. IC interfaced by ESI to the MS allowed the determination of EDTA in µg L^{-1} quantities without any pre-concentration [397]. Metal complexes of EDTA were very stable and could be observed after ESI-LC-MS as [M+metal]$^+$ ions in positive ionisation mode [398]. ESI CE-MS was applied to separate and to quantify the stabile Ni-EDTA complexes [399].

Drugs and diagnostic agents

ESI was the most common interface for monitoring drugs and their metabolites in the aquatic environment. Industrial effluents from pharmaceutical industries, STP effluents, groundwater and surface water samples were studied. Low concentrations of pollutants of concern made pre-concentration necessary before compounds were detected and identified by ESI-LC-MS and MS/MS [400–404].

Environmental water samples were examined by ESI-LC-MS/MS to analyse the antibiotic penicillin, sulfonamide, nitroimidazole, nitrofuran, 2,4-diaminopyrimidine and macrolide compounds and chloramphenicol. SRM and MRM techniques were applied [401]. To elaborate a method for determining drugs by (ion spray) ESI-LC-MS and MS/MS 22 Different neutral and weakly basic drugs (e.g. antiphlogistics, βblockers, β2-sympathomimetics, lipid regulators, antiepileptic agents, psychiatric drugs and vasodilators) were spiked into wastewater, river water and drinking water samples. The determination of phenazone, carbamazepine, cyclophosphamide, ifosfamide and pentoxifylline was affected by organic matrix compounds which made application of ESI-LC-MS/MS essential. MS/MS detection limits of five neutral drugs were down to 10 ng L^{-1} [402]. Methods for the determination of drug residues in water by means of API-LC-MS or API-CE-MS applying ESI or APCI interfaces were elaborated and drugs such as paracetamol, clofibric acid, penicilin V, naproxen, bezafibrate, carbamazepine, diclofenac, iboprufen and mefenamic acid could be separated and determined. The method was then applied to river water samples, where naproxen, bezafibrate, diclofenac, iboprufen and clofibric acid were determined in ng L^{-1} concentrations [295]. ESI-LC-MS and MS/MS were used to analyse and confirm 18 antibiotics in water samples after SPE (Lichrolute EN and C_{18}) or after lyophilization. The group of antibiotics analysed included penicillins, tetracyclines (TETs), sulfonamides and macrolide antibiotics. Quantification limits were 50 ng L^{-1} for TETs and 20 ng L^{-1} for all other antibiotics examined [404]. ESI-LC-MS, MS/MS and HPLC/UV were applied to determine 13 sulfonamide drugs in environmental water samples after LiChrolut EN SPE. Detection limits ranged from 0.2–3.7 µg L^{-1} for all sulfonamides while recoveries were 50–90%. Sulfamethoxazole and sulfadiazine were detected at 30–2000 and 10–100 ng L^{-1}, respectively [400]. Five tetracycline antibiotic derivatives in ground and waste water were analysed by ESI-LC-MS(+) after C_{18} or polymeric (Oasis) phase SPE. CID spectra were presented [405].

The qualitative and quantitative analysis of iodine containing X-ray contrast media and their metabolites in environmental waters of the city of Berlin (Germany), in raw and treated sewage, in surface waters, bank filtrate and raw drinking water, was performed by ESI-LC-MS and MS/MS in positive mode [406, 407]. Concentrations observed were 1.6–20.7 µg L^{-1} [403].

An ESI-LC-MS and MS/MS method for the determination of neutral drugs e.g. caffeine, propyphenazone, 4-aminoantipyrin, diazepam, glibenclamide, nifedipine, omeprazole, oxyphenbutazone and phenylbutazone in groundwater, surface and wastewater was presented. Concentration levels of these compounds in 14 STP effluents and 11 German rivers were reported [408]. Salicylic acid, ketoprofen, naproxen, diclofenac, iboprufen and genfibrozil were determined by ESI-LC-MS after SPE in water samples taken from several Spanish rivers and STP effluents. Results were compared with results from toxicity testing [409]. Sixty different pharmaceuticals covering analgesics, antipyretics, antiphlogistics, antirheumatics, lipid reducing compounds, antiepileptics, vasodilatators, tranquillizers, β-blockers, antineoplastic drugs, iodated X-ray contrast media and antibiotics were determined by GC-MS or ESI-LC-MS/MS and maximum contrations were reported [410]. Sixteen aromatic sulfonamides were monitored in the effluents of municipal STPs applying ESI-LC-MS and MS/MS in positive mode. Concentrations observed in secondary effluents or surface waters ranged from 5 to 1700 ng L^{-1} [411]. The very polar phenylsulfonamides and their metabolites, both relevant for water works, were determined by ESI-LC-MS(-) and by GC-MS after derivatization. Real environmental samples and samples from a testfilter were analysed [412]. ESI-FIA- and LC-MS and MS/MS besides GC-MS were applied to follow polar and nonpolar organic pollutants in wastewater treatment process. Total ion and mass chromatograms as well as MS/MS spectra of pharmaceuticals were presented. Substance-specific scans in the MS/MS mode were used for identification (cf. APCI) [296].

The extent of exposure to cyclophosphamide and ifosfamide in 24 workers in two hospitals was monitored in air and from wipe and pad samples as well as from gloves and urine using ESI-LC-MS (ion spray) and MS/MS. Because of incorrect use of the vertical laminar airflow hoods an increased contamination was observed [413].

Different drugs were submitted to aerobic biodegradation and results were monitored in FIA-ESI-MS mode [414]. The degradation of the pharmaceutical compound 4-fluorocinnamic acid in wastewater treatment process over a period of 149 h was followed by ESI-LC-TOFMS in the negative mode to recognize precursor compound and biodegradation products. The tentative biodegradation pathway was presented [415]. An ESI-LC-TOFMS in positive and negative mode was applied to perform "exact mass" measurements of aromatic sulfonamides and sulfonates in spiked samples and anaerobically treated textile wastewater [416].

A library of ESI and APCI product ion mass spectra of a number of steroids, morphine and drugs of abuse based on wideband excitation in an ion trap mass spectrometer was assembled and was applied to real environmental samples [292].

Dyes

Dyes are used for manifold applications, predominantly in the textile industries but the food industry is also uses some of these compounds. The strongly polar compounds can be ionised by ESI interface in negative mode without fragmentation [201]. Ion chromatography (IC) coupled with MS-LC separation in the ion-pairing mode by addition of volatile amines allowed the separation and determination of the very strongly polar metabolites of dyes [416, 417].The determination of sulfonated azo dyes in water and wastewater [43] or dye stuffs, PAHs, surfactants and pesticides in environmental matrices [28] were the topics of papers which compared the results of application of the different interface types, ESI, TSP and APCI [43] or ESI, APCI, TSP and PBI [28], respectively. In very early applications of ESI-LC-MS Bruins et al. [418, 419] reported on the analysis of dyes. The behavior of these compounds in the wastewater treatment process was elucidated by several groups [200], biodegradation processes were observed and followed [201, 420, 421] and even on-line ESI-LC-MS coupled with NMR was performed for identification [422]. Substance specific scans, loss of SO_3, in CID mode helped to improve the confirmation [201, 420].

SPE followed by CZE/UV and optimized capillary zone electrophoresis (CZE) with ESI-MS detection was used to determine monosulfonated (Mordant Yellow 8) and a series of disulfonated azo dyes (Acid Red 1, 13, 14 and 73, Mordant Red 9, Acid Yellow 23 and Acid Blue 113) quantitatively in spiked (3 mg l^{-1} of each compound) groundwater samples and industrial effluents [423]. Azo dyes besides pesticides and herbicides were determined by ion trap MS interfaced by a commercial ESI to the LC-device. By adjusting the repeller voltage (in-source CID) and doing MS/MS in the ion trap for $[M+H]^+$ ions CID spectra were obtained. With online ESI-LC-ITMS, detection limits of 0.1–1.0 ng could be easily achieved. IT-MS/MS and ESI-CID data were provided and compared [424]. ESI was used for coupling CZE and MS or LC and MS to analyse SPE concentrated sulfonated azo dyes and LAS in industrial effluents. CZE-MS offered higher separation power and was less affected by matrix components than LC-MS. Detection limits for sulfonated azo dyes in SIM mode were observed with 100–800 µg L^{-1} for CZE-MS [425]. Capillary electrophoresis (CE) with UV-DAD detection or MS interfaced by ESI in negative mode were applied for the identification of five reactive vinylsulfone dyes and their hydrolysis products in spent dyebaths and raw and treated wastewaters after C_{18} SPE. Concentrations of the different dyes and hydrolysis products in sewage effluents ranged from 23 to 42 µg L^{-1} [426]. Anthraquinone dyes (e.g. Disperse Blue 3) was characterised besides other pollutants in the effluents of a textile company by ESI-LC-MS and -MS^n in positive ion mode and in addition using the in-line data obtained from LC/NMR [422]. Sulfophthalimide, sulfophthalamide, sulfophthalamic acid and sulfophthalic acid as metabolites of sulfophthalocyanine textile dyes were determined qualitatively and quantitatively by ion pairing ESI-LC-MS and MS/MS resulting in quantification limits of 2–10 µg L^{-1} [417]. Eighteen polysulfonated anionic dyes and their degradation products were analysed by HPLC-UV and ESI-LC-MS(-) in the form of $[M\text{-}xA]^{x-}$ (A = H or Na) using volatile ion-pairing reagents. MS spectra were presented and structures were proposed according to the MS spectra obtained [427]. Aromatic sulfonic acids and sulfonated azo dyes were analysed by ESI-LC-MS. For improvement of LC separation non-volatile TBA ion pairing agents were used and then removed on-line prior to ESI-MS analysis [297]. The coupling of ion-exchange and ion-pairing chromatography by ESI with tandem mass spectrometry (MS/MS) was studied by Siu et al. who performed LC separations of a mixture of six permitted food dyes [428].

Quaternized cellulose is used as a sorbent to remove azo dyes from wastewater samples. For cellulose recycling purposes reductive degradation was performed and products of Orange II and Remazol Red F3B were determined by ESI-LC-MS and MS spectra of azo dyes and reduction products were reported [429]. As major oxidative degradation products of the azo dye Uniblue A four reaction products were identified under UV irradiation (254 nm)/ peroxydisulfate treatment. Possible reaction pathways were discussed [421].

Estrogenic compounds
The metabolic degradation products of natural and synthetic estrogens, present as glucuronids or sulfates, are strongly polar compounds in order to make their renal elimination possible. Therefore LC-MS is the method of choice to determine these compounds in underivatized form. For the analyses of the the precursor compounds, however, GC-MS is amenable and widely used because of the high sensitivity observed (cf. 15.3.3.2 APCI, estrogens). Comprehensive studies with ESI and APCI methods were performed [298], raw and treated wastewaters were examined quantitatively [300, 301].

The most prominent ions under ESI(+) ionisation observed in natural water extracts were [M+Na]$^+$ adduct ions [298]. Estrogens and progestogens of natural and anthropogenic origin from environmental samples were analysed using different SPE materials prior to ESI-LC-MS. For the determination of estrogens MS was operated in negative mode while progestogens were determined in positive ion mode. [M-H]⁻ or [M+Na]⁺ ions were obtained [430].

Explosives

World War II wastes, such as explosives may cause considerable problems because of the mobility of their metabolites generated under anaerobic conditions, the anilines, which are under suspicion as carcinogens. Results of ESI and APCI were taken to compare TSP-LC-MS(-) results in the negative mode for the explosives TNT, RDX and hexyl and their degradation products in groundwater samples from ammunition hazardous waste sites. Applying SPE (LiChrolut® EN) 31 compounds could be identified, such as nitramines and their by-products, TNT and partially nitrated toluenes, 1,3,5-trinitrobenzene (cf. 15.3.3.1 TSP, explosives) [205]. Cyclic nitramine explosives were examined in real and spiked soil samples to recognize degradation pathways. Mononitroso, dinitroso and trinitroso metabolites and ring cleavage products were determined by ESI-LC-MS in negative mode, resulting in [M-H]⁻ ions [431]. The explosive 1,3,5-trinitro-1,3,5-triazacyclohexane (RDX) and its nitroso-RDX metabolites were analysed by ESI and APCI-LC-MS in groundwater samples after SPE (Sep-Pak Porapak RDX) (cf. 15.3.3.2 APCI, explosives). APCI provided a 20-fold greater signal for nitroaromatics than ESI. Detection limits of 0.03–0.14 mg L^{-1} and recoveries of 71–130% were found [302]. EI-MS or ESI-MS spectra of explosive residues contained in a drainage water extract of an ammunition plant were published and the list of compounds examined was reported. The comparability and the limitations of the application of these libraries for identification were discussed [291]. A two-step (anaerobic/aerobic) composting was performed in a reactor system containing 2,4,6-trinitrotoluene-contaminated soil. Three new TNT metabolites, 4-acetylamino-2-hydroxylamino-6-nitrotoluene, 4-formamido-2-amino-6-nitrotoluene and 4-acetylamino-2-amino-6-nitrotoluene, were observed as [M+H]⁺ ions, which arose first under anaeriobic conditions and then were degraded aerobically to > 99% [432].

Haloacetic acids and disinfection byproducts

The main source of haloacetic acids in the environment is free chlorine or bromine. In the disinfection process of drinking water chlorine is applied, which may result in the generation of halogenated carboxylic acids besides a broad spectrum of volatile halogenated compounds amenable for electron capture detection (GC-ECD) or GC-MS. The application of ESI-LC-MS meanwhile has been established for quality control of drinking water to determine these contaminants substance-specifically [433, 434]. A method for the quantitative determination of nine chlorinated and brominated haloacetic acids at ppt-levels was elaborated and put into practice with a flow injection ESI-FAIMS/MS (high field asymmetric waveform ion mobility) device [435]. Another method applying ESI-LC-MS after SPE (Li-Chrolut EN, HR-P, Isolute ENV+ and Oasis HLB) for the quantitative determination of several mono- up to tri-halogenated acetic acids containing chloro or bromo substituents or a mixture of both in tap water, drinking and swimming pool water was presented [436]. The problem of separating haloacetic acids from matrix compounds present in the extracts and of determining them was solved by applying three different approaches: FAIMS [435, 437, 438, 439], MS after generation of organic complexes with high m/z ratios by coupling the analytes with perfluoroheptanoic acid or by application of high resolution MS using TOF instruments, all MS interfaced by ESI. New disinfection byproducts generated by applying ozone with either chlorine or chloramine determined by ESI-LC-MS were reported [440].

Organoarsenic compounds
The coupling of ion-exchange (IEC) and ion-pairing chromatography using ESI coupled with MS/MS allowed the determination of several environmentally important organoarsenic compounds in complex mixtures. An arsenobetaine could be confirmed by MS/MS after IEC [428]. Arsenc species were determined by ESI-LC-MS applying post column methanol addition [441].

Polycyclic aromatic hydrocarbons (PAH)
The results of PAH analysis with different types of interfaces (e.g. ESI, APCI, PBI and TSP - were reported by Clench et al. reviewing the state of the art of various mass spectral techniques [28]. For more polar PAHs pneumatically assisted ESI-LC-MS was used to determine mixtures of hydroxy polycyclic aromatic hydrocarbons. The abundance of ions dependent on flow rates was shown. ESI ionization was found to be less sensitive compared to APCI ionisation [304]. PAH analysis with ESI-LC-MS combined with RP-LC with post-column addition of silver nitrate was applied for the determination of 10 PAHs in river water. PAHs resulted in $[M]^+$ and $[M+Ag]^+$. The detection limits of different PAHs in spiked samples ranged from 0.001 to 0.03 µg L^{-1} [442].

The gas-phase reactions of PAHs with OH and NO_3 radicals generated in situ via photolysis of methyl nitrite were followed by ESI-MS. Naphthalene and d_8-labeled naphthalene were used as PAH components [443].

Phenols
The determination of phenols was preferentially performed using GC-MS with analytes in underivatized or derivatized form, but LC-MS methods were also developed. API methods for the analysis of phenols in aqueous matrices were applied [315, 316, 317]. APCI-LC-MS was found to be more sensitive than ESI application despite the possibility of improving ESI-sensitivity by a post-column addition of diethylamine [317]. Detection limits were observed with 0.02–20 ng injected onto the column. The determination of alkylphenols and bisphenol A as compounds with endocrine disruptor potential was also performed by ESI-LC-MS from aqueous [318, 319] and sediment samples with detection limits in the low µg L^{-1} range [346].

With the use of ESI-CE-MS a total of 11 priority phenols could be easily separated and quantitatively determined [444].

Sulfonic acids

Aromatic sulfonic acids are hardly degradable and, because of their polarity, very mobile during the drinking water treatment process resulting in the need for a reliable determination method. Twenty two aromatic sulfonates were extracted using four different SPE phases before the compounds were separated, confirming the excellent performance and separation power of ESI-CE-MS (cf. Fig. 15.6) [395]. Twenty aromatic sulfonic acids and their metabolites observed in industrial wastewaters were determined by ESI-LC-MS(-) and MS/MS using volatile amines for ion-pairing purposes in LC separation [445]. Several sulfonic acids such as p-toluenesulfonic acid, naphthalene-2,6-disulfonic acid and 2-aminonaphthalene-1,5-disulfonic acid were used as markers for studying processes in the leachate plume and in the groundwater downstream of a landfill. ESI-LC-MS(-) mass chromatograms and spectra were shown and the feasibility of ESI in the monitoring was discussed [446]. Benzene and naphthalenesulfonates in leachates and plumes of landfills were determined by ESI-LC-MS(-) after ion-pairing LC [447]. A sensitive ESI-LC-MS method for the determination of poly(naphtalenesulfonate) (PNS) contaminants in water after ion-pairing SPE extraction using ammonium acetate was elaborated and validated by naphthalenesulfonate–formaldehyde condensates. Environmental relevance of these compounds was confirmed by the analyses of grab samples of waste, river and ground water, containing PNS type compounds levels between 53 ng L^{-1} and 32 μg L^{-1} [448]. Naphthalenesulfonate–formaldehyde condensates were analysed by ESI-LC-MS(-) after ion-pairing LC [449], supporting the results reported by Crescenzi et al. [448]. Aromatic sulfonates known as hardly degradable pollutants in the aqutic environment were determined in landfill leachates and groundwater by ESI-LC-MS in negative mode [450]. Eighteen monomeric aromatic sulfonates which were contained in heavily loaded industrial wastewaters from textile industry were examined by ESI-LC-MS(-) before and after anaerobic and aerobic degradation. Six selected compounds were quantified in SRM mode [451]. Quadrupole mass spectrometers in positive and negative ESI-LC-TOFMS mode were applied to perform "exact mass" measurements of aromatic sulfonates and sulfonamides in spiked samples and anaerobically treated textile wastewater [416].

Surfactants

Several review papers have reported on the application of ESI in the analysis of surfactants. Di Corcia reviewed LC-MS methods for the unequivocal identification of isomers, oligomers and homologues of the technical blends of surfactants and their biodegradation intermediates in environmental samples at trace levels with particular attention to ESI and TSP applications [40]. Clench et al. [28] described the applications of LC-MS in environmental analysis using the interfaces PBI, TSP, APCI and ESI in use or just coming into use in th early 1990s.

Reports obtained from ESI interfaced LC-MS on non-ionic surfactants as medium polar compounds compared to the strong polar anionics predominantly covered the AEOs, NPEOs and some up-coming surfactants.

When APCI and ESI-LC-MS were compared in their ionisation efficiency for AEOs spiked into raw wastewater the more effective APCI technique was preferentially applied because low detection limits could be obtained (cf. 15.3.3.2 APCI, surfacants) [334]. Non-ionic polyethoxylate surfactants such as aliphatic alkylethoxylates (AEOs) and nonylphenol polyethoxylates (NPEOs) in water were determined by ESI-LC-MS after SPE using GCB. Recoveries were 85–97% and concentrations in municipal wastewaters were observed at ppt levels [452]. Surface water from the river Elbe (Germany) was qualitatively analysed, particularly for polar organic pollutants, using ESI and APCI-LC-MS and MS/MS. For identification, diagnostic scans in the MS/MS mode or neutral loss (NL) mode were performed, which confirmed non-ionic and anionic surfactants (cf. 15.3.3.2 APCI, surfactants) [226]. A mixture of 2-butyl branched AEOs, each containing an average of five ethoxy units was biodegraded prior to analysis by ESI-LC-MS. MS and MS/MS spectra and mass chromatograms of neutral and acidic metabolized branched surfactants were reported proving that the ethoxylate chain was shortened or oxidized resulting in highly polar intermediates [453]. ESI-FIA-MS/MS(+), LC-MS and MS/MS analyses were performed to characterise non-ionic surfactants in the wastewater extracts of Thessakloniki (Greece) STP after SPE using C_{18} or LiChrolut EN. MS and MS/MS spectra of non-ionics as dominating pollutants (AEOs and NPEOs) in inflow and effluent samples were presented (cf. 15.3.3.2 APCI, surfactants) [336].

ESI-LC-MS and MS/MS were used to identify the components of three non-ionic surfactant mixtures, NPEOs, secondary alcohol ethoxylates (SAE) and primary AEOs [454] used in wool scouring [455] and to monitor intermediates in the photocatalytic degradation of SAEs over a suspension of TiO_2 particles. Typical spectra at various stages of degradation were obtained, indicating the initial preferential cleavage of ethoxyl groups over OH radical reactions with the aliphatic alkyl chains resulting in cleavage at the secondary carbon in the molecules [454]. The photocatalytic decomposition products were the same as observed in the MS under CID conditions [455]. A robust LC-determination of NPEOs and octylphenol ethoxylates (OPEOs) in sewage plant effluents by fluorescence detection was confirmed by ESI-LC-MS(+) using 1-(4-methoxyphenyl)hexan-1-ol as internal standard. The limit was 5 µg L^{-1} applying 100 mL of effluent, which was extracted by GCB-SPE. Use of MS/MS in neutral loss (NL) mode allowed an increased selectivity in detection [456]. NPEOs, OPEOs and AEOs of different alkyl chain length were routinely determined by ESI-LC-MS(+) in wastewater and sludges, treated in different ways, in gradient elution mode [457]. Nonylphenols (NPs) and NPEOs extracted from wastewater treatment plant influent and effluent and surface water sediment were determined by ESI-LC-MS in negative or positive ionisation mode, respectively, using stable isotope-labeled standards [458]. The short and long EO-chain APEOs, besides halogenated NPEOs and nonylphenolpolyether carboxylates (NPECs), in estuarine water and sediment samples after high-temperature sonicated extraction were determined by ESI-LC-MS in positive or negative mode. Limitations of the method were discussed [459]. A method for the determination of NPEO homologues by ESI-LC-MS in positive mode in river water samples was described by Takino et al. [460]. ESI-LC-MS and MS/MS were used to characterise recalcitrant intermediate species generated from biotransformation of the branched alkyl chain of industrial blends of NPEOs. After a biodegradation period of 2 weeks, species oxidized in both side-chains, alkyl (CAPEO) and PEG (APEC) chain, could be confirmed. Less abundant metabolites having only the alkyl chain carboxylated (CAPEs) were also formed and before degradation proceeded in a slow transformation to alkyl and polyether chain carboxylated compounds (CAPECs) [461]. A novel rapid screening method, combined precursor ion scanning and multiple reaction monitoring, using ESI MS/MS or APCI MS/MS (cf. 15.3.3.2 ESI, surfactants) was elaborated to monitor homologue mixtures of NPEOs contained in the inflows and effluents of STPs. The method proved to be more selective and specific than current methods for NPEO profiling [340].

ESI-LC-MS and MS/MS were applied to elucidate the metabolism of 4-NPs, the metabolites of NPEOs. The metabolites observed were identified as 4-hydroxy- and 4-(dihydroxy)-NPs, glucosylated at the phenolic OH group and further glucosylated or glucuronidated [462]. Anionic and non-ionic surfactants (e.g. LAS and short and long chain NPEOs, respectively) were characterised in the effluents of a textile company by ESI-LC-MS and MS^n using an ion trap in positive ion mode. The combination of MS^n and stop-flow-LC/NMR in-line data allowed identification to a great extent [339, 422]. Halogenated APEOs and their metabolites were quantitatively analysed and identified in drinking, surface and wastewater as well as in river sediments and STP sludges by ESI or APCI-LC-MS(+/-) under optimized conditions [345]. Brominated and chlorinated compounds of NPEOs and OPEOs were confirmed. Quantitative determination of NPEO surfactant homologues in marine sediment using normal-phase ESI-LC-MS was performed resulting in detection limits of 2–10 ng g^{-1} with linear calibration graphs from 0.5–500 ng [463]. Popenoe et al. [464] described an ESI (ion spray) LC-MS method for the quantitative analysis of the anionic surfactants, alkyl sulfates (AS) and alkyl ethoxysulfates (AES) in natural waters. The method was validated with spiked samples, using 36 AES homologue species. Linear calibration curves and recoveries > 90%, except for highly spiked effluents, (75%), were observed. Comparing ESI and APCI for the analysis of AES APCI in positive mode at low cone voltages proved to be the method of choice because its fragment ion spectra revealed the alkyl chain length and the number of ethoxylate moieties [326]. This wa, in contradiction, reported after a comparison of API techniques in positive or negative mode as presented in Fig. 15.3 [21]. The analysis of selected AES homologues was performed by ESI-FIA-MS/MS(-) or ESI-LC-MS(-), homologue distributions in industrial blends were presented [329]. Solid-phase micro-extraction (SPME) was used for concentration of LAS homologues (C_{10}–C_{13}) from municipal wastewaters. Homologues were then determined with high selectivity and sensitivity by ESI-LC-MS under in-source-CID conditions [465]. SPME was also applied for the concentration of water-soluble components of sludges and sediments before the pollutants (e.g. phthalates, fatty acids, nonionic surfactants, chlorinated phenols and carbohydrate derivatives) were detected by ESI-LC-MS [466]. LAS in wastewater inflows and effluents and coastal waters (Cadiz, Spain) were determined by automated SPE followed by ESI-LC-MS(-) and ESI-CE-MS(-) and results were compared to CE-UV. LAS in concentrations > 990 µg L^{-1}, > 136 µg L^{-1} or > 739 µg L^{-1} were observed in inflows, effluents and coastal waters [467]. LAS together with dialkyltetralinsulfonates (DATS) in aqueous environmental samples could be detected and quantified by LC-FL and subsequently were confirmed by ESI-LC-MS. Compounds were extracted by graphitized carbon black (GCB) [468]. The long-chain intermediates from the biodegradation of LAS in the marine environment were monitored by ESI-LC-MS(-) after SPE (C$_{18}$ followed by SAX). The metabolite pattern observed indicated an (-oxidation resulting in C$_{11}$-chain molecules in seawater samples and C$_{13}$-chain molecules in interstitial waters [469]. Carboxylic degradation products, the so-called sulphophenyl carboxylates (SPCs), of LAS in coastal waters were concentrated by SPE and analysed by ESI-LC-MS(-). Total ion chromatograms (TICs) and SIM traces were presented. Fragment ions obtained by source CID were tabulated. Recoveries of 51–96% were observed, with with a general improvement parallel to increasing alkyl chain length [470]. Testfilter experiments with LAS resulted in SPCs determined by ESI-LC-MS(-) applying an ion suppressor module to improve sensitivity [394]. The quantitative determination of LAS in sea water and sediment samples from the German Bight of the North Sea or Waddensea marinas and estuaries applying ESI-FIA and LC-MS and MS/MS in the negative mode were performed. Concentrations observed for LAS compounds were <39–106 ng g^{-1} dry matter and < 30 ng L^{-1} in water samples from estuaries [333].

Elimination efficiencies for surfactants as predominantly observed pollutants in a conventional and in parallel three bio-membrane assisted wastewater treatment plants were studied. ESI-FIA and LC-MS and MS/MS in negative mode were used to qualify and quantify LAS. Diagnostic scans were applied for confirmation [343]. The elimination efficiency of LAS in a wet air oxidation reactor by chemical treatment of a wastewater was monitored by ESI-FIA-MS(-) applying a pattern recognition [471].

The determination of perfluorinated anionic surfactants, perfluorinated alkanesulfonates and perfluorocarboxylates, in surface water samples after an accidental release of perfluorosurfactant contaminated fire-fighting foam was performed qualitatively and quantitatively by ESI-LC-MS and MS/MS. The ESI-LCMS and MS/MS(-) TICs of the determination of perfluorooctanesulfonate (PFOS), perfluorohexanesulfonate (PFHxS), perfluorooctanoic acid (PFOA), perfluoroheptanoic acid (PFHpA), perfluorododecanoic acid in water samples were presented [472]. Anionic and non-ionic surfactants and their degradation products were determined qualitatively and quantitatively by ESI-LC-MS in sewage sludge samples. Ultrasonification was applied for extraction while ESI interface in negative mode was used for more polar compounds (e.g. NPECs). APCI (cf. APCI) was used predominantly in positive mode to ionise the AEOs, NPEOs and PEGs [346]. Already in 1990 organic ammonium, sulfate and sulfonate surfactant compounds were determined by Conboy et al. after ion chromatography by ESI-IC-MS in positive mode. MS and MS/MS spectra were provided (see Fig. 15.7) [473]. The qualitative and quantitative analysis of the cationic surfactants ditallow-dimethylammonium chloride (DTDMAC), diethylester dimethylammonium chloride and diesterquat by microbore LC-MS under ESI ionisation was examined using LAS as ion-pairing reagent. STP inflows and effluents and river water samples were studied. Cationic surfactant concentrations of 0.4 up to 140 µg L^{-1} were found in river and sewage waters [474]. After SFE using CO_2, DTDMAC was determined by HPLC-UV or HPLC-fluorescence and confirmed by ESI-LC-MS(+) in anaerobically stabilized sewage sludges to study their presence in the phasing-out period. It became obvious that DTDMAC levels in sludge had dropped by 94% from 1991–1994 due to the producers' voluntary ban on its use in Europe [475]. An industrial blend of C_{12} and C_{14}-N-methylglucamides and their biodegradation under aerobic conditions was examined by ESI-LC-MS to determine precursor and metabolite compounds in municipal sewage plant inflow and effluent samples. Degradation involved ω-oxidation of the alkyl chain followed by a ω-oxidation. The C_4 glucamic acid could be observed as a degradation intermediate, but higher homologues were not found, probably due to rapid breakdown [476]. To determine non-ionic surfactants of alkyl polyglucoside type (APGs) in wastewater effluents an ESI-LC-MS method was elaborated [477, 478]. Despite the fact that monitoring also covered potentially arising metabolites as observed in an APG degradation study monitored under TSP ionisation conditions [212] no metabolites were detected here. The qualitative determination of so-called gemini surfactants with highly improved surface activity by APCI or ESI-FIA-MS and LC-MS and MS/MS was reported. Mass spectra, fragment ion spectra and fragmentation pathways were presented [479]. Results of LC techniques and enzyme-linked immunosorbent assays for the determination of surfactants in wastewaters were obtained in an interlaboratory comparison study performed in seven laboratories. The non-ionic NPEOs, AEOs, coconut diethanol amides and the anionic LAS, NPEO-sulfates and secondary alkane sulfonates (SAS) were quantitatively determined by ESI and APCI-LC-MS and MS/MS (cf. 15.3.3.2 APCI, surfactants) [327]. In parallel to stability studies of pesticides the stability of surfactant compounds contained in water matrices or adsorbed on SPE phases was examined using ESI or APCI-LC-MS. NPEOs, alcohol ethoxylates (AEOs), coconut diethanol amides (CDEAs) and LAS were examined. After storage at 4 °C or 20 °C the compounds were analysed with SIM detection limits at 0.05–0.70 µg L^{-1} (cf. 15.3.3.2 APCI, surfactants) [328]. Advantages and drawbacks of the application of added non-ionic, anionic and cationic surfactant mixtures to nitroaromatic spiked sediments for an improved extraction of these nitroaromatic compounds were discussed after results of ESI-LC-MS of fractions from SFE were compared [480]. Polar Fenton oxidation products of surfactants (lauryl sulfate) were analysed by ion spray (ESI) MS. ESI-LC-MS provided information on the oxidation mechanism which resulted in mainly hydroxyl and epoxide group or aldehyde compounds [481]. MS spectra were presented [482].

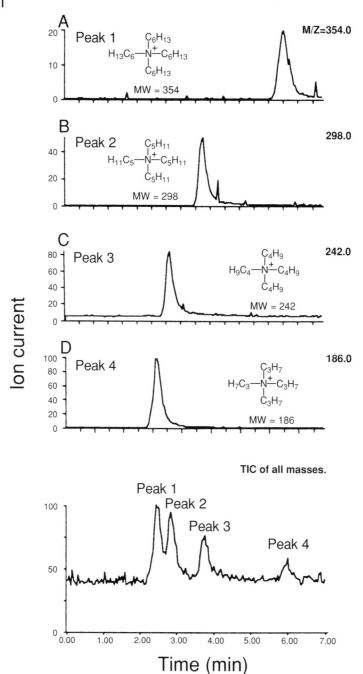

Fig. 15.7 One of the first published ion chromatography application interfaced by ESI with MS. ESI-IC-MS(+) total ion current trace (TIC) and selected mass traces of an industrial blend of quaternary ammonium compounds. Reproduced with permission from [473] © American Chemical Society, 1990.

Toxins

No reports were found of the analyses of free biogenic toxins in the environment, i.e. in surface or seawaters, but on the seafood toxins contained in shellfish and other seafood many papers are available since the invention of the soft ionizing interfaces. A review paper from Quilliam [396] reported the different types of toxins. Domoic acid is an amnesic shellfish-poisoning toxin originating from blue mussels which could be contained in shellfish. Its quantitative determination was studied by ESI-LC-MS(+) and MS^n. MRM results were presented which allowed an unambigous identification and quantitation combined with a minimum of sample clean-up [483]. ESI-LC-MS was used to handle thermolabile, polar seafood toxins such as domoic acid and isodomoic acids. LC eluent compositions were optimized for detection [410]. Even the very lipophilic shellfish toxins such as okadaic acid could be studied by ESI-LC-MS in positive mode, resulting in $[M+Na]^+$ and $[M+NH_4]^+$ adduct ions [484]. ESI or an ion spray interfacing technique operated in MS and MS/MS mode were found to be useful tools to determine and identify the shellfish toxins ciguatoxin and maitotoxin [485]. Cyanobacterial hepatotoxins in drinking water were analysed by CE-MS [486] and new pectenotoxins could be identified by LC-MS and MS/MS [487]. ESI-LC-MS(+) with micro-LC was applied to study SPE conditions of microcystins [488].

Pesticides (incl. herbicides, fungicides etc.)

Anilides, quaternary amines, toluidines and thiocyanate compounds.
The quaternary ammonium herbicides paraquat and diquat were first separated by CZE and then detected by ESI on a laboratory-built TOF instrument by means of volatile buffers. Depending on the buffer the ESI-TOF mass spectra of paraquat and diquat under these conditions showed singly and doubly charged molecular ions [489]. Moyano et al. also applied ESI-CZE-MS(+) to the determination of the herbicides mepiquat, chlormequat, diquat, paraquat and difenzoquat in water. MS/MS provided the structural information essential for the confirmation of identity [490]. For product quality check-up of formulations of the herbicide chlormequat ESI-CZE-MS was used, allowing the determination of contaminants contained in the formulation [491]. The quaternary ammonium pesticides paraquat; diquat, difenzoquat, chlormequat and mepiquat were analysed by ESI-LC-MS(+). Prior to analysis ion-pairing extraction was applied to concentrate compounds prior to analysis on different SPE materials (e.g. C_8, C_{18}, and PS-VDB) [492]. Separation of chlormequat on ODS1 combined with optimised ESI-LC-MS(+) allowed direct quantification on an ion trap instrument at levels lower than those required for residue analysis in foods and also in drinking water [493]. The determination of diquat and paraquat in water samples was performed by ion-pairing ESI-LC-MS using labeled diquat-d_4 dibromide or paraquat-d_8 dichloride and trifluoroacetic acid for ion-pairing purposes. The ions observed were $[M^{2+}\text{-}H]^+$ and $[M^{2+}+OOCCF_3]^+$ for diquat and paraquat, respectively. Detection limits

were < 0.2 µg L^{-1} [494]. The addition of heptafluorobutyric acid (HFBA) as ion pairing agent allowed the determination of paraquat and diquat in water without sample preparation by ESI-LC-MS(+). Detection limits for paraquat and diquat were observed with 5 and 1 µg L^{-1} [495]. ESI-FIA and ESI-CZE-MS(+) resulted in quantitation limits (S/N=3) of 200 µg L^{-1} for paraquat and diquat. [496]. Ion-pairing LC was applied to analyse the quaternary ammonium herbicides diquat, paraquat, difenzoquat, mepiquat and chlormequat by ESI and APCI-LC-MS(+). Detection limits down to 0.1–4 µg l^{-1} were obtained [347]. From solid samples chlormequat and mepiquat residues in grain were quantitatively determined by ESI-LC-MS and MS/MS after extraction with MeOH–H$_2$O–acetic acid (75:24:1, v/v/v) prior to an SPE and elution step. Detection limits were observed with 2 and 6 µg kg^{-1} for mepiquat and chlormequat [497].

Anilide pesticides were analysed by Ferrer et al. [498] who reported an ESI-LC-MS(-) method for the determination of the oxanilic and sulfonic acid derivatives of acetochlor, alachlor and metolachlor after C$_{18}$ SPE or liquid/liquid extraction using dichloromethane from surface and groundwater samples. Quantitation limits of 0.01 µg L^{-1} combined with recoveries of 88% for both oxanilic and sulfonic acid derivatives were obtained (cf. triazines) [240]. An ESI (ion spray) LC-MS/MS method was developed for the qualitative and quantitative analysis of the very polar sulfonic acid metabolites of acetochlor, alachlor, metolachlor and dimethenamid observed in groundwater at trace levels. MS/MS spectra were reported and calibration graphs were linear at 0.25–10 ng injected on-column [499]. Anilides and N-substituted amine pesticides in natural waters after SPE (Carbograph 4) were simultaneously determined qualitatively and quantitatively using ESI-LC-MS [500]. An automated SPE procedure combined with ESI-LC-MS(-) (ion spray) was used for the detection and quantification of the acidic herbicides benazolin, bentazone and 6- and 8-hydroxybentazone in environmental waters. Good fragmentation was observed under ESI conditions providing more information than TSP-LC-MS. SIM detection limits of 0.01–0.03 µg L^{-1} were observed, dependent on the extraction voltages [501]. Bentazone and alachlor were examined when an on-line dual-precolumn-based trace enrichment step applying the SPE materials (PLRS-S, Hyshere-l, LiChrolut EN and Isolute ENV +) and two different pHs was studied (cf. 13.3.3.2 APCI, triazines) [502]. ESI [503], FAB and APCI-LC-MS and MS/MS [175] were used for the multiresidue determination of a broad range of pesticides and to explain fragmentation mechanisms observed in a TSP study of anilides [175]. The methods were validated for selectivity, linear dynamic range, detection limit, precision and ruggedness. Biodegradation of alachlor in river water over a period of 28 d was performed. ESI-LC-MS besides GC/MS after derivatization were applied to follow the disappearance of alachlor. Several degradation products could be confirmed by comparison with synthetic standards [504]. ESI-LC-MS and MS/MS were performed on an orthogonal-acceleration (oa) TOF to characterise the photodegradation products of alachlor in water after on-line SPE. MS and product ion MS spectra of precursor and degradation products were presented and structures proposed for the observed product ions were shown [505] or degradation was followed by applying ESI-LC-MS and MS/MS. Effective monitoring of the degradation process was possible and unknowns could be identified using MS/MS [506].

Carbamates. A fast, sensitive and selective method for the concentration and analysis of 9 N-methylcarbamate pesticides was reported by Volmer et al. [507]. Three different SPME fibres combined with short-column ESI-LC-MS(+) and MS/MS were applied. The detection limits observed were 0.3–1.9 µg L^{-1}. Signal intensities increasing by a factor of 2–7 were observed [508] using non-volatile buffers in the separation process prior to ESI-MS. After LC removal of the non-volatile buffers was essential. The results obtained by ESI and APCI-LC-MS and MS/MS for the analysis of the eight N-methylcarbamate pesticides and their degradation products were compared with results obtain with the application of TSP or PBI (cf. 15.3.3.1 TSP, carbamates) [108]. ESI-LC-MS and TSP-LC-MS were used for quantitative determination of 10 different carbamate pesticides which showed a broad variety in polarity. ESI-SIM detection limits were typically 10–60 pg which was 10–150 times better than using TSP-MS (cf. 15.3.3.1 TSP, carbamates) [509]. Interfacing a commercial ESI source to an ITMS allowed the determination of carbamates as well as triazines and azo dyes. Identification could be performed either by IT-MS/MS or by ESI-CID [424].

The determination of carbamates by ESI-MS in environmental waters was reported in various papers. In one study the simultaneous determination of 26 non-acidic and 13 acidic pesticides in natural waters was performed. Recoveries observed were > 80% except for carbendazim, butocarboxim, aldicarb and molinate which were all better than 67% [500]. Forty five different pesticides of thiocarbamate, carbamate and carbamoyloxime type e.g. molinate or mercaptodimethur, carbaryl, ethifencarb, primicarb, propoxur, carbofuran, butocarboxim, aldicarb, carbetamide, methomyl, or oxamyl, aldicarbsulfone and butoxycarboxim besides other types were analysed SPE (Carbograph 4) by ESI-LC-MS(+) (cf. 15.3.3.2 ESI, triazines) [510]. For the multiresidue determination of carbamates and thiocarbamates in environmental samples parallel to a broad range of N- and P-containing pesticides ESI [503], FAB, PD and APCI-LC-MS and MS/MS interfacing techniques besides TSP [175] were used. Carbofuran and promecarb traces in water were determined quantitatively by means of ESI-LC-MS(+) (ion spray).TIC traces and mass spectra were presented. Detection limits of < 25 pg were observed [511]. Seven N-methylcarbamate [357] and 12 carbamate pesticides [512] were quantitatively analysed by ESI-LC-MS in river water or surface water from a lake, a groundwater well, a cistern, a farm pond and from drinking water samples, respectively. Compared to APCI-LC-MS ESI was found to be more sensitive [357]. Applying SPME coupled to ESI-LC-MS the carbamate pesticides aminocarb, asulam, barban, chlorpropham, methomyl, oxamyl, promecarb, propham and carbofuran in leachates of soil samples could be determined by applying positive ionisation. [513].

For the analysis of 10 carbamates in fruits and vegetables a comparison of ESI and APCI-LC-MS switching from positive to negative ionisation was performed (cf. APCI; carbamates). APCI was more efficient than ESI. Detection limits were equivalent to 0.002–0.033 mg per kg of crop [358]. Nevertheless, DiCorcia applied ESI-LC-MS(+) to analyse 12 N-methylcarbamate insecticides in 10 types of fruits and vegetables. Recoveries were > 80% for all matrices with detection limits of a few hundred pg per g sample [514]. For the routine quantification of phenoxycarb, pyrimicarb and other types of pesticides e.g., carbendazim, thiophanate methyl, phosmet, thiabendazole in apples and pears ESI-LC-MS was applied, showing limits of detection and quantitation of 0.01–0.02 and 0.02–0.05 mg kg^{-1}, respectively [515]. Thirteen carbamate pesticides were monitored by a newly established ESI and APCI-LC-MS procedure in oranges, grapes, onions and tomatoes after concentration by solid-phase dispersion (MSPD). Quantitative results were reported for both interface types (cf. 15.3.3.2 APCI, carbamates) [355].

The degradation of the carbamates carbofuran and methiocarb in estuarine waters was monitored by TSP- and ESI-LC-MS in positive and negative mode. TSP- and ESI-LC-MS in positive and negative mode were used for product monitoring and identification of hydrolytic and microbial degradation products (cf. 15.3.3.1 TSP, carbamates) [257]. When carbofuran, together with diuron, isoproturon, atrazine and alachlor, dissolved in surface waters at low levels was treated by UV light the physicochemical degradation products were analysed and characterised by applying ESI-LC-MS and MS/MS (cf. 16.3.3.2 ESI, phenylurea) [506].

Organophosphorus compounds.
An ESI-LC-MS(+) and MS/MS method was elaborated for the detection and characterization of phosphorothioate and -dithioate herbicides in environmental matrices. Compounds studied were diazinon, chlorpyrifos-methyl, chlorpyrifos, azinphos-ethyl, azinphos-methyl and phosmet. MS and MS/MS spectra were reported which confirmed that low-energy CID of [M+H]+ enabled identification by characteristic fragmentation patterns of the phosphorodithioates [516]. ESI-LC-MS was used to elucidate fragmentation behavior observed in a TSP-LC-MS study of the phosphorus pesticides butenoate, dichlorovos and trichlorfon [175]. A test mixture of 17 pesticides containing the organophosphorus compounds dimethoate, fenamiphos, coumaphos, fenchlorphos, chlorpyriphos and bromophos-ethyl at low ng L^{-1} levels were analysed by on-line SPE-ESI-LC-MS and MS/MS. A pesticide MS/MS library was successfully applied for identification [322, 353]. An ESI-LC-MS procedure was elaborated for the analysis of the polar and/or thermally labile organophosphorus pesticides trichlorfon, dichlorvos, dimethoate, oxydemeton-methyl, cis- and trans-mevinphos, demeton-S-methyl, fenamiphos, fenitrothion, fenthion and diazinon in groundwater [517] and surface water samples [518]. A variety of SPE materials (Amberchrom, LiChrolut EN, cyclohexyl, SDB, C18 and Isolute ENV) was examined. In contrast to TSP-LC-MS no thermal degradation was observed for trichlorfon [518]. Under these conditions transformation products such as fenthion sulfoxide were also observed. Detection limits in SIM mode were observed with 0.01–0.20 µg L^{-1} [517] or 0.001 µg L^{-1} [518], respectively.

A study with environmental samples was performed analyzing 26 non-acidic (e.g. organophosphorus, carbamate, triazine, anilide, N-substituted amine and phenylurea type) and 13 acidic (urea and phenoxy acid type) target pesticides simultaneously by ESI-LC-MS with recoveries of about 80% [500]. Drinking, ground and river water samples containing the phosphorus compounds omethoate, demeton sulfoxide, demeton sulfone, monocrotophos, trichlorphon, dimethoate, azinophos-methyl, malathion, ethoprophos, diazinon, phoxim, primiphos-methyl were concentrated by SPE using Carbograph 4 prior to ESI-LC-MS [510]. Several phosphorus target pesticides could be monitored using on-line SPE and ESI-LC-MS(+) detection. Results were confirmed by MS/MS [519]. The analysis of groundwaters spiked with fenamiphos was performed by ESI-LC-MS. The stability of samples adsorbed by Lichrolut-EN or styrene divinylbenzene cartridges and stored at −20 °C, 4 °C and ambient temperature for 80 d was studied [520]. ESI-IC-MS was successfully applied for glyphosate determination [397]. An automated SPE-LC-MS and MS/MS method interfaced by ESI was described for determination of the very polar phosphorus pesticides glyphosate, glufosinate and the glyphosate biodegradation product aminomethylphosphonic acid in drinking and surface waters [397, 521]. The detection in negative mode was selective, reliable and applicable to wastewater, drinking and surface water. Detection and quantitation limits were 0.03 and 0.05 µg L^{-1}, respectively with recoveries of 96% [521]. Samples were analysed by ESI-LC-MS at 1 µg L^{-1} levels and the background was reduced by about 2 orders of magnitude and salt adduction could be prevented [397]. For the routine quantification of the phosphorus pesticides dimethoate, thiophanate methyl, phosmet and other types of pesticides e.g. phenoxycarb, pyrimicarb, carbendazim, thiabendazole in apples and pears an ESI-LC-MS method was elaborated. Limits of detection and quantitation were 0.01–0.02 and 0.02–0.05 mg kg^{-1}, respectively [515]. A novel metabolite of profenofos could be identified by ESI-LC-MS. Spectra of the metabolite which was identified as the glucosyl-sulfate conjugate of 4-bromo-2-chlorophenol were presented [522]. Ionic compounds contained in agricultural chemicals were determined by ESI-ion chromatography-MS and TSP-ion chromatography-MS/MS applying a solid-phase chemical suppressor. The method was tested on a standard solution. ESI spectra give intense [M-H]- peaks for methyl phosphate and methyl sulfate [523]. To follow the physicochemical degradation products of pirimiphos methyl after ozone treatment, different MS methods, ion-trap EI or PICI-GC/MS, or ESI or APCI interfaced LC-MS, were applied. Illustrative ESI and APCI mass chromatograms of the pirimiphos methyl oxidation products and fragmentation data were reported and an ozonolysis degradation pathway was proposed (cf. 15.3.3.2 APCI, organophosphorus compounds) [155].

Phenoxycarboxylic acids.
In many contributions reporting on acidic pesticides in environmental samples ESI applied as ion spray was predominantly performed to analyse these pollutants. APCI, however, was not as effective as ESI as studies with standard solutions of the pesticide mixtures made obvious [325] when phenoxy acid compounds were determined using both types of interface. MSn quantitative results were used for confirmation. Mass detection after CZE-MS interfaced by ESI was successfully applied to analyse drinking water spiked with chlorinated acid herbicides. Selected-ion elec-

tropherograms of the 16 analytes provided detection limits (S/N = 3) at 8–250 µg L^{-1} [524]. To improve the ionisation efficiency the analysis of different types of so-called post-emergence herbicides of 2-(4-aryloxyphenoxy) propionic acid type (fluazifop, haloxyfop, fenoxaprop and quizalofop) in spiked groundwater and drinking water samples by ESI-LC-MS (ion spray) was studied [525]. Separation of enantiomers was performed on a chiral phase under CE conditions combined with ESI-MS in negative ionisation mode. [M-H]- ions for the pesticide compounds mecoprop, dichlorprop and fenoprop [526].

Eighteen acidic herbicides such as phenoxy acids, sulfonylureas and phenols or their biodegradation products in groundwater samples were quantified by ESI-LC-MS(-) and confirmed by MS/MS. Recoveries of phenoxy acid derivatives were >80%. [527]. Di Corcia et al. quantified 13 acidic pesticides of phenoxy acid and urea type and 26 non-acidic (base and neutral) pesticides of carbamate, triazine, anilide, N-substituted amine and organophosphorus type in natural waters after Carbograph 4 SPE. Recoveries observed for acidic compounds were better than 80% [500]. SPE applying GCB was also used successfully to concentrate and to determine acidic pesticides in aqueous samples by ESI-LC-MS [528, 529]. Screening examinations applying four different SPE materials (PLRS-S, Hyshere-l, LiChrolut EN and Isolute ENV +) were performed for concentration of a wide range of polar and acidic pesticides (e.g. 2,4,5-T and MCPA) in river water prior to analysis applying ESI-LC-MS(-) (cf. 15.3.3.2 ESI, triazines) and confirmation by MS/MS detection [502]. A protocol for the automated SPE in combination with ESI-LC-MS(-) for the quantitative determination of the phenoxy acid derivatives 2,4-D, MCPA, MCPP and MCPB in environmental waters was established [501]. A broad spectrum of pesticides contained in shallow groundwater samples from 2 sandy and 2 clay catchment areas were analysed using ESI-LC-MS or APCI-LC-MS. MCPA was observed as the predominant acidic pesticide compound when > 300 samples were analysed [351]. Phenoxyalkanoic pesticide compounds were among the target compounds in surface water samples from Southeastern regions of France and from the St. Lawrence River in Canada. ESI-LC-MS results confirmed by MS/MS in the negative mode and ion trap LC-MS/MS were found to be comparable [374].

The metabolisation of MCPA in water and soil samples was followed by ESI-LC-MS(-) and MS/MS. The main metabolite of MCPA, 4-chloro-2-methylphenol was determined with a detection limit of 40 ng L^{-1} for ground and surface water samples [530].

To optimize analytical results in the separation process TBA was used as the ion pairing agent in the quantitative analyses of phenoxy acid derivatives such as MCPA, 2,4-D, mecoprop, dichlorprop, 2,4,5-T, MCPB, 2,4-DB and 2,4,5-TP in drinking and surface water samples [529]. Under the same conditions, ion-pairing with TBA fluoride [531] or ammonium acetate [532], the arylphenoxypropionic herbicides fluazifop, fenoxaprop, quizalofop, haloxyfop and diclofop and some of their Me, Et and Bu-esters were determined in surface, ground water or drinking water [531] or in soil [532] by ESI-LC-MS in SIM mode. Recoveries and detection limits were reported.

Phenylureas, thioureas and sulfonylureas.
Three different types of urea pesticides were used on a large scale: phenylureas, thioureas and sulfonylureas. Therefore these and their degradation products will be observed in the environment. For optimization ESI and FAB ionisation results in combination with LC-MS were compared on the same magnetic sector instrument. The mass spectra of several herbicides (e.g. bensulfuron methyl, bromacil and degradation products) were compared for each technique and high resolution results were obtained [533]. Different types of so-called post-emergence herbicides (sulfonylureas, imidazolines and 2-(4-aryloxyphenoxy) propionic acids) were studied by ESI-LC-MS (ion spray) after Carbograph 4-SPE. Recoveries were reported to be > 85%. Fragmentation and signal intensities, dependent on orifice plate voltage, were examined [525]. For improvement of LC separations in the analysis of phenylurea pesticides non-volatile buffers were applied which were removed prior to ESI-MS by a postcolumn removal technique [508]. To achieve good LC separations an eight component mixture of sulfonylurea crop protection chemicals were analysed on a capillary electrophoresis (CE) system adapted to an ESI-MS system (ion spray) [534]. CZE-UV results from the determination of 12 sulfonylurea herbicides in water were confirmed by ESI-LC-MS in positive mode combined with in-source CID [535]. Phenylurea herbicides in water samples could be determined with detection limits in the low pg range using both API-techniques, ESI and APCI, in MS and MS^n (15.3.3.2 APCI, phenylureas) [325]. For a high throughput combined with a substance-specific determination of the polar urea pesticides metoxuron, monuron, monolinuron, chlortoluron, metobromuron, metabenzthiazuron, isoproturon, diuron, linuron, ESI-FIA-MS without prior chromatographic separation was performed successfully (cf. 15.3.3.2 APCI, phenylureas) [373]. ESI-LC-MS in parallel to TSP-LC-MS in SIM mode was used in a multiresidue method for the determination of the sulfonylurea herbicides chlorsulfuron and the methyl esters of sulfometuron, metsulfuron, tribenuron, bensulfuron, chlorimuron and primisulfuron. Mass spectra with three structure-significant ions essential for unambiguous identification in environmental samples were obtained with both interfaces, ESI as well as TSP (cf. 15.3.3.1 TSP, phenylureas) [271]. From both interface types characteristic advantages and drawbacks (cf. 15.3.3.2 ESI, triazines) could be reported if they were applied to the analysis of phenylurea pesticides in estuarine waters [240]. The optimization of SPE and ESI-LC-MS and MS/MS conditions for ultratrace analysis of polar pesticides like urea pesticides 3,4-dichlorophenylurea (DPU) and 3,4-dichlorophenylmethylurea (DPUM) in water were described. Detection limits were in the range of 0.2 to 2 ng L^{-1}. The applicability was demonstrated by the analyses of surface and estuarine water [536]. For a study of a wide range of polar and acidic pesticides in river water by LC-UV/DAD and ESI-LC-MS and MS/MS detection, an on-line dual-precolumn-based trace enrichment step resulted in reliable results if MS/MS was used for determination and confirmation [502]. In a study the simultaneous determination of 13 (6) acidic and 26 (10) pesticides in natural waters (in waters of the Calabria region, Italy [537]) after SPE (Carbograph 4) using ESI-LC-MS was performed. Confirmation was obtained by MS applying different extraction voltages for fragmentation purposes (cf. 15.3.3.2

ESI, triazines) [537]. Recoveries observed were > 80% except for carbendazim, butocarboxim, aldicarb and molinate, all better than 67% [500]. An aoTOF-MS interfaced by ESI was used to screen and identify unknown compounds and pesticides in water samples by MS and MS/MS. Structures for compounds observed besides pesticides were proposed [538]. Traces of the phenylurea pesticides linuron and monolinuron in water were determined quantitatively. Calibration graphs obtained after Supelclean ENVI-18 SPE were linear with detection limits < 25 pg [511]. Large numbers of phenylurea herbicide analyses led to the elaboration of on-line preconcentration techniques coupled to ESI-LC-MS. The procedure was demonstrated and validated with several pesticides using 10 ml of sample, resulting in detection limits of about 10 ng L^{-1} [539]. ESI-LC-MS and MS/MS were applied to quantify and to confirm 16 different herbicides of sulfonylurea [527] type in surface water samples. Surface water samples were extracted by SPE (Spe-ed RP-102). As confirmation criteria RT, molecular ion and two fragment ions besides ion abundance ratios were defined. Quantitation at 0.1 and 1.0 ppb level was demonstrated [540].

Diuron and isoproturon were analysed by ESI-LC-MS(+) and MS/MS in surface water samples from Southeastern regions of France and from the St. Lawrence River in Canada. Confirmation was performed by MS/MS [374]. The determination of one thiourea and four urea pesticides in environmental waters was arried out by ESI-LC-MS(+) and MS/MS (cf. 15.3.3.2 ESI, carbamates) [512]. The seven sulfonylurea herbicides thifensulfuron methyl, metsulfuron methyl, triasulfuron, chlorsulfuron, rimsulfuron, tribenuron methyl and bensulfuron methyl in environmental waters were quantitatively analysed by ESI-LC-MS(+) after SPE. Recoveries under SIM were reported [541]. Phenylurea herbicides and their degradation products in water were examined by ESI-LC-MS(+) resulting in a separation of 22 compounds within a 50 min run. Detection limits were in ng L^{-1} ranges. The effect of the LC mobile phase on the MS response was studied [542]. In water and food phenylureas and sulfonylureas could be determined quantitatively by ESI-LC-MS and MS/MS at low pg and fg levels. The ESI approach proved to have better sensitivity and precision than existing TSP-LC-MS methods [503]. ESI-LC-MS(-) and MS/MS were applied for the determination of the mothproofing agents sulcofuron and flucofuron in environmentals waters [543, 544]. Product ion mass spectra of mitins and MRM chromatograms were reported. Absolute detection limits in MRM mode for sulcofuron and flucofuron were reported [543]. Results of C_{18} SPE and LLE were comparable [544]. The reactions of ^{14}C-labelled isoproturon and some of its metabolites, including [^{14}C]-4-isopropylaniline, in aqueous solutions with humic monomer catechol was followed by ESI-LC-MS(+/-). It was observed that aniline-derived pesticides covalently bound in soil may not be fully undegradable, nor fully immobile [545].

A hot phosphate-buffered water extraction system (buffer, water at 90 °C or soxhlet with methanol) followed by subsequent C_{18}-SPE coupled on-line to ESI-LC-MS was used for the analysis of monolinuron, metabenzthiazuron, linuron and neburone in naturally aged agricultural soil. [546]. The comparison of ESI and APCI-LC-MS for the determination of diflubenzuron in fruits, proved that APCI was more efficient than ESI. The application of positive–negative mode switching MS allowed detection limits equivalent to 0.002–0.033 mg per kg crop [358]. A novel screening method elaborated for the automated detection and identification of isotopically labelled pesticide compounds using ESI-LC-MS(+) and MS/MS on an ion trap and isotope pattern recognition software. Identification of degradation products could be achieved and metabolic pathways were elucidated [547]. The sulfonylurea pesticide triasulfuron was determined in soil samples by ESI-LC-MS(+) after MeOH–phosphate buffer extraction and SPE (Supelclean LC-Si) clean up with overall recoveries of > 83% [548] while chlorsulfuron, metsulfuron-methyl, thifensulfuron-methyl and triasulfuron in soil could be quantified by ESI-LC-MS after acidic extraction with recoveries > 80% [549]. The same sulfonylurea herbicides and in addition tribenuron-methyl were analysed in soil water. Quantification was performed, based on the integrated abundance of [M-H]$^-$ ions. SRM was used for confirmation [550]. The sulfonylurea herbicides nicosulfuron, thifensulfuron methyl, metsulfuron methyl, sulfometuron methyl, chlorsulfuron, bensulfuron methyl, tribenuron methyl and chlorimuron methyl were quantitatively determined in soil extracts by ESI-LC-MS(+) and MS/MS in SRM mode, proving a 400-fold increase in quantification [551]. ESI-LC-MS in positive and negative mode was applied to monitor biodegradability of thifensulfuron methyl [552] and tribenuron-methyl in soil [381]. Five and four ions of soil-induced degradation products were identified as metabolites [552] or [381] (cf. 15.3.3.2 APCI, ureas), respectively. Diuron and atrazine or diuron and isoproturon dissolved in water were submitted to photolysis by UV light performed under conditions selected to be close to those found in the environment. To follow the photolysis products SPE-LC using two short columns (cf. 15.3.3.2 ESI, triazines) on-line coupled with ESI-MS [553] or ESI-LC-MS and MS/MS [506] was applied, respectively. Two degradation products of isoproturon could be confirmed [506]. The chemical degradation of chlortoluron observable during the water disinfection process with HOCl/ClO$^-$ was elucidated by ESI-LC-MS and MS^n. The product ion spectra of chloro-hydroxylated and hydroxylated by-products resulted in a quite complex fragmentation pattern. A fragmentation scheme was proposed [554].

Triazines.
ESI-LC-MS was used to characterise and differentiate the triazine herbicides atrazine, terbuthylazine, propazine and prometryn. In source CID spectra were reported. Low-energy CID of [M+H]$^+$ ions confirmed the characteristic fragmentation patterns and permitted distinction of isomeric triazines [555]. Results obtained from ESI-LC-MS and APCI-LC-MS and MS/MS for the analysis of polar triazine pesticides in water confirmed a better performance of APCI compared to ESI (cf. 15.3.3.2 APCI, triazines) [325].

To optimize the whole analytical procedure, extraction and detection SPE and ESI-LC-MS and MS/MS were studied with polar pesticides like atrazine and some of their transformation products such as deisopropylatrazine, hydroxyatrazine and deethylatrazine. Detection limits in surface and estuarine water were in the range 0.2 to 8 ng L^{-1}. [536]. The application of ESI and APCI developed as techniques to solve the interpretation of fragmentation mechanisms was performed in parallel in a comprehensive TSP study of triazine pesticides with N-heterocyclic, phenylureas, sulfonylureas, thioureas, anilides, carbamates, thiocarbamates, and organophosphorus compounds [175]. The dependence of detection limits on the eluent flow-rate and signal response of different types of pesticides such as triazines was studied by ESI and APCI. Dramatic losses in sensitivity for the hydrophobic pesticide compounds were observed with increasing flow rates, triazines were less sensitive with both ionisation techniques (cf. 15.3.3.2 APCI, triazines) [385]. Ion-trap MS (IT-MS) was also applied for analysis using a commercial ESI interface. Environmental contaminants such as pesticides of triazine and carbamate type and azo dyes were analysed and identified by IT-MS^n or ESI-CID (cf. 15.3.3.2 ESI, dyes). Detection limits were reported by Lin et al. [424]. While LC-MS interfaced by ESI had been developed as a common routine method for the determination of triazine pesticides the analytical approach using SFC-MS interfaced by ESI was reported. Compared to TSP-LC-MS applied in parallel, the ESI-SFC-MS method using SPE by RP-C$_{18}$ or by SAX for concentration was one order of magnitude more sensitive than TSP but was suitable only for the less polar chlorotriazines [276]. Even micellar electrokinetic chromatography (MEKC) was combined with ESI-MS(+) for the analysis of a mixture of the triazine herbicides atrazine, propazine, ametryn and prometryn using sodium dodecylsulfate for electrophoresis [556].

In real environmental water samples traces of atrazine and hydroxyatrazine were determined quantitatively after SPE (Supelclean ENVI-18) by means of ESI-LC-MS(+).Calibration graphs were linear at 0.025–10 ng on-column combined with detection limits < 25 pg [511]. The six major very polar degradation products of atrazine, amino and hydroxyatrazines, were determined in spiked river, drinking and groundwaters samples by ESI-LC-MS in SIM mode after SPE (Carbograph 4) in ng L^{-1} ranges [557]. Under trace level analysis conditions using ESI-LC-MS detection, limits for atrazine and hydroxyatrazin of 10 and 30 pg, respectively, were observed [558]. Samples of drinking, ground and river waters were spiked with a mixture of 45 different pesticides e.g. the triazine derivatives cyanazine, simazine, atrazine and terbutylazine before SPE (Carbograph 4) and ESI-LC-MS(+) was performed (cf. 15.3.3.2 ESI, carbamates) [510].

In a study covering a wide range of polar and acidic pesticides deethylatrazine and atrazine besides anilide, phenoxy acid, phenylurea, carbamates and other types of specific pesticides in river water were determined by ESI-LC-MS and MS/MS. Recoveries, depending on preconcentration steps, obtained with different SPE materials (PLRS-S, Hyshere-1, LiChrolut EN and Isolute ENV +) and at different pH values were reported [502]. Sixteen of the most widely used pesticides in Southern Italy were monitored in surface water samples taken in the Calabria region. Triazines were determined quantitatively by LC-UV and ESI-LC-MS(+) and were confirmed by MS [537]. In another study the simultaneous determination of 26 non-acidic (base and neutral: e.g. triazine, carbamate, anilide, N-substituted amine, urea and organophosphorus type) and 13 acidic (sulfuron and phenoxy acid type) pesticides in natural waters was performed using ESI-LC-MS. Recoveries were reported (cf. 15.3.3.2 ESI, carbamates) [500]. Triazines (e.g., simazine, atrazine, terbutylazine) besides other pesticides in soils were extracted by a hot phosphate-buffered water extraction with subsequent C_{18}-SPE prior to ESI-LC-MS analysis. In this way 35 target pesticides could be monitored and results were confirmed by MS/MS [519, 546]. An ESI interfaced LC-MS (oaTOF) was used for the identification of polar organic microcontaminants in surface waters (the river Rhine) performing accurate mass determination. The triazine derivatives propazine and terbutylazine could be identified by their positively recorded ion masses and the losses of propene and butene, repectively [559]. ESI-LC-MS/MS was applied to monitor and confirm the hydroxylated degradation products of atrazine in the stream water from Goodwater Creek watershed samples collected over a period of 2 years [275]. When ESI and TSP were compared with respect to their feasibility for the analysis of pesticides such as atrazine, simazine, ametryne, cyanazine, deethylatrazine and deisopropylatrazine and triazine metabolites or chlortoluron, isoproturon, diuron, linuron and diflubenzuron in estuarine waters, TSP was found to offer greater sensitivity for triazines than for phenylurea herbicides, whereas ESP was more sensitive for phenylurea herbicides [240].

SPME coupled to ESI-LC-MS was used to determine the triazine pesticides simazine, atrazine, propazine and prometryn in leachates obtained from soil samples. SPME compared with other extraction methods showed less interference from the matrix compound [513].

Bonding studies to examine the strength of adsorbance of the fungicide anilazine to humic substances and dissolved organic matter (DOM) were performed applying ESI and APCI ionisation. If anilazine bound residues could be observed, a high release of the main metabolite, the dihydroxy-anilazine, was found [390]. Several degradation studies were performed which were accompanied by ESI-LC-MS and MS/MS in order to monitor results. The physicochemical degradation applying UV radiation to atrazine contained in surface water at low levels was examined [506]. The applicability of SPE-LC using two short columns and/or single-short-column LC combined on-line with ESI-MS was demonstrated for the photolysis of the pesticide atrazine dissolved in water under conditions selected to be close to those found in the environment [553]. Applying ESI-LC-MS allowed one to follow the rapid metabolization of atrazine by an enzyme from Pseudomonas strain ADP resulting in carbon dioxide, ammonia and chloride. The precursor compound and the active enzyme could be identified [560]. A Rhizobium sp. strain was applied to biodegrade atrazine and ^{14}C-labeled atrazine. The metabolic process was quantitatively followed by ESI-LC-MS using simazine as an internal standard. Hydroxy-atrazine was the only metabolite which was detected after 8 d [389]. Atrazine degradation products generated by Fenton's reagent (Fe^{2+} and H_2O_2) were analysed by ESI-LC-MS combined with an in-line radioisotope detector. The 14C-labeling procedure made it obvious that a derivatization of the more polar atrazine products is necessary [561]. For a reductive dechlorination of atrazine, fine-grained zero-valent Fe was applied and degradation was followed by ESI-LC-MS and MS/MS. The dechlorinated product 2-ethylamino-4-isopropylamino-1,3,5-triazine was characterised. Detection limits observed were 0.15 ng mL^{-1} for triazines and 6.2 ng mL^{-1} for the degradation product [562].

Antifouling pesticides
Organotin compounds have been observed in manifold surface water samples because of their widespread use in antifouling paintings and in agricultural chemicals. Their determination could be performed by ESI-LC-MS after SPE. Compared to LC-ICP-MS (inductive coupled plasma) ESI-LC-MS showed a decreased sensitivity [563].

Miscellaneous
An ESI-LC-MS and MS/MS method was described for the quantitative determination of tebufenozide in water samples. The in-source CID spectra were reported. The detection and quantification limits were 0.001 and 0.005 ppb, respectively [564]. To avoid signal suppression in the determination of tebufenozide and hydroxy- tebufenozide in wheat hay a method for postcolumn introduction of an internal standard was elaborated for ESI-LC-MS applications [565]. The insecticide avermectin B1 was determined in oranges using ESI-LC-MS(+). Different fragmentor voltages were applied and fragmentation patterns observed were tentatively identified [566]. The degradation products from the aerobic and anaerobic incubation of emamectin benzoate in microbially active soil were characterised by ESI-LC-MS and MS/MS proving that emamectin benzoate is biodegradable in soil [567]. ESI-LC-MS and APCI-LC-MS were applied for the determination of clofentezine besides others pesticides in fruits. APCI was found to be more efficient than ESI [358].

Traces of imidazolinone herbicides in natural waters (river, ground and drinking water) could be determined by ESI-LC-MS. The product-ion mass spectrum of the [M+H]$^+$ ion of imazethapyr was presented. [568]. The imidazolinone herbicides imazapyr, *m*-imazamethabenz, *p*-imazamethabenz, *m,p*-imazamethabenz-methyl, imazethapyr and imazaquin were determined in water and soil. Detection limits of 4–7 ng L^{-1} in groundwater, 9–13 ng L^{-1} in river water and 0.1–0.05 ng g^{-1} in soil were observed [569]. For the examination of the extraction efficiency of the same imidazolinone herbicides contained in soil samples ESI-LC-MS was used. Detection limits were < 14 ng g^{-1} (S/N = 3) [570]. Mean recoveries for imazethapyr were 92% [571]. In the same way six imidazolinone herbicides in five different soil types were analysed. Extraction profiles were reported. Recoveries of 95–105% were obtained [572]. Stout et al. described an ESI-LC-MS and MS/MS method for the determination of the six imidazolinone herbicides imazapyr, imazamethabenz, imazmethapyr, imazamethabenz-methyl, imazethapyr and imazaquin in in tap, lake and well water samples which made sample clean-up obsolete [573]. The advantages of ESI-LC-MS and MS/MS for pesticide and herbicide residue analysis with respect to the determination of imidazolinones in H$_2$O, and imazethapyr and its metabolites in plants were discussed from a historic, scientific and economic viewpoint [574]. A microwave-assisted extraction (MAE) of the herbicide imidazolinone and its 1-hydroxyethyl as well as glucosyl(1-hydroxyethyl) metabolite was elaborated using ESI-LC-MS. Recoveries observed were about 100% for all compounds. Compared with conventional extraction procedures sample throughput could be increased six-fold [575]. Sixteen different herbicides of imidazolinone, sulfonylurea and sulfonamide type in surface water were determined quantitatively and confirmed by ESI-LC-MS. Confirmation was performed by MS/MS, reaching quantitation levels of 0.1 and 1.0 ppb [540].

To concentrate the fungicide carbendazim extracted from spiked water, environmental water and soil extracts immunoaffinity extraction was performed coupled to RP-LC-ESI-MS. Quantitation limits were 100 ppb and 25 ppt in soil and lake water, respectively [576]

15.4
Conclusions

In the past two decades, LC-MS has become a generally accepted analytical technique in many fields of analyses. In the environmental analysis of organic pollutants besides GC-MS LC-MS has gained importance for the determination and identification of polar organic compounds (cf. Tab. 15.1–15.3), though the role of the application to the analysis of inorganics should not be forgotten [577–581].

However, its full significance was not recognized until after LC-MS had become established as a hyphenated research technique. This method revealed its special potential when, in order to determine the degree of pollution in aquatic matrices, sum parameter analyses, such as chemical oxygen demand (COD) or total organic carbon (TOC), came to be replaced by substance-specific LC-MS methods which afforded identification of polar contaminants. The importance of LC-MS analysis comes to light especially if we take into account that, apart from all those anthropogenic polar substances released into the environment that can be determined in a reliable way only since the eighties, still far more biogenic polar compounds exist in the environment. The significance of these substances in the environment may of course be relativized, since only a small number of them are of any ecotoxicological relevance. However, those polar substances that hardly degrade in the environment, are extremely mobile in aquatic systems, and make their way to drinking water treatment plants, are not to be neglected in their number. Although they could be removed from drinking water to a great extent, yet the substances applied for hygienisation can, in combination with dissolved organic matter, lead to new undesirable polar drinking water byproducts.

Previously, GC-MS analysis could do no more than help, dependent on the origin of the water, to account for between 5 and a maximum of 25% of the carbon water contained in the shape of defined chemical compounds – and this was achieved only after prior expensive derivatisation of a part of the non-volatile pollutants. With the introduction of LC-MS analysis, the number of identified polar substances has increased considerably, but the potential of the method could not be fully exploited, since the transferability and dissemination of mass spectral data and libraries of product ion spectra to another instrument, different in type, is quite restricted. The availability of standards for identification is limited, too, and is a question of costs. The use of the expensive high-resolution instruments such as magnetic instruments, time-of-flight (TOF) or Fourier transform ion cyclotron resonance (FTICR) mass spectrometers in combination with MS/MS, or the possibility of MS^n studies on cheaper ion trap instruments will improve the capacity for identifying especially metabolites or physicochemical degradation products.

It should not go unmentioned, as also documented in the applications reviewed before, that no interface is able to ionise each compound with the same selectivity and sensitivity. Moreover nearly every type of interface has its advantages in the analysis of polar compounds and only the application of several interfaces provides the maximum of desired information.

Nevertheless LC-MS with the most common API interfaces applied has undoubtedly become the most powerful tool for those environmental chemists who are working with samples taken from aquatic systems. Chemists working in this research area had already recognized this perspective in the late 1970s [2–5], other users, however, realised it only about 10 years later [128, 129, 152–154, 215, 582]. Today, this technique sees essential improvements and the perspective it opens up now is its combination with other powerful techniques, e.g. NMR [339, 422]. This combination has already led to some profound results and more amazing results can be expected in future with the combination of new analytical techniques.

References

1 Niessen, W.M.A: Developments in Interface Technology for Combined Liquid Chromatography, Capillary Electrophoresis, Supercritical Fluid Chromatography-Mass Spectrometry, in *Applications of LC-MS in Environmental Chemistry; J. Chromatogr. Library*, Vol. 59, ed. D. Barceló, D. Elsevier, Amsterdam **1996**, p. 1.

2 Talroze, V.L., Skurat, V.E., Gorodetskii, I.G. et al., *Russian J. Phys. Chem.* **1972**, 46, 456.

3 Horning, E.C., Carroll, D.I., Dzidic, I. et al., *J. Chromatogr.* **1974**, 99, 13.

4 Scott, R.P.W., Scott, C.G., Munroe, M. et al., *J. Chromatogr.* **1974**, 99, 395.

5 Arpino, P.J., Baldwin, M.A., McLafferty, F.W. *Biomed. Mass Spectrom.* **1974**, 1, 80.

6 Clement, R.E., Eiceman, G.A., Koester, C.J. *Anal. Chem.* **1995**, 67, 221R.

7 Gordon, D.B., Lord, G.A., Jones, D.S. *Rapid Commun. Mass Spectrom.* **1994**, 8, 544.

8 Clement, R.E., Yang, P.W., Koester, C.J. *Anal. Chem.* **1999**, 71, 257R.

9 Clement, R.E., Eiceman, G.A., Koester, C.J. *Anal. Chem.* **1993**, 65, 85R.

10 Clement, R.E. Langhorst, M.L., Eiceman, G.A. *Anal. Chem.* **1991**, 63, 270R.

11 Clement, R.E. Yang, P.W., Eiceman, G.A. *Anal. Chem.* **1997**, 69, 251R.

12 *Application of LC-MS in Environmental Chemistry; J. Chromatogr. Library*, Vol. 59, ed. D. Barceló, Elsevier, Amsterdam **1996**, p.564.

13 Stan, H.J., Heberer, Th.: in *Analysis of Pesticides in Ground, Surface Water*, ed. H.J. Stan, (Ed.), Springer, Berlin **1995**, p.143.

14 Lin, H.-Y., Voyksner, R.D.: Practical Aspects of Ion Trap Mass Spectrometry, in *CRC Series Modern Mass Spectrometry* Vol. 111, Chemical Environmental, Biomedical Applications, eds.R.E. March, J.F.J. Todd, CRC, Boca Raton, FL **1995**, ch. 14.

15 Yergey, A.L., Edmonds, Ch.G., Lewis I.A.S. et al.: *Liquid Chromatography/Mass Spectrometry Techniques, Applications*, Plenum Press, New York, **1990**, p. 1.

16 Carpioli, R.M.: *Continous-Flow Fast Atom Bombardment Mass Spectrometry*, Wiley, Chichester **1990**, p.1.

17 Rosen, J.D.: *Applications of New Mass Spectrometry Techniques in Pesticide Chemistry*, Wiley, New York **1987**, p. 1.

18 Brown, M. *ACS Symp. Ser.*, **1990**, 420, 1.

19 Niessen, W.M.A.: Liquid Chromatography – Mass spectrometry, in *Chromatographic Science Series*, Vol. 79, Marcel Dekker, New York **1998**, p. 648.

20 Barceló, D.: Sample handling, analysis of pesticides, their transformation products in water matrices by liquid chromatographic techniques, in *Techniques, Instrumentation in Analytical Chemistry*, Vol. 21; Sample Handling, Trace Analysis of Pollutants - Techniques, Applications, Quality Assurance, ed. Barceló, D., Elsevier, Amsterdam **2000**, p.155.

21 Schröder, H.Fr., Ventura, F: Applications of liquid chromatography-mass spectrometry in environmental chem-

istry; Characterization, determination of surfactants, their metabolites in water samples by modern mass spectrometric techniques, in *Techniques, Instrumentation, in Analytical Chemistry*, Vol. 21; Sample Handling, Trace Analysis of Pollutants - Techniques, Applications, Quality Assurance, ed. Barceló, D., Elsevier, Amsterdam **2000**, p. 828.
22. Slobodnik, J., Brinkman, U.A.Th. LC/MS interfacing systems in environmental analysis: Applications to polar pesticides, in *Analytical Chemistry*, Vol. 21, Sample Handling, Trace Analysis of Pollutants - Techniques, Applications, Quality Assurance, ed. Barceló, D., Elsevier, Amsterdam **2000**, p. 935.
23. Schröder, H.Fr.: Separation, Identification, Quantification of Surfactants, their Metabolites, in Waste Water, Surface Water, Drinking Water by LC-TSP-MS, FIA-TSP-MS, MS-MS, in *Applications of LC-MS in Environmental Chemistry*, ed. Barceló, D., Elsevier, Amsterdam **1996**, p. 263.
24. Barceló, D. *Anal. Chim. Acta* **1992**, *263*, 1.
25. Barceló, D. *Analyst* **1991**, *116*, 681.
26. Lopez-Avila, V. *J. AOAC Int.* **1999**, *82*, 217.
27. Moder, M., Popp, P.: in *Applications of Solid Phase Microextraction*, ed. Pawliszyn, J., Royal Society of Chemistry, Cambridge **1999**, p. 311.
28. Clench, M.R., Scullion, S., Brown, R., et al., *Spectrosc. Eur.* **1994**, *6*, 16, 21.
29. Ferrer, I., Barceló, D. *Analusis* **1998**, *26*, M118.
30. Stan, H.-J, Fuhrmann, B.: Schriftenreihe Biologische Abwasserreinigung zum Kolloquium im Sonderforschungsbereich 193 der DFG an der Technischen Universität Berlin, Biologische Behandlung Industrieller und Gewerblicher Abwässer, Anwendung der LC-MS in der Wasseranalytik, **2001**, Vol. 11, p. 3.
31. Slobodnik, J., van Baar, B.L.M., Brinkman, U.A.Th. *J. Chromatogr. A* **1995**, *703*, 81.
32. Barceló, D., Hennion, M.-C., *Anal. Chim. Acta* **1995**, *318*, 1.
33. Dean, J.R., Wade, G., Barnabas, I.J. *J. Chromatogr. A* **1996**, *733*, 295.
34. Pacakova, V., Stulik, K., Jiskra, J. *J. Chromatogr. A* **1996**, *754*, 17.
35. Cserhati, T., Forgacs, E. *J. Chromatogr. B* **1998**, *717*, 157.
36. Pico, Y., Font, G., Molto, J.C. et al., *J. Chromatogr. A* **2000**, *885*, 251.
37. Jones, T.L., Betowski, L.D., Lopez-Avila, V. *Trends Anal. Chem.* **1994**, *13*, 333.
38. Berger, U., Kölliker, S., Oehme, M. *Chimia* **1999**, *53*, 492.
39. Kiewiet, A.T., de Voogt, P. *J. Chromatogr. A* **1996**, *733*, 185.
40. DiCorcia, A. *J. Chromatogr. A* **1998**, *794*, 165.
41. Marcomini, A., Zanette, M. *J. Chromatogr. A* **1996**, *733*, 193.
42. Hites, R.A. *Int. J. Mass Spectrom. Ion Processes* **1992**, *118/119*, 369.
43. Riu, J., Schonsee, I., Barceló, D. et al., *Trends Anal. Chem.* **1997**, *16*, 405.
44. Reemtsma, T. *Trends Anal. Chem.* **2001**, *20*, 500.
45. Reemtsma, T. *Trends Anal. Chem.* **2001**, *20*, 533.
46. Krost, K.J., *Appl. Spectrosc.* **1993**, *47*, 821.
47. Wright, L.H. *J. Chromatogr. Sci.* **1982**, *20*, 1.
48. Stamp, J.J., Siegmund, E.G., Cairns, T. et al., *Anal. Chem.* **1986**, *58*, 873.
49. Games, D.E., Lant, M.S., Winwood, S.A. et al., *Biomed. Mass Spectrom.* **1982**, *9*, 215.
50. Cairns, T., Siegmund, E.G., Doose, G.M. *Biomed. Mass Spectrom.* **1983**, *10*, 24.
51. Cairns, T., Siegmund, E.G., Stamp, J.J. *Rapid Commun. Mass Spectrom.* **1987**, *1*, 90.
52. Barefoot, A.C., Reiser, R.W. *J. Chromatogr.* **1987**, *398*, 217.
53. Barefoot, A.C., Reiser, R.W. *Biomed. Environ. Mass Spectrom.* **1989**, *18*, 77.
54. Barefoot, A.C., Reiser, R.W., Cousins, S.A. *J. Chromatogr.* **1989**, *474*, 39.
55. Levsen, K., Wagner-Redeker, W., Schäfer, K.H. et al., *J. Chromatogr.* **1985**, *323*, 135.
56. Krost, K.J. *Anal. Chem.* **1985**, *57*, 763.

57 Sim, P.G., Boyd, R.K., Gershey, R.M. et al., *Biomed. Environ. Mass Spectrom.* **1987**, *14*, 375.
58 Perreault, H., Ramaley, L., Sim, P.G. et al., *Rapid Commun. Mass Spectrom.* **1991**, *5*, 604.
59 Games, D.E., Rontree, J.A., Fowlis, I.A. *J. High Res. Chromatogr.* **1994**, *17*, 68.
60 Garcia, J.F., Barceló, D. *J. High Res. Chromatogr.* **1993**, *16*, 633.
61 Parker, C.E., Haney, C.A., Harvan, D.J. et al., *J. Chromatogr.* **1982**, *242*, 77.
62 Voyksner, R.D., Bursey, J.T. *Anal. Chem.* **1984**, *56*, 1582.
63 Voyksner, R.D., Bursey, J.T., Pellizzarri, E.D. *J. Chromatogr.* **1984**, *312*, 221.
64 Parker, C.E., Haney, C.A., Hass, J.R. *J. Chromatogr.* **1982**, *237*, 233.
65 Parker, C.E., Yamaguchi, K., Harvan, D.J. et al., *J. Chromatogr.* **1983**, *319*, 273.
66 Shalaby, L.M. *Biomed. Mass Spectrom.* **1985**, *12*, 261.
67 Honing, M., Barceló, D., Jager, M.E. et al., *J. Chromatogr. A* **1995**, *712*, 21.
68 Kleintop, B.L., Eades, D.M., Yost, R.A. *Anal. Chem.* **1993**, *65*, 1295.
69 Bellar, T.A., Budde, W.L., Kryak, D.D. *J. Am. Soc. Mass Spectrom.* **1994**, *5*, 908.
70 Willoughby, R.C., Browner, R.F. *Anal. Chem.* **1984**, *56*, 2626.
71 de la Guardia, M., Garrigues, S. *Trends Anal. Chem.* **1998**, *17*, 263.
72 Aguilar, C., Borrull, F., Marce, R.M. *Chromatographia* **1996**, *43*, 592.
73 Ho, J.S., Budde, W.L. *Anal. Chem.* **1994**, *66*, 3716.
74 Cappiello, A., Famiglini, G., Palma et al., *Environ. Sci. Technol.* **1995**, *29*, 2295.
75 Cappiello, A., Famiglini, G., Berloni, A. *J. Chromatogr. A* **1997**, *768*, 215.
76 Jimenez, J.J., Bernal, J.L., Del Nozal, M.J. et al., *J. Assoc. Off. Anal. Chem.* **2000**, *83*, 756.
77 Cappiello, A., Famiglini, G., Bruner, F. *Anal. Chem.* **1994**, *66*, 1416.
78 Rezai, M.A., Famiglini, G., Cappiello, A. *J. Chromatogr. A* **1996**, *742*, 69.
79 Aguilar, C., Borrull, F., Marce, R.M. *J. Chromatogr. A* **1998**, *805*, 127.
80 Slobodnik, J., Hogenboom, A.C., Louter, A.J.H. et al., *J. Chromatogr. A* **1996**, *730*, 353.
81 Louter, A.J.H., Hogenboom, A.C., Slobodnik, J. et al., *Analyst* **1997**, *122*, 1497.
82 Bruner, F., Berloni, A., Palma, P. *Chromatographia* **1996**, *43*, 279.
83 Kim, I.S., Sasinos, F.I., Stephens, R.D. et al., *Anal. Chem.* **1991**, *63*, 819.
84 Mattina, M.J.I. *J. Chromatogr.* **1991**, *542*, 385.
85 Cappiello, A., Famiglini, G. *Anal. Chem.* **1995**, *67*, 412.
86 Brown, M.A., Stephens, R.D., Kim, I.S. *Trends Anal. Chem.* **1991**, *10*, 330.
87 Jones, T.L., Betowski, L.D., Lesnik, B. et al., *Environ. Sci. Technol.* **1991**, *25*, 1880.
88 Betowski, L.D., Pace, C.M., Roby, M.R. *J. Am. Soc. Mass Spectrom.* **1992**, *3*, 823.
89 Slobodnik, J., Hoekstra-Oussoren, S.J.F., Jager, M.E. et al., *Analyst* **1996**, *121*, 1327.
90 Slobodnik, J., Jager, M.E., Hoekstra-Oussoren et al., *J. Mass Spectrom.* **1997**, *32*, 43.
91 Jedrzejewski, P.T., Taylor, L.T. *J. Chromatogr. A* **1995**, *703*, 489.
92 Bagheri, H., Slobodnik, J., Marce Recasens, R.M. et al., *Chromatographia* **1993**, *37*, 159.
93 Minnaard, W.A., Slobodnik, J., Vreuls et al., *J. Chromatogr. A* **1995**, *696*, 333.
94 Slobodnik, J., Louter, A.J.H., Vreuls, J.J. et al., *J. Chromatogr. A* **1997**, *768*, 239.
95 Mattina, M.J.I. *J. Chromatogr.* **1991**, *549*, 237.
96 White, J., Brown, R.H., Clench, M.R. *Rapid Commun. Mass Spectrom.* **1997**, *11*, 618.
97 Prosen, H., Zupancic-Kralj, L., Marsel, J. *J. Chromatogr. A* **1995**, *704*, 121.
98 Schutz, S., Hummel, H.E., Duhr, A. et al., *J. Chromatogr. A* **1994**, *683*, 141.
99 Atienza, J., Jimenez, J.J., Herguedas, A. et al., *J. Chromatogr. A* **1996**, *721*, 113.
100 Kambhampati, I., Roinestad, K.S., Hartman, T.G. et al., *J. Chromatogr. A* **1994**, *688*, 67.
101 Marcé, R.M., Prosen, H., Crespo, C. et al., *J. Chromatogr. A* **1995**, *696*, 63.
102 Dijkstra, R.J., Van Baar, B.L.M., Kientz, C.E. et al., *Rapid Commun. Mass Spectrom.* **1998**, *12*, 5.

103 Primus, T.M., Tawara, J.N., Johnston, J.J. et al., *Environ. Sci. Technol.* **1997**, *31*, 346.
104 Bagheri, H., Slobodnik, J., Brinkman, U.A.Th. *Anal. Lett.* **2000**, *33*, 249.
105 Petrier, C., David, B., Laguian, S. *Chemosphere* **1996**, *32*, 1709.
106 Acher, A.J., Hapeman, C.J., Shelton, D.R. et al., *J. Agric. Food Chem.* **1994**, *42*, 2040.
107 Hapeman, C.J., Anderson, B.G., Torrents, A. et al., *J. Agric. Food Chem.* **1997**, *45*, 1006.
108 Pleasance, S., Anacleto, J.F., Bailey, M.R. et al., *J. Am. Soc. Mass Spectrom.* **1992**, *3*, 378.
109 Mattina, M.J.I.: in *Applications of LC-MS in Environmental Chemistry*, ed. D. Barceló, Elsevier, Amsterdam **1996**, p.325.
110 Hsu, J., D. Barceló (Ed.), in *Applications of LC-MS in Environmental Chemistry*, ed. D. Barceló, Elsevier, Amsterdam **1996**, p. 345.
111 Clark, L.B., Rosen, R.T., Hartman, T.G. et al., *Int. J. Environ. Anal. Chem.* **1991**, *45*, 169.
112 Clark, L.B., Rosen, R.T., Hartman, T.G. et al., *Int. J. Environ. Anal. Chem.* **1992**, *47*, 167.
113 Straub, R., Voyksner, R.D., Keever, J.T. *J. Chromatogr.* **1992**, *627*, 173.
114 Turnipseed, S.B., Roybal, J.E., Rupp, H.S. et al., *J. Chromatogr. B* **1995**, *670*, 55.
115 Maguire, R.J. *Water Sci. Technol.* **1992**, *25*, 265.
116 Voyksner, R.D., Straub, R.F., Keever, J.T. et al., *Environ. Sci. Technol.* **1993**, *27*, 1665.
117 Singh, R.P., Brindle, I.D., Jones, T.R.B. et al., *J. Am. Soc. Mass Spectrom.* **1993**, *4*, 898.
118 Doerge, D.R., Clayton, J., Fu, P.P. et al., *Biol. Mass Spectrom.* **1993**, *22*, 654.
119 Bonfanti, L., Careri, M., Mangia, A. et al., *J. Chromatogr. A* **1996**, *728*, 359.
120 Anacleto, J.F., Ramaley, L., Benoit, F.M. et al., *Anal. Chem.* **1995**, *67*, 4145.
121 Pace, C.M., Betowski, L.D. *J. Am. Soc. Mass Spectrom.* **1995**, *6*, 597.
122 Gremm, T.J., Frimmel, F.H. *Chromatographia* **1994**, *38*, 781.
123 Ventura, F.: Characterization of surfactants in water by desorption ionization methods, in *Environmental Analysis: Techniques, Applications, Quality Assurance, Vol. 13*, ed. D. Barceló, Elsevier, Amsterdam **1993**, p.481.
124 Rivera, J., Fraisse, D., Ventura, F. et al., *Fresenius' J. Anal. Chem.* **1987**, *328*, 577.
125 Rivera, J., Fraisse, D., Ventura, F. et al., *Biomed. Environ. Mass Spectrom.* **1988**, *16*, 403.
126 Rivera, J., Caixach, J., Espadaler, I. et al., *Water Supply* **1989**, *7*, 97.
127 Crathorne, B., Fielding, M., Steel, C.P. et al. *Environ. Sci. Technol.* **1984**, *18*, 797.
128 Watts, C.D., Crathorne, B., Fielding, M. Et al., in *Org. Micropollut. Aquat. Environ.*, eds. A. Bjørseth, G. Angeletti, Comm. Eur. Communities, EUR 8518, Reidel, Dordrecht **1984**, p. 120.
129 *Org. Micropollut. Aquat. Environ.*, eds. Rivera, J., Ventura, F., Caixach, J. et al., Comm. Eur.Communities, EUR 10388, Reidel, Dordrecht **1986**, p.77.
130 Ventura, F., Fraisse, D., Caixach, J. et al., *Anal. Chem.* **1991**, *63*, 2095.
131 Rivera, J., Ventura, F., Caixach, J. et al., *Int. J. Environ. Anal. Chem.* **1987**, *29*, 15.
132 Rockwood, A.L., Higuchi, T. *Tenside Surfact. Deterg.* **1992**, *29*, 6.
133 Takeuchi, T., Watanabe, S., Kondo, N. et al., *Chromatographia* **1988**, *25*, 523.
134 Lawrence, D.L. *J. Am. Soc. Mass Spectrom.* **1992**, *3*, 575.
135 Borgerding, A.J., Hites, R.A. *Anal. Chem.* **1992**, *64*, 1449.
136 Ventura, F., Caixach, J., Figueras, A. et al., *Water Res.* **1989**, *23*, 1191.
137 Ventura, F., Figueras, A., Caixach, J. et al., *Water Res.* **1988**, *22*, 1211.
138 Rivera, J., Caixach, J., Ventura, F. et al., *Chemosphere* **1985**, *14*, 395.
139 Schneider, E., Levsen, K., Daehling, P. et al., *Fresenius' J. Anal. Chem.* **1983**, *316*, 488.
140 Fernandez, P. Valls, J.M. Bayona, J. Albaigés *Environ. Sci. Technol.* **1991**, *25*, 547.
141 Valls, M., Bayona, J.M., Albaigés, J. *Int. J. Environ. Anal. Chem.* **1990**, *39*, 329.

142 Field, J.A., Barber, L.B., Thurman, E.M. et al., *Environ. Sci. Technol.* **1992**, *26*, 1140.
143 Simms, J.R., Keough, T., Ward, S.R. et al., *Anal. Chem.* **1988**, *60*, 2613.
144 Gilliam, J.M., Landis, P.W., Occolowitz, J.L. *Anal. Chem.* **1983**, *55*, 1531.
145 Ventura, F., Figueras, A., Caixach, J. et al., *Org. Mass Spectrom.* **1988**, *23*, 558.
146 Siegel, M.M., Tsao, R., Oppenheimer, S. et al., *Anal. Chem.* **1990**, *62*, 322.
147 Schneider, E., Levsen, K., Boerboom, A.J.H. et al., *Anal. Chem.* **1984**, *56*, 1987.
148 Shiraishi, H., Otsuki, A., Fuwa, K. *Bull. Soc. Chem. Jpn.* **1992**, *55*, 1410.
149 Ventura, F., Caixach, J., Romero, J. et al., *Water Sci. Technol.* **1992**, *25*, 257.
150 Gilliam, J.M., Landis, P.W., Occolowitz, J.L. *Anal. Chem.* **1984**, *56*, 2285.
151 Goad, L.J., Prescott, C.M., Rose, M.E. *Org. Mass. Spectrom.* **1984**, *19*, 101.
152 Schneider, E., Levsen, K. in *Org. Micropollut. Aquat. Environ.*, eds. A. Bjørseth, G. Angeletti, Comm. Eur. Communities, EUR 10388, Reidel, Dordrecht **1986**, p. 14.
153 Schneider, E., Levsen, K. *Fresenius' J. Anal. Chem.* **1987**, *326*, 43.
154 Rivera, J., Ventura, F., Caixach, J. et al., in *Org. Micropollut. Aquat. Environ.*, eds. A. Bjørseth, G. Angeletti, Comm. Eur. Communities, EUR 11350, Reidel, Dordrecht **1988**, p. 329.
155 Coran, S.A., Bambagiotti-Alberti, M., Giannellini, V. et al., *Rapid Commun. Mass Spectrom.* **1998**, *12*, 281.
156 Kalinoski, H.T., Hargiss, L.O. *J. Am. Soc. Mass Spectrom.* **1992**, *3*, 150.
157 Parees, D.M., Hanton, S.D., Cornelio Clark, P.A. et al., *J. Am. Soc. Mass Spectrom.* **1998**, *9*, 282.
158 Yakimov, M.M., Timmis, K.N., Wray, V. et al., *Appl. Environ. Microbiol.* **1995**, *61*, 1706.
159 Grange, A.H., Donnelly, J.R., Brumley, W.C. et al., *Anal. Chem.* **1994**, *66*, 4416.
160 Ventura, F., Caixach, J., Figueras, A. et al., *Fresenius' J. Anal. Chem.* **1989**, *335*, 222.
161 Reemtsma, T. *J. Chromatogr. A* **1996**, *733*, 473.
162 Yang, Y., Egestad, B., Jernströmn, B. et al., *Rapid Commun. Mass Spectrom.* **1991**, *5*, 499.
163 Teffera, Y., Baird, W.M., Smith, D.L. *J. Chromatogr.* **1992**, *577*, 69.
164 Enya, T., Kawanishi, M., Suzuki, H. et al., *Chem. Res. Toxicol.* **1998**, *11*, 1460.
165 Corless, C., Reynolds, G., Graham, N. et al., *Water. Res.* **1989**, *23*, 1367.
166 Calvosa, L., Monteverdi, A., Rindone, B. et al., *Water. Res.* **1991**, *25*, 985.
167 Borrett, V., Dagley, I.J., Kony, M. et al., *Eur. Mass Spectrom.* **1995**, *1*, 59.
168 *Applications of New Mass Spectrometry Techniques in Pesticide Chemistry*, eds. Fujiwara, H., Wratten, S.J. Rosen, J.D., Wiley, New York **1987**. p.128.
169 Kotrebai, M., Tyson, J.F., Block, E. et al., *J. Chromatogr. A* **2000**, *866*, 51.
170 Bruchet, A., Legrand, M.F., Arpino, P. et al., *J. Chromatogr. Biomed. Appl.* **1991**, *562*, 469.
171 Reiser, R.W., Barefoot, A.C., Dietrich, R.F. et al., *J. Chromatogr.* **1991**, *554*, 91.
172 Bourgeois, G., Benicha, M., Azmani et al., *Analusis* **1995**, *23*, 23.
173 Ohkura, T., Takechi, T., Deguchi, S. et al., *Jpn. J. Toxicol. Environ. Health* **1994**, *40*, 266.
174 Tondeur, Y., Sovocool, G.W., Mitchum, R.K. et al., *Biomed. Environ. Mass Spectrom.* **1987**, *18*, 733.
175 Volmer, D., Levsen, K. *J. Am. Soc. Mass Spectrom.* **1994**, *5*, 655.
176 Schröder, H. Fr. *Environm. Monitoring, Assessment* **1997**, *44*, 503
177 Onisko, B.C., Barnes, J.P., Staub, R.E. et al., *Biol. Mass Spectrom.* **1994**, *23*, 626.
178 Melnyk, M.C., Carlson, R.E., Busch, K.L. et al., *J. Mass Spectrom.* **1998**, *33*, 75.
179 Soubayrol, P., Dana, G., Bolbach, G. et al., *Analusis* **1996**, *24*, M34.
180 Chivukula, M., Spadaro, J.T., Renganathan, V. *Biochemistry* **1995**, *34*, 7765.
181 Borgerding, A.J., Hites, R.A. *Environ. Sci. Technol.* **1994**, *28*, 1278.
182 Brumley, W.C., Brownrigg, C.M., Grange, A.H. *J. Chromatogr. A* **1994**, *680*, 635.
183 Day, B.W., Skipper, P.L., Zaia, J. et al., *Chem. Res. Toxicol.* **1994**, *7*, 829.

184 Wolf, S.M., Vouros, P. *Chem. Res. Toxicol.* **1994**, *7*, 82.
185 Kostiainen, R., Tuominen, J., Luukkanen, L. et al., *Rapid Commun. Mass Spectrom.* **1997**, *11*, 283.
186 Teffera, Y. *Diss. Abstr. Int. B* **1992**, *53*, 240B.
187 Wolf, S.M., Vouros, P., Norwood, C. et al., *Anal. Chem.* **1995**, *67*, 891.
188 Tsuruda, L.S., Lame, M.W., Jones, A.D. *Drug Metab. Dispos.* **1995**, *23*, 129.
189 Wolf, S.M., Annan, R.S., Vouros, P. et al., *Biol. Mass Spectrom.* **1992**, *21*, 647.
190 Wolf, S.M., Vouros, P., Norwood, C. et al., *J. Am. Soc. Mass Spectrom.* **1992**, *3*, 757.
191 Tyler, A.N., Romo, L.K., Frey, M.H. et al., *J. Am. Soc. Mass Spectrom.* **1992**, *3*, 637.
192 Voyksner, R.D., Pack, T. *Rapid Commun. Mass Spectrom.* **1991**, *5*, 263.
193 Perkins, J.R., Parker, C.E., Tomer, K.B. *J. Am. Soc. Mass. Spectrom.* **1992**, *3*, 139.
194 Lane, S.J., Brinded, K.A., Taylor, N. et al., *Rapid Commun. Mass Spectrom.* **1993**, *7*, 953.
195 Bitsch, F., Shackleton, C.H.L., Ma, W. et al., *Rapid Commun. Mass Spectrom.* **1993**, *7*, 891.
196 Niessen, W.M.A. *J. Chromatogr. A* **1998**, *794*, 407.
197 Blakley, C.R., Vestal, M.L. *Anal. Chem.* **1983**, *55*, 750.
198 Arpino, P. *Mass Spectrom. Rev.* **1992**, *11*, 3.
199 Barceló, D., Honing, M., Chiron, S., in *Applications of LC-MS in Environmental Chemistry,* ed. D. Barceló, Elsevier, Amsterdam **1996**, p. 219.
200 Yinon, J., Betowski, L.D., Voyksner, R.D., in *Applications of LC-MS in Environmental Chemistry,* ed. D. Barceló, Elsevier, Amsterdam **1996**, p.187.
201 Rafols, C., Barceló, D. *J. Chromatogr. A* **1997**, *777*, 177.
202 Groeppelin, A., Linder, M.W., Schellenberg, K. et al., *Rapid Commun. Mass Spectrom.* **1991**, *5*, 203.
203 Verweij, A.M.A., De Bruyn, P.C.A.M., Choufoer, C. et al., *Forensic Sci. Int.* **1993**, *60*, 7.
204 Gates, P.M., Furlong, E.T., Dorsey, T.F. et al., *Trends Anal. Chem.* **1996**, *15*, 319.
205 Astratov, M., Preiss, A., Levsen, K. et al., *Int. J. Mass Spectrom. Ion Processes* **1997**, *167/168*, 481.
206 Godejohann, M., Astratov, M., Preiss, A. et al., *Anal. Chem.* **1998**, *70*, 4104.
207 Koeber, R., Bayona, J.M., Niessner, R. *Environ. Sci. Technol.* **1999**, *33*, 1552.
208 Greaves, J., Bieri, R.H. *Int. J. Environ. Anal. Chem.* **1991**, *43*, 63.
209 Sigman, M.E., Schuler, P.F., Ghosh, M.M. et al., *Environ. Sci. Technol.* **1998**, *32*, 3980.
210 Evans, K.A., Dubey, S.T., Kravetz, L. et al. *Anal. Chem.* **1994**, *66*, 699.
211 Schröder, H. Fr. *Vom Wasser* **1992**, *79*, 193.
212 Schröder, H. Fr. Proceedings of the 4th World Surfactant Congress, eds. Comité Espanol de la Detergencia (C.E.D.), Royal Society of Chemistry, Cambridge **1996**, Vol. 3, pp. 121–135.
213 Schröder, H.Fr. *Vom Wasser,* **1992**, *78*, 211.
214 Schröder, H. Fr. *Vom Wasser* **1991**, *77*, 277.
215 Schröder, H. Fr. *Vom Wasser* **1989**, *73*, 111.
216 Clark, L.B., Rosen, R.T., Hartman, T.G. et al., *Res. J. Water Pollut. Control Fed.* **1991**, *63*, 104.
217 Schröder, H. Fr. *J. Chromatogr.* **1993**, *647*, 219.
218 Schröder, H. Fr. *Water Sci. Technol.* **1991**, *23*, 339.
219 Schröder, H. Fr. *Korr. Abwasser* **1992**, *39*, 387.
220 Paxéus, N., Schröder, H.Fr. *Water Sci. Technol.,* **1996**, *33*, 9.
221 Schröder, H. Fr. *Water Sci. Technol.* **1998**, *38*, 151.
222 Schröder, H. Fr. *Vom Wasser* **1993**, *80*, 323.
223 Schröder, H. Fr. *Vom Wasser* **1993**, *81*, 299.
224 Schröder, H. Fr. *Water Sci. Technol.* **1992**, *25*, 241.
225 Schröder, H. Fr. *Water Sci. Technol.* **1996**, *34*, 21.
226 Schröder, H.Fr. *J. Chromatogr. A* **1997**, *777*, 127.
227 Schröder, H. Fr. *J. Chromatogr. A* **1995**, *712*, 123.
228 Schröder, H. Fr. *Fresenius' J. Anal. Chem.* **1995**, *353*, 93.

229 Schröder, H. Fr. *J. Chromatogr.* **1991**, *554*, 251.
230 Schröder, H. Fr. *J. Chromatogr.* **1993**, *643*, 145.
231 Schröder, H. Fr. *Water Sci. Technol.* **1996**, *33*, 331.
232 Schröder, H. Fr. *Trends Anal. Chem.* **1996**, *15*, 349.
233 Schröder, H. Fr., in *DVGW-Schriftenreihe zum 7. BMFT-Statusseminar "Neue Technologien in der Trinkwasserversorgung"* Nr. 108, Eschborn **1990**, p. 121.
234 Schröder, H. Fr. *Vom Wasser* **1994**, *83*, 187.
235 Wils, E.R.J., Hulst, A.G. *Rapid Commun. Mass Spectrom.* **1993**, *7*, 413.
236 Mirocha, C.J., Cheong, W., Mirza, U. et al., *Rapid Commun. Mass Spectrom.* **1992**, *6*, 128.
237 Maruyama, J., Noguchi, T., Matsunaga, S. et al., *Agric. Biol. Chem.* **1984**, *48*, 2783.
238 White, K.D., Sphon, J.A., Hall, S. *Anal. Chem.* **1986**, *58*, 562.
239 Barceló, D., Durand, G., Bouvot, V. et al., *Environ. Sci. Technol.* **1993**, *27*, 271.
240 Molina, C., Durand, G., Barceló, D. *J. Chromatogr. A* **1995**, *712*, 113.
241 Kochany, J., Maguire, R.J. *J. Agric. Food Chem.* **1994**, *42*, 406.
242 Vreeken, R.J., Van Dongen, W.D., Ghijsen, R.T. et al., *Int. J. Environ. Anal. Chem.* **1994**, *54*, 119.
243 Barceló, D., Durand, G., Vreeken, R.J. *J. Chromatogr.* **1993**, *647*, 271.
244 Barceló, D., Durand, G., Vreeken, R.J. et al., *J. Chromatogr.* **1991**, *553*, 311.
245 Volmer, D., Levsen, K., Wünsch, G. *J. Chromatogr. A* **1994**, *660*, 231.
246 Chiron, S., Dupas, S., Scribe, P. et al., *J. Chromatogr. A* **1994**, *665*, 295.
247 Bagheri, H., Brouwer, E.R., Ghijsen, R.T. et al., *J. Chromatogr.* **1993**, *647*, 121.
248 Moore, K.M., Jones, S.R., James, C. *Water Res.* **1995**, *29*, 1225.
249 Chiron, S., Valverde, A., Fernandez-Alba, A. et al., *J. Assoc. Off. Anal. Chem.* **1995**, *78*, 1346.
250 Liska, I., Slobodnik, J. *J. Chromatogr. A* **1996**, *733*, 235.
251 Salau, J.S., Honing, M., Tauler, R. et al., *J. Chromatogr. A* **1998**, *795*, 3.
252 Honing, M., Barceló, D., van Baar, B.L.M. et al., *J. Am. Soc. Mass Spectrom.* **1994**, *5*, 913.
253 Volmer, D., Levsen, K., Honing, M. et al., *J. Am. Soc. Mass Spectrom.* **1995**, *6*, 656.
254 Lopez-Avila, V., Jones, T.L. *J. Assoc. Off. Anal. Chem.* **1993**, *76*, 1329.
255 Liu, C.-H., Mattern, G.C., Yu, X. et al., *J. Agric. Food Chem.* **1991**, *39*, 718.
256 Durand, G., de Bertrand, N., Barceló, D. *J. Chromatogr.* **1991**, *554*, 233.
257 Chiron, S., Torres, J.A., Fernandez-Alba, A. et al., *Int. J. Environ. Anal. Chem.* **1996**, *65*, 37.
258 Kodama, S., Yamamoto, A., Matsunaga, A. *J. Agric. Food Chem.* **1997**, *45*, 990.
259 Lacorte, S., Ehresmann, N., Barceló, D. *Environ. Sci. Technol.* **1996**, *30*, 917.
260 Sabik, H., Jeannot, R. *J. Chromatogr. A* **1998**, *18*, 197.
261 Lacorte, S., Barceló, D. *J. Chromatogr. A* **1995**, *712*, 103.
262 Lacorte, S., Lartiges, S.B., Garrigues, P. et al., *Environ. Sci. Technol.* **1995**, *29*, 431.
263 Halarnkar, P.P., Leimkuehler, W.M., Green, D.L. et al., *J. Agric. Food Chem.* **1997**, *45*, 2349.
264 Chiron, S., Martinez, E., Barceló, D. *J. Chromatogr. A* **1994**, *665*, 283.
265 Vreeken, R.J., Ghijsen, R.T., Frei, R.W. et al., *J. Chromatogr. A* **1993**, *654*, 65.
266 Barceló, D., Durand, G., Albaiges, J., in *Org. Micropollut. Aquat. Environ.*, eds. A. Bjørseth, G. Angeletti, Comm. Eur. Communities, EUR 13152, Reidel, Dordrecht **1991**, p.132.
267 Geerdink, R.B., Kienhuis, P.G.M., Brinkman, U.A.Th. *J. Chromatogr.* **1993**, *647*, 329.
268 Geerdink, R.B., Kienhuis, P.G.M., Brinkman, U.A.Th. *Chromatographia* **1994**, *39*, 311.
269 Ramanand, K., Nagarajan, A., Suflita, J.M. *Appl. Environ. Microbiol.* **1993**, *59*, 2251.
270 Mattern, G.C., Liu, C.-H., Louis, J.B. et al., *J. Agric. Food Chem.* **1991**, *39*, 700.
271 Volmer, D., Wilkes, J.G., Levsen, K. *Rapid Commun. Mass Spectrom.* **1995**, *9*, 767.

272 Brady, J.F., Turner, J., Skinner, D.H. *J. Agric. Food Chem.* **1995**, *43*, 2542.
273 Geerdink, R.B., Attema, A., Niessen, W.M.A. et al., *LC.GC Int.* **1998**, *11*, 361.
274 Geerdink, R.B., Berg, P.J., Kienhuis, P.G.M. et al., *Int. J. Environ. Anal. Chem.* **1996**, *64*, 265.
275 Lerch, R.N., Donald, W.W., Li, Y.-X. et al., *Environ. Sci. Technol.* **1995**, *29*, 2759.
276 Nelieu, S., Stobiecki, M., Sadoun, F. et al., *Analusis* **1994**, *22*, 70.
277 Abian, J., Durand, G., Barceló, D. *J. Agric. Food Chem.* **1993**, *41*, 1264.
278 Meesters, R.J.W., Forge F., Schröder, H. Fr. *Vom Wasser* **1995**, *84*, 287.
279 Schmitt, P., Freitag, D., Sanlaville, Y. et al., *J. Chromatogr. A* **1995**, *709*, 215.
280 Nelieu, S., Stobiecki, M., Kerhoas, L. et al., *Rapid Commun. Mass Spectrom.* **1994**, *8*, 945.
281 Mascolo, G., Lopez, A., Foldenyi, R. et al., *Environ. Sci. Technol.* **1995**, *29*, 2987.
282 Mascolo, G., Lopez, A., Passino, R. et al., *Water Res.* **1994**, *28*, 2499.
283 Fischer, J.B., Michael, J.L. *J. Chromatogr. A* **1995**, *704*, 131.
284 Papilloud, S., Haerdi, W., Chiron, S. et al., *Environ. Sci. Technol.* **1996**, *30*, 1822.
285 Salau, J.S., Alonso, R., Batllo, G. et al., *Anal. Chim. Acta* **1994**, *293*, 109.
286 Schrap, S.M., van den Heuvel, H., van der Meulen, J. et al., *Chemosphere* **2000**, *40*, 1389.
287 Eisert, R., Levsen, K., Wunsch, G. *Int. J. Environ. Anal. Chem.* **1995**, *58*, 103.
288 Fernandez-Alba, A.R., Valverde, A., Aguera, A. et al., *J. Chromatogr. A* **1996**, *721*, 97.
289 Stout, S.J., Peterson, R.P., daCunha, A.R. et al., *J. Assoc. Off. Anal. Chem.* **1995**, *78*, 862.
290 Maycock, R., Hastings, M., Portwood, D. *Int. J. Environ. Anal. Chem.* **1995**, *58*, 93.
291 Schreiber, A., Efer, J., Engewald, W. *J. Chromatogr. A* **2000**, *869*, 411.
292 Baumann, C., Cintora, M.A., Eichler, M. et al., *Rapid Commun. Mass Spectrom.* **2000**, *14*, 349.
293 Kienhuis, P.G.M., Geerdink, R.B., *Trends Anal. Chem.* **2000**, *19*, 460.
294 Aguera, A., Fernandez-Alba, A.R. *Analusis* **1998**, *26*, M123.
295 Ahrer, W., Scherwenk, E., Buchberger, W. *J. Chromatogr. A* **2001**, *910*, 69.
296 Schröder, H. Fr. *Waste Management* **1999**, *19*, 111.
297 Sacher, G., Nussbaum, R., Rissler, K. et al., *Chromatographia* **2001**, *54*, 65.
298 Ma, Y.-C., Kim, H.-Y. *J. Am. Soc. Mass Spectrom.* **1997**, *8*, 1010.
299 Lagana, A., Bacaloni, A., Fago, G. et al., *Rapid Commun. Mass Spectrom.* **2000**, *14*, 401.
300 Baronti, C., Curini, R., d'Ascenzo, G. et al., *Environ. Sci. Technol.* **2000**, *34*, 5059.
301 de Alda, M.J.L., Barceló, D. *J. Chromatogr. A* **2000**, *892*, 391.
302 Cassada, D.A, Monson, S.J., Snow, D. et al., *J. Chromatogr. A* **1999**, *844*, 87.
303 Ahrer, W., Buchberger, W. *Fresenius J. Anal. Chem.* **1999**, *365*, 604.
304 Galceran, M.T., Moyano, E. *J. Chromatogr. A* **1996**, *731*, 75.
305 Schröder, H. Fr. *Proceedings of the IWA-Conference, Critical F12Technologies to the World in 21st Century: Pollution Control, Reclamation in Process Industries*, Beijing, P.R.China, **2000**, p.125.
306 Thomas, D., Sim, P.G., Benoit, F.M. *Rapid Commun. Mass Spectrom.* **1994**, *8*, 105.
307 Mansoori, B.A. *Rapid Commun. Mass Spectrom.* **1998**, *12*, 712.
308 Letzel, T., Poschl, U., Rosenberg, E. et al., *Rapid Commun. Mass Spectrom.* **1999**, *13*, 2456.
309 Letzel, T., Rosenberg, E., Wissiack, R. et al., *J. Chromatogr. A* **1999**, *855*, 501.
310 Koeber, R., Bayona, J.M., Niessner, R. *Int. J. Environ. Anal. Chem.* **1997**, *66*, 313.
311 Koeber, R., Niessner, R., Bayona, J.M. *Fresenius' J. Anal. Chem.* **1997**, *359*, 267.
312 Marvin, C.H., Smith, R.W., Bryant, D.W. et al., *J. Chromatogr. A* **1999**, *863*, 13.
313 Charles, L., Pepin, D. *J. Chromatogr. A* **1998**, *804*, 105.
314 Thomas, D., Crain, S.M., Sim, P.G. et al., *J. Mass Spectrom.* **1995**, *30*, 1034.
315 Puig, D., Silgoner, I., Grasserbauer, M. et al., *Anal. Chem.* **1997**, *69*, 2756.

316 Wissiack, R., Rosenberg, E., Grasserbauer, M. *J. Chromatogr. A* **2000**, *896*, 159.
317 Jauregui, O., Moyano, E., Galceran, M.T. *J. Chromatogr. A* **1997**, *787*, 79.
318 Pedersen, S.N., Lindholst, C. *J. Chromatogr. A* **1999**, *864*, 17.
319 Motoyama, A., Suzuki, A., Shirota, O. et al., *Rapid Commun. Mass Spectrom.* **1999**, *13*, 2204.
320 Aguilar, C., Ferrer, I., Borrull, F. et al., *Anal. Chim. Acta* **1999**, *386*, 237.
321 Castillo, M., Domingues, R., Alpendurada, M.F. et al., *Anal. Chim. Acta* **1997**, *353*, 133.
322 Slobodnik, J., Hogenboom, A.C., Vreuls, J.J. et al., *J. Chromatogr. A* **1996**, *741*, 59.
323 Azevedo, D.A., Lacorte, S., Viana, P. et al., *Chromatographia* **2001**, *53*, 113.
324 de Almeida Azevedo, D., Lacorte, S., Vinhas, T. et al., *J. Chromatogr. A* **2000**, *879*, 13.
325 Baglio, D., Kotzias, D., Larsen, B.R. *J. Chromatogr. A* **1999**, *854*, 207.
326 Jewett, B.N., Ramaley, L., Kwak, J.C.T. *J. Am. Soc. Mass Spectrom.* **1999**, *10*, 529.
327 Castillo, M., Riu, J., Ventura, F. et al., *J. Chromatogr. A* **2000**, *889*, 195.
328 Petrovic, M., Barceló, D. *Fresenius J. Anal. Chem.* **2000**, *368*, 676.
329 Bachus, H., Stan, H.-J. Schriftenreihe Biologische Abwasserreinigung zum Kolloquium im Sonderforschungsbereich 193 der DFG an der Technischen Universität Berlin, Biologische Behandlung industrieller und gewerblicher Abwässer, Anwendung der LC-MS in der Wasseranalytik, **2001**, Vol. 11, p. 248.
330 Billian, P., Hock, W., Doetzer, R. et al., *Anal. Chem.* **2000**, *72*, 4973.
331 Scullion, S.D., Clench, M.R., Cooke, M. et al., *J. Chromatogr. A* **1996**, *733*, 207.
332 Yamagishi, T., Hashimoto, S., Kanai, M. et al., *Bunseki Kagaku* **1997**, *46*, 537.
333 Bester, K., Theobald, N., Schröder, H.Fr. *Chemosphere*, **2001**, *45*, 817.
334 Cretier, G., Podevin, C., Rocca, J.-L. *Analusis* **1999**, *27*, 758.
335 Li, H.-Q., Schröder, H. Fr. *Water. Sci. Technol.* **2000**, *42*, 391.
336 Schröder, H. Fr., Fytianos, K. *Chromatographia* **1999**, *50*, 583.
337 Efer, J., Schreiber, A., Ceglarek, U., Engewald, W. Schriftenreihe Biologische Abwasserreinigung zum Kolloquium im Sonderforschungsbereich 193 der DFG an der Technischen Universität Berlin, Biologische Behandlung industrieller und gewerblicher Abwässer, Anwendung der LC-MS in der Wasseranalytik, **1999**, Vol. 11, p. 125.
338 Efer, J., Ceglarek, U., Anspach, T. et al., *Wasser Abwasser* **1996**, *138*, 577.
339 Preiss, A., Levsen, K. Schriftenreihe Biologische Abwasserreinigung zum Kolloquium im Sonderforschungsbereich 193 der DFG an der Technischen Universität Berlin, Biologische Behandlung industrieller und gewerblicher Abwässer, Anwendung der LC-MS in der Wasseranalytik, **1999**, Vol. 11, p. 169.
340 Plomley, J.B., Crozier, P.W., Taguchi, V.Y. *J. Chromatogr. A* **1999**, *854*, 245.
341 Castillo, M., Barceló, D., in *Techniques, Instrumentation in Analytical Chemistry*, Vol. 21, Sample Handling, Trace Analysis of Pollutants - Techniques, Applications, Quality Assurance, ed. Barceló, D., Elsevier, Amsterdam **2000**, p.537.
342 Schröder, H.Fr. *J. Chromatogr. A.*, **2001**, *926*, 127.
343 Li, H.-Q., Jiku, F., Schröder, H. Fr. *J. Chromatogr. A* **2000**, *889*, 155.
344 Schröder, H.Fr. *Water Sci. Technol.*, **2002**, *46*, 57.
345 Petrovic, M., Diaz, A., Ventura, F. et al., *Anal. Chem.* **2001**, *73*, 5886.
346 Petrovic, M., Barceló, D. *Anal. Chem.* **2000**, *72*, 4560.
347 Castro, R., Moyano, E., Galceran, M.T. *J. Chromatogr. A* **1999**, *830*, 145.
348 Santos, T.C.R., Rocha, J.C., Barceló, D. *J. Chromatogr. A* **2000**, *879*, 3.
349 Santos, T.C.R., Rocha, J.C., Alonso, R.M. et al., *Environ. Sci. Technol.* **1998**, *32*, 3479.
350 Spliid, N.H., Koppen, B. *J. Chromatogr. A* **1996**, *736*, 105.
351 Spliid, N.H., Koppen, B. *Chemosphere* **1998**, *37*, 1307.
352 Pothuluri, J.V., Evans, F.E., Doerge, D.R. et al., *Arch. Environ. Contam. Toxicol.* **1997**, *32*, 117.

353 Hogenboom, A.C., Slobodnik, J., Vreuls, J.J. et al., *Chromatographia* **1996**, *42*, 506.
354 Doerge, D.R., Bajic, S. *Rapid Commun. Mass Spectrom.* **1992**, *6*, 663.
355 Fernandez, M., Pico, Y., Manes, J. *J. Chromatogr. A* **2000**, *871*, 43.
356 Thomas, D.H., Lopez-Avila, V., Betowski, L.D. et al., *J. Chromatogr. A* **1996**, *724*, 207.
357 Takino, M., Yamagami, T., Daishima, S. *Bunseki Kagaku* **1997**, *46*, 555.
358 Barnes, K.A., Fussell, R.J., Startin, J.R. et al., *Rapid Commun. Mass Spectrom.* **1997**, *11*, 117.
359 Nunes, G.S., Marco, M.P., Ribeiro, M.L. et al., *J. Chromatogr. A* **1998**, *823*, 109.
360 Itoh, H., Kawasaki, S., Tadano, J. *J. Chromatogr. A* **1996**, *754*, 61.
361 Aguilar, C., Ferrer, I., Borrull, F., Marce et al., *J. Chromatogr. A* **1998**, *794*, 147.
362 Lacorte, S., Vreuls, J.J., Salau et al., *J. Chromatogr. A* **1998**, *795*, 71.
363 Lacorte, S., Barceló, D. *Anal. Chem.* **1996**, *68*, 2464.
364 Lacorte, S., Molina, C., Barceló, D. *J. Chromatogr. A* **1998**, *795*, 13.
365 Fernandez-Alba, A.R., Aguera, A., Contreras, M. et al., *J. Chromatogr. A* **1998**, *823*, 35.
366 Lacorte, S., Jeanty, G., Marty, J.-L. et al., *J. Chromatogr. A* **1997**, *777*, 99.
367 Oubina, A., Ferrer, I., Gascon, J. et al., *Environ. Sci. Technol.* **1996**, *30*, 3551.
368 Ferrer, I., Barceló, D. *J. Chromatogr. A* **1999**, *854*, 197.
369 Ferrer, I., Barceló, D. *J. Chromatogr. A* **1997**, *778*, 161.
370 Kawasaki, S., Ueda, H., Itoh, H. et al., *J. Chromatogr.* **1992**, *595*, 193.
371 Katagi, T. *J. Agric. Food Chem.* **1993**, *41*, 496.
372 Chiron, S., Rodriguez, A., Fernandez-Alba, A. *J. Chromatogr. A* **1998**, *823*, 97.
373 Dobberstein, P., Münster, H. *J. Chromatogr. A* **1995**, *712*, 3.
374 Jeannot, R., Sabik, H., Sauvard, E. et al., *J. Chromatogr. A* **2000**, *879*, 51.
375 Hogenboom, A.C., Speksnijder, P., Vreeken, R.J. et al., *J. Chromatogr. A* **1997**, *777*, 81.
376 Hogenboom, A.C., Niessen, W.M.A., Brinkman, U.A.Th. *J. Chromatogr. A* **1998**, *794*, 201.
377 Ferrer, I., Ballesteros, B., Marco, M.P. et al., *Environ. Sci. Technol.* **1997**, *31*, 3530.
378 Thomas, K.V. *J. Chromatogr. A* **1998**, *825*, 29.
379 van der Heeft, E., Dijkman, E., Baumann, R.A. et al., *J. Chromatogr. A* **2000**, *879*, 39.
380 Yarita, T., Sugino, K., Ihara, T. et al., *Anal. Commun.* **1998**, *35*, 91.
381 Bossi, R., Vejrup, K., Jacobsen, C.S. *J. Chromatogr. A* **1999**, *855*, 575.
382 Ferrer, I., Hennion, M.-C., Barceló, D. *Anal. Chem.* **1997**, *69*, 4508.
383 Ferrer, I., Pichon, V., Hennion, M.-C. et al., *J. Chromatogr. A* **1997**, *777*, 91.
384 Barnes, K.A., Fussell, R.J., Startin, J.R. et al., *Rapid Commun. Mass Spectrom.* **1995**, *9*, 1441.
385 Asperger, A., Efer, J., Koal, T. et al., *J. Chromatogr. A* **2001**, *937*, 65.
386 Hogenboom, A.C., Brinkman, U.A.Th., Niessen, W.M.A. *LC.GC Int.* **1997**, *10*, 669.
387 Geerdink, R.B., Niessen, W.M.A., Brinkman, U.A.Th. *J. Chromatogr. A* **2001**, *910*, 291.
388 Lerch, R.N., Blanchard, P.E., Thurman, E.M. *Environ. Sci. Technol.* **1998**, *32*, 40.
389 Bouquard, C., Ouazzani, J., Prome, J.-C. et al., *Appl. Environ. Microbiol.* **1997**, *63*, 862.
390 Klaus, U., Pfeifer, T., Spiteller, M. *Environ. Sci. Technol.* **2000**, *34*, 3514.
391 Bobeldijk, I., Broess, K., Speksnijder, P. et al., *J. Chromatogr. A*, **2001**, *938*, 15.
392 Aguera, A., Almansa, E., Malato, S. et al., *Analusis* **1998**, *26*, 245.
393 Mol, H.G.J., van Dam, R.C.J., Vreeken, R.J. et al., *J. Chromatogr. A* **1999**, *833*, 53.
394 Barnes, K.A., Fussell, R.J., Startin, J.R. et al., *Rapid Commun. Mass Spectrom.* **1997**, *11*, 159.
395 Loos, R., Alonso, M.C., Barceló, D. *J. Chromatogr. A* **2000**, *890*, 225.
396 Quilliam, M.A., in *Applications of LC-MS in Environmental Chemistry*, ed. D. Barceló, Elsevier, Amsterdam **1996**, p. 414.

397 Bauer, K.-H., Knepper, T.P., Maes, A. et al., *J. Chromatogr. A* **1999**, *837*, 117.
398 Baron, D., Hering, J.G. *J. Environ. Qual.* **1998**, *27*, 844.
399 Sheppard, R.L., Henion, J. *Electrophoresis* **1997**, *18*, 287.
400 Hartig, C., Storm, T., Jekel, M. *J. Chromatogr. A* **1999**, *854*, 163.
401 Lange, F.T., Sacher, F., Metzinger, M., Wenz, M. *Schriftenreihe Biologische Abwasserreinigung zum Kolloquium im Sonderforschungsbereich 193 der DFG an der Technischen Universität Berlin, Biologische Behandlung industrieller und gewerblicher Abwässer, Anwendung der LC-MS in der Wasseranalytik*, **2001**, Vol. 11, p. 193.
402 Ternes, T.A., Hirsch, R., Mueller, J. et al., *Fresenius J. Anal. Chem.* **1998**, *362*, 329.
403 Putschew, A., Jekel, M. *Schriftenreihe Biologische Abwasserreinigung zum Kolloquium im Sonderforschungsbereich 193 der DFG an der Technischen Universität Berlin, Biologische Behandlung industrieller und gewerblicher Abwässer, Anwendung der LC-MS in der Wasseranalytik*, **2001**, Vol. 11, p. 213.
404 Hirsch, R., Ternes, T.A., Haberer, K. et al., *J. Chromatogr. A* **1998**, *815*, 213.
405 Zhu, J., Snow, D.D., Cassada, D.A. et al., *J. Chromatogr. A* **2001**, *928*, 177.
406 Putschew, A., Wischnack, S., Jekel, M. *Sci. Total Environ.* **2000**, *255*, 129.
407 Hirsch, R., Ternes, T.A., Lindart, A. et al., *Fresenius J. Anal. Chem.* **2000**, *366*, 835.
408 Ternes, T., Bonerz, M., Schmidt, T. *J. Chromatogr. A* **2001**, *938*, 175.
409 la Farré, M., Ferrer, I., Ginebreda, A. et al., *J. Chromatogr. A* **2001**, *938*, 187.
410 Sacher, F., Lange, F.T., Brauch, H.-J. et al., *J. Chromatogr. A* **2001**, *938*, 199.
411 Hartig, C., Jekel, M. *Schriftenreihe Biologische Abwasserreinigung zum Kolloquium im Sonderforschungsbereich 193 der DFG an der Technischen Universität Berlin, Biologische Behandlung industrieller und gewerblicher Abwässer, Anwendung der LC-MS in der Wasseranalytik*, **1999**, Vol. 11, p. 93.
412 Knepper, T.P., Kirschhöfer, F., Lichter, I. et al., *Environ. Sci. Technol.* **1999**, *33*, 945.
413 Minoia, C., Turci, R., Sottani, C. et al., *Rapid Commun. Mass Spectrom.* **1998**, *12*, 1485.
414 Möhle, E., Kempter, C., Kern, A. et al., *Acta Hydrochim. Hydrobiol.* **1999**, *27*, 430.
415 New, A.P., Freitas dos Santos, L.M., Biundo, G. L. et al. *J. Chromatogr. A* **2000**, *889*, 177.
416 Storm, T., Hartig, C., Reemstma, T. et al., *Anal. Chem.* **2001**, *73*, 589.
417 Reemtsma, T. *J. Chromatogr. A.* **2001**, *919*, 289.
418 Bruins, A.P., Covey, T.H., Henion, J.D. *Anal. Chem.* **1987**, *59*, 2642.
419 Bruins, A.P., Weidolf, L.O.G., Henion, J.D. et al., *Anal. Chem.* **1987**, *59*, 2647.
420 Edlund, P.O., Lee, E.D., Henion, J.D. et al., *Biomed. Environ. Mass Spectrom.* **1989**, *18*, 233.
421 McCallum, J.E.B., Madison, S.A., Alkan, S. et al., *Environ. Sci. Technol.* **2000**, *34*, 5157.
422 Preiss, A., Sanger, U., Karfich, N. et al., *Anal. Chem.* **2000**, *72*, 992.
423 Riu, J., Schonsee, I., Barceló, D. *J. Mass Spectrom.* **1998**, *33*, 653.
424 Lin, H.-Y., Voyksner, R.D. *Anal. Chem.* **1993**, *65*, 451.
425 Barceló, D., Riu, J. *Biomed. Chromatogr.* **2000**, *14*, 14.
426 Poiger, T., Richardson, S.D., Baughman, G.L. *J. Chromatogr. A* **2000**, *886*, 271.
427 Holcapek, M., Jandera, P., Zderadicka, P. *J. Chromatogr. A* **2001**, *926*, 175.
428 Siu, K.W.M., Guevremont, R., Le Blanc, J.C.Y. et al., *J. Chromatogr.* **1991**, *554*, 27.
429 Laszlo, J.A. *Environ. Sci. Technol.* **1997**, *31*, 3647.
430 de Alda, M.J.L., Barceló, D. *J. Chromatogr. A* **2001**, *938*, 145.
431 Groom, C.A., Beaudet, S., Halasz, A. et al., *J. Chromatogr. A* **2001**, *909*, 53.
432 Bruns-Nagel, D., Drzyzga, O., Steinbach, K. et al., *Environ. Sci. Technol.* **1998**, *32*, 1676.
433 Hashimoto, S., Otsuki, A. *HRC-J. High Resolut. Chromatogr.* **1998**, *21*, 55.
434 Takino, M., Daishima, S., Yamaguchi, K. *Analyst* **2000**, *125*, 1097.
435 Ells, B., Barnett, D.A., Purves, R.W. et al., *Anal Chem.* **2000**, *72*, 4555.

436 Loos, R., Barceló, D. *J. Chromatogr. A* **2001**, *938*, 45.
437 Ells, B., Barnett, D.A., Froese, K. et al., *Anal. Chem.* **1999**, *71*, 4747.
438 Magnuson, M.L., Kelty, C.A. *Anal. Chem.* **2000**, *72*, 2308.
439 Debre, O., Budde, W.L., Song, X. *J. Am. Soc. Mass. Spectrom.* **2000**, 11, 809.
440 Richardson, S.D., Thruston, A.D., Jr., Caughran, T.V.et al., *Proc.-Water Qual. Technol. Conf.* **1998** 2223.
441 Shimizu, N., Inoue, Y., Daishima, S. et al., *Anal. Sci.* **1999**, *15*, 685.
442 Takino, M., Daishima, S., Yamaguchi, K. et al., *J. Chromatogr. A* **2001**, *928*, 53.
443 Sasaki, J., Aschmann, S.M., Kwok, E.S.C. et al., *Environ. Sci. Technol.* **1997**, *31*, 3173.
444 Tsai, C.Y., Her, G.R. *J. Chromatogr. A* **1996**, *743*, 315.
445 Storm, T., Reemtsma, T., Jekel, M. *J. Chromatogr. A* **1999**, *854*, 175.
446 Suter, M.J.F., Riediker, S., Zipper, C. et al., *Analusis* **1997**, *25*, M23.
447 Riediker, S., Suter, M.J.F., Giger, W. *Water Res.* **2000**, *34*, 2069.
448 Crescenzi, C., Di Corcia, A., Marcomini, A. et al., *J. Chromatogr. A* **2001**, *923*, 97.
449 Wolf, C., Storm, T., Lange, F.T. et al., *Anal. Chem.* **2000**, *72*, 5466.
450 Suter, M.J.-F., Riediker, S., Schwoerer, V.G. *Schriftenreihe Biologische Abwasserreinigung zum Kolloquium im Sonderforschungsbereich 193 der DFG an der Technischen Universität Berlin, Biologische Behandlung industrieller und gewerblicher Abwässer, Anwendung der LC-MS in der Wasseranalytik*, **1999**, Vol. 11, p. 41.
451 Storm, T., Reemtsma, T., Jekel, M. *Schriftenreihe Biologische Abwasserreinigung zum Kolloquium im Sonderforschungsbereich 193 der DFG an der Technischen Universität Berlin, Biologische Behandlung industrieller und gewerblicher Abwässer, Anwendung der LC-MS in der Wasseranalytik*, **1999**, Vol. 11, p. 57.
452 Crescenzi, C., DiCorcia, A., Samperi, R. et al., *Anal. Chem.* **1995**, *67*, 1797.
453 DiCorcia, A., Crescenzi, C., Marcomini, A. et al., *Environ. Sci. Technol.* **1998**, *32*, 711.
454 Sherrard, K.B., Marriott, P.J., Amiet, R.G. et al., *Environ. Sci. Technol.* **1995**, *29*, 2235.
455 Sherrard, K.B., Marriott, P.J., McCormick, M.J. et al., *Anal. Chem.* **1994**, *66*, 3394.
456 Mackay, L.G., Croft, M.Y., Selby, D.S. et al., *J. Assoc. Off. Anal. Chem.* **1997**, *80*, 401.
457 Cohen, A, Klint, K., Bøwadt, S. et al., *J. Chromatogr. A* **2001**, *927*, 103.
458 Ferguson, P.L., Iden, C.R., Brownawell, B.J. *J. Chromatogr. A* **2001**, *938*, 79.
459 Ferguson, P.L., Iden, C.R., Brownawell, B.J. *Anal Chem.* **2000**, *72*, 4322.
460 Takino, M., Daishima, S., Yamaguchi, K. *J. Chromatogr. A* **2000**, *904*, 65.
461 DiCorcia, A., Costantino, A., Crescenzi, C. et al., *Environ. Sci. Technol.* **1998**, *32*, 2401.
462 Bokern, M., Nimtz, M., Harms, H.H. *J. Agric. Food Chem.* **1996**, *44*, 1123.
463 Shang, D.Y., Ikonomou, M.G., MacDonald, R.W. *J. Chromatogr. A* **1999**, *849*, 467.
464 Popenoe, D.D., Morris, S.J., III, Horn, P.S. et al., *Anal. Chem.* **1994**, *66*, 1620.
465 Ceglarek, U., Efer, J., Schreiber, A. et al., *Fresenius J. Anal. Chem.* **1999**, *365*, 674.
466 Moder, M., Popp, P., Pawliszyn, J. *J. Microcol. Sep.* **1998**, *10*, 225.
467 Riu, J., Eichhorn, P., Guerrero, J.A. et al., *J. Chromatogr. A* **2000**, *889*, 221.
468 Crescenzi, C., DiCorcia, A., Marchiori, E. et al., *Water Res.* **1996**, *30*, 722.
469 González-Mazo, E., Honing, M., Barceló, D. et al., *Environ. Sci. Technol.* **1997**, *31*, 504.
470 Riu, J., González-Mazo, E., Gómez-Parra, A. et al., *Chromatographia* **1999**, *50*, 275.
471 Mantzavinos, D., Burrows, D.M.P., Willey, R. et al., *Water Res.* **2001**, *35*, 3337.
472 Moody, C.A., Kwan, W.C., Martin, J.W. et al., *Anal. Chem.* **2001**, *73*, 2200.
473 Conboy, J.J., Henion, J.D., Martin, M.W. et al., *Anal. Chem.* **1990**, *62*, 800.
474 Radke, M., Behrends, T., Foerster, J. et al., *Anal. Chem.* **1999**, *71*, 5362.
475 Fernandez, P., Alder, A.C., Suter, M.J.-F. et al., *Anal. Chem.* **1996**, *68*, 921.

476 Eichhorn, P., Knepper, T.P. *J. Mass Spectrom.* **2000**, *35*, 468.

477 Eichhorn, P., Knepper, T.P. *J. Chromatogr.* A **1999**, *854*, 221.

478 Eichhorn, P., Knepper, T.P. Schriftenreihe Biologische Abwasserreinigung zum Kolloquium im Sonderforschungsbereich 193 der DFG an der Technischen Universität Berlin, Biologische Behandlung industrieller und gewerblicher Abwässer, Anwendung der LC-MS in der Wasseranalytik, **1999**, Vol. 11, p. 156.

479 Schröder, H.Fr. Schriftenreihe Biologische Abwasserreinigung zum Kolloquium im Sonderforschungsbereich 193 der DFG an der Technischen Universität Berlin, Biologische Behandlung industrieller und gewerblicher Abwässer, Anwendung der LC-MS in der Wasseranalytik, **2001**, Vol. 11, p. 223.

480 Porschmann, J., Blasberg, L., Mackenzie, K. et al., *J. Chromatogr.* A **1998**, *816*, 221.

481 Cuzzola, A., Raffaelli, A., Saba, A. et al., *Rapid Commun. Mass Spectrom.* **1999**, *13*, 2140.

482 Cuzzola, A., Raffaelli, A., Saba, A. et al., *Rapid Commun. Mass Spectrom.* **2000**, *14*, 834.

483 Furey, A, Lehane, M., Gillman, M. et al., *J. Chromatogr.* A **2001**, *938*, 167.

484 Quilliam, M.A. *J. Assoc. Off. Anal. Chem. Int.* **1995**, *78*, 555.

485 Lewis, R.J., Holmes, M.J., Alewood, P.F.et al., *Nat. Toxins*, **1994**, *2*, 56.

486 Siren, H., Jussila, M., Liu, H. et al., *J. Chromatogr.* A **1999**, *839*, 203.

487 James, K.J., Bishop, A.G., Draisci, R. et al., *J. Chromatogr.* A **1999**, *844*, 53.

488 Rivasseau, C., Hennion, M.-C. *Anal. Chim. Acta* **1999**, *399*, 75.

489 Hiraoka, K., Saito, S., Katsuragawa, J. et al., *Rapid Commun. Mass Spectrom.* **1998**, *12*, 1170.

490 Moyano, E., Games, D.E., Galceran, M.T. *Rapid Commun. Mass Spectrom.* **1996**, *10*, 1379.

491 Wycherley, D., Rose, M.E., Giles, K. et al., *J. Chromatogr.* A **1996**, *734*, 339.

492 Castro, R., Moyano, E., Galceran, M.T. *J. Chromatogr.* A **2000**, *869*, 441.

493 Evans, C.S., Startin, J.R., Goodall, D.M. et al., *J. Chromatogr.* A **2000**, *897*, 399.

494 Taguchi, V.Y., Jenkins, S.W.D., Crozier, P.W. et al., *J. Am. Soc. Mass Spectrom.* **1998**, *9*, 830.

495 Marr, J.C., King, J.B., *Rapid Commun. Mass Spectrom.* **1997**, *11*, 479.

496 Song, X., Budde, W.L. *J. Am. Soc. Mass Spectrom.* **1996**, *7*, 981.

497 Juhler, R.K., Vahl, M. *J. Assoc. Off. Anal. Chem.* **1999**, *82*, 331.

498 Ferrer, I., Thurman, E.M., Barceló, D. *Anal. Chem.* **1997**, *69*, 4547.

499 Vargo, J.D. *Anal. Chem.* **1998**, *70*, 2699.

500 DiCorcia, A., Nazzari, M., Rao, R. et al., *J. Chromatogr.* A **2000**, *878*, 87.

501 Chiron, S., Papilloud, S., Haerdi, W. et al., *Anal. Chem.* **1995**, *67*, 1637.

502 Hogenboom, A.C., Hofman, M.P., Jolly, D.A. et al., *J. Chromatogr.* A **2000**, *885*, 377.

503 Volmer, D.A., Vollmer, D.L., Wilkes, J.G. *LC.GC* **1996**, *14*, 216.

504 Mangiapan, S., Benfenati, E., Grasso, P. et al., *Environ. Sci. Technol.* **1997**, *31*, 3637.

505 Hogenboom, A.C., Niessen, W.M.A., Brinkman, U.A.Th. *Rapid Commun. Mass Spectrom.* **2000**, *14*, 1914.

506 Hogenboom, A.C., Niessen, W.M.A., Brinkman, U.A.Th. *J. Chromatogr.* A **1999**, *841*, 33.

507 Volmer, D.A., Hui, J.P.M. *Arch. Environ. Contam. Toxicol.* **1998**, *35*, 1.

508 Gardner, M.S., Voyksner, R.D., Haney, C.A. *Anal. Chem.* **2000**, *72*, 4659.

509 Honing, M., Riu, J., Barceló, D. et al., *J. Chromatogr.* A **1996**, *733*, 283.

510 Crescenzi, C., DiCorcia, A., Guerriero, E. et al., *Environ. Sci. Technol.* **1997**, *31*, 479.

511 Giraud, D., Ventura, A., Camel, V. et al., *J. Chromatogr.* A **1997**, *777*, 115.

512 Wang, N., Budde, W.L. *Anal. Chem.* **2001**, *73*, 997.

513 Moder, M., Popp, P., Eisert, R. et al., *Fresenius J. Anal. Chem.* **1999**, *363*, 680.

514 DiCorcia, A., Crescenzi, C., Lagana, A. et al., *J. Agric. Food Chem.* **1996**, *44*, 1930.

515 Lacassie, E., Dreyfuss, M.-F., Daguet, J.L. et al., *J. Chromatogr.* A **1999**, *830*, 135.

516 Banoub, J., Gentil, E., Kiceniuk, J. *Int. J. Environ. Anal. Chem.* **1995**, *61*, 143.

517 Molina, C., Grasso, P., Benfenati, E. Et al., *J. Chromatogr. A* **1996**, *737*, 47.
518 Molina, C., Honing, M., Barceló, D. *Anal. Chem.* **1994**, *66*, 4444.
519 Hernandez, F., Sancho, J.V., Pozo, O. et al., *J. Chromatogr. A* **2001**, *939*, 1.
520 Molina, C., Grasso, P., Benfenati, E. et al., *Int. J. Environ. Anal. Chem.* **1996**, *65*, 69.
521 Vreeken, R.J., Speksnijder, P., Bobeldijk-Pastorova, I. et al., *J. Chromatogr. A* **1998**, *794*, 187.
522 Capps, T.M., Barringer, V.M., Eberle, W.J. et al., *J. Agric. Food Chem.* **1996**, *44*, 2408.
523 Mohsin, S.B. *Anal. Chem* **1999**, *71*, 3603.
524 Song, X., Budde, W.L. *J. Chromatogr. A* **1998**, *829*, 327.
525 D'Ascenzo, G., Gentili, A., Marchese, S. et al., *Environ. Sci. Technol.* **1998**, *32*, 1340.
526 Otsuka, K., Smith, C.J., Grainger, J. et al., *J. Chromatogr. A* **1998**, *817*, 75.
527 Koppen, B., Spliid, N.H. *J. Chromatogr. A* **1998**, *803*, 157.
528 D'Ascenzo, G., Gentili, A., Marchese, S. et al., *Chromatographia* **1998**, *48*, 497.
529 Crescenzi, C., Di Corcia, A., Marchese, S. et al., *Anal. Chem.* **1995**, *67*, 1968.
530 Pozo, O., Pitarch, E., Sancho, J.V. et al., *J. Chromatogr. A* **2001**, *923*, 75.
531 D'Ascenzo, G., Gentili, A., Marchese, S. et al., *J. Chromatogr. A* **1998**, *813*, 285.
532 Lagana, A., Fago, G., Marino, A. et al., *Anal. Chim. Acta* **1998**, *375*, 107.
533 Reiser, R.W., Fogiel, A.J. *Rapid Commun. Mass Spectrom.* **1994**, *8*, 252.
534 Garcia, F., Henion, J.D. *J. Chromatogr.* **1992**, *606*, 237.
535 Krynitsky, A.J., *J. Assoc. Off. Anal. Chem.* **1997**, *80*, 392.
536 Steen, R.J.C.A., Hogenboom, A.C., Leonards, P.E.G. et al., *J. Chromatogr. A* **1999**, *857*, 157.
537 Curini, R., Gentili, A., Marchese, S. et al., *Chromatographia* **2001**, *53*, 244.
538 Bobeldijk, I., Vissers, J.P.C., Kearney, G. et al., *J. Chromatogr. A* **2001**, *929*, 63.
539 Baltussen, E., Snijders, H., Janssen, H.-G. et al., *J. Chromatogr. A* **1998**, *802*, 285.

540 Rodriguez, M., Orescan, D.B. *Anal. Chem.* **1998**, *70*, 2710.
541 DiCorcia, A., Crescenzi, C., Samperi, R. et al., *Anal. Chem.* **1997**, *69*, 2819.
542 DiCorcia, A., Costantino, A., Crescenzi, C. et al., *J. Chromatogr. A* **1999**, *852*, 465.
543 Hancock, P.M., White, S.J.G., Catlow, D.A. et al., *Rapid Commun. Mass Spectrom.* **1997**, *11*, 195.
544 Hancock, P.M., Walsh, M., White, S.J.G. et al., *Analyst* **1998**, *123*, 1669.
545 Scheunert, I., Reuter, S. *Environ. Pollut.* **2000**, *108*, 61.
546 Crescenzi, C., Di Corcia, A., Nazzari, M. et al., *Anal. Chem.* **2000**, *72*, 3050.
547 Drexler, D.M., Tiller, P.R., Wilbert, S.M. et al., *Rapid Commun. Mass Spectrom.* **1998**, *12*, 1501.
548 Gennari, M., Ferraris, L., Negre, M. et al., *J. Assoc. Off. Anal. Chem.* **2000**, *83*, 1076.
549 Marek, L.J., Koskinen, W.C. *J. Agric. Food Chem.* **1996**, *44*, 3878.
550 Bossi, R., Koppen, B., Spliid, N.H. et al., *J. Assoc. Off. Anal. Chem.* **1998**, *81*, 775.
551 Li, L.Y.-T., Campbell, D.A., Bennett, P.K. et al., *Anal. Chem.* **1996**, *68*, 3397.
552 Brown, H.M., Joshi, M.M., Van, A.T. et al., *J. Agric. Food Chem.* **1997**, *45*, 955.
553 Hogenboom, A.C., Steen, R.J.C.A., Niessen, W.M.A. et al., *Chromatographia* **1998**, *48*, 475.
554 Zambonin, C.G., Losito, I., Palmisano, F. *Rapid Commun. Mass Spectrom.* **2000**, *14*, 824.
555 Banoub, J., Gentil, E., Kiceniuk, J. *Int. J. Environ. Anal. Chem.* **1995**, *61*, 11.
556 Nelson, W.M., Tang, Q., Harrata, A.K. et al., *J. Chromatogr. A* **1996**, *749*, 219.
557 DiCorcia, A., Crescenzi, C., Guerriero, E. et al., *Environ. Sci. Technol.* **1997**, *31*, 1658.
558 Cai, Z., Cerny, R.L., Spalding, R.F. *J. Chromatogr. A* **1996**, *753*, 243.
559 Hogenboom, A.C., Niessen, W.M.A., Little, D. et al., *Rapid Commun. Mass Spectrom.* **1999**, *13*, 125.
560 Boundy-Mills, K.L., de Souza, M.L., Mandelbaum, R.T. et al., *Appl. Environ. Microbiol.* **1997**, *63*, 916.
561 Arnold, S.M., Talaat, R.E., Hickey, W.J. et al., *J. Mass Spectrom.* **1995**, *30*, 452.

562 Monson, S.J., Ma, L., Cassada, D.A. et al., *Anal. Chim. Acta* **1998**, *373*, 153.

563 González-Toledo, E., Companó, R., Dolors Prat, M. et al., *J. Chromatogr. A* **2002**, *946*, 1.

564 Banoub, J.H., Martin, R.C., Hodder, H. et al., *Analusis* **1997**, *25*, M15.

565 Choi, B.K., Gusev, A.I., Hercules, D.M. *Anal. Chem.* **1999**, *71*, 4107.

566 Valenzuela, A.I., Redondo, M.J., Pico, Y. et al., *J. Chromatogr. A* **2000**, *871*, 57.

567 Chukwudebe, A.C., Atkins, R.H., Wislocki, P.G. *J. Agric. Food Chem.* **1997**, *45*, 4137.

568 D'Ascenzo, G., Gentili, A., Marchese, S. et al., *J. Chromatogr. A* **1998**, *800*, 109.

569 Lagana, A., Fago, G., Marino, A. *Anal. Chem.* **1998**, *70*, 121.

570 D'Ascenzo, G., Gentili, A., Marchese, S. et al., *Analusis* **1998**, *26*, 251.

571 Stout, S.J., DaCunha, A.R., Safarpour, M.M. *J. Assoc. Off. Anal. Chem.* **1997**, *80*, 426.

572 Krynitsky, A.J., Stout, S.J., Nejad, H. et al., *J. Assoc. Off. Anal. Chem.* **1999**, *82*, 956.

573 Stout, S.J., daCunha, A.R., Picard, G.L. et al., *J. Agric. Food Chem.* **1996**, *44*, 2182.

574 Stout, S.J., daCunha, A.R., Picard, G.L. et al., *J. Assoc. Off. Anal. Chem.* **1998**, *81*, 685.

575 Stout, S.J., daCunha, A.R., Picard, G.L. et al., *J. Agric. Food Chem.* **1996**, *44*, 3548.

576 Bean, K.A., Henion, J.D. *J. Chromatogr. A* **1997**, *791*, 119.

577 Ahrer, W., Buchberger, W. *J. Chromatogr. A* **1999**, *854*, 275.

578 Koester, C.J., Beller, H.R., Halden, R.U. *Environ. Sci. Technol.*, **2000**, *34*, 1862.

579 Charles, L., Pepin, D., Casetta, B. *Anal. Chem.* **1996**, *68*, 2554.

580 Buchberger, W., Ahrer, W. *J. Chromatogr. A* **1999**, *850*, 99.

581 Magnuson, M.L., Urbansky, E.T., Kelty, C.A. *Anal. Chem.* **2000**, *72*, 25.

582 Rivera, J., Ventura, F., Caixach, J. et al., in *Organic Micropollutants in the Aquatic Environment*, Vienna **1985**, p. 77.

16
Gas Chromatography/Ion Trap Mass Spectrometry (GC/ITMS) for Environmental Analysis

Michel Sablier and Toshihiro Fujii

16.1
Introduction

The development of the concept of linking analytical instrumentation is motivated by the need for chemists to unequivocally detect, characterize, and quantify compounds in unknown mixtures. Mass spectrometry has long been recognized as a valuable analytical tool because of its superior reproducibility, repeatability, specificity, and limits of detection. The coupling of the "universal" detection power of mass spectrometry (MS) with the relatively inexpensive and versatile separation capabilities of gas chromatography (GC), together with the technological advances that have been achieved over the past 30 years, has led to widespread use of GC/MS instrumentation, some 80% of which is represented by GC/MS quadrupole-based systems.

Quadrupole ion trap mass spectrometry is a recent development in mass spectrometry that appears to be particularly well suited for high-sensitivity chromatographic applications. An ion trap, or Paul trap, is an extraordinary device that functions both as an ion store and as a mass spectrometer. The development of GC/MS using an ion trap mass spectrometer has recently culminated in the realization of low-cost instruments capable of classical modes of ionization associated with MS/MS, ranking ion traps in the domain of both high-specificity mass spectrometers and benchtop mass spectrometers. Advances in ion trap technology for GC/MS, such as improved trapping efficiency and resolution, have led to numerous applications in many areas of analytical chemistry. The ion storage capability of ion traps provides a sensitivity advantage over that of traditional quadrupole mass spectrometers and has spurred rapid growth of the GC/MS ion trap market, with probably over 5000 ion trap units sold worldwide.

Although ion trap mass spectrometry is a relatively new technique, the use of ion traps for GC coupling has been widespread, with the main driving force residing in the ability to obtain full-scan mass spectra with a very high sensitivity, often one order of magnitude larger than with traditional quadrupole mass spectrometers.

Many aspects of ion traps are reported in the literature [1–3], and these features are not covered here. Additionally, Chapter 10 of this Handbook covers the general principles and applications of mass spectrometry. Rather, this chapter focuses on the use of GC/ion trap mass spectrometry (ITMS) in environmental analysis, describing the principles of operation, giving examples of applications of ion traps in the environmental field, and highlighting recent developments in this area.

16.2
Practical Aspects of GC/ITMS

16.2.1
Historical survey

The first commercial GC ion trap mass spectrometer, the ITD700®, was launched by Finnigan in 1984. It was capable of acquiring full-scan and selected ion electron ionization (EI) mass spectra over a mass range of 650 u for samples eluting from a GC column coupled to an ion trap via a heated open-split interface. The ITD800® ion trap detector was then introduced as an upgraded version of the ITD700® and included chemical ionization (CI) capabilities and automatic ionization control routine procedures. ITMS®, the next ion trap mass spectrometer marketed by Finnigan, was a multipurpose research instrument providing MS/MS custom scan procedures through its scan editor software. The ITS40® ion trap, also marketed by Finnigan, was introduced as a benchtop GC/MS system especially designed for routine analysis. With the introduction of MS/MS technology, offering the potential of performing selected-ion monitoring analysis on an instrument priced competitively with current benchtop GC/MS linear quadrupole instruments, there is little doubt that ion trap technology is becoming of increasing interest to manufacturers of mass spectrometers (cf. Appendix for a list of manufacturers and representative products).

16.2.2
Principles of Operation

GC/MS coupling has been the driving force behind the commercial development of the quadrupole ion trap, as evidenced by its ability to obtain full-scan mass spectra with a high sensitivity. The introduction of ion trap mass spectrometers occurred during a period of growth in the application of mass spectrometry to analysis of compounds of environmental interest, with commercial systems consisting mainly of a capillary column GC combined with a conventional quadrupole analyzer. The problem of introducing a capillary column effluent into a low-pressure device with a limited pumping speed was, in fact, solved about 20 years ago from ex-

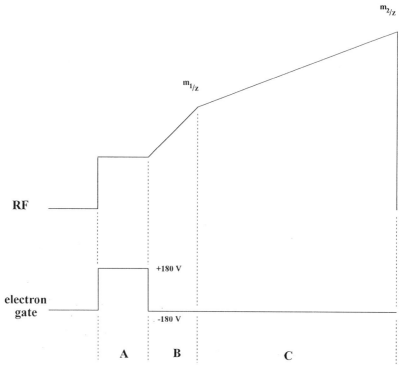

Fig. 16.1 Scan function of the ion trap for obtaining a mass spectrum. The scan function consists of an ionization period A, where the electron gate is biased at a positive potential (cf. text), a period B for ejection of the background ions, and an analytical ramp C, where ions are acquired from mass m_1 to mass m_2.

pertise gained with coupling such columns to quadrupolar instruments. The main advance was the development of the mass-selective instability mode of mass analysis, which revolutionized the use of the ion trap as a mass analyzer. In this mode of operation, a radiofrequency RF voltage is applied to the central ring electrode of the ion trap, while the two end-cap electrodes are held at ground potential [4]. By definition, each ion confined within the ion trap is associated with a value of the q_z parameter on the q_z axis of the stability diagram correlated with a particular mass-to-charge m/z ratio. To record a mass spectrum, the RF voltage is increased with time so that ions of successively greater m/z ratio are ejected through holes in one of the end-cap electrodes as they develop unstable trajectories, ions are then detected with a classical electron multiplier. The scanning sequence of the ion trap consists of three steps performed repeatedly: (1) ionization and storage of ions above a chosen m/z ratio, (2) ejection of background ions below the m/z range of interest, and (3) ejection and detection of ions in order of increasing m/z ratio (Fig. 16.1). Full-scan mass spectra covering the desired range of analysis are then recorded. The execution of a single scan program is called a microscan.

During a GC/MS acquisition, the data for each microscan are summed, and the analytical spectrum obtained from the data system results from the averaging of this sum. The ion trap is operated with helium buffer gas to provide adequate cooling of the ions generated; the operating pressure (ca. 1 mTorr) is usually provided by the carrier gas flow through the GC column, avoiding the necessity for an additional helium gas cylinder.

16.2.3
Ionization and Scanning Modes

16.2.3.1 Electron Ionization

The most common configuration for GC/ITMS in its first stage of development was a single capillary GC column coupled to an ion trap. Electron ionization (EI) in the trap is achieved by passing a current of several tens of microamperes through a filament. Injection of the electrons from the filament is controlled by a gate lens biased at a positive potential during the ionization period and a negative potential during the analytical ramp. In routine use, the validity of ion trap for its sensitivity as an analytical tool must be correlated to a reliable spectra library searching procedure for rapid identification of unknowns in a chromatogram. However, in the ion trap, the affectors of spectral characteristics depend on ion chemistry and instrumental design. Specific compounds can undergo ion–molecule reactions under EI conditions, and a large amount of ions can result in space-charge effects; these factors are related to sample concentration and are a limitation for rapid and efficient identification of unknowns.

The number of ions that are created and trapped at a given sample pressure is controlled by the ionization time and the filament emission current. A longer ionization period leads to a greater number of trapped ions, and consequently a poorer resolution for preponderant ions of low m/z compared to high m/z ions as the ions are scanned out from the ion trap, since the total number of ions is decreased during the analytical ramp. In contrast, with quadrupole mass analyzers, space-charge effects are more noticeable because all of the ions above a particular m/z ratio are trapped at the same time. It is noteworthy that, in most commercial instruments, recognizing space-charge effects in GC/ITMS data is complicated by the acquisition of centroided peaks for reconstructed mass spectra which tends to hinder the characterization of changes in resolution, peak shape, and mass assignment. These space-charge effects, which impair the performance of GC/ITMS, have been addressed by using modified control routines to reduce the number of ions generated. For this purpose, automatic gain control (AGC) has been introduced to reduce space-charge effects by limiting the number of ions generated at different points during the elution of a sample during a chromatographic run. When AGC returns a large total ion current, the ionization time for subsequent microscans is reduced (conversely, the ionization time for the following microscans is increased when the measured total ion current is low). Along a chromatographic peak, the total ion current changes with the concentration profile during elution, AGC responds to the change by decreasing the ionization time at the top of the peak, where the concentration of the sample is highest, and increasing the ionization time along the sides of the eluting peak, where the concentration of the sample is lower. To summarize, AGC minimizes space-charge effects when the sample pressure is high, and acts to enhance the sensitivity when the sample pressure is low. In addition, reduction of the ionization time limits the extent of perturbing ion–molecule reactions between newly generated ions and neutral precursors. A compromise between AGC settings and filament emission current usually gives optimized sensitivity [5].

EI spectra covering the full-scan mass range are normally acquired with a segmented EI scan program. The EI scan program of commercial ion traps normally produces a complete mass spectrum by combining the spectra obtained from four scan segments and includes: (1) a variable ionization time, (2) a rapid RF ramping for ejection of low mass ions, and (3) an RF analytical ramp for the defined segment (Fig. 16.2). The main purpose of the segmented scan program is to reduce self-chemical ionization and ion–molecule processes by shortening the residence time of the ions in the ion trap between ionization and detection. Moreover, the segmented scan program allows adjustment of the ionization settings of each segment (ionization time, RF level). For GC/MS experiments, the scan repetition rate (number of scans acquired each second) can be increased by reducing the number of scan segments and/or reducing the mass range of analysis. A larger scan repetition rate is obtained if a single scan segment is used.

A supplementary RF voltage, the axial modulation, is applied during the scan program to improve the resolution by increasing the efficiency with which the ions are resonantly ejected from the ion trap during the RF analytical ramp.

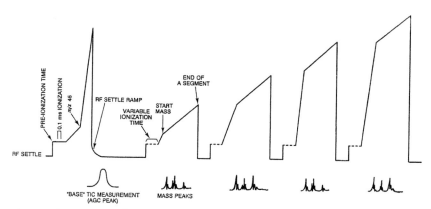

Fig. 16.2 Representative four-segment scan function for acquiring EI mass spectra. Each AGC scan function consists of a pre-ionization step, a fixed short ionization time related to the AGC settings, and a rapid rf scan to get a rough measurement of the total ionic current. All segments include the steps described in the text. (Reprinted with permission from [5])

16.2.3.2 Chemical ionization

Chemical ionization (CI), a less energetic ionization method than EI, is often used in conjunction with EI to aid in identifying the molecular weights of samples. CI has the advantage of allowing better control of transferred internal energy (through the choice of reagent gases), which favors the formation of protonated molecules through proton transfer or radical cations through charge exchange processes. The flexibility of being able to switch from EI to CI in the same experiment has contributed to the growing interest in ion trap analyzers. In addition, the CI potentialities of ion traps have played a major role in establishing ion traps as promising devices for analytical applications. CI operation of the ion trap was reported in 1987, only 3 years after its commercial introduction [6]. The first drawbacks encountered under CI conditions, space-charge effects and ion–molecule reactions, were circumvented by using modified routine sequences similar to the AGC procedure described below.

The reagent gas and sample pressures employed for CI operation of ion traps are intermediate between those used in conventional mass spectrometers and those used in ion cyclotron resonance (ICR) spectrometers. CI in an ion trap is effected by using 10^{-6} to 10^{-4} Torr of reagent gas compared to the 1 Torr range of pressure used in conventional ion sources. If the sample to reagent gas ratio is drastically decreased, even down to 1:10 to limit the space-charge effects, this feature precludes however effective limitation of sample electron ionization processes. As a consequence, fragments resulting from EI can contribute significantly to the resulting CI mass spectrum. What is accomplished under classical CI conditions by using a high pressure ion source is achieved in the ion trap by using long reaction times. Ion traps are generally operated in the mass-selective instability mode. During the timing sequence, the ion trap is first held at a low RF voltage to eliminate residual ions (Fig. 16.3). After the ionization period, the RF voltage is raised to an RF value in accordance with the m/z ratio of the stored reagent ions. At this value, lower mass ions show unstable trajectories, while higher mass ions are inefficiently trapped. This first reaction period allows an effective population of reagent ions to be established through ion–molecule processes, typically CH_5^+ and $C_2H_5^+$ under methane CI conditions (pressure in the $1,0 \cdot 10^{-5}$ Torr range and reaction time of about 15 ms). The nature of the reagent population can be controlled through the length of the first reaction period; for example, shortening this period induces an increase in the CH_5^+ ion population at the expense of $C_2H_5^+$ ions under methane CI conditions. This controls the ongoing ion–molecule reactions between reagent ions and sample molecules, and reduces EI interferences, since trapping is inefficient for high-mass ions. The desired reagent ion population is subjected to a fast RF voltage ramp to achieve the sample reaction period RF level. Ion–molecule reactions between the sample molecules and reagent ions take place during this interval, typically 50 to 100 ms, in which the RF value settles, allowing efficient trapping of the resulting product ions. This value of 50 to 100 ms should be compared to the 10^{-4} s ion residence time in conventional high pressure ion sources. The RF voltage is subsequently scanned for mass analysis. The entire sequence is repeated and the acquired microscan spectra are added and averaged to obtain the analytical spectrum.

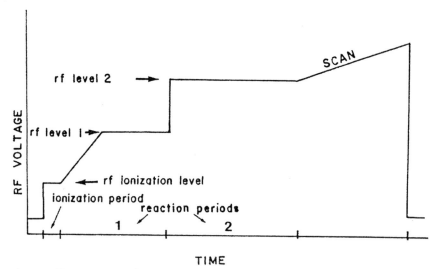

Fig. 16.3 Timing sequence for acquiring a mass spectrum under CI ion trap conditions. A low RF voltage is first applied to clear the ion trap of residual ions; the RF voltage is then allowed to settle at RF level 1 and RF level 2 to allow reagent ion formation and the CI process, respectively; and the RF analytical ramp is scanned for data acquisition. (Reprinted with permission from [6]).

Experimental variables do, however appear to be critical in obtaining reproducible CI spectra with ion traps, this is a feature common to all mass spectrometer instruments. But improper selection of RF voltages, time intervals, or operating pressures can yield unexpected results, and could be at the origin of some criticisms concerning this mode of ionization in ion trap mass spectrometry. Obviously, and as a matter of fact, under methane CI conditions, ion traps present characteristic properties such as a smaller m/z ratio for the $CH_5^+/C_2H_5^+$ ions, and a reduced yield of adduct ions $[M+29]^+$, $[M+41]^+$. The lack of effective stabilizing conditions for the removal of internal energy has been proposed to account for the absence of such adducts [6].

Among the parameters influencing the CI ion trap conditions, reaction time, temperature and pressure are of most importance. For example, the time allowed for the reagent ions to react with the sample molecules (reaction period 2) has been shown to alter the appearance of CI spectra and could certainly explain the main differences noted in comparing ion trap mass spectrometry with linear quadrupole mass spectrometry. However, prolonging the reaction time increases the total number of sample ions; this feature can be used to enhance weak signal ions within the confines of sampling time limitation to avoid space-charge effects. The ratio of the helium buffer gas pressure to the CI reagent gas pressure is typically in the range 100:1. Collisional deactivation by helium has a dramatic effect on the CI mass spectrum in limiting subsequent fragmentations from the protonated molecules generated under the CI process. Thus, there are several major differences between conditions employed in CI ion trap operations and those used with conventional mass spectrometers.

Accumulation of a large concentration of reagent ions, which accounts for the large population of sample ions, is the main characteristic of the ion trap under CI conditions. Optimum conditions for CI experiments are controlled through a pre-scan procedure to determine the appropriate ionization and reaction times: automatic reaction control (ARC), analogous to AGC used in EI experimentss. During chromatographic elution of a sample, ARC acts to accommodate these parameters to provide a full mass spectral scan in accordance with the concentration profile. The two major types of ion–molecule reactions, that take place are charge transfer and proton transfer. The exothermicity of these ion–molecule reactions is such that there exists a risk that EI-like ions will appear in the CI mass spectrum; for example, under methane CI conditions and in the presence of reagent gas ions generated during the first reaction period, the radical ion $CH_4^{+\cdot}$ is likely to undergo high-energy charge transfer reactions while the protonated ion–molecule reaction product CH_5^+ undergoes mainly low-energy proton transfer reactions. Particular attention has been paid to these features of ion traps in the development of modified ARC procedures now available in commercial instruments. These new ARC-CI functions are based on extended control over the composition of the reagent ion population.

16.2.3.3 Full Scan Versus Selected-Ion Monitoring

The selected-ion monitoring (SIM) scanning mode is used to increase sensitivity by collecting continuously the ion current for a single m/z ratio. For beam-type mass spectrometers, the SIM mode is usually more sensitive than the full scan mode because of the likely signal-to-background ratio (S/B) enhancement provided by this continuous single mass monitoring mode. However, in ion traps, the S/B gain obtained by switching from full scan mass spectrum acquisition to SIM is lower than with quadrupole analyzers where continuous scanning for one mass is carried out. Indeed, reducing the acquisition range to record a single mass shortens one part of the scan program, affecting only the number of microscans, inducing only a relatively modest increase in S/B. This major drawback for GC/MS is counterbalanced by the improved quantitation obtained with the SIM acquisition procedure, since the GC peak profile is more accurately reproduced due to the increase in the scan repetition rate.

16.2.4
Advances in GC/ITMS

Although the ion trap is considered by many analysts to be the most or one of the most sensitive instruments for GC/EI MS and positive ion GC/CI MS, new techniques to improve instrument performance are being developed. We briefly describe here new scanning sequences, the fitting of external ion sources, and applications of GC/MS/MS procedures.

16.2.4.1 Methods for Improving Performances: Increasing the Signal-to-Background Ratio

One advantage of the ion trap is its large dynamic range. However, a high level of background ions may prevent the detection of dilute analytes. The problem to solve is to increase S/B by employing shorter ionization times to lower the background ions and avoid space-charge effects. Some researchers have proposed using SIM methods to selectively eject background ions prior to mass analysis to reduce background interferences and to enhance sensitivity.

Since ions are physically trapped according to their secular frequency, single frequency resonance ejection methods have been used to remove ions of a particular m/z ratio by applying an additional potential to the end-cap electrodes at the fundamental axial secular frequency of the ion to be ejected. Broadband excitation and combination of RF voltages and resonant ejection frequencies have been employed for this purpose. Nonlinear relationships between the ionization time and the resulting ion signal have led to the conclusion that single frequency resonance ejection methods are more suited to the elimination of a single ion species due to the step-by-step procedure of these ejection techniques [7].

Multiple-frequency resonance ejection methods have been developed to eject specifically one or more ion species simultaneously. Digital waveform generators have been used for this purpose and have been shown to provide greater control over the excitation processes [7]. Most of these methods are derived from ICR mass spectrometry.

Commercial introduction of ion isolation/ejection techniques to reduce background interferences by the use of waveform generators coupled to digital function generators and personal computer systems has greatly contributed to customization of GC/ITMS instruments. These instruments are now providing with an extended simplicity numerous methods for the utilization of customized sequences for specific analytical applications.

16.2.4.2 External Ion Sources

Although ions can be generated by EI and CI of compounds introduced directly into an ion trap, it might be desirable to apply other methods of ionization. Moreover, the use of an external ion source may solve the problem of self-CI during GC/EI-ITMS analyses. Such considerations have been the basis of studies on the injection of externally generated ions into an ion trap [8].

In the external ion source configuration of ion traps, a dual EI/CI ion source is directly coupled to the ion trap through a focusing lens system. The externally generated ions are accelerated from the ion source, focused through a lens system and injected into the ion trap. A simple gated electronic lens system is used to correlate the injection step to the ionization period of the classical scanning sequence. Ion injection into ion traps is facilitated by the presence of helium bath gas at a relatively high pressure (1.0 (10^{-3} Torr), which removes kinetic energy of the entering ions and permits efficient trapping.

External ion sources have been developed principally to address the perturbing effect of self-CI processes that form $[M+1]^+$ adduct ions during EI process. Such reactions result in an increased $[M+1]^+/M^+$ ratio or in the appearance of artifact peaks due to adduct formation, both of which lead to ambiguity in identification of samples, especially for samples containing isotopes. Self-CI effects are easily observed at relatively low partial pressure of samples for compounds having a high gas-phase acidity, or possessing proton-donor or proton-acceptor functions. The proposed solution lies in reducing the density of neutral compounds (analytes or contaminants) that are involved in the ion-molecule reactions leading to these adducts.

Besides reducing ion-molecule reaction processes, injection of externally generated ions offers the possibility of conducting negative ion chemical ionization (NICI) experiments. NICI is not efficient in the usual configuration of ion traps because of the lack of thermal electrons due to the presence of the RF storage field. Consequently, under EI conditions, positive ions are estimated to be formed 10^3 times more efficiently in ion traps than negative ion. In addition, nascent negative ions are subjected to recombination reactions with the predominant positive ions to form neutral species. The negative ion mode of operation is useful for the analysis of compounds bearing electronegative functional groups such as halogenated pesticides. Consequently to use the NICI mode, the appropriate species must be formed in an external ion source and subsequently injected into the ion trap.

The major drawbacks to using an external ion source was related to the complexity of operation in switching from EI to CI in contrast to the usual configuration, and the requirement for an additional vacuum system. Nevertheless, external ion source GC/ITMS is undergoing commercial development, and manufacturers currently provide external ion source designs for the new generations of ion-trap mass spectrometers.

16.2.4.3 GC/MS/MS

In general, SIM procedures can be useful for enhancing sensitivity for target analytes in clean matrices, but at the expense of mass spectral information. In addition, as the sample matrix becomes more complex, it becomes more difficult to obtain the necessary sensitivity to distinguish the analyte signal from any interfering chemical background signal. Therefore, when only SIM procedures are used, low detection limits are rarely attainable in complex matrices.

For improved sensitivity and selectivity, GC/MS/MS in the ion trap meets the criteria for obtaining low detection limits. MS/MS enhances selectivity by separating the target compound from the chemical background (due to the matrix), and, in essence, the matrix is virtually eliminated. The result is a better signal/background ratio (and lower detection limits). Collision induced dissociation (CID) of isolated ion species in an ion trap has thus become a powerful technique.

In comparison to linear quadrupole instruments, the CID process in an ion trap is more complex since the selected ionic species are resonantly excited at their secular frequency of motion. This frequency is subject to some slight changes when the ions move away from the center of the trap and when the number of ions confined within the trap changes. In addition, a large set of parameters needs to be adjusted (ionization time, RF level for ionization, RF and/or DC voltages for excitation, isolation times, RF level for CID, excitation frequency, excitation voltage) and, finally, tuning procedures are time consumpting. Application of CID for GC coupling, for which elution peak width might not exceed a few seconds, has only been made possible by the development of sophisticated set of programs. Multi-frequency irradiation methods are usually employed to perform CID with high efficiency and short periods of irradiation. However, compared to beam-type mass spectrometers, ion traps offer three main advantages when used in the MS/MS mode. First, the ion trap operates in the pulsed mode, and allows accumulation of ions mass selectively over time in such a way that a rather constant target ion number can be selected over varying concentration profiles. Second, the CID process in the ion trap involves a large number of collisions between the mass selected ion and light helium buffer gas atoms. Under these conditions, the energy transferred per collision is very low ; as a result, only the lowest dissociation pathways are attained, which simplifies the CID spectra. Third, it is possible to totally dissociate the mass-selected ion and, confine within the ion trap nearly all of the fragment ions (this accounts for the high efficiency of dissociation generally reported in the literature; efficiency of dissociation is defined as 100 times the ratio of the sum of the product ion signal intensities to the parent ion signal intensity). The ion isolation CID sequence can be repeated one or several times to provide $(MS)^3$ and $(MS)^n$ sequences, a process known as multiple-stage mass-selective operation. Clearly, the development of benchtop GC/MS/MS ion traps over the past few years has expanded the use of MS/MS for analytical purposes.

16.3
Examples of Applications of GC/ITMS

16.3.1
Requirements for Environmental Analysis

In environmental analyis, strict criteria are highly desirable. The detection of identified pollutants and toxic material must be certain, and original analysis should be supported by a confirmatory procedure. Moreover, recording a mass spectrum in the low parts per million (ppm) to parts per trillion (ppt) concentration range is much more complex than recording a mass spectrum of a reference standard. Although confirmatory procedures are well-documented for a variety of analytical methods in the regulatory arena, including mass spectrometry techniques after the elaboration of good laboratory practices, confirmatory procedures for ion trap mass spectrometry are not so clearly defined. However, the examples of the following section show that ion traps have the outlined criteria of reliability, practicality, accuracy, limit of detection, and specificity.

Of fundamental importance are the instrumental tune-up and calibration tests. Originally, decafluorotriphenylphosphine (DFTPP; bis(perfluorophenyl)-phenyl phosphine) was recommended as one of the most appropriate reference compounds, although perfluorobutylamine (PFTBA; FC 43) has a large popularity among the mass spectrometrists. A second test compound, 4-bromo-fluorobenzene (BFB) was introduced in 1984 for the tuning procedure in methods for the analysis of volatile organic compounds (VOCs). In both cases, relative peak abundance measurements in the ion trap from repetitive injections of small quantities of DFTPP and BFB through a capillary GC column have been shown to be good, with relative standard deviation of 3 to 26% [9].

Good mass spectra quality indices must be provided. These indices consist of criteria defining the characteristics for an acceptable mass spectrum including: the absence of ions in excess of the values expected for the molecular ion (except adduct ions under CI conditions), the absence of non-logical neutral losses, molecular and fragment ions within the theoretical range of variation for the isotopic abundances relative to the parent ions. Comparison with standard MS spectra from the reference NIST (National Institute of Standards and Technology) library usually provides good results, even at low levels of concentration during GC injection into the ion trap.

The ability to detect a residue at a level one order of magnitude below the regulatory level is normally required. Careful attention must be paid to extrapolating spectral data of standard reference spectra, which can be misleading with respect to the true level of detection available in the real sample.

Coupling GC to mass spectrometry implies criteria directly related to the GC separation, and retention time criteria are of paramount importance to the mass spectrometric identification. One advantage of combining capillary column technology with mass spectrometry is that only a small quantity of sample is required for full mass analysis. However, scan repetition rate must be restricted to the values specified in the protocol employed and retention time correlation must be bracketed in a window within 15 scans of the reference standard. EI spectra have been widely used for identification due to their "fingerprinting" capabilities allowing identification through comparison with published data and library references or through the fragmentation patterns of the sample. One of the main strengths of GC/ITMS lies in the ability of ion traps to reliably produce full mass spectra at low concentration levels. For confirmatory purposes, as processed usually in mass spectrometry, selected-ion monitoring procedures can be satisfactory only if conducted on more than three ions of structural significance and observed in the expected relative abundance ratio with respect to the precursor molecular ion. Monitoring a smaller number of ions can lead to misleading identification of residues, and this approach is generally used for screening purposes only.

The following sections report successful use of GC/ITMS for the detection and characterization of compounds of environmental concerns (see Tab. 16.1).

Table 16.1 Compounds of environmental concern characterized by GC/ITMS and reported sensitivities.

Compound	Reported Sensitivity	Reference
Benzene, carbon tetrachloride, 1,2-dichloroethane, trichloroethene, 1,4-dichlorobenzene, 1,1-dichloroethene, 1,1,1-trichloroethane, vinyl chloride	0.1–10 µg L^{-1} [a]	10
Acetone, acrolein, acrylonitrile, allyl chloride	1 µg L^{-1} [a]	11
2,3,7,8-Tetrachlorodibenzodioxin	100 fg µL^{-1} [b]	16
(Tri-n-butylmethyl)tin	1.3 pg[c]	18
(Di-n-butylmethyl)tin	1.9 pg[c]	18
Tebufelone	0.27 ng[d]	19
Diazepam	2 ng	20

a Corresponding to 0.5 to 50 ng of each compound in a sample of 5 mL of water; purge-and-trap extraction technique; fluorobenzene as internal standard at 1 µg L^{-1}.
b Reported as a current detection limit.
c Under CI acetonitrile ionization conditions; presented as minimal detection amounts of Sn for S/B = 3.
d GC/MS/MS analysis; daughter ion spectrum of the m/z 248 fragment ion.

16.3.2
Determination of Volatile Organic Compounds in Drinking Water; EPA Methods

The monitoring of volatile organic compounds (VOCs) in drinking water has led to their regulation in drinking water in several countries, especially in the United States where the Environmental Protection Agency (EPA) has set such regulations since 1979. Distinct approaches have emerged, focusing on either identifying one or more listed analytes or targeting a specific analyte for which a methodology is optimized. These regulations have established maximum VOC contaminant levels. In the early 1990s, a combination of a capillary GC column and a benchtop ion trap was used to evaluate EPA Methods 524.2 [10, 11]. A standard purge-and-trap extraction system was used, and AGC and full mass range scanning of the ion trap were employed for complete identification of compounds. Optimum purge-and-trap GC/MS conditions for the detection of eight regulated (benzene, carbon tetrachloride, 1,2-dichloroethane, trichloroethene, 1,4-dichlorobenzene, 1,1-dichloroethene, 1,1,1-trichloroethane, vinyl chloride) and 51 unregulated volatile compounds in drinking water were determined. At the 2 µg L^{-1} level with a 5 mL water sample, the grand mean measurement accuracy for 54 compounds was 95% of the true value, with a mean relative standard deviation of 4% [10]. At the 0.2 µg L^{-1} level, the grand mean measurement accuracy for 52 compounds was 95%, with a mean relative standard deviation of 3% [10]. At this level, the failure to detect chloromethane and dichlorodifluoromethane was attributed to their high vapor pressure and poor retention. Extension of the method to 28 additional compounds gave reliable results for a maximum contaminant level of 1 µg L^{-1} or lower [11]. Standardization and quality control proposals for VOCs in drinking water and ambient air, and for less volatile compounds were subsequently summarized [12]. During these investigations, the ITD operating in the full scan mode of acquisition was demonstrated to be a good compromise between minimum analysis time and complete separation of compounds of different origin in the same mixture. This is in accordance with cost-effectiveness in analytical laboratories, which requires the optimization of the number of samples that can be processed in a given work period without loss of quality of analysis.

16.3.3
Detection of Dioxins and Furans

Polychlorodibenzodioxins (PCDDs) and polychlorodibenzofurans (PCDFs) are of great environmental concern, and their determination by GC/ITMS is a good example of the application of GC/ITMS in analytical chemistry. These compounds are determined mainly in complex mixtures, where they may be present at trace level. Because of their suspected toxicity and their persistence in the environment, PCDDs and PCDFs are classified as priority pollutants. The methods of detection specified by environmental agencies usually call for monitoring the total concentration of all the congeners of these compounds, with a particular emphasis on the overall isomers of the most potent 2,3,7,8-tetrachloro-substituted congeners of PCDD (2,3,7,8-T_4CDD).

High-resolution gas chromatography (HRGC) coupled with high-resolution mass spectrometry (HRMS) is currently prescribed by most regulatory agencies. However, such determinations are costly due to the time required for extensive sample preparation, the use of ^{13}C-labeled internal standards, and expensive high-resolution mass spectrometers. The main problem in detecting dioxins lies in the likely presence of interferents, such as the matrix itself or PCDFs, and the overlap of isotopic clusters during the analysis requiring HRMS and resolving power of several tens of thousands. Tandem mass spectrometry (MS/MS) has been proposed to circumvent the costly use of HRMS, with detection limits approaching those obtained under HRMS conditions; however, neither technique can completely remove all interferences, making the two techniques complementary [13]. Recently, a rapid screening technique for the detection and quantitation of 2,3,7,8-TCDD using GC/ITMS operated in the MS/MS mode has been reported [14]. Although the sensitivity of the ion trap MS/MS technique appeared to be fairly good and comparable to that of tandem quadrupole mass spectrometry, the scanning functions available at that time in ITMS were a limiting factor in the development of GC/ITMS for the quantitation of dioxins. Software advances since then have overcome this limitation. Commercial software is now available for deconvoluting mass spectra obtained from coeluting compounds and for conducting MS/MS experiments using multiple-reaction monitoring with several daughter ions, which is required for the analysis of tetra- to octa-PCDDs/PCDFs during a single chromatographic run. Since then, the CID procedure has been used to optimize both molecular ion isolation and directed fragmentation of the PCDDs and PCDFs with an ion trap [15]. The multifrequency irradiation mode of excitation was found to be the best compromise in terms of conversion efficiencies for the parent ion to the [M–COCl] daughter ion and for compatibility of irradiation duration on a gas chromatographic time scale.

A comparative study of three mass spectrometric methods for the determination of tetra- to octa-chlorodibenzo-p-dioxins/furans using HRMS, triple-stage quadrupole mass spectrometry, and ion trap mass spectrometry has been recently published [16]. The following aspects of the determination of PCDDs/PCDFs were examined : (1) tuning procedures, (2) calibration curve preparation, (3) 2,3,7,8-T_4CDD detection limits, (4) examples of ion signals from 2,3,7,8-T_4CDD obtained at 50 times the HRMS detection limit, (5) relative response factors, (6) ionization cross sections, and (7) comparison of signal ions due to hexachloro-p-dibenzodioxin (H_6CDD) congeners from real samples. The current 2,3,7,8-T_4CDD detection limits of the three methods (coupled to GC) are 10 fg L^{-1} by HRMS, 150 fg L^{-1} by triple-stage quadrupole, and 100 fg L^{-1} by ion trap. Examples of the ion signals obtained with each technique for a low concentration of 2,3,7,8-T_4CDD are shown in Fig. 16.4, where the signal-to-background ratios for HRMS and the ion trap are seen to be comparable. For HRMS and triple-stage quadrupole mass spectrometry, the approved methods called for the observation of two m/z ratios, whereas for the ion trap, there is no approved method and additional ion species were monitored to increase sensitivity and selectivity. These investigations permit one to state that, while the HRMS detection limit for T_4CDD is lower than that of triple stage quadrupoles and ion traps, there is evidence that all interferences are not eliminated by using HRMS alone. Consequently, there is a need for instruments that can achieve high specificity by MS/MS operation and ion traps are ideally suited for this purpose.

16.3.4
Other Examples

The advantages and disavantages of the ion trap approach to multi-residue pesticide analysis have been assessed for the purpose of conducting routine analysis for the determination of pesticide residues in food [17]. The driving force behind this approach is to reduce analytical costs. Chemical ionization is currently used in GC/MS analysis of trace levels of pesticides to reduce interferences of matrix components and to provide higher specificity through the production of protonated molecular ions for the residues of interest. The precision and accuracy of ion trap for trace level detection of pesticides is within acceptance ranges (relative standard deviation of less than 10% with a correlation coefficient of 0.995 or greater). The ion trap has been found to be sensitive for the detection of nitrogen-containing compounds. The use of a single ion area measurement is acceptable for trace level quantification. Finally, the ion trap can be used in multi-residue pesticide analysis as a replacement detection system capable of automatic confirmation and quantification in a shortened analytical step [17].

A growing concern about the presence of organotin in the environment has created a need for faster, more sensitive and more accurate analytical methods for its

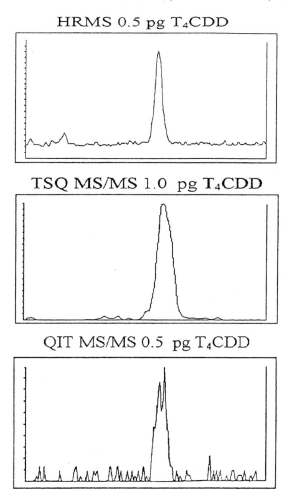

Fig. 16.4 Ion signals obtained for low concentrations of 2,3,7,8-T$_4$CDD: (top) HRMS, 0.5 pg injected in 1 µL (signal obtained at 50 times the detection limit), sum of m/z 320 [M]$^{+\bullet}$ and 322 [M + 2]$^{+\bullet}$; (middle) triple-stage quadrupole MS, 1.0 pg injected in 1 µL, sum of m/z 257 and 259[M − CO^{37}Cl]$^+$; and (bottom) ion trap, 0.5 pg in 1 µL, sum of m/z 257, 259, 194, and 196. [M]$^{+\bullet}$, and [M+2]$^{+\bullet}$ are the molecular ion with ^{35}Cl atoms only and the molecular ion with a single ^{37}Cl atom respectively; m/z 257, 259 and 194, 196 correspond to loss of COCl$^\bullet$ and 2COCl$^\bullet$, respectively. (Reprinted with permission from [16].)

detection in environmental samples. GC/ITMS has been shown to be a powerful technique for determining trace and ultratrace quantities of tributyltin and its degradation products in water after hydride derivatization and Grignard methylation, which provides lower detection limits under EI conditions [18]. The wide linear dynamic range and picogram sensitivity of the ion trap operating in the EI mode make this GC/MS configuration suitable for routine trace and ultratrace analysis of organotin.

The performance of GC/ITMS for the analysis of a model drug, tebufelone, has been studied to evaluate the selectivity, linear range, accuracy, and precision of this method for sampling drugs in biological matrices [19]. Compared to GC linear quadrupole mass spectrometry, the SIM mode of operation of the ion trap provided a higher degree of selectivity for the analysis of tebufelone spiked rabbit plasma samples, and gave linear standard curves over three orders of magnitude of concentration, with an associated detection limit of 100 pg mL^{-1}.

Diazepam has been used as a test compound for a comparative study of GC/ITMS versus a GC linear benchtop quadrupole mass spectrometer in both the full mass scanning mode and the SIM mode [20]. A major concern involved evaluating whether GC/ITMS provides mass spectra in a concentration-dependent way and whether this technique yields mass spectra that can be searched against conventional mass spectral data libraries. In the full scan mode, the ion trap had a signal/background ratio of 1400 for a 2 ng injection of diazepam, with an ion ratio precision varying from 5 to 13%. In the SIM mode, the ion trap had an average signal/background ratio of 14,000, with an ion ratio precision of 6 to 15% [20]. Overall, compared to quadrupole mass spectrometry, GC/ITMS in the full scan mode provided an equivalent precision in ion ratio at a greater signal/background ratio, but was 5 to 10-fold less accurate in the SIM mode.

16.4
Future Prospects in GC/Chemical Ionization-ITMS

16.4.1
Chemical Ionization in Environmental Analysis

CI techniques are used favorably to generate molecular ion species for the unequivocal identification of compounds. Determining the molecular weights of the compounds under investigation is an important step in the analytical process. One of the main advantages of CI is that the degree of fragmentation, and therefore the amount of energy deposited into the analyte, can be controlled by the choice of reagent gas and provides direct information on the molecular weight of the analytes under these soft ionization conditions. This is certainly why CI mass spectrometry has become a powerful analytical technique in various disciplines, including pharmacology, medicinal chemistry, forensic science, petroleum exploration, and environmental analytical chemistry. Moreover, listings of the molecular weights of the majority of drugs, toxic substances, and their related metabolites are now accessible. In addition, CI is characterized by the relative facility with which switching from one reagent gas to another allows one to perform confirmatory experiments to reveal the presence of specific ions, whatever their polarity, by the interchange of the reactant gas, methane for ammonia for example. However, due to the limited extent of fragmentation, CI must generally be coupled to MS/MS procedures to ensure complete and accurate characterization.

Fundamental investigations of the ionization mechanism for CI have been extensive, but novel GC/MS applications are still being developed. The criteria for developing a CI reagent are its availability in a highly pure state, its relative inertness toward many substrate molecules, and, consequently, its ability to give only limited reactive ions confined to the low mass region of the acquired mass spectrum. Among the potential positive ion reagents useful for CI, examples with acetonitrile and pentafluorobenzyl alcohol as CI reagents are reported here (Section 16.4.2).

As stated earlier, the CI process takes place during the reaction period of the CI scan mode of the ion trap. Since the duration of this period is long compared to the short residence time of ions in a classical high-pressure CI ion source, a gain in specificity can be expected when this mode of ionization is used in ion traps. This feature has recently permitted the development of ion attachment mass spectrometry in the ion trap with sodium ion as a reagent ion (Section 16.4.3.2).

16.4.2
Examples of Unusual Reagents for Chemical Ionization

Acetonitrile has been reported to be an effective reactant for the positive ion CI of long-chain hydrocarbons, as well as in localizing the unsaturated hydrocarbon double-bond position [21]. More recently, a rapid method has been presented for determining the location of double bonds in polyunsaturated fatty acid methyl esters by ITMS [22]. EI is known to cause double bond migration in fatty acid methyl esters, resulting in ambiguous spectra, and CI with the usual reagent gases does not yield useful fragments. Methods based on derivatization to induce fragmentation from charged or radical sites remote from the double bonds have the disadvantage of requiring an additional chemical modification step prior to analysis. CI-based methods with vinylamine, and vinyl methyl ether as reagent gases are limited to special cases and to determining the locations of a minimum number of double bonds. The acetonitrile mass spectrum under ion trap CI conditions includes major ions at m/z 40 (loss of H) and m/z 54 (identified as the 1-(methyleneimino)-1-ethenylium ion), inducing adduct formation with the fatty acid methyl ester at M+40 and M+54, respectively. The $[M+54]^+$ adduct ion observed in the CI mass spectrum is a superposition of isomers corresponding to reaction across each double bond and is indicative of the degree of unsaturation. Tandem mass spectrometry conducted on the isolated $[M+54]^+$ ion results in diagnostic ions that include the hydrocarbon end, and ions that include the methyl ester end. Locations of double bonds have been demonstrated for fatty acid methyl esters with up to six double bonds when considering these ions together [22].

A significant contribution to the measurement of hydroxy carbonyls in air has been recently demonstrated by using pentafluorobenzyl alcohol as a chemical ionization reagent monitoring the intensity of the products $[M+H]^+$ and $[M+181]^+$ ions [23]. Hydroxy carbonyls and other carbonyls are first derivatized with (pentafluorobenzyl) hydroxylamine, and then silylated with bis(trimethylsilyl)trifluoroacetamide to improve resolution and sensitivity in the chromatograms. Pentafluorobenzyl alcohol CI-mass spectra are straightforward for the identification of glycolaldehyde and hydroxyacetone in the presence of coeluting interferences by monitoring both the [M+H]$^+$ and the [M+181]$^+$ ion signals, which are severely enhanced compare to pure methane-CI. The first measurements of hydroxyacetone and 3-hydroxy-2-butanone in ambient air have been reported, and ultratrace concentrations (pptv levels) of methyl vinyl ketone, methacrolein, methylglyoxal, glycolaldehyde, and hydroxyacetone have been measured [23]. Extension of the method to the measurement of water-soluble carbonyls, for which no or little ambient air data exist, is straightforward.

16.4.3
Ion Attachment Mass Spectrometry

16.4.3.1 Principle

Thermal alkali-metal ion association reactions are described by the following simplified expression:

$$A^+ + M + N \rightarrow [A+M]^+ + N$$

Where A denotes a positively charged alkali metal ion, M is a neutral species, and N acts as a third-body. The binding energy of the molecule M, considered as a Lewis base, to the alkali metal cation A$^+$ is defined by the enthalpy change for the preceding reaction. Deriving primarily from electrostatic forces, the binding energy of the reagent alkali metal ion A$^+$ to the molecule M must be high enough to permit a significant number of adducts to be formed at the partial pressure used in the experiments. Cationized molecular adducts are generally stable and, apart from their intrinsic interest as a chemical process, association reactions would therefore be potentially useful for determining the molecular weights of the neutral species M by monitoring the m/z ion ratio of the [A+M]$^+$ adduct. Such use of alkali metal cations as reagent species for CI mass spectrometry has been investigated previously [24, 25]. The relative binding energies of a large number of molecules have been measured and correlated with theoretical calculations. These results indicate that the ion binding energies of a wide range of alkali metal ion complexes are high enough to be detectable at low concentrations as long as the attachment process is kinetically efficient. Consequently, the chemical ionization process of alkali metal ion association reactions (ion attachment mass spectrometry, IAMS) offers a unique and interesting potential in analytical chemistry [26]. Currently IAMS is available commercially in a complete form (Anelva Co.).

Alkali metal ions are generated by thermionic emission externally to the ion source and injected into a chamber containing a reagent gas with a trace amount of sample [24]. The ions bind to the sample according to the termolecular process described above. Typically, molecules that have intrinsically high alkali metal ion chemical ionization sensitivities are molecules that are polar or polarizable species. Binding energies are usually in the 50 kcal mol^{-1} range, or less, for effective binding. Then, it has been demonstrated that trace amounts of alkenes can be detected in the presence of alkanes [24]. More recently, ion attachment mass spectrometry has been developed for continuous measurement of perfluoro compounds (PFCs) of environmental concern in semiconductor manufacturing [27]. Five greenhouse gases, CF_4, CHF_3, C_2F_6, SF_6, and c-C_4F_8, were studied with the intention of developing improved methods for PFC analysis at the trace level (ppb range). The results demonstrate the feasibility of real-time measurements for PFC trace monitoring by generating only adduct ions from Li^+ ion attachment process.

16.4.3.2 Sodium Ion Attachment Reactions with GC/ITMS

The first direct demonstration of the applicability of alkali ion attachment reactions using a sodium cation emitter as a novel and sensitive technique of ionization for ion trap mass spectrometry was made with an ion trap mass spectrometer equipped with an external ion source to generate the reagent Na^+ ions [28]. The combination of alkali metal ion attachment with an ion trap may well represent a noticeable improvement in the selectivity and sensitivity of current GC ion trap technology, and may enlarge the scope of use of this type of mass spectrometer.

Sodium adduct formation has been applied to the detection and characterization of derivatives of explosives, and the detection of phthalate samples. A signal-to-background ratio of ca. 10 : 1 was obtained with a mixture of dimethyl, diethyl, and di-n-octyl phthalates injected into a GC column at the 100 pg level. The major peaks in all of the spectra corresponded to the molecular adducts $[M+Na]^+$, giving straightforward data on the molecular weights of the samples. For detection of explosives, sodium ion attachment reactions offer the advantages of direct determination of molecular weights with a very low level of fragmentation. This simplifies the interpretation of the mass spectra and offers the possibility of distinguishing between pre-ionization decomposition and ion fragmentation. Two nitramine derivatives, 1,3,5-trinitro-1,3,5-triazacyclohexane (RDX) and 1,3,5,7-tetranitro-1,3,5,7-tetraazacyclooctane (HMX), could be deduced from the m/z 121 adduct observed under solid probe introduction conditions associated with the Na^+ ion attachment reaction. Analysis of pentaerythrityl tetranitrate (PETN) permits unequivocal identification of decomposition products as well as the molecular adduct.

Sodium ion attachment reactions have been investigated for commercial GC ion trap mass spectrometers [29]. The alkali metal ion method was shown to be particularly suitable for ion traps by simply replacing the electron filament with a sodium emitter, and inverting the gate lens potential to allow the injection of positive ions. Figure 16.5 compares the relative response in the EI and sodium ion attachment modes of ionization under the same chromatographic separation conditions for a mixture of 10 organic compounds bearing different functional groups. The optimized conditions for both cases lead to the conclusion that the sensitivity for the mixture components is satisfactorily large under sodium ion attachment mass spectrometry. With the detection of solely molecular adducts, it is unlikely that any confusion in assignment of the molecular weights of the detected species would result. It appears that the detection may be sensitive to the structure of the analyte in the formation of the molecular adducts since the Na^+ affinity favors bind-

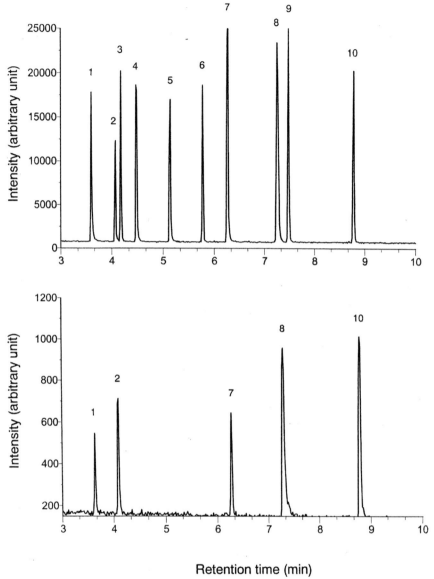

Fig. 16.5 Reconstructed total ion current (RTIC) chromatograms of a test mixture of 10 compounds, each at a concentration of 0.1 g L^{-1} (1 µL injected; split 1:40). Conditions: (top) EI and (bottom) Na$^+$ ionization conditions. Elution order: 1, benzonitrile; 2, n-octyl aldehyde; 3, 1-decene; 4, o-dichlorobenzene; 5, thioanisole; 6, iodooctane; 7, 2,6-dimethyl aniline; 8, o-anisaldehyde; 9, n-undecane; 10, methyl caprate.

ing to certain compounds under the ion trap conditions as shown by the disappearance of peaks corresponding to 1-decene, o-dichlorobenzene, thioanisole, iodooctane and n-undecane under the Na^+ ionization conditions. EI and sodium ion attachment ionization modes in the ion trap gave a similar signal/background ratio for functionalized compounds in the test mixture (Fig. 16.5). Methyl caprate, which was chosen as a model compound to define some of the figures of merit for the technique, provided a detection limit in the 10^{-4} g L^{-1} range.

Ion trap mass spectrometers demonstrate complete applicability to these alkali metal ion attachment reactions due to their *in-time* confinement capability, which provides a high efficiency for alkali ion attachment reactions compared to conditions in a classical quadrupole ion source. This method should be of general interest to those needing to solve problems involving organic pollutants, samples diluted in a matrix, pyrolytic products and samples that are of environmental concern, in situations where classical chemical ionization techniques are not always satisfactory due to their structure-dependent responses. One of the main advantages of sodium ion attachment reactions for such analysis is that the ionization conditions are not pressure dependent, unlike conventional CI ionization processes.

16.5
Conclusion

It was recently recognized that "the quadrupole ion trap has undergone a renaissance from a novel ion storage device to a conventional mass analyzer used for GC/MS" [7]. Clearly, the coupling of gas chromatography to ion traps has been a driving force for the development of commercial instruments, but recent advances in software capabilities have greatly contributed to the customization of operating procedures for analytical purposes. Consequently, gas chromatography/ion trap mass spectrometry (GC/ITMS) has become increasingly important for both qualitative and quantitative analysis of a wide range of organic compounds, and a large number of applications are being found for ion trap mass spectrometry in analytical chemistry.

The development of ion trap systems, first targeted to compete with GC benchtop quadrupole mass spectrometers, has led to sophisticated high performance instruments that are simple to operate and accessible to a wider range of users. The reasonable cost of current commercial GC/ITMS instruments, with additional CI and MS/MS capabilities, opens up the possibilities for environmental analysis to a broader range of laboratories. New analytical methods appear to be very promising for the future of the ion trap, taking advantage of what are the ion trap's unique abilities, namely, ion storage and mass selectivity.

16.6
Appendix: List of Main Manufacturers and Representative Products for GC/ITMS

- Thermoquest (San Jose, CA; http://www.thermo.com), Finnigan PolarisQ Benchtop ion trap GC/MS

- Varian Inc. (Palo Alto, CA; http://www.varianinc.com), Saturn 2000 ion trap GC/MS

- Hitachi Ltd. (Tokyo, Japan; http://www.hitachi-hitech.com), 3DQ ion trap GC/MS, until recently, Hitachi Ltd. and Teledyne Tech. were developing the 3DQ quadrupole ion trap mass spectrometer.

References

1 March, R.E.; Hugues, R.J.; Todd, J.F.J. *Quadrupole Storage Mass Spectrometry.* Wiley Interscience, New York **1989**.
2 *Practical Aspects of Ion Trap Mass Spectrometry*, eds. March, R.E.; Todd, J.F.J. CRC Press, Boca Raton, FL **1995**, Vol. 1–3.
3 Todd, J.F.J. *Mass Spectrom. Rev.* **1991**, *10*, 3.
4 Stafford, G.C.; Kelley, P.E.; Syka, J.E.P. et al. *Int. J. Mass Spectrom. Ion Processes* **1984**, *60*, 85.
5 Huston, C.K. *J. Chromatogr.* **1992**, *606*, 203.
6 Brodbelt, J.S.; Louris, J.N.; Cooks, R.G. *Anal. Chem.* **1987**, *59*, 1278.
7 Yates, N.A.; Booth, M.M.; Stephenson, J.L.; Yost, R.A., in *Practical Aspects of Ion Trap Mass Spectrometry*, vol. III, chap. 6, eds. R.E. March and J.F.J. Todd, CRC Press, Boca Raton, FL **1995**.
8 Louris, J.N.; Amy, J.W.; Ridley, T.Y. et al. *Int. J. Mass Spectrom. Ion Processes* **1989**, *88*, 97.
9 Eichelberger, J.W.; Budde, W.L. *Biomed. Environ. Mass Spectrom.* **1987**, *14*, 357.
10 Eichelberger, J.W.; Bellar, T.A.; Donnelly, J.P.; Budde, W.L. *J. Chromatogr. Sci.* **1990**, *28*, 460.
11 Munch, J.W.; Eichelberger, J.W. *J. Chromatogr. Sci.* **1992**, *30*, 471.
12 Budde, W.L. in *Practical Aspects of Ion Trap Mass Spectrometry*, vol. III, chap. 12, eds. R. E. March and J.F.J. Todd, CRC Press, Boca Raton, FL **1995**.
13 Reiner, E.J.; Schellenberg, D.H.; Taguchi, V.Y. et al. *Chemosphere* **1990**, *10*, 1385.
14 Plomley, J.B.; Koester, C.J.; March, R.E. *Org. Mass Spectrom.* **1994**, *29*, 372.
15 Plomley, J.B.; March, R.E.; Mercer, R.S. *Anal. Chem.* **1996**, *68*, 2345.
16 March, R.E.; Splendore, M.; Reiner, E.J. et al. *Int. J. Mass Spectrom.* **2000**, *197*, 283.
17 Cairns, T.; Chiu, K.S.; Navarro, D. et al., in *Practical Aspects of Ion Trap Mass Spectrometry*, vol. III, chap. 13, eds. R.E. March and J.F.J. Todd, CRC Press, Boca Raton, FL **1995**.
18 Plzak, Z.; Polanska, M.; Suchanek, M. *J. Chromatogr. A* **1995**, *699*, 241.
19 Wehmeyer, K.; Knight, P.M.; Parry, R.C. *J. Chromatogr. B* **1996**, *676*, 53.
20 Fitzgerald, R.L.; O'Neal, C.L.; Hart, B.J. et al., *J. Anal. Toxicol.* **1997**, *21*, 445.
21 Monetti, G.; Pieraccini, G.; Favretto, D. et al. *J. Mass Spectrom.* **1998**, *33*, 1148.
22 Van Pelt, C.; Brenna, J.T. *Anal. Chem.* **1999**, *71*, 1981.
23 Spaulding, R.S.; Frazey, P.; Roa, X. et al. *Anal. Chem.* **1999**, *71*, 3420.
24 Hodges, R.V.; Beauchamp, J.L. *Anal. Chem.* **1976**, *48*, 825.
25 Fujii, T.; Ogura, M.; Jimba, H. *Anal. Chem.* **1989**, *61*, 1026.
26 Fujii, T. *Mass Spectrom. Rev.* **2000**, *19*, 111.
27 Fujii, T.; Arulmozhiraja, S.; Nakamura, M.; Shiokawa, Y. *Anal. Chem.* **2001**, *73*, 2937.
28 Faye, T.; Brunot, A.; Sablier, M. et al. *Rapid Commun. Mass Spectrom.* **2000**, *14*, 1066.
29 Sablier, M.; Fujii, T.; Rolando, C. *Proceedings of the 48th ASMS Conference on Mass Spectrometry*, Long Beach CA, **2000** June 11–15.

Section IX
Application 3: Process Control

Introduction

John Green

The quality of life currently enjoyed by a large proportion of the world's population is ever more dependent upon the multitude of various materials we are able to produce and use. Fabrics, plastics, prepared and preserved foodstuffs and pharmaceuticals exemplify broad categories of products that are used everyday by millions. All these commodities are the result of manufacturing and conversion processes that are controlled for quality, consistency and economic viability.

Processes can usually be categorised as either continuous or batch operations. A continuous process accepts a continuous supply of feedstocks and produces product continuously. A batch process accepts a charge of feedstocks that is converted to product and subsequently removed from the process equipment for storage and sale. Process control is important to both these types of operation.

Process control is commonly included as a part of process engineering and has been dealt with in many engineering textbooks. An introduction to the fundamental concepts can be found in several references [1–4].

In a more general sense process control encompasses those procedures used to ensure that products are manufactured according to a previously agreed specification with a guarantee of purity and a fitness for purpose whilst the process used operates efficiently. Process control in its simplest form can involve inspection of the final product of a process which, if satisfactory, leads to a decision to continue operating the process in a similar manner whilst, if the product is unsatisfactory, the process must be altered to bring the product back to the agreed specification. Operating a process out of control results in waste or at least involves a need for reprocessing, reduces the economic viability of the operation and can have a serious environmental impact.

A schematic illustrating basic process control is presented in Fig. 1. The process is shown as a simple conversion of feedstocks to products. Measurements can be made on either the feedstocks or the products and the results are used by the process control system to modify the processing conditions, if necessary. In cases where the feedstocks are the major source of variation then the measurements are used in a feed forward mode to change the process. Crude oil refining is a case where this type of control is desirable because the feedstock varies with the

Introduction

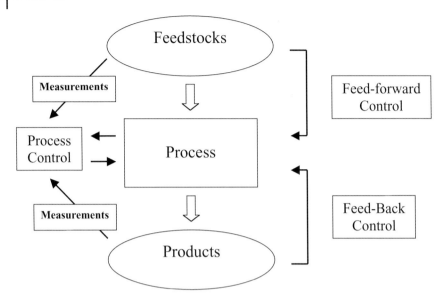

Fig. 1 A schematic view of process control.

origin of the material. In cases where the feedstock is essentially invariant, for instance air separation plants, and the product is to be produced to stringent specifications then feed back control is often more useful. In other cases a combination of these can be the most effective scheme.

In practice, many processes which are used to produce today's complex materials to exacting specifications are very complex and involve multiple stages each of which needs to be in control and each of which has an impact on the following stages and the final product. The processes are commonly subject to external influences such as a change in the feedstock characteristic (for example the industrial production of hydrogen and the drying of harvested grain with different moisture and protein contents) or changes in the temperature or humidity of the ambient conditions. There are also internal influences, such as changes to the equipment being used, causing changes in flow rates, or the progressive deactivation of a catalyst used in the process. Good process control is designed to cope with these changes and either indicates the process changes that are needed to rectify the problems or initiates the necessary changes automatically.

Process control is achieved by taking account of a selection of inputs, which involve some form of process measurement, and comparing these to a known set of acceptable values. Modifications to the process conditions are made accordingly. Process control can be manual or automatic. Manual operation involves manufacturing personnel taking note of the measurements or alarms activated by measurements and acting accordingly. Automatic operation utilises the measurement signals by feeding them into an electronic or computerised system that compares measured values with a database of values representative of acceptable or optimised operation and then initiates automatic corrective action on the basis of an established model.

The inputs or incoming signals to a process control scheme can be from a variety of measurement sources. In general these can either be measurements of the process conditions or measurements of the process materials involved in the operation. Temperature, pressure and flow are the most common process conditions used. As process control becomes more demanding then chemical and physical measurements of the materials being processed or produced can be used. These include for example colour, density and chemical composition. This is the important link to chemical measurements and the spectroscopic applications, which are the subject of this section of this handbook.

Concepts of Process Analysis

Process analysis is the general term applied to analytical techniques used to provide information for controlling processes. Such methods can provide a wide range of information relating to the physical and chemical properties of the material being examined and also the properties and characteristics of the products that will result as a consequence of the process conditions being used. This information, when used effectively, can provide a significant input into process control schemes which impact upon the quality, consistency and usefulness of the product as well as the efficiency of the operation.

Process analysis can be achieved in different ways including traditional laboratory methods carried out in a central laboratory, using methods in localised 'at-plant' laboratories or automated on-line methods. The terms at-line and in-line are also used to describe different approaches to the application of the analytical methods. In large manufacturing processes, which can be exemplified by integrated petrochemical complexes or, to a slightly lesser degree, large food manufacturing operations, transporting samples to central laboratories and waiting for results brings with it unacceptable delays. Using techniques located near the process operations and arranged so that relatively non-specialised staff can carry out the tests provides immediate results on-demand that can be acted upon quickly. Even better, if the implementation costs can be justified, is to install equipment directly on the operating process so that measurements are made in real time. Coupling such on-line measurements to appropriate control mechanisms results in tight control of the process to provide product that is consistent and on specification.

Methods of monitoring processes as they proceed changes the out-dated philosophy of quality control as an 'end of pipe' activity to a quality assurance philosophy where a process being continuously monitored and kept within prescribed conditions will be assured of producing acceptable product. The out-dated 'end of pipe' quality control examination could lead to products being discarded, sold at a loss, or recycled for reprocessing. In contrast, continuous process control can identify deviations from acceptable operating conditions before off-specification product is produced and process changes can be made to bring the process back to optimum operating conditions. Even within specification limits process analysis and control can improve the consistency of a product, which is an advantage to customers with stringent processing criteria.

Data generated from process analysis techniques are commonly displayed on control charts and the term statistical process control (SPC) is often used to describe the use of such data visualisation [5]. As the amount of data available increases, due both to different measurements and greater frequency of measurements, then combinations of different data provide improved methods of monitoring the processes concerned. The procedures and concepts of multivariate SPC incorporating principal component analysis (PCA) and partial least squares (PLS) analysis then become important [6]. The different SPC approaches are all aimed at providing better process control and improved process understanding.

Processes and processing equipment or plant are now commonly used for producing a range of materials including different grades of one product or entirely different materials. Process control and the associated primary measurements have their contributions to make to this type of operational regime. Changing the grade of a material produced can involve a reduction in production rate and the production of unusable material during the period of change. Ensuring that the change can be made as rapidly and efficiently as possible can have important economic consequences. In cases where equipment is used for different products cleaning procedures must be effective to ensure good manufacturing practice, here again the effective application of process analysis can be beneficial.

Continuous processes during normal, in-control operation, operate in a state of equilibrium and fluctuations should be at a minimum, however, there are significant periods of operation where non-normal conditions are experienced. Such conditions include start-up after for example a maintenance shutdown, abnormally high or low production requirements and catalyst regeneration or replenishment operations. It is here that process control using well-designed process analysis techniques can be particularly beneficial by reducing the time that the plant is not operating according to the requirements. With batch processes each period of operation, feedstock charging, reaction and completion are to some extent different conditions and can need different process control and associated analysis protocols.

Processes for producing chemicals, materials and goods are complex and varied and to ensure the correct product is produced effectively process control procedures are vital but underlying these procedures process analysis techniques, including spectroscopies can be used as an important source of primary data.

Practical Considerations for Process Analysis

Consideration must be given to the sample for which results are sought, the requirements of the analysis, the available equipment, its capability and suitability and the feasibility of linking to a control system that can bring about the necessary changes to the process.

Samples may be solid, liquid, gaseous or multiphase, at high or low temperatures, under non-ambient pressures, with variable flow rates and as a result of their composition may be corrosive or abrasive. All these factors affect the feasibility of applying process analysis techniques.

Analytical requirements vary according to the process control that is required. Chemical composition may be needed at percentage levels or trace levels with single components or complete analyses being required. Physical characteristics of the sample may be important as with many polymer-processing operations. The accuracy and precision needed will depend upon the application of the results. The speed of response is also important, for example in some process applications of leak detection of explosive materials analysis is required every few seconds whereas in processes that change slowly a rapid response would be unnecessary as process modifications could not be made in such a short time.

The analytical equipment that is chosen needs to be capable of satisfying the requirements. It often needs to be robust to withstand the chemical environment. Cost is an important factor, which is linked to whether equipment may be multiplexed so as to carry out a number of similar analyses. Operationally, simple, reliable and long-term calibration is important, the capability of automatic fault diagnosis can be an advantage and the benefits of a non-invasive technique are considerable.

If analytical equipment can be identified to do the required task then the links to process control must be considered, for the analytical results to be of value appropriate changes to the process must be possible.

Spectroscopy and Process Analysis

Spectroscopy is only one of the general methods of analysis used to monitor processes. Chromatography, electrochemical techniques and a broad range of physical measurements are commonly used [7] but are obviously beyond the scope of both this section and indeed this Handbook. The role of spectroscopy in process analysis has recently been reviewed [8].

Process analysis can be carried out in a traditional analytical laboratory with samples being transported from the process of interest to the laboratory. This is still a commonly used procedure but has serious drawbacks, notably the delays that inevitably result. These delays can mean that the process is producing off-specification material for a considerable time before the fact is realised. This is especially important if large volume or high value materials are being produced. An alternative is to locate the laboratory close to the process operations [9] and avoid the serious delays involved in transporting samples and waiting for analytical results. Automating the analysis so that process operators can use the equipment quickly and efficiently helps to reduce delays even further. On-line or in-line analysis where no extraction of sample is required is the preferred procedure although the technology costs involved sometimes dictate that this is not economically viable. In other situations, especially where safety considerations are involved, rapid analyses can be essential and automated on-line analysis with a minimum delay in obtaining the result is the only acceptable procedure.

Introduction

Using spectroscopic techniques for process control can present special requirements. A laboratory environment being used for process analysis is technically no different from the use of such equipment as described elsewhere in this Handbook. Operational considerations may dictate that the equipment is available at specific times to accept the process samples according to a prescribed schedule but this is an organisational matter for the laboratory and manufacturing management. The use of spectroscopy in on-line and in-line environments is usually very different. Process conditions commonly experienced include chemical vapours, dust and vibration, whilst some equipment may be open to the elements, located in zoned areas and used by operations staff unaccustomed to such equipment. The siting of equipment is therefore important. Siting options include a controlled cabinet, an analyzer house or a designated part of the process area or control room.

The interface between the sample and the spectrometer is vital wherever a spectrometer is sited. The sample can be piped into the spectrometer or, in some cases; the radiation used by the spectrometer can be transmitted to a convenient sample or probe location point using optical fibres or other light-pipe devices. The sample presented to the spectrometer must be representative of the material from which the measurement is required and the interaction between the radiation of the spectrometer and the sample must be suitable for the measurement to be made (sufficient power and suitably clean interface).

Sample lines can be a serious cause of malfunction of process analysis techniques, which will subsequently adversely affect the process control procedures. Long sample lines or those having high pressure drops will cause delays in the sample transport and therefore the analysis will relate to process conditions already passed. Sample lines can become blocked as a result of entrained material or as a result of insufficient heat tracing causing condensation or solidification. Consequently results will either be unavailable or will be erroneous. Practical sampling procedures and sample pretreatments have been briefly summarised and discussed [10].

Consideration needs to be given to the sample interface and the measurement technique used. Some techniques relate to the whole sample whereas others are very much surface measurements. For example, microwave spectroscopy and infrared transmission measurements provide values on the bulk sample whereas X-ray fluorescence and Raman spectroscopy are very much surface techniques, only penetrating the sample to a limited degree.

Common Spectroscopies for Process Analysis & Control

The spectroscopic techniques most commonly used for process analysis involve the use of infrared or UV/visible radiation. Mass spectroscopy has a considerable number of varied applications especially in the area of gas analysis. NMR technology is being increasingly used for a range of applications. Atomic spectroscopies, used extensively from a laboratory base for process control, are finding applications in automatic on-line measurement where the sampling systems can be suitably

Introduction | 275

adapted. Details of these spectroscopies in process analysis and control and their applications are specifically dealt with in the following parts of this section of the Handbook.

Specialised Spectroscopies and Emerging Techniques

There are a number of spectroscopies that are not commonly used for process analysis but do have specialised applications. Other techniques are emerging as possible methods for the future. Some of these techniques and their applications are discussed briefly below.

Microwave spectroscopy

Although the most well known use of microwave spectroscopy is in the fundamental studies of the rotational structure of free molecules and as a method of determining dipole moments the technique is finding applications in the area of process analysis. Its use for process samples depends upon the analyte having different permittivity characteristics to that of the sample as a whole and so the technique is admirably suited to the determination of water or moisture in solids and liquids.

Microwave spectroscopy has found applications in the food processing industry as well as in the polymer and chemicals manufacturing industries. Water, fat and protein determinations have been demonstrated together with reported applications for the measurement of polymerisation rates and the water content of organic and mineral acids. Determinations have also been reported of the percentage solids in lime slurry and, in another application, the water, phenol and diphenol oxide content of a reaction mixture [11]. A varied range of solid and liquid samples has been examined including a range of meat and dairy products, pet foods, doughs and coal. The technique is applicable to the bulk properties of a sample and is therefore not restricted to that section of the sample in proximity to the cell windows as is sometimes the case in some optical and X-ray fluorescence spectroscopies. The equipment is suitable for use in the form of a bench-top instrument or as an in-pipe installation.

It has been suggested that developments in microwave spectrometers could provide sensors for particular chemical compounds with high specificity and sensitivity [12]. Indeed microwave spectroscopy has been used to specifically monitor ethylene oxide concentrations in medical sterilisation units [13].

REMPI Spectroscopy

The use of resonance enhanced multiphoton ionisation (REMPI) spectroscopy linked to time of flight mass spectroscopy has been demonstrated in on-line monitoring of combustion by-products in industrial flue gases [14] and in a research project dedicated to the analysis of coffee roasting processes [15]. REMPI is a highly selective and sensitive technique and depends upon selective ionisation of target species in a flowing gas stream. In the examples cited this is of course a hyphenated technique.

Ion Mobility Spectrometry

Ion mobility spectrometry (IMS) is a form of mass analysis and is most commonly associated with military use for the portable detection of nerve gases and the detection of explosives and drugs at international borders and airports. Applications in process analysis are being found in the analysis of gas streams. It is particularly suited to trace analysis (ppb) of easily ionisable species. The sample is drawn over a semi-permeable membrane through which the compounds of interest pass into an ionisation chamber. Typically a nickel 63 source provides a means for ionization, the ions then pass down a drift tube to the detector, small ions arriving before larger ions. Examples of continuous analysis of ammonia, and organic amines are described in application notes provide by Molecular Analytics [16]. Other companies providing IMS equipment have concentrated more on the military and security applications.

Acoustic Emission Spectroscopy and Ultrasound Techniques

Acoustic emission spectroscopy has been applied to process analysis to detect physical rather than chemical changes. However its use appears less widespread than it maybe deserves. Detection of the movement of particles or bubbles in reactors can, when coupled with process knowledge, provide a means of monitoring the extent of process changes. Applications include the monitoring of granulation, fluidisation, agglomerization, milling and drying procedures. A reported application is in the detection of particle entrainment during a solvent removal process, which shows that the process has reached completion. A primary benefit realised from such a solvent removal application is the rapid turnaround of processing equipment that can be achieved. However a secondary benefit is an improved process, as further processing causes the removed solvent, which is subsequently recycled, to become contaminated with the entraining material. Eventually this results in the solvent being unsuitable for recycling or the product of the primary process becoming impure and out of specification. A number of application notes are available from Process Analysis and Automation [17]. Acoustic techniques have been used to monitor crystallisation processes [18].

Active ultrasound uses a source of sound radiation, which is applied to a process sample, with a detector placed such that modification to the signal can be detected and related to changes in the sample. Signal attenuation, velocity measurements and wavelength selective absorption provide the means of probing the sample. This approach promises to provide both chemical and physical information but as yet has not been used extensively. A number of on-line polymer-related studies have been reported in which polymer flow behaviour, viscosity, blend characterisation, and foaming-process monitoring have been examined [19].

Inferential Analysis

Inferential analysis [20, 21] is not a spectroscopy but could have a bearing upon the use of all process analysis techniques. It is a term being used to describe measurements that are not made but are inferred from other properties of the process under scrutiny. These methods rely upon process models being available for the process concerned. The value of this approach, quite apart from the fact that no expensive equipment is needed, is that it can give an indication of a measurement when it is impossible to extract a sample without it undergoing change or where inserting a probe is impractical. Inferential methods can also be useful to provide values between the frequency of the installed measurement devices or indeed when the measurement devices are off-line for maintenance purposes. The quality of an intermediate or a product, can in some instances be inferred from the values of temperature, pressure and flow rates in the area of the process under consideration.

Summary

Process control, achieved in part as a result of good process measurements, can provide a number of operational benefits including:

- Operating cost improvements
- Increased rate of production
- Reduced energy use
- Reduced laboratory costs
- Reduced production of off-specification material
- Less re-working of material
- Less waste production and disposal
- Reduced environmental impact
- Improved health and safety as a result of reduced exposure to chemicals
- Continuous control of product quality
- Improved process understanding

Effective process control needs reliable, robust, often rapid inputs from measurement techniques. Spectroscopy can provide such inputs if the right technique is selected, the sample interface is well designed and the data analysis is carried out effectively.

References

1 *Coulson and Richardson's Chemical Engineering*, Volume 3, eds. J. F. Richardson, D. G. Peacock, Pergamon Press, 1994, Vol. 3, pp. 560–731.
2 K. Dutton, S. Thompson, B. Barraclough, *The Art of Control Engineering*, Addison Wesley Longman, London 1997.
3 F. G. Shinsky, *Process Control Systems, Application, Design and Tuning*, 3rd Edition, McGraw Hill, New York, 1988.
4 Chem. Eng., **1999**, *106* (8), 76.
5 J. S. Oakland, R. F. Followell, *Statistical Process Control*, 2nd Edition, Heinmann Newnes, 1990.
6 M. H. Kaspar, W. H. Ray, *AIChE J.*, **1992**, *38*, 1593.
7 K. J. Clevett, *Process Analyzer Technology*, Wiley, New York 1986.
8 *Spectroscopy in Process Analysis*, ed J. Chalmers, Sheffield Academic Press, 2000.
9 T. Lynch, *CAST*, **1999**, June/July, 4.
10 J. R. P. Clarke, *Process Control Quality*, **1992**, *4* (1), 1.
11 Epsilon Industrial Inc., 2215 Garnd Avenue, Parkway, Austin, Texas 78728; www.epsilon-gms.com
12 H. D. Rudolph, Microwave Spectroscopy – Instrumentation and Applications, in *Encyclopedia of Analytical Science*, ed. A. Townshend, Academic Press, New York 1995, Vol. 6, p. 3271.
13 Z. Zhu, I. P. Matthews, W. Dickinson, *Rev. Sci. Instrum.*, **1997**, *68*(7) 2883.
14 R. Zimmermann, H. J. Heger, A. Kettrup et al., *Rapid Commun. Mass Spectrom.*, **1997**, *11*, 1095.
15 R. Zimmermann, H. J. Heger, R. Dorfner et al., *SPIE*, **1997**, *3108*, 10.
16 Molecular Analytics, 25 Loveton Circle, PO Box 1123, Sparks, MD 21152-1123; www.ionpro.com
17 Process Analysis & Automation, Fernhill Road, Farnborough, Hampshire, GU14 9RX, UK; www.paa.co.uk
18 J. G. Bouchard, M. J. Beesley, J. A. Salkeld, J. A., *Process Control Quality*, **1993**, *4*(4), 261.
19 L. Piche, R. Gendron, A. Hamel et al., *Plastics Eng.*, **1999**, October, 39.
20 M. Tham, Department of Chemical & Process Engneering, University of Newcastle upon Tyne, UK; http://lorien.ncl.ac.uk/ming/infer/infer.htm
21 J. V. Kresta, T. E. Marlin, J. F. McGregor, *Comput. Chem. Eng.*, **1994**, *18* (7), 597.

17
Optical Spectroscopy

John Green

17.1
Introduction

Optical spectroscopy can be defined in terms of the electromagnetic spectrum and includes those regions from the vacuum UV to the mid- or far-IR. The wavelengths associated with the different spectroscopies are shown in Fig. 17.1 together with an expanded view of the optical spectroscopy region including the different wavelength units commonly used.

The commonest form of optical spectroscopy measurement involves simple absorption governed by the well-known Beer–Lambert law. However, reflection, light scattering, fluorescence and chemiluminescence methods are also employed in applications relating to process control.

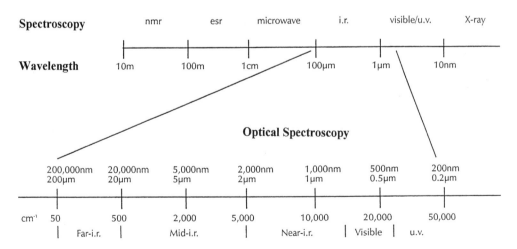

Fig. 17.1 The electromagnetic spectrum showing the wavelength units commonly associated with the different spectroscopies used in chemical analysis. The wavelength range relating to the optical spectroscopies is expanded in the diagram. Adapted from a diagram supplied by Clairet Scientific [1].

17.1 Introduction

The interaction of radiation of the UV, visible and IR regions of the electromagnetic spectrum with molecules causes different perturbations to occur in the molecules concerned. UV and visible light generally cause electronic transitions whilst interaction with IR radiation results in vibration modes being excited. Simple absorption measurements involve the differential measurement of the radiation on passing through the sample. Reflection measurements use the same general principle but the detector is positioned to measure reflected radiation. Scattering of radiation results from some interactions and the resulting shift in radiation frequency is a measure of the species present in the sample. Raman spectroscopy makes use of this phenomenon. Fluorescence measurements monitor the radiation emitted from a species within the sample that has previously been excited by the radiation source. Chemiluminescence is the measurement of emitted light from a species, usually as a result of the interaction with a second chemical. The chemical interaction produces a product in an electronically excited state that, on relaxation, emits chemiluminescence radiation.

Instruments vary widely in their design depending upon the purpose for which they are built. Common features include a source of radiation, a means of bringing the radiation and the sample of interest together in a cell or probe, and a detector. In applications to process measurement perhaps the most distinctive feature is the sample interface. The source of radiation used and the detectors are similar and often identical to laboratory-based instrumentation. Almost all of today's instruments include data acquisition and control electronics together with a user interface in a computerized form. To obtain the optimum performance from analytical and control systems, links to distributed control systems for feed-back and feed-forward control are vital.

Optical spectroscopies are appropriate for many process analysis applications because they can provide a rapid analysis of process stream composition in an industrially robust form requiring little regular maintenance. On-line implementation means that extracting manual samples is avoided and results obtained using computerized data handling methods can be linked to automatic control schemes. Recently a review of spectroscopy in process analysis has been published which contains descriptions and describes applications of optical spectroscopies [2].

The optical spectroscopies having established or potential process analysis applications described in this section include:

- Mid-infrared
- Non-dispersive infrared analysers
- Near-infrared
- Ultraviolet/visible
- Raman spectroscopy
- Fluorescence techniques
- Chemiluminescence
- Laser techniques
- Optical sensors

The emphasis will be on the applications of these techniques rather than a detailed description of the techniques themselves that are covered in other sections of the Handbook.

17.2
Mid-infrared

By far the most common use of mid-infrared radiation for process analysis is in the non-dispersive infrared analysers that are discussed below. The widespread use of FTIR spectrometers in the mid-IR has yet to be fully realized in process analytical applications. The requirements for the optical components and the wavelength stability of the instruments available have, until recently, detracted from the use of this region of the spectrum in on-line process analysis. Optical fibers that provide such a benefit to the applications of NIR (see below) are not available for the mid-IR in robust forms or forms that are capable of transmitting over more than a few tens of metres. Improvements and developments to sample cells, particularly designs of attenuated total reflectance (ATR) cells, for use with mid-IR are being made and will influence the application of the technique. An impressive list of applications including both FTIR and the NDIR approaches has been compiled [2, 3].

17.3
Non-dispersive Infrared Analysers

In terms of the number of optical spectroscopic instruments used in process analysis non-dispersive infrared is perhaps the most common and has an established record of success. The mid-infrared section of the spectrum provides a region rich in chemical information having absorptions specific to individual chemical species where interferences can be avoided.

These NDIR instruments are mainly used for gas samples although multiple reflection cells are available in a variety of materials that are applicable for the analysis of liquids. Filters are chosen according to the analyte to be determined and the other components of the matrix. Carbon dioxide, carbon monoxide, ethylene, nitrogen oxides and sulfur dioxide are typical analytes.

There is not an extensive recent scientific literature devoted to these methods but many of the instrument suppliers provide technical details and specifications in their literature and websites [4, 5].

A variety of designs are commercially available and include combinations of single and dual beam and single and multiple wavelength.

The instruments, see Fig. 17.2, comprise a sample cell through which the radiation is passed together with a reference cell to compensate for radiation source drift and output. Contamination of the cell walls can be compensated for using selected dual beam instruments in which one wavelength is used for the measurement and a second wavelength, not absorbed by the sample is a measure of contamination.

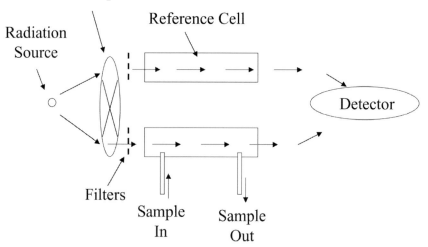

Fig. 17.2 Typical arrangement of components of a dual beam NDIR process analyser.

Narrow bandpass filters can improve the selectivity of such analysers and can be a means of using the instruments for multiple components. Using a filter chamber filled with an interfering sample component can increase the utility of such instruments.

17.4
Near-infrared Spectroscopy

Whilst infrared spectroscopies all involve vibrational mode excitation of molecules different regions of the infrared interact in different ways with the species present. Mid-infrared cause fundamental vibrations to occur whereas near-infrared results in the excitation of overtone and combination modes of vibration. These modes are so-called forbidden transitions and they result in the weak absorptions that give NIR spectroscopy some of its unique properties.

Although near-infrared radiation was discovered by Herschel in 1800, NIR spectroscopy has only recently become an established technique for process analysis [6, 7]. The increasing success and widespread application of NIR spectroscopy in this area is a result of several advantageous features and technical developments.

Cell path lengths need to be longer for NIR than for mid-IR or UV/visible spectroscopy because the absorptions are weaker. This has two advantages, firstly the process stream often contains particulates that can block narrower cells and secondly the pressure drops of longer pathlength cells are considerably reduced.

NIR radiation can be transmitted down readily available, relatively cheap optical fibers thus allowing spectrometers to be located remotely from the process streams they are monitoring, away from potentially hazardous areas. Optical fibers also provide a convenient means of multiplexing a single spectrometer to several streams, thus reducing the overall cost of the analyses. Optical fibers, optical switching and sample probes have been crucial to the development of NIR spectroscopy for process analytical applications [8, 9].

The improved use and availability of chemometrics allows the overlapping spectra of multicomponent mixtures that are characteristic of NIR absorptions to be analysed successfully and the components quantified. A series of calibration samples of known composition are analysed to establish a calibration model, which is subsequently used to determine the composition of unknown samples. The same approach can be used to correlate NIR spectra with non-compositional properties of samples so the technique is regularly used to determine and estimate operational and quality properties of process streams, for example octane numbers, distillation points and viscosity.

As a consequence of the above factors NIR spectroscopy is now applied to a wide range of process applications in the oil refining, petrochemical, polymer, pharmaceutical, food, environmental and agricultural industries.

Instrumentation for NIR process analysis can be of several forms. Perhaps most simply, but less commonly, fixed filter photometers may be used to pass several chosen wavelengths of light through a sample extracted from a process stream. Interference filters located on a rotating disc allow light of certain frequencies to pass through the sample. The absorption measurements achieved allow sample composition and properties to be estimated as a result of previous calibrations. More commonly, scanning spectrometers, Fourier transform, photodiode array and spectrometers utilizing an acousto-optic tunable filter are used, allowing a continuous spectrum of a chosen sample to be obtained for subsequent analysis using chemometrics and previously determined calibration routines.

NIR radiation, at least those wavelengths suitable for probing the overtone bands, is not absorbed by silica optical fibers and so the spectrometer can be remotely linked to the sample. In hazardous areas such as many petrochemical applications this allows the spectrometer to be located in a safe area without having to transport the sample from the chemical plant. Furthermore it is possible to switch the NIR radiation between fibers so that one source can be used sequentially to excite and analyse several sample points. Both these advantages, resulting from the transparency of silica to NIR radiation can reduce the cost of such applications considerably.

Various probes are available for introducing the radiation to the sample. Transmission probes can be either of an insertion or flow-through type. Reflectance probes utilise a reflective surface to return the radiation to the detector. Internal reflectance probes employ attenuated total reflectance at a sample surface to obtain an absorption value.

A typical arrangement is shown in Fig. 17.3 illustrating a single spectrometer multiplexed to three process stream transmission cells by optical fibers. Spectral

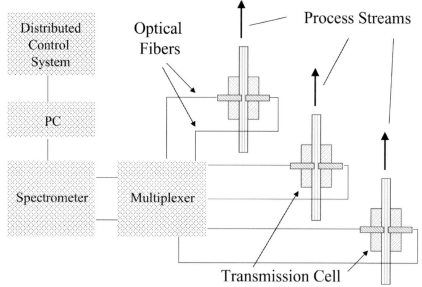

Fig. 17.3 Schematic of typical multiplexed NIR process analysis arrangement showing spectrometer, multiplexer, optical fibers and transmission cells together with the connections to the PC and distributed control system.

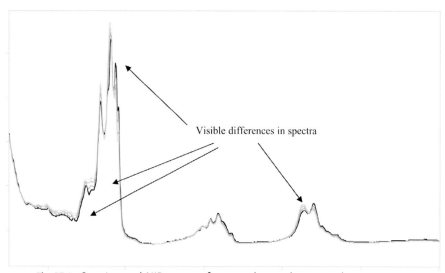

Fig. 17.4 Superimposed NIR spectra of pentane, hexane, heptane and octane.

data are then analysed by the computer and transmitted to the distributed control system.

Because NIR absorptions are typically broad, individual components of mixtures overlap very considerably. This is illustrated in Fig. 17.4 that shows the superimposed spectra of pentane, hexane, heptane and octane. In order to obtain quantitative information from such spectra prior calibration using PLS calibration routines is required. When the individual components have significant differences in their spectra calibration is not difficult and several samples prepared to cover the expected range of samples concentrations will serve to provide a calibration. In cases where the sample to be analysed contains similar components (e.g. hydrocarbons) then more samples will be required to build an appropriate calibration. Transferring calibrations from one instrument to another is possible, sometimes without any mathematical transformation routine. This is especially true if the spectrometers concerned are of the same design and manufacture.

Various probes are available for introducing the radiation to the sample. Transmission probes can be either of an insertion or flow-through type. Reflectance probes utilise a reflective surface to return the radiation to the detector. Internal reflectance probes employ attenuated total reflectance at a sample surface to obtain an absorption value.

The applications of NIR spectroscopy to process analysis and control are far too numerous to comprehensively cover in a text of this type. A regular review of literature in this area is given in NIR News [10]. Examples in Tab. 17.1 illustrate the breadth of NIR applications to process analysis. Table 17.2 summarises the use of NIR in process analysis.

Table 17.1 Examples of the application of NIR spectroscopy to on-line process analysis

Industrial Sector	Example Reference from Recent Literature
Petrochemicals	Feed-forward control of a steam cracker for production of ethylene and propylene 11
	Fast on-line analysis of process alkane gas mixtures by NIR spectroscopy 12
Pharmaceuticals	On-line measurement of moisture and particle size in the fluidized-bed processing with NIR spectroscopy 13
	Automated system for the on-line monitoring of powder blending processes using NIR spectroscopy 14
Polymers	On-line NIR sensing of CO_2 concentration for polymer extrusion foaming processes 15
Environment	
Food	In-line measurement of tempered cocoa butter and chocolate by means of NIR spectroscopy 16
Agriculture	On-line cane analysis by NIR spectroscopy 17

Table 17.2 Near-IR spectroscopy for process analysis

Molecular parameter used in measurement:	Interaction with sample:
Overtone and combination bands of molecular vibrations	Liquids, transmission depth Solids, surface reflectance correlation to bulk property Gases, transmission path length
Typical information sought:	
Quantitative information on sample components	**Limits of determination:** Absorbing materials, percentages to, at very best, 10 ppm levels
Measurement environment:	**Time needed for analysis:**
Liquids, gases and solids can be investigated	Calibration, hours/days depending on sample availability and optical differences in sample components Sample preparation, not generally applicable to on-line analysis. Measurement, seconds/minutes Evaluation, seconds using predetermined calibration routines.
Equipment:	
Near-IR spectrometer Sample cells and probes Optical fiber connections (optional) Fiber multiplexer	
Type of laboratory:	**Cost:**
Optimal use as an on-line technique. Can be used in QC/QA laboratory after initial development work	Large on-line installation: 250–500 K GBP Laboratory spectrometer: 30–50 K GBP
Skill needed:	**Sample requirement:**
Specialist to arrange/develop analysis Competent technical skill to operate.	Liquids, mm/cm pathlength Gases, adequate pathlength depending on pressure, e.g. 10 cm Solids, surface exposure to radiation
Techniques yielding similar information:	
Optical spectroscopies	

17.5
Ultraviolet/Visible Spectroscopy

UV and visible photometers are commonly used for a range of process stream components. Hydrogen sulfide and sulfur dioxide are common examples. The approach is similar to the use of NDIR measurements and several manufacturers supply equipment based on this principle [4, 5]. The source of radiation is split into two beams (sample and reference) and passed alternately through filters. The 'sample' filter allows light to pass that is absorbed by the analyte of interest the 'reference' filter provides a beam that is not absorbed by the analyte. The ratio of the absorptions is a measure of the analyte concentration. The choice of filters is crucial if potential interferences are to be avoided. Equipment design allows such analyzers to be used at elevated temperatures and pressures and depending upon the analyte concerned ppm values are achieved.

Alternatively the whole UV/Visible spectrum may be recorded and by using appropriate calibration and chemometric techniques it is often possible to determine severely overlapping peaks.

In a similar way to NIR, UV/Visible can be used in absorption and reflection modes. Spectra that are due to electronic transitions are generally broad, as in the NIR, because several vibrational levels will be populated by the primary electronic transition.

The applications of UV/Visible spectroscopy depend upon the analyte of interest having absorptions in the relevant region of the spectrum. For organic compounds aromatic unsaturation provides an excellent absorption chromophore. Thus UV spectroscopy has been used to determine antioxidant levels in polymers [18] during the process of extrusion, the aromatic contents of petroleum [19], the purity of organic products, effluent control of fluorine in the nuclear processing industry [4], determination of methylhydroquinone inhibitor in the storage of acrylonitrile [4], aromatics in process water to minimize environmental impact and as a means of assessing the chemical oxygen demand.

Colour is an important quality control property in many industries including the water and petrochemical industries, each uses visible spectrometry methods to assess quality according to comparisons with different standards. ASTM and Saybolt colour measurements are reported using process UV photometers [4].

UV/Visible spectroscopy is also used as the detection means for a number of process analyses following wet-chemical sample treatment with a selected reagent. Systems are available for the analysis of ammonium, nitrate, nitrite and phosphate in a range of aqueous process streams following appropriate clean-up and reagent colour development [20].

17.6
Raman Spectroscopy

The Raman effect was first predicted and then demonstrated early in the 20th century. Until recently it has been a specialist tool with niche applications very much confined to laboratory environments. Recent developments and improvements in laser light sources, detector technology, optical fibers and optical filters have led to an increase in its use. This is now having its impact upon process analysis as robust equipment becomes available and the potential advantages of the technique become widely appreciated.

Raman spectroscopy is a scattering technique in that it utilizes the scattered radiation when light impinges on a sample. Most incident light is scattered elastically, having the same wavelength as the original light (Raleigh scattering) but a very small proportion is scattered inelastically having a frequency shifted from

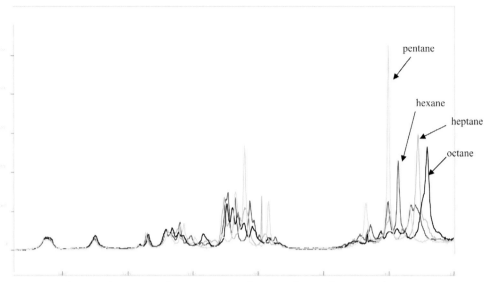

Fig. 17.5 Superimposed Raman spectra of pentane, hexane, heptane and octane

that of the original frequency. The shift is caused by interaction with the sample molecules and relates directly to their molecular vibrations. Raman spectroscopy gives information akin to and complementary to mid-infrared spectroscopy.

Raman is particularly sensitive to unsaturation in symmetrical environments and is therefore an appropriate method for many aromatic species. Generally it is the sensitivity to large changes in polarizability that provides strong Raman signals. Infrared in contrast responds more sensitively to vibrations with large changes in dipole moments. With simple molecules Raman and IR are completely complementary, with for example oxygen, nitrogen and hydrogen Raman gives a signal whilst for IR these molecules are inactive. However the spectral characteristics of Raman resemble mid-IR in that sharp features are dominant in contrast to the NIR where bands are broad, diffuse and largely overlapping for different chemicals.

Raman spectroscopy has a number of benefits for process analysis applications. For appropriate samples Raman spectra are rich in information content. Figure 17.5 shows the superimposed Raman spectra of pentane, hexane, heptane and octane which can be compared with the corresponding NIR spectra in Fig. 17.4. The spectra and resultant information are similar to those obtained from mid-IR.

The wavelength of the incident exciting radiation can however be in the NIR, visible or UV regions of the spectra thus making the use of optical fibers possible. Optical fibers bring their own advantages including multiplexing capabilities and remote location of spectrometer and sample points. Options for the sample interface include a window, or site glass, into the process providing an essentially noninvasive method or an insertion probe incorporating a ruggedised bundle of fibers. Both the exciting radiation and the scattered radiation can enter and exit the sample through the same window and use the same fiber optic cable. In practice the arrangements may not be quite as simple if only because of the precautions that need to be taken as a result of using laser light in operational areas [21, 22]. Using either window or probe the sample interface is therefore a flow-past device, as compared to the flow through devices common with absorption and transmission measurements. This can be very important for viscous polymers or liquids and for multiphase streams that may otherwise block the flow path.

The choice of Raman instrumentation for process analysis is dependent upon the particular application. Laser light sources of different wavelengths: 532 nm, 633 nm 785 nm and 1064 nm being commonly used with 785 nm being perhaps the most common. The light source chosen will depend upon the sample composition, the sensitivity needed, the existence of potential fluorescing materials and the availability of a suitable detector. Most spectrometers used for process applications are dispersive with charge coupled device (CCD) detectors although in some applications where fluorescence presents problems the use of Fourier transform (FT) instruments is advantageous. Depending upon the probes used, optical filtering may be required. With the fiber bundle probes optical filtering is not used and silica artifacts are minimized by the fiber end geometries. With the site glass interfaces optical filtering is incorporated into the optics to remove the Raman signal originating from the silica and also to remove the Raleigh scattered light, otherwise the Raman signals from the sample would be obscured. Holographic filters have become the commonly used choice in process applications. Whilst CCD detectors are most common the use of charge induction device (CID) detectors is reported to offer some advantages [23]. Array detectors offer an alternative means of multiplexing the system by using different locations on the CCD chip to provide the signal from different sample points.

A schematic diagram showing the main components of a typical process analysis arrangement for Raman spectroscopy is shown in Fig. 17.6.

Fluorescence is commonly regarded as a major problem in the use of Raman spectroscopy and indeed fluorescence of species in the samples of interest can mask the Raman signal of the analyte molecules. However it is often possible to choose a laser wavelength that avoids the problem or to use techniques such as shifted excitation Raman difference spectroscopy [24] or subtracted shifted Raman spectroscopy [25].

Quantification of Raman spectra requires the use of internal standardization and this can sometimes be achieved using an essentially invariant feature of the sample. For example in the analysis of styrene in a polymerization reaction the styrene phenyl peak was used [26].

Applications of Raman for process analysis have been reported in the open literature and in the patent literature which is clearly a sign that some of the applications are regarded as commercially sensitive.

Distillation column monitoring was one of the first reports of using on-line Raman spectroscopy in real time for process analysis. This application used FT Raman coupled to a remote sample cell with optical fibers [27] to determine the 'bottoms' product of a solvent recovery column. The system was linked to a control

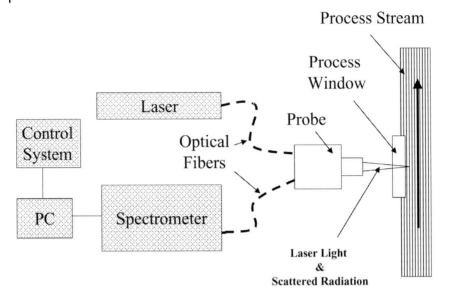

Fig. 17.6 Schematic diagram of a typical arrangement using Raman spectroscopy for process analysis.

strategy using statistical process control with defined upper and lower action limits.

The monitoring of the calcination process in the production of the rutile-structured titanium dioxide from the anatase structure is recorded as a notable success [28]. A fiber optic probe appropriately positioned in the calcination kiln monitors the composition of the powder undergoing processing. The Raman bands of the two forms of titanium dioxide are quite distinct and as a result the operating conditions of the kiln (fuel and air-flows) can be adjusted to give optimum production.

The production of phosphorus trichloride from phosphorus and chlorine is an important step in the manufacture of a number of agrochemical products. Raman spectroscopy has been used to monitor the reaction and control the raw material feed rates. This maximises production of phosphorus trichloride, minimizes the formation of phosphorus pentachloride and ensures safe operation when plant shut-down periods are needed. Remote analysis is achieved using optical fibers to provide a safer operation and a more rapid analysis than was previously possible [29, 30].

Hydrocarbon streams have also been investigated and Raman-using systems have been patented. The monitoring of emulsion polymerizations by Raman spectroscopy is also proposed. Pharmaceutical process applications of Raman spectroscopy include the monitoring of the active ingredient in a product without sample preparation. The identification of gemstones and the determination of the synthetic/natural origins are well known.

Gas analysis can be achieved using Raman spectroscopy and is of particular value when the gases and vapours concerned are infrared inactive. In a pharmaceutical application oxygen can be detected in the headspace of sealed vials [31]. Anaesthetic compounds have been determined using the laser cavity as a cell [32]. A trace hydrogen sensor using Raman scattering has been reported [33]. Gas–liquid equilibria for a cyclohexane–toluene mixture have been investigated using sequential Raman measurements of the gas and liquid phases [34].

A variety of commercial polymerisation reactions have been followed using on-line Raman spectroscopy in which the progressive disappearance of the unsaturation provides a measurement of the extent of reaction. Emulsion co-polymerisation of methyl methacrylate, butyl acrylate and styrene has been followed in this way [28].

A number of model in situ and on-site applications of low resolution, and therefore low cost, Raman spectroscopy have been reported including the quantitative monitoring of synthetic rubber and polystyrene emulsion polymerisations, detection of illicit drugs and explosives and detection of cyanide in wastewater using a surface enhanced Raman system [26].

It has been noted elsewhere that Raman and infrared techniques provide complementary information and therefore it is not surprising that these different approaches are commonly considered for the same application. In some cases it has been concluded that either Raman or infrared techniques can be used to obtain the information required. Comparisons of the techniques for aromatic hydrocarbon determinations have been reported from laboratory studies [35, 36].

Developments in SERS and SERRS have yet to make an impact upon process analysis but some potential means of using these enhanced sensitivity methods have been proposed.

Table 17.3 summarizes the use of Raman in process analysis.

17.7
Laser Diode Techniques

Laser diodes are providing analytical capabilities in the area of process analysis, particularly for the analysis of gases. A range of diodes is available covering the NIR wavelength range of 0.6 to 2 µm, suitable for detecting overtone and combination bands. Individual diodes have very narrow wavelength outputs and can only be tuned over a narrow range of several nm. The high spectral resolution of diode laser sources provides the specificity of the technique as the radiation is tuned selectively to correspond to the absorption features of the analyte molecule. Oxygen, carbon monoxide, hydrogen chloride, hydrogen fluoride and ammonia are amongst the gases that can be analysed using this approach. The monitoring of ammonia in stack gas has also been achieved using laser diode techniques with a reported detection limit of 2 ppm at 100 Torr with a 1 m path length [37]. Commercial equipment is available for these applications [38]. Laser diode methods for monitoring atmospheric gases have been reviewed [39]. In a later paper this was extended to on-line process monitoring [40].

Table 17.3 Raman spectroscopy for process analysis

Molecular parameter used in measurement:	Interaction with sample:
Molecular bond vibrations from scattered radiation	Laser radiation focused within sample meaning different depths can be probed for appropriate samples. Information obtained from μm depths of field
Typical information sought:	**Limits of determination:**
Quantitative information on sample components. Supplementary molecular structural information available if needed.	Raman active materials, percentages to ppm levels. Lower detection levels with surface enhanced methods but not generally used for process analysis.
Measurement environment:	
Liquids, gases and solids can be investigated	**Time needed for analysis:**
Equipment:	Calibration, depends on analyte and availability of distinct Raman absorption differences in sample components
Laser source Raman spectrometer Sample probes Raleigh scattering filters Optical fiber connections (optional) Fiber multiplexer (optional)	Sample preparation, not generally applicable to on-line analysis. Measurement, seconds/minutes Evaluation, seconds using predetermined calibration routines.
Type of laboratory:	**Cost:**
Optimal use for process analysis as an on-line technique. Can be used in QC/QA/research laboratory	Large on-line installation: 250–500 K GBP Laboratory spectrometer: 50–80 K GBP Lower cost spectrometers becoming available
Skill needed:	**Sample requirement:**
Specialist to arrange/develop analysis Competent technical skill to operate.	Access to sample through non-contaminated optically transparent window. Use of probe incorporating optical fibers.
Techniques yielding similar information:	
Other optical spectroscopies Raman is a complementary technique to mid-IR	

The application of NIR laser diode spectroscopy to the on-line analysis of atmospheric pressure chemical vapour decomposition has been reported. In the process of depositing a thin layer of tin on a glass surface the monitoring of methane in the presence of oxygen, water and dichloromethyl tin dichloride in the reactor is important. Laser diode spectroscopy has been used at high sensitivity (detection limit of 0.01%) and at high frequency (5 Hz) for this purpose [40].

The methods are robust withstanding temperatures of several hundred degrees and elevated pressures [38].

17.8
Fluorescence

Fluorescence spectroscopy has similar applications to UV/Visible spectroscopy as it originates from radiation emission from an excited electronic state of the species to be analysed. The most common application in process terms is the determination of hydrocarbons in water. A range of hydrocarbons including fuels, oils, aromatic chemicals and PAHs [41] has been determined in industrial, process and potable waters. An on-site laser probe for the detection of petroleum products in water and soil has been described although it is not known whether such systems are in regular use for this application [42]. Although not the most common method for determining sulfur dioxide commercial equipment has been available based upon fluorescence methods. A reported specialized use of fluorescence in process control involved investigations of the residence time of polymeric materials in extruders in which a fluorescence additive at a low concentration was traced as it passed through the extruder entrained within the polymer [43]. There are also considerable prospects for the use of on-line fluorescence measurements linked to control in the food industry and an application to sugar samples has been described [44].

17.9
Chemiluminescence

Chemiluminescence principles are used in commercial analyzers for the determination of nitric oxide for process control purposes [45]. The nitric oxide is reacted with oxygen to produce nitrogen dioxide in an electronically excited state; this then decays with the emission of light. Detection of the emitted light with a photomultiplier tube provides a measure of the original concentration of the nitric oxide. By suitably treating process samples containing both nitrogen dioxide and nitric oxide both species can be determined. In as much as optical means are used to detect the chemiluminescence this is another use of optical spectroscopy in process analysis.

17.10
Optical Sensors

Sensors based upon optical spectroscopy principles have been developed for the process measurement of pH and dissolved oxygen as well as the physical characteristics of turbidity. The pH and oxygen sensors rely upon sol–gel immobilized reagents that respond to changes in the process stream in which they are immersed. Evanescent wave absorption is used to detect pH related colour changes and fluorescent quenching of an oxygen sensitive ruthenium compound provides a means of detecting dissolved oxygen. The determination of hydrocarbons in water has been reported using evanescent wave absorption into a hydrophobic coating on an optical fiber that serves to concentrate the analyte [46, 47]. Commercialisation of such devices is still in its infancy but a colour and turbidity monitor is available from Siemens Environmental Systems [48].

17.11
Cavity Ringdown Spectroscopy

This technique has yet to find an established place in process analysis although the prospects look interesting. The fundamentals of the technique have been described [49] and the use for gas analysis discussed by several authors [50, 51]. Furthermore the development to applications involving condensed systems using a form of evanescent wave has been described [52].

The story of the application of optical spectroscopy to process analysis is not complete, there are developments and improvements being made such that, in the future, chapters will be added as the capabilities are extended to enable us to gain a better appreciation of our manufacturing processes.

References

In the area of process analysis useful information is often contained within the literature and on the websites of commercial companies supplying equipment. Consequently, the references below contain some such details as examples of the equipment and information available. In this respect the details are not intended to be comprehensive. The inclusion or otherwise of specific companies does not imply endorsement or otherwise of the products described in the websites.

1 Clairet Scientific, 17 Scirocco Close, Moulton Park Industrial Estate, Northampton, NN3 6AP. www.clairet.co.uk
2 *Spectroscopy in Process Analysis*, ed. Chalmers, J. M., Sheffield Academic Press, 2000.
3 Coates, J. P., Shelley, P. H., in *Encyclopedia of Analytical Chemistry*, ed. Meyers, R. A., Wiley, New York 2000, p.8217.
4 Teledyne Analytical Instruments, 16830, Chestnut Street, City of Industry, California 91748, USA. www.teledyne-ai.com
5 Servomex, Servomex Group Ltd, Jarvis Brook, Crowborough, East Sussex, UK. www.servomex.co.uk
6 Goldman, D. S., in *Encyclopedia of Analytical Chemistry*, ed. Meyers, R. A., Wiley, New York 2000, Vol. 9, p.8256.
7 McClure, W. F., *Anal. Chem.*, 66(1), **1994**, 43A.
8 Workman, J., *NIR News*, 6(4), **1995**, 8.
9 Workman, J., *NIR News*, 6(6), **1995**, 7.
10 www.nirpublications.com
11 Ganorieau, J. P., Riberi, E., Loublier, M. et al., *Entorphie*, 34 (210), **1998**, 23.
12 Boelens, H. F. M., Kok, W. T., De Noord, O. et al., *Appl. Spectrosc.*, 54(3), **2000**, 406.
13 Goebel, S. G., Steffens K-J., *Pharm. Ind.*, 60(10), **1998**, 889.
14 Sekulic, S. S., Wakeman, J, Doherty, P. et al., *J. Pharm. Biomed. Anal.*, 17(8), **1998**, 1285.
15 Nagata, T., Ohshima, M., *Polym. Eng. Sci.*, 40(8), **2000**, 1843.
16 Bolliger, S., Zeng, Y., Windhab, E. J., *J. Am. Oil Chem. Soc.*, 76(6), **1999**, 659.
17 Staunton, S. P., Lethbridge, P. J., Grimley et al., *Proceedings of the 21st Conference of the Australian Society of Sugar Cane Technology*, 1999, p. 20.
18 Herman, H., Hope, P., *Polymer Process Engineering 97*, ed. Coates, P. D., Institute of Materials, London 1997.
19 NovaChem BV van Rensselaerweg 4, 6956 AV Spankeren/Dieren, The Netherlands. www.novachembv.com
20 Bran+Luebbe GmbH, PO Box 1360, D-22803 Norderstedt. www.bran-luebbe.de
21 Carleton, F. B., Weinberg, F. J., *Proc. R. Soc. London, Ser. A*, 447, **1994**, 513.
22 Adler, J., Carleton, F. B., Weinberg, F. J., *Proc. R. Soc. London, Ser. A*, 440, **1993**, 443.
23 Bonner-Denton, M., Gilmore, D. A., *SPIE*, 2388, **1995**, 121.
24 Shreve, A. P., Cherepy, N. J., Mathies, R. A., *Appl. Spectrosc.*, 46, **1992**, 707.

25 Bell, S. E. J., Bourguignon, E. S. O., Dennis, A, *Analyst*, *123*, **1998**, 1729.
26 Clarke, R. H., Londhe, S., Premasiri, W. R. et al., *J. Raman Spectrosc.*, *30*, **1999**, 827.
27 Martin, M. Z., Garrison, A. A., Roberts, M. J. et al., *Process Control Quality*, *5*, **1993**, 187.
28 Everall, N., King, B., Clegg, I, *Chem. Br.*, *36*(7), **2000**, 40.
29 Jobin Y., *Application Note, Raman Applications in the Agri-Chemical Industry*, Instruments SA, Jobin Yvon/Spex Division, 16–18 rue du Canal, 91165 Longjumeau.
30 Gervasio, G. J., Pelletier, M. J., *At-Process.*, *3*, **1997**, 7.
31 Gilbert et al., *SPIE*, *2248*, **1994**, 391.
32 Gregoris et al., *SPIE*, *1336*, **1990**, 247.
33 Alder-Golden, S. M., et al., *Appl. Opt.*, *31*(6), **1992**, 831.
34 Kaiser et al, *Ber. Bunsen-Ges., Phys. Chem.*, *96*, **1992**, 976.
35 Gresham, C. A., Gilmore, D. A., Bonner Denton, M., *Appl. Spectrosc.*, *53*(10), **1999**, 1177.
36 Cooper, J. B., Wise, K. L., Welch, W. T. et al., *Appl. Spectrosc.*, *51*(11), **1997**, 1613.
37 Martin, P. A., Feher, M., *NIR News*, *7*(3), **1996**, 10.
38 Norsk Elektro Optikk A/s PO box 384, N-1471, Skarer, Norway. www.neo.no
39 Feher, M., Martin, P. A., *Spectrochim. Acta, Part A*, *51*, **1995**, 1579.
40 Holdsworth, R. J., Martin, P. A., Raisbeck, D. et al., *Topical Issues in Glass*, *3*, **1999**.
41 Turner Designs, http://oilinwatermonitors.com
42 Schade, W., Bublitz, J., *Environ. Sci. Technol.*, *30*, **1996**, 1451.
43 Senoucci, A., Hope, P. S., Hilliard, L. A. et al., *Annual Meeting of the Polymer Processing Society*, Manchester, 1993.
44 Christensen, J., Norgaard, L., Carsten, L., *Spectrosc. Eur.*, *11*(5), **1999**, 20.
45 Eco Physics, 3915 Research Park Drive, Ann Arbor, Michigan 48108–1600. http://ic.net/~ecophys
46 Burck, J., Roth, S., Kramer, K. et al., *J. Hazardous Mater.*, *83*, **2001**, 11.
47 Forschungszentrum Karlsruhe, Institute for Instrumental Analysis. www.ifia.fzk.de/home_en.htm
48 http://www.siemens.co.uk/env-sys/
49 Paul, J. B., *Anal Chem.*, **1997**, 287A.
50 Wen-Bin, Y., *Ultra Sensitive Trace Gas Detection using Cavity Ringdown Spectroscopy*, IFPAC January 21–24, 2001, Amelia Island, FL, USA; www.Meeco.com; www.tigeroptics.com;
51 Crosson, E et al, *Cavity Ringdown Spectroscopy: Developing a Simple and Rugged Trace Isotope Analyzer*, IFPAC January 21–24, 2001 Amelia Island, FL, USA; Informed Diagnostics Inc, Sunnyvale, CA
52 Pipino, A. C. R., *Phys. Rev. Lett.*, *83*(15), **1999**, 3093.

18
NMR

Loring A. Weisenberger

18.1
Introduction

Nuclear magnetic resonance (NMR) offers process control a wide variety of information. NMR has gained the reputation of being very expensive in terms of capital, maintenance and personnel and is therefore generally regarded as being limited to the research laboratory. However, depending on the information sought, a NMR analyzer for process control is easier to apply and thus less expensive than many of the common process instruments. Details on the theory and execution of modern NMR techniques are located elsewhere in this book and are not duplicated in this chapter. An excellent review by Maciel [1] on the application of NMR to process control and quality control is available. More recently, Nordon et al. [2] offer a review of the past as well as a look to the future for process NMR spectrometry.

18.2
Motivations for Using NMR in Process Control

The driving force behind the desire to apply NMR spectroscopy to process control lies in its inherent quantitative nature and its ability to differentiate between chemical structures. In common with many spectroscopic techniques, NMR can be simply explained as a means of applying energy to a sample, measuring the energy absorbed and relating this to some property of the sample. For Fourier transform NMR, the input energy is in the form of a radio frequency (RF) pulse. The output energy is a RF pulse, at the same frequency, that decays with time and originates from the nuclei in the sample. The difference reveals information about the sample. The magnet serves to produce energy levels in the nuclei that will respond to the RF pulse.

The quantitative nature of NMR comes from the response of the nuclei in the sample. The amplitude of the response from any nucleus being excited is the same as the amplitude of the response from all other individual nuclei excited under the same conditions. Therefore, the contribution to the overall signal from any single nucleus is a simple fraction of all those excited. In words more commonly associated with optical spectroscopy the extinction coefficient for all excited nuclei are the same. This leads to the inherently simple quantitative nature of NMR spectroscopy.

Differentiating between chemical structures comes from a second effect. The exact frequency at which a nucleus absorbs energy is determined by the electrons surrounding, and therefore shielding, the nucleus from the external magnetic field. This shielding is therefore a function of the associated chemical structure. Different chemical structures shift the frequency at which different nuclei respond. Fortunately, this frequency difference is consistent for similar chemical structures and is known as the chemical shift. Chemical shifts are referenced to a peak in the spectrum of an agreed standard compound. The chemical shift relative to the reference chemical shift indicates what types of chemical structures are in the sample. The chemical shifts, relative to tetramethyl silane (TMS), of various chemical structures are well known and documented for many nuclei and especially for ^{1}H and ^{13}C NMR spectroscopy. Other nuclei use different chemical shift reference compounds. NMR signals have both amplitude and frequency. Together these two characteristics allow the determination of the chemical structures to be determined and a quantitative estimate of the sample components to be made. The actual experiment is a little more complicated than described here. In practice the time response of the nuclei must be considered and the magnetic coupling between nuclei can further complicate the spectrum. Both coupling and relaxation times can however be used to gain additional information about the sample. The utilization of relaxation rates as a means of analyzing process samples is considered later in this section.

When a spectrum is collected and the chemical structures of the individual components are known, a simple computation allows the compounds in the sample to be quantified. In general, organic compounds give a number of peaks in their NMR spectra. However, in order to quantify the components of a sample only one peak from each compound is normally, needed. Each identified peak is integrated and divided by the number of nuclei associated with the chemical structure that it represents, e.g. for a methyl group with three protons the peak area is divided by three, for a methylene group with two protons the number is two and so forth. Each normalized area is divided by the sum of all of the normalized areas of all of the relevant peaks. This provides a mole fraction for the compounds. The weight fraction for that compound is determined when the normalized areas are multiplied by their respective molecular weights and divided by the sum of the weights for all of the relevant peaks. The quantitative determination of a complex mixture of chemicals is only limited by the resolution between peaks and the ability to assign a chemical structure or compound to a peak or set of peaks. Furthermore, even when detailed chemical structures are not available, useful process information can be obtained from NMR spectra by considering more generalized structures such as aliphatics and aromatics which give signals in specific regions of the spectrum.

In comparison to optical spectroscopies calibration of NMR has a fundamental simplicity because the concept of an extinction coefficient is not required. For optical techniques different chemical structures in a given molecule respond differently in amplitude to the excitation. A correction factor or extinction coefficient is used to correct the amplitude in order to correctly quantify the sample. The extinction coefficients are determined by running a series of samples with differing amounts of the compounds, measuring the response amplitude at a given frequency and generating a calibration curve. Such calibrations are required for FTIR, NIR or UV but not for NMR. In the case of very complex mixtures, optical techniques require training sets for complex mathematical algorithms. These training sets are generated by varying operating parameters, measuring the response of the optical technique and analyzing the sample by independent techniques. NMR does not require training sets since complex mathematical algorithms are not generally applied. One NMR spectrum acquired under the proper conditions can provide a detailed analysis of a complex mixture in a straightforward algebraic fashion. On the other hand, chemometrics used with NMR data on highly complex systems such as fuels will prove to be a powerful combination. The simple relationship between the NMR signal and quantity of material will simplify the application of chemometrics to systems which yield highly complex NMR spectra with severely overlapping peaks.

Similarly, chromatographic techniques require calibrations of retention times and response factors. Calibrations are usually accomplished in advance by injecting known mixtures into the system and measuring retention times and response factors. Chromatographic techniques are notoriously time consuming, of the order of several minutes even with a short column. The faster the chromatogram is acquired, generally the poorer the resolution and the greater the artifacts such as tailing. Gas chromatography also suffers from temperature problems. If the inlet or column temperature is too high, the sample might decompose. If the inlet or column temperature is too low, a higher molecular weight compound may not vaporize and will eventually foul the inlet or column. NMR does not suffer from these kinds of constraints. Usually a single pulse with an acquisition time of less than 10 s is all that is required for good data. The temperature of the sample compartment, be it a flow probe or static cell can be regulated to the sample temperature.

As stated, the quantitation from a NMR spectrum is relative only to those components measured, but in many cases, absolute quantitation is required. Absolute quantitation is the determination of the amount of a substance with regard to all of the substances present. It is usually expressed as the amount of a substance per unit volume or weight of the sample but can also be expressed as a percentage. NMR provides a convenient means for demonstrating the difference between relative and absolute measurements. Assume a solvent system is composed of water, acetone and acetonitrile in a 5:3:2 ratio by weight. ^{13}C NMR would only detect the acetone and acetonitrile. And the relative result would be 60% acetone and 40% acetonitrile i.e. it only represents the carbon-containing solvents. However, ^{1}H NMR would detect all three solvents. The result would reflect the exact ratio of the three solvents. The absolute result is equivalent to the relative result in this case. The ^{13}C spectrum could be made into an absolute quantitation method simply by calibrating the NMR signal against a known amount of either solvent. Some kind of calibration is required for most (but not all) uses of absolute quantitation. For process NMR, this usually involves measuring the absolute response against *at least* two standards. Generally, the amplitude (area) calibration is linear against concentration. Also, it only needs to be calibrated for one compound since all others are relative to that compound. For single unknown samples, an internal standard compound can be weighed into a known amount of sample to provide direct calibration. In the above example, one might choose ^{13}C NMR over ^{1}H NMR when the acetone or acetonitrile are much less than water. By eliminating the water signal, the dynamic range for the compounds of interest is increased.

Conceptually, from an engineering perspective, NMR for process control is quite simple. The sample is brought to the instrument from the process via a side stream. The sample flows through a quartz, sapphire or zirconia tube positioned in the magnetic field and returns to the process via another side stream. The design of the sample system depends upon the conditions of the process sample, including temperature, pressure and flow rate. The data are acquired and transferred to the process control room without operator intervention. The sample handling system can include a manifold so that one analyzer can monitor and provide process control information for more than one position in the process. Generally, the unwanted deposits on windows or elsewhere in the instrument associated with the optical techniques are not a problem. No waste is generated as is the case with LC or titrations. No column compatibility or fouling issues are present that are normally associated with GC. No one has to collect samples.

For on-line applications the effects of sample flow on the NMR signal must be considered. First, the flow rate must be slow enough to allow the magnetization to come to equilibrium. That means the time in the magnet must be in the range from one to five times the spin–lattice relaxation time, T_1. Whilst this can be a long time, the advantage to the flowing system is that the time inside the RF coil is short so that the RF pulse rate can be faster than the normal five times T_1 since each pulse will excite a new set of molecules. The second issue is that the nuclei excited by the RF pulse must be in the coil long enough for their signal to be detected by the coil. The third issue is that the flow introduces some errors which cause T_2 to be shorter than in a static sample leading to broader peaks. In the case of broadline NMR that has little effect. In the case of high resolution NMR, it can have a detrimental effect, depending on the analysis.

Another driving force behind using NMR for process control is the non-destructive nature of the technique. In some cases, the material is packed into a suitable container, usually a tube or vial, analyzed and then returned to the bulk, or if the material flows, it can be piped through the instrument. For other methods the sample is analyzed neat or in a solvent. In both cases the sample can usually be recovered with minimal effort. No waste is generated.

NMR provides several alternative approaches to a problem. If one nucleus does not prove effective, then an alternative may work. If one pulse sequence does not provide the best information, another may prove more effective. If one option is too expensive, a less expensive version might work. If the molecular dynamics is not in the correct range, try a higher or lower temperature. If the frequency domain does not provide enough resolution, the time domain might. In any event, some of these choices are obvious and can be made prior to proceeding. Others require some experimentation and analysis. This versatility is demonstrated in the two basic methods of NMR for process control. Broadline NMR is a time domain technique and FT-NMR, also called (Fourier transform) high resolution NMR, is a frequency domain technique. While based on the same general principles, the information available from either technique is very different.

18.3
Broadline NMR

The least expensive process NMR option is a small 10–20 MHz (0.2–0.4 T) magnet with a desktop computer containing most of the necessary electronics for a complete pulsed NMR experiment. This is commonly referred to as broadline NMR. With a low field strength, the linewidth of any peak is large compared to the chemical shift scale. Separating peaks into meaningful chemical shifts and the appropriate chemical structures is not attempted. The time domain is used exclusively to provide information on the sample.

Physically, the magnet is small, $20 \times 20 \times 20$ cm^3. The temperature control unit can be considerably larger, depending on the temperature range and control desired. The RF electronics are usually contained within a personal computer. A RF pulse is used to excite the sample and the response of the sample is recorded as a function of time.

The response signals from the nuclei decay exponentially with time. Many factors affect this decay, including field homogeneity and the motions of the molecules. When considering only the pure relaxation processes, the relaxation is characterized by the spin–spin relaxation time, T_2. When external factors such as magnetic field inhomogeneity are considered, the apparent spin–spin relaxation time is referred to as T_2^*. Since homogeneity will influence all molecules in the same way, only the molecular motions are considered here. The resolution between molecular species is a function of the difference between the characteristic times of the molecules. Typically, T_2^* is directly related to the "shape" of the free induction decay (FID). In the case of a single component, the T_2^* is related to the time response by the following equation:

$$M(t) = M_0 \exp(-t/T_2^*) \tag{1}$$

where $M(t)$ is the magnetization at time t and M_0 is the initial magnetization. In the case of a multi-component system the response is the simple sum of all of the components as a function of time.

$$M_{total}(t) = \Sigma M_{n0} \exp(-t/T_2^*{}_n), \; n = 1 \text{ to the number of components} \tag{2}$$

where $M_{total}(t)$ is the total magnetization at time t and n is the component index. With the appropriate curve fitting software, all of the variables can be determined with several iterations. The limitation to this method is the differences between the T_2^*'s. In general if the ratio of the T_2^*s is not greater than two, the fitting algorithms will have difficulty accurately discerning the variables for each component. If the T_2^*s can be determined independently, then the restriction is relaxed. Each species may also have a Gaussian component which can be accounted for in a similar fashion to the previous equations.

In general, when the components and their T_2^*'s have been previously determined, the initial magnetization for each component can be calculated. Knowing the initial magnetization for each of the components allows the mole fraction of each component to be determined as its initial magnetization divided by the total initial magnetization.

The drawback to this approach is relating the time domain information to a chemical species or physical phenomena. Many broadline systems are used in an empirical fashion in which the results are correlated with a process control parameter or measured physical property. However, this approach alone does not indicate which chemical species is being measured. The signals could be due to water or fat content, bound water versus free water in a slurry or different phases of a polymer, crystalline, amorphous or interfacial regions. To ensure that these correlations are due to specific chemical species broadline NMR data must be related to some other primary analytical technique such as high resolution NMR.

An even more simplistic approach is used when doing spin counting which is simply "counting" all of the nuclei (spins) contributing to the signal. In spin counting, a broadline system is calibrated using a known volume of a standard solution. Immediately after the RF pulse, the initial response of the sample is recorded. This is mathematically compared to the initial response of the standard. The result is a correlation of the number of protons in the given volume. The technique requires precise temperature and pressure control. An example is measuring the total hydrogen content of a petroleum fraction and comparing that to an equal weight of a standard such as n-octane.

Other NMR parameters, such as the spin–lattice relaxation time, T_1, and spin–lattice relaxation time in the rotating frame, $T_{1\rho}$, are used in a similar fashion although the equations are different. Each uses a unique pulse sequence to probe a specific dynamic property of the molecule. T_1s are related to the short range, high frequency motions of a molecule whereas, $T_{1\rho}$s are related to longer range motions in the kHz region. Chemical compounds with similar T_2^*'s might have vastly different T_1s or $T_{1\rho}$s and therefore alternative methods are available for determining the different components in a solution. Diffusion times add yet another option to the range of motions that can provide differences between compounds. Diffusion times are measured by altering the static magnetic field such that the position of a molecule in the sample is encoded in the signal. As the molecule moves throughout the volume, the encoding changes the signal. More diffusion results in faster decay of the signal. Thus, the diffusion weighted analysis favors slow moving and generally larger molecules. The magnetic field is altered using linear magnetic field gradients added to the probe or the magnet.

Broadline NMR made its mark in the food industry [3–12] with substantial work done in Europe [13–16]. Originally, Chapman and others [17] applied wideline NMR to the determination of solid fat in oil. By heating a fat in oil sample, the fat is melted and the total intensity of the FID is measured. Then a second measurement is taken at a lower temperature where the fat has solidified. The ratio of the two measurements indicates the solid fat content. This method is quick and more precise than other methods such as differential scanning calorimetry (DSC).

More simplistically, several methods exist to determine the amount of water or oil in foodstuffs. One is the spin counting method described previously. The absolute signal amplitude of the FID is simply measured as a function of the weight of the material. When measured against a carefully calibrated set of standards, the amount of material is determined. The drawback to this method is that it only works well when only one measurable component is present and the other components are not detected by the NMR instrument.

Guillou and Tellier [18] used a 20 MHz wideline instrument to measure ethanol in alcoholic beverages. A Carr–Purcell–Mieboom–Gill [19] pulse sequence is used to encode the scalar coupling of the methyl and methylene in ethanol. By examining the difference between the modulated and unmodulated echoes at time $t=1/(2J)$ it is possible to determine the volume percent of ethanol in an aqueous solution. The proportion is given by the equation

$$\Delta(1/(2J)) \propto (5d_a/m)t_v \tag{3}$$

Where the right-hand side of the proportionality represents the hydrogen content of ethanol per unit volume, where d_a is the density of pure ethanol, t_v is the percentage of alcohol, m is the molecular mass and J is the scalar coupling of the methylene and methyl protons. The correlation is very good in the 0–70% (v/v) range. With no sample preparation, they measured several wines. Based on the differing characteristics of the wines, they were able to show also that glucose does not interfere with the ethanol measurements.

Engelbart and others [20] provided a detailed molecular description of the curing process of towpreg materials and its relationship to changes in the NMR signal in a broadline system. Figure 18.1 shows how these materials, which are carbon fibers coated by "towing" them through epoxy resin, exhibit a change in T_2 of four orders of magnitude as the material cures. By monitoring the T_2 of the material, the degree of cure is monitored and the optimum temperature program can be chosen. Additionally, Fig. 18.2 shows that the degree of cure can be monitored successfully while strips of towpreg are pulled through the NMR magnet and probe. Fig. 18.3 shows the set-up of the fibers being pulled through the magnet and probe. By using four strips of towpreg, cured between 10 and 40%, they found good agreement between DSC measurements and the on-line NMR system.

NMR work on polyethylene (PE) is extensive, not surprisingly because its production exceeds all other plastics by an order of magnitude. The physical properties vary with composition. Most producers use high resolution NMR to characterize

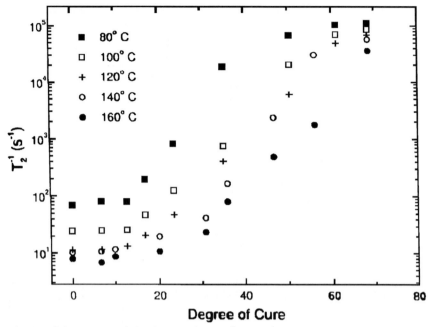

Fig. 18.1 Laboratory correlation between degree of cure and spin–spin relaxation time. Reprinted with permission from [20].

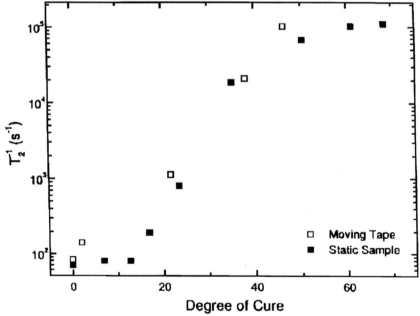

Fig. 18.2 On-line simulation: Correlation between degree of cure and spin–spin relaxation time for both static and moving towpreg. Reprinted with permission from [20].

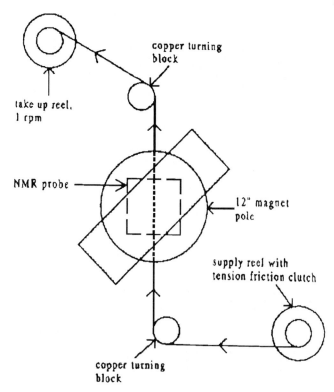

Fig. 18.3 Laboratory set-up for NMR dynamic measurements. Reprinted with permission from [20].

the chain branching of their different grades of PE. The process of obtaining NMR spectra is slow and difficult, depending on the solvents and conditions used and on the grade of the polymer. However, most of the PE producers use NMR for quality control of their product lines and many use it for process control and monitoring of their reactors, albeit with a large time delay. Many are also using this information for blending different grades of PE to provide the desired properties of a particular product line.

With this much interest in polyethylene, many attempts have been made to use low frequency NMR in either the time domain or frequency domain to monitor and control the production more rapidly. Auburn International (now part of Oxford Instruments) developed a widely adopted system based on the time domain spectrometers [21]. In this case, sample preparation is no longer an issue since the system accepts either powder or pellets and no solvent is used. The Auburn systems determine crystalline and amorphous ratios, viscosity, melt index and molecular weight. For other types of polymers, the list of advertised measurements include tacticity, rubber content, copolymer analysis, and various rheological properties. These values are determined by correlating several routine but laborious methods with the decay of the NMR signal under various pulse sequences. The man-hours saved by using this technique are an added value to the fact that the information is returned quickly enough to control the process.

The manufacturers of these kinds of instruments all have a long list of applications. Most of the applications are similar and many are pre-packaged as specific analyzers so that they simply unpack, set up and data is acquired within an hour or so after the magnet temperature equilibrates. The instrument vendors for broadline systems are Bruker Instruments [22], Oxford Instruments [23], Praxis [24], Process Control Technology (PCT) [25] and Resonance Instruments [26]. Determination of oil and/or water content dominates the applications. Oil and water analyses are established for seeds and soil in the agriculture industry, catalysts and detergents in the chemicals industry, capsules, tablets and powders in the pharmaceutical and cosmetic industries as well as a wide variety of foodstuffs.

Although the determination of oil and water dominates the use of NMR spectroscopy there is a wide range of other imaginative applications. In the polymer industry, in addition to the PE characteristics discussed previously, many other properties are analyzed using broadline NMR systems. The details of these applications are generally available from the manufacturers of broadline NMR systems mentioned in the previous paragraph. For fibers, the amount of added spin finish on the outside layer of the polymer is determined based on the distinct NMR signals resulting from differences in the mobility of components in the finish and the polymer. In a similar analysis, plasticizer content can be determined in bulk polymer or final product samples, utilizing the higher mobility of the lower molecular weight plasticizer to distinguish it from the polymer. The composition of blended materials can be determined to give the rubber or filler content. Polymer blend compatibility can also be assessed by utilizing the different relaxation rates of the various blend components. Like the towpreg example described previously, the extent of polymerization and the degree of crosslinking are other properties that can be monitored as a result of the changes in NMR relaxation rates as the mobility of the polymer changes.

The food industry continues to provide some interesting and unusual applications of process control beyond water and oil measurements, such as the study of a cooking process [27] or of a freezing process. Catalysis is another area with unique applications. The characteristics of a catalyst that can be measured by broadline NMR include activity and selectivity.

Many analyses easily cross industry lines. The chemical and pharmaceutical industries are interested in the analyses that determine the coating weight of a coated particle such as with a time-released drug. Like the spin finish on fibers, the coating has a different relaxation rate than the encapsulated material. Fluorine analysis is another unique application of NMR technology since the signal derives from the fluorine nuclei rather than hydrogens. The materials for which this technique for process analysis and control might be used include toothpaste, fluoropolymers and fluorochemicals.

Droplet size analysis is of particular interest to both the food industry, for margarine, and to the cosmetic industry, for various make-up emulsions. Droplet size is also a good example of the use of magnetic field gradients to encode spatial in-

Table 18.1 Broadline NMR.

Characterized parameter:	Surface specificity	
Energy of radio frequency absorption	Information depth: Bulk	Detectability: >1%
Type of information:	**Resolution**	
Composition or property based on specific nuclei relaxation	Depth: Lateral: Other: Not used for spatially resolved analysis	
Measurement environment: Difficulties	**Time needed for analysis:**	
Strong magnetic field RF excitation Air or nitrogen	Preparation Measurement Evaluation 0.5 to 5 min 0.5 to 5 min 1 to 5 min	
Equipment:	**Cost [ECU]:**	**No. of facilities:**
Broadline NMR analyzer	$ 40–80,000	Common
Type of laboratory: User skill needed:	**Sample**	
Small Unskilled	Form type Solid, liquid or slurry	Size: 0.5 to 150 ml
Techniques yielding similar information		
FTIR, GC-MS		

formation into the NMR analysis. By applying a magnetic field gradient, the rate that a mobile molecule moves through a sample is monitored. With the appropriate model, the size and distribution of the sizes of the domains are determined. The complement to this analysis is the determination of particle size for pigments or fillers in liquids such as paint. In this case, the signal does not generally originate from the component of interest. Instead the matrix provides the NMR signal necessary to determine the average particle size. This is similar to the way NMR is used to determine the pore sizes and their distribution in catalysts or in rocks for oil recovery.

The use of broadline NMR in process control is summarized in Tab. 18.1.

18.4
FT-NMR

As the field strength of the magnets increases, the resolution in the frequency domain increases. Peaks representing different chemical structures are resolved. A direct correlation exists between the chemical shift of a peak and the chemical species that it represents. The relative mole and weight fractions of an identified chemical are easily calculated.

18.4 FT-NMR

At the lower end of the magnet strengths, around 40–60 MHz (0.9–1.4 T), the systems are made using small permanent magnets typically measuring $30 \times 30 \times 30$ cm^3. Depending on the design, the fringe field of the magnet is usually contained inside the magnet enclosure. The electronics are considerably larger and more complex than the broadline systems possibly occupying a couple of 50 cm racks 1 m tall. The electronics are generally more robust than for the broadband NMR analyzer. The RF, temperature control and shim power supply are all housed in the rack with the magnet on top. Plenty of space and fans are added for cooling the electronics in a hostile industrial environment. The hardware is often driven by a personal computer that is separate from the other electronics. These systems are ideal for placing on-line or near the process. They are easily fitted to an industry standard protective enclosure while the computer can be located in the control room. There is little maintenance beyond the initial installation. A sample delivery system is simply a pipe running from the process stream to the instrument with a tube running through the magnet made of a more specialized non-magnetic material such as glass, quartz, sapphire or zirconia. The choice of which flow tube is used is determined by the harshness of the sample, pressure, temperature and cost. The sample is returned to the process without alteration. In general, after the analysis is initially established, these systems are run automatically as an analyzer with the required information being transmitted to the process control panel. Alternatively, they can also be used as small bench-top analyzers in a QA/QC laboratory.

The use of low field FT-NMR in process control is summarized in Table 18.2.

On the high resolution side of magnet strengths, the systems range from 100 MHz up to 900 MHz (2.3–21 T). The magnets are made of superconducting wire that is cooled to –269 (C by liquid helium. The dewar of liquid helium is blanketed with liquid nitrogen to reduce the boil-off rate of the liquid helium. These designs provide high field magnets without the large cost of the electricity required by electromagnets. The maintenance costs are higher since liquid nitrogen must be replaced weekly or bi-weekly and liquid helium must be topped quarterly or semi-annually but this is more than offset by the electrical savings.

These systems also occupy a lot of space. The fringe field of a magnet is the magnetic field outside the physical magnet itself. The stronger the magnetic field, the larger the fringe field will be. This can be dangerous since ferromagnetic objects are easily pulled towards the field, creating uncontrollable projectiles. Untrained or poorly supervised personnel may inadvertently wheel a gas cylinder past the magnet with catastrophic effects, possibly to both the magnet and the personnel. It is also dangerous because persons with pacemakers or metal plates can be adversely affected by any strong magnetic field. The extent of the fringe field for a superconducting magnet is typically measured in meters, so that a safety zone around the magnet is necessary. Furthermore, the electronics should be outside this zone as well, thereby occupying additional space. Some systems now come with shielded magnets that have substantially reduced fringe fields. Nevertheless, it is difficult to imagine a superconducting system anywhere near the process line. Although in some extreme cases where the costs are justified, superconducting systems could be used on-line [1].

Table 18.2 Low field FT-NMR.

Characterized parameter:	Surface specificity:	
Energy of radio frequency absorption	Information depth: Bulk	Detectability: >1%
Type of information:	**Resolution:**	
Composition based on specific nuclei	Depth: Lateral: Other: Not normally used for spatially resolved analysis	
Measurement environment: Difficulties	**Time needed for analysis:**	
Strong magnetic field RF excitation Air or Nitrogen	Preparation 0.5 to 5 min	Measurement Evaluation 0.5 to 5 min 1 to 5 min
Equipment:	**Cost [ECU]:**	**No. of facilities:**
Low field FT-NMR Spectrometer	$80–120,000	Less common
Type of laboratory: **User skill needed:**	**Sample**	
Small Unskilled	Form type Solid, liquid or slurry	Size: 0.5 to 5 ml
Techniques yielding similar information:		
FTIR, GC-MS		

Both types of FT-NMR system can contribute to process control whether on-line or in the laboratory. The smaller systems are ideal for monitoring solutions or mixtures with a detection limit of about 1%. This involves the integration of peaks from the materials of interest in the sample. In some cases, the important parameter is the exact chemical shift of a particular peak. As conditions in a process change, the position of a sensitive peak will shift in a predictable manner. Changes in the chemical shift of a peak can be related to changes in pH or complex formation. In some cases, the appearance or disappearance of a peak can be used to monitor the progression of a reaction.

Haw and others [28] demonstrated the utility of low field frequency domain NMR for determining the amount of oxygenates in gasoline. By using LC pumps, they blended methyl *tert*-butyl ether (MTBE) and/or ethanol with gasoline and monitored the process using a 42 MHz NMR system with a flow probe. Figure 18.4 shows only minor degradation in the signal of ethyl benzene when the flow is increased from static to 6 ml min^{-1}. From the NMR spectrum, they calculated the amount and type of added oxygenate. A ^{1}H NMR spectrum (Fig. 18.5) of a typical gasoline shows features including the aromatic (8–6 ppm), olefinic (6.5–5 ppm) and aliphatic protons (3–0 ppm). It also demonstrates that with a sample flow of 2 ml min^{-1} the MTBE (3.2 ppm) can be monitored up to 20% v/v. In a more demanding analysis, ethanol can be measured from 2% v/v in the presence of MTBE (Fig. 18.6).

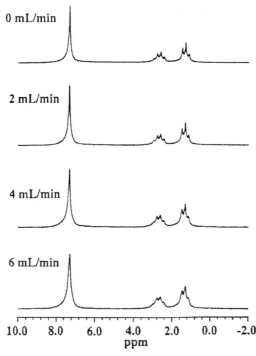

Fig. 18.4 The 42 MHz proton spectra of ethyl benzene flowing at various rates. The spectra were not sensitive to flow rates in the range shown. Reprinted with permission from [28].

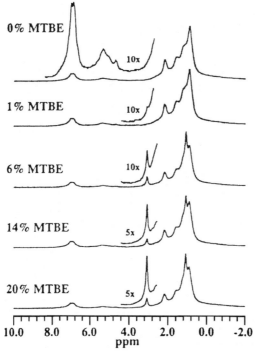

Fig. 18.5 The 42 MHz proton spectra of various blends of MTBE and regular, unleaded gasoline. The highlighted areas show signals from MTBE or the aromatic and olefinic protons of gasoline. All spectra were acquired with 12 scans on samples flowing at 2.00 ml min^{-1}. Reprinted with permission from [28].

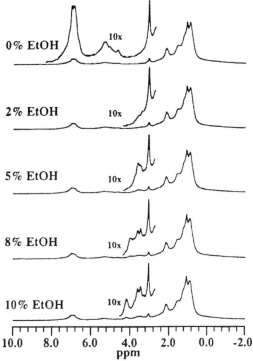

Fig. 18.6 The 42 MHz proton spectra of various blends of ethanol in 10% MTBE in gasoline. All spectra were acquired with 12 scans on samples flowing at 2.00 ml min^{-1}. The concentration dependence of the hydroxyl proton chemical shift is apparent in the expanded plots. Reprinted with permission from [28].

Edwards and Giammateo [29] (Process NMR Associates [30]) have reported the application of an on-line NMR system to monitor the sulfuric acid alkylation process at a refinery. This is an excellent example of using the full capabilities of a low-field frequency-domain NMR analysis. Figure 18.7 shows the full spectrum which includes the acid/water peak at about 10 ppm and the hydrocarbons from about 8 to 0 ppm. In this example, the weight percent of the acid/water fraction is determined against the hydrocarbon fraction by simply measuring the area beneath each peak. This ratio is an indication of the quality of the emulsion. Similarly, the alkane to alkene ratio is determined and used to monitor feed compositions. This is more clearly demonstrated in Fig. 18.8 where the olefins are in the region from 6 to 4 ppm. In a more complex analysis, but demonstrating the versatility of the NMR measurements, the acid strength is determined from the position of the acid/water peak. The position is entered into a third order equation that is related to the acid strength. All of this information is acquired from a single measurement in less than 2 min and returned to the controller for optimizing the process. The previous method for determining the acid strength was a titration, where the emulsion was separated into acid and hydrocarbon fraction. Samples for the titration method were taken every 4 h. The titration method was slow, labor intensive and hazardous since samples had to be collected from the alkylation unit, separated into fractions and poured into the apparatus. The NMR system simply

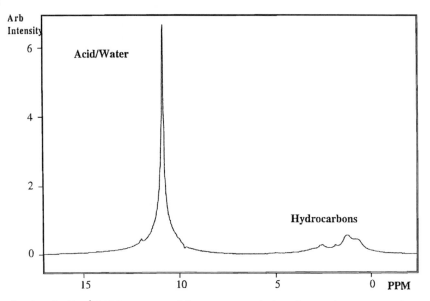

Fig. 18.7 On-Line ^{1}H NMR spectrum of the intact H_2SO_4/hydrocarbon emulsion. Reprinted with permission from [29].

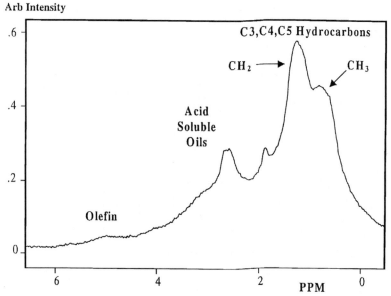

Fig. 18.8 ^{1}H NMR spectrum of the hydrocarbon region of the intact emulsion. Reprinted with permission from [29].

takes a sample via a side stream to the instrument and returns the sample to the unit with another pipe.

The Foxboro Company [31] purchased the process NMR division of Elbit ATI in 1997 and turned a process/portable NMR system into a series of analyzers for the petroleum refining industry. By using a variety of modeling and chemometric algorithms, the Foxboro NMR analyzers convert NMR data into traditional refining measurements including API gravity, viscosity, and reed vapor pressure (RVP). These analyzers touch on nearly every aspect of the refining process from crude oil blending through distillation, fluid catalytic cracking and sulfuric acid alkylation to product blending.

One of the most complex examples of using NMR for process control is its application to transforming growth factor-β_3. Blommers and Cerletti [32] developed a methodology using high resolution NMR. The 2D NOE spectrum of the TGF-β3 serves as a fingerprint to monitor and control batch-to-batch variations of the material. The NMR spectrum for a given batch is compared to the TGF-β3 as well as the reference spectra for fully reduced TGF-β3 and for a mutant version. This is a challenging application since the differences between the 25 kDa structures lie in the placement and number of disulfide bonds and hence the three-dimensional structure.

In addition to Process NMR Associates and Foxboro, Anasazi [33] and Hitachi [34] both produce low field FT-NMR systems although the instruments are designed for laboratory use rather than as process analyzers. High field FT-NMR systems are produced by three manufacturers, Bruker Instruments [35], JOEL [36] and Varian [37].

The use of high field FT-NMR in process control is summarized in Tab. 18.3.

Table 18.3 High field FT-NMR.

Characterized parameter:	Surface specificity:	
Energy of radio frequency absorption	Information depth: Bulk	Detectability: 10 ppm
Type of information:	**Resolution:**	
Composition based on specific nuclei	Depth: 100 µm Lateral: 10 µm Other:	
Measurement environment: Difficulties	**Time needed for analysis:**	
Strong magnetic field RF excitation Air or nitrogen	Preparation Measurement Evaluation 0.5 to 5 min 0.5 to 24 h 1 to 5 min	
Equipment:	**Cost [ECU]:**	**No. of facilities:**
FT-NMR Spectrometer	$ 200 K–$ 4 MM	Common
Type of laboratory: **User skill needed:**	**Sample**	
Large Experienced technician or professional	Form type Solid, liquid or slurry	Size: Large 0.5 to 5 ml
Techniques yielding similar information:		
FTIR, GC-MS		

18.5
Conclusion

With the ever-increasing demands for efficiency and quality in production, analyzers and controllers of all types will slowly gain acceptance in many industries. NMR offers a wide variety of options for on-line analysis as well as at-line plant laboratory analysis. Information generated by NMR analyzers can be immediately fed back into the process, providing more accurate and rapid control. Implementation is relatively easy. Installation requires a fast loop from the process, possible sample conditioning and running the communications lines to the control room. Some software development is required to convert the NMR measurements into meaningful data for the operators or control units. Such procedures are no different than for most on-line analyzers. The benefits of NMR analyzers depend upon their capability of providing quantitative chemical information without complex models or calibrations. As the NMR analyzer becomes more valued, the cost will seem less of an obstacle.

References

1 Maciel, G., *NATO ASI Ser., Ser. C*, **1994**, *447*, 225–75.
2 Nordon, A., McGill, C., Littlejohn, D., *Analyst*, **2001**, *126*, 260–72.
3 Wiggall, P., Ince, A., Walker, E., *J. Food Technol.*, **1970**, *5*(4), 353–62.
4 Wettstrom, R., *J. Am. Oil Chem. Soc.*, **1971**, *48*(1), 15–17.
5 Tessier, A., Linard, A., Delaveau, P. et al., *Z. Lebensm.-Uters. Forsch.*, **1983**, *176*(1), 12–15.
6 Sleeter, R., *J. Am. Oil Chem. Soc.*, **1983**, *60*(2), 343–9.
7 Schmidt, S., *Adv. Exp. Med Biol.*, **1991**, *302*, 599–613.
8 Rutledge, D., *J. Chim. Phys. Phys.-Chim. Biol.*, **1992**, *89*(2), 273–85.
9 Simoneau, C., McCarthy, M., Reid, D. et al., *Trends Food Sci. Technol.*, **1992**, *3*(8-9), 208–11.
10 Davenel, A., Marchal, P., *Dev. Food Sci.*, **1994**, *36*, 35–42.
11 McDonald, P., *Food Process*, ed. Gaonkar, A., Elsevier, Amsterdam 1995, 23–36.
12 Sun, X., Moreira, R., *J. Food Process. Preserv.*, **1996**, *20*(2), 157–67.
13 Sambuc, E., *Rev. Fr. Corps Gras*, **1974** 21(12), 689–98.
14 Ribaillier, D., *Inf. Technol.*, **1980**, *67*, 19–28.
15 Rutledge, D., Khaloui, M., Ducauze, C., *Rev. Fr. Corps Gras*, **1988**, *35*(4), 157–62.
16 Bezecna, L., Tenkl, L., Konradova, M. Slechtitelska, S., *Czech. Rostl. Vyroba*, **1991**, *37*(1), 75–80.
17 Chapman, D., Richards, R. E., York, R. W., *JOACS*, **1960**, *37*, 243–6.
18 Guillou, M., Tellier, C., *Anal. Chem.*, **1988**, *60*, 2182–2185.
19 Meiboom, S., Gill, D., *Rev. Sci Instrum.*, **1958**, *29*, 688–691.
20 Engelbart, R., Conradi, M., Stoddard, R., *43rd Int. SAMPE Symp.*, **1998**, *43*, 928–36.
21 Roy, A., Marino, S., *Am. Lab.*, **1999**, *31*(21), p. 32–33.
22 Bruker Analytik GmbH, Rheinstetten, Germany.
23 Oxford Instruments Analytical, Buckinghamshire, UK.
24 Praxis, San Antonio, TX, USA.
25 Process Control Technologies, Fort Collins, CO, USA.
26 Resonance Instruments Ltd., Witney, UK.
27 Stapley, A., Goncalves, J., Gladden, L. et al., *IChemE Res. Event–Eur. Conf. Young Res. Chem. Eng.*, **1995**, *2*, 1064–6.
28 Skloss, T., Kim, A., Haw, J., *Anal. Chem.*, **1994**, *66*, 536-42.
29 Edwards, J., Giammateo, P., *Proc. Annu. ISA Anal. Div. Symp.*, **1998**, *31*, 73–77.
30 Process NMR Associates, Danbury, CT, USA.
31 The Foxboro Company, Foxboro, MA, USA.
32 Blommers, M., Cerletti, N., *Pharm. Sci.*, **1997**, *3*(1), 29–36.
33 Anasazi Instruments, Inc., Indianapolis, IN, USA.
34 Hitachi Instruments, Inc., San Jose, CA, USA.
35 Bruker Analytik GmbH, Rheinstetten, Germany.
36 JOEL Ltd, Akishima, Japan.
37 Varian, Inc., Palo Alto, CA, USA.

19
Process Mass Spectrometry

Christian Hassell

19.1 Introduction

Process mass spectrometry (MS) is a very powerful technique for process monitoring and control, providing a unique combination of speed, selectivity, dynamic range, accuracy, precision and flexibility. The technique has become a standard for gas-phase analysis in several industrial applications, including steel manufacturing, fermentation off-gas analysis, and the production of ethylene oxide and ammonia. Among its attributes:

- Speed: of the order of 1 s per analyte per sample stream, thus permitting true real-time analysis.
- Sensitivity: low pp (10^{-6}) detection limits are routine for many applications, and low ppb (10^{-9}) or even ppt (10^{-12}) levels are achievable for certain applications.
- Wide dynamic range: MS detection limits are not dependent on pathlength, as with most optical techniques, thus the same MS analyzer can be used for measurements from ppb to nominal 100%.
- Easily multipointed: most commercial systems are supplied with multiple sample valves, accommodating up to 64 or more sample streams.
- Cost: although the system hardware is quite expensive, the ability to multipoint often results in lower cost per sample point.
- Flexibility: new analytes are easily added to the sample analysis matrix.
- Ideal for process diagnostics/information rich: mass spectral interpretation is relatively straightforward, especially in conjunction with spectral libraries.

On the other hand, MS has its limitations, and each of the above attributes has its caveats, as will be described later in this chapter. Among its limitations are:

- Most commercially available instrumentation is limited to the analysis of gas-phase sample streams, although membrane technology is sometimes employed for the analysis of volatile organics in certain liquid streams.

- Overlap of fragmentation patterns can be severe for some applications, making analysis of some stream components very difficult.

As with any process analyzer, thorough application review is the key to matching the process analysis need to the most appropriate measurement technology.

Wide acceptance of process mass spectrometry has been hindered by a perception that it is too complicated and delicate. This perception may have its roots in many undergraduate chemistry laboratories, where it is quite common to have hands-on experience with optical spectrometers and gas chromatographs, whereas MS is often introduced as the large complex instrumentation in the basement, run by a sort of "priesthood" of specialists. (This same situation may be one of the barriers to acceptance of nuclear magnetic resonance for process monitoring.) While it is true that process MS involves moving parts and somewhat delicate source materials, such items have been ruggedized in current commercial instrumentation to a degree that MS reliability and ease of use are equivalent to, or in some cases better than, other common process analysis technologies.

This chapter seeks to clarify the current state of process MS instrumentation and dispel some of the common misperceptions. It is aimed at the chemist or engineer who needs a practical solution to solve a process problem. As such, it is biased toward practical instrumentation that is commercially available. Review articles [1] are available that describe the latest instrumental developments that utilize more exotic mass analyzers and novel ionization sources.

19.2
Hardware Technology

Figure 19.1 is a block diagram of a typical process analyzer system, consisting of a sample collection and conditioning system, sample manifold, sample inlet, ion source, mass analyzer, detector, and a data analysis and output system that interfaces with the process control system. The dashed line indicates the parts of the overall system that are considered to comprise the analyzer itself (i.e., what is normally included when one purchases a process MS). Figure 19.2 is a photograph of a commercial process MS that incorporates these components. Aspects of these various components are described below, with emphasis on how they are applied in a process mass spectrometer.

Some issues are common to all components in the analyzer system, including consideration of materials of construction and heating/cooling requirements. Many options also exist for installing a process MS in toxic, corrosive or explosive environments. For example, a general purpose analyzer can be placed in an analyzer house or shelter that is temperature-controlled and designed to protect from such hostile environments. Alternatively, the components within the dashed line of Fig. 19.1 can be placed within a cabinet that is purged with an inert atmosphere and that includes an integral temperature control system, thereby eliminating the need for a separate shelter.

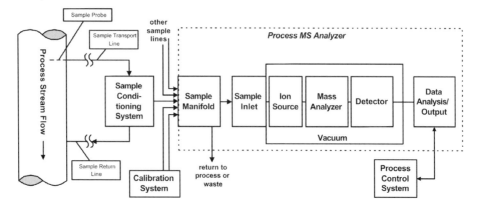

Fig. 19.1 Block diagram of a process mass spectrometer analyzer system.

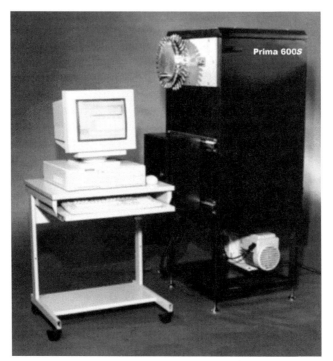

Fig. 19.2 A commercial process mass spectrometer. Note rotary valve on the upper part of the analyzer, facing the computer. (Photograph used with permission of ThermoOnix Inc.).

19.2.1
Sample Collection and Conditioning

Although a thorough treatment of sample collection and conditioning is beyond the scope of this chapter, a few aspects merit mention. Because mass spectrometers are often used to monitor multiple sample points throughout a process, they are usually positioned in a central location to which samples are transported from distant sample points. Often it is possible to transport samples several hundred meters, but the system design must ensure representative samples by considering temperature, flow and pressure changes throughout the run. Particulates should be excluded with filters to approximately 0.5 µm; several stages of filtering are often employed along the sample system to reduce pluggage of the final smallest filter. Generally, most sample line problems can be avoided by keeping all sample lines as short as possible, keeping them flowing continuously, and ensuring that lines are heated if condensation is possible. Figure 19.1 illustrates a common method of ensuring representative sampling and rapid continuous flow of sample. With this method, known as a fast-loop sampling system, the sample is rapidly transported to and from the sample conditioning system, with a small bypass fraction drawn into the sample manifold.

A sample manifold is usually used at the mass spectrometer to connect several sample lines and calibration gases to the analyzer. Many designs are available, usually incorporating either solenoid or rotary valves; a rotary valve can be seen on the side of the analyzer in Fig. 19.2. Most modern manifold designs permit all samples not being analyzed to flow continuously via a bypass (this bypass loop is commonly referred to as the *slow-loop*), thus permitting better representative sampling while preventing condensation that might occur in a line that is not actively being transported. Several proprietary manifold designs are very good at minimizing cross-contamination between samples, while achieving rapid stream switching and reducing design complexity in sample conditioning systems.

Water, in the form of vapor or droplets, is a common element of many sample streams. Too much water vapor can either overwhelm the mass spectrometer, or present a significant sample transport problem if condensation should occur in the sample line. It is possible to remove water vapor with driers, but one must ensure that other analytes of interest are not being removed as well. Fine water droplets can be removed with coalescing filters, and in some cases large water droplets can be excluded by simply designing the sample line with several right-angle bends.

19.2.2
Sample Inlet

Several different techniques can be employed to introduce the sample into the mass spectrometer. These inlets must transition the sample from the higher pressure of the sample line to the vacuum required for the mass spectral analysis, yet this transition must occur without affecting the concentrations of the analytes of interest. Specialized inlets are occasionally used for liquid sample analysis, and it is also possible to vaporize or sparge liquid streams for analysis; however, this discussion will only focus on the far more common application to gas stream analysis.

19.2.2.1 Direct Capillary Inlets

These are used on most commercial process mass spectrometers. Often a 1 m long capillary of 10–100 µm inner diameter is sufficient to provide the necessary pressure drop. Deactivated fused silica is the most common capillary material, since it is reasonably inert and does not exhibit significant memory effects with most sample streams. Heating the capillary further reduces memory effects and pluggage due to condensation. Capillaries can also be made of other materials such as stainless steel or nickel if silica is problematic. Molecular leaks (pinhole orifices) are sometimes used, either in conjunction with or in place of the capillary. Porous frits, either of sintered glass or metal, are sometimes used to avoid pluggage problems with a capillary or molecular leak, but these can often exhibit greater memory effects.

19.2.2.2 Membrane Inlets

These are used for many sample introduction systems. They are useful precisely because they violate the normal requirement that the sample introduced into the MS be representative of the sample stream, in that they exclude certain portions of the sample in order to reduce interferents or to preconcentrate the sample to increase the sensitivity. Membranes are sometimes used to directly sample volatile organics in liquid streams (e.g., VOCs in wastewater), permitting the analyte to permeate to the low pressure side of the membrane for direct introduction into the MS ion source without requiring vaporization or sparging of the liquid. Membrane inlets are also frequently used for ambient air monitoring for volatile organics, achieving a degree of preconcentration of analytes while minimizing the introduction of normal air gases. Calibration of membrane inlet systems can be more difficult than direct inlet systems, and standard membrane materials can be problematic for analysis of very polar compounds, but resources [2] are available to assist in the selection of the best membrane material and design of the system.

19.2.2.3 Gas Chromatography (GC)

This is sometimes employed as a process MS inlet. Process GC–MS presents more fault and routine maintenance issues, and the delay associated with a chromatographic separation often negates the significant speed advantage of process MS. However, few analytical techniques are as powerful as GC-MS, especially for pilot plants and processes with very complex matrices or with frequent production of unknown byproducts. Commercial process GC-MS instrumentation is available, and is being ruggedized and made more rapid with further advances in the area of "fast-GC".

19.2.3
Ionization

Ionization is the process by which the analyte of interest gains a charge, thus allowing it to interact with the electromagnetic fields of the mass analyzer. Electron impact (EI) ionization is used almost universally in process MS due to its stability and relative simplicity. In EI, sample molecules pass through an energetic beam of electrons emitted from a hot filament. This beam imparts a charge on the molecule, and frequently fragments the molecule into smaller charged particles. The profile of fragments for a given compound is very reproducible between instruments, and is known as a fragmentation pattern or cracking pattern. The electron beam energy is usually set to 70 eV, which provides very stable production of ions, even with minor variations in electron beam energies. Lower energies are occasionally used to achieve less fragmentation, which can be useful if the fragmentation patterns of stream components are very similar; however, this approach should be used with caution, since ion production can be erratic and the resultant component quantitation more variable. Use of 70 eV EI also permits comparison of unknown spectra with various EI-MS libraries, which are usually composed of 70 eV spectra.

Various filament materials are available for process MS. Tungsten is rugged and inexpensive, but not suitable for applications requiring accurate CO and CO_2 measurements due to outgassing and hydrocarbon reactions on the reactive W surface. Tungsten has a higher work function and thus operates at a higher temperature, which can help to heat the source to reduce adsorption and contamination. Rhenium is not commonly used for process MS, although it is useful for hydrocarbon measurement applications due to the reduced formation of CO and CO_2. Thoriated iridium filaments have a low work function and hence produce large quantities of electrons at relatively low temperatures, which can contribute to filament lifetime; these are also very good for hydrocarbon and CO and CO_2 measurement applications. Choice of filament material is largely application-dependent, but thoriated iridium filaments usually offer the best balance of ruggedness and freedom from interference.

EI sources are available in either closed or open designs. In a closed design, the sample is introduced into a confined region that contains the ionizing electron beam. This confinement results in a higher concentration of sample (due to the limited conductance of the design) and higher degree of ionization, resulting in greater sensitivity. The disadvantage to this closed design is that the increased residence time of the gas components can result in greater source contamination, requiring more frequent cleaning. The open design permits better pumping and clearing of the source region, but this in turn causes a reduction in sensitivity. In either case, materials of construction for the source can affect the degree of contamination, especially in the case of polar molecules interacting with relatively active stainless steel surfaces. Either type can be heated with an auxiliary heater to reduce contamination from streams that contain high concentrations of hydrocarbons or polar ("sticky") molecules; if the analysis requirements are suited to tungsten filaments, the higher work function of W often provides sufficient radiant heating of the source to reduce contamination.

In addition to ion production and fragmentation, ion sources contain a number of ion lens elements that focus and accelerate the ion beam for injection into the mass analyzer. Contamination from the sample can alter the electrostatic fields of these lens elements, resulting in changes in performance that require more frequent recalibration or cleaning of the source elements. A careful assessment of the application at hand will lead to the best source design, often leading to analyzers that require routine maintenance only on an annual basis. For example, in the analysis of high concentrations of polar organics, a heated open source with Pt components might be warranted. For analysis of trace nonpolar compounds in ambient air, an unheated closed source would result in excellent sensitivity with minimal risk of source contamination.

Electron impact ionization is rugged, reliable and provides excellent precision and accuracy for most applications. However, the fragmentation of the parent molecules in the analysis stream can sometimes lead to spectra that are severely overlapped or even indistinguishable. For example, in certain hydrohalocarbon production processes, single halogens and hydrogens are easily cleaved from the parent molecules, resulting in virtually identical mass spectra. In such cases, it would be advantageous to utilize a softer ionization technique (i.e., one that predominantly produces charged parent molecules, with minimal fragmentation), such as chemical ionization (CI); indeed, both EI and CI mass spectra are often required for confirmation of compound identification in laboratory mass spectrometry. In the case of the hydrohalocarbons, the parent molecules would be easily distinguished with CI. However, in practice, CI is very difficult to utilize on-line, since the technique requires additional reagent gases, and is not as stable or rugged as EI. Softer ionization, either with CI or other techniques such as electrospray, glow discharge or field ionization, is an area of current research that promises to extend the utility of mass spectrometry for a number of difficult applications.

19.2.4
Mass Analyzers

After the sample molecules have received a charge, and possibly undergone fragmentation, the charged particles are injected into the mass analyzer, sometimes referred to as the mass filter, which separates these particles according to their mass. In actuality, this separation is carried out on the basis of the mass-to-charge ratio (m/z) of the particles. Many different types of mass analyzer designs are available, but magnetic sectors and quadrupoles are the most common choices for most process applications.

19.2.4.1 Sector Mass Analyzers

These employ either an electric or a magnetic field to separate the charged particles that are injected from the ion source. The field bends the ions into a circular trajectory, with the heavier particles (higher m/z) undergoing less of a bend than the lighter particles (lower m/z). A static field can be used, with several detectors

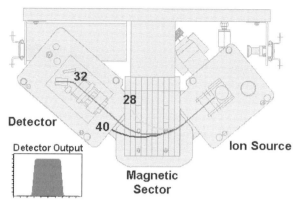

Fig. 19.3 Magnetic sector mass analyzer, with m/z 32 impinging on the detector; note flat-topped peak profile. (Used with permission of ThermoOnix Inc.)

placed in the path of the individual trajectories; this design offers good stability at the expense of flexibility, since a new detector would need to be installed if a new species needed to be monitored. A more common approach is to use a single detector and scan the field such that each trajectory falls on the detector as the field is swept; this offers a greater degree of flexibility, and modern electronics permit stability on a par with the static field design. Although either electric or magnetic fields can be utilized for sector MS, magnetic sector instruments are usually preferred for process MS because of their better long-term stability compared to electric sector instruments. Figure 19.3 illustrates a magnetic sector analyzer.

19.2.4.2 Quadrupole Mass Analyzers

These consist of four parallel rods with the ends arranged in a square pattern as shown in Fig. 19.4. The rods at opposite corners of the square are electrically connected. An rf signal is applied to one set, and an equivalent but inverted rf signal is applied to the opposite set. Likewise, opposing positive and negative DC offset voltages are applied to the rod sets. This combination creates a resonance condition that permits particles with a specific m/z to travel down the length of the rods, while all other m/z particles are annihilated by collision with one of the rods or some other surface. By scanning the DC offset, different m/z particles will resonate to reach the detector, thus resulting in a mass spectrum. Although the small diameter quadrupole rods in residual gas analyzers are sufficient for some process monitoring applications, better stability for more complex processes is usually achieved with larger diameter rods designed for process mass analyzers.

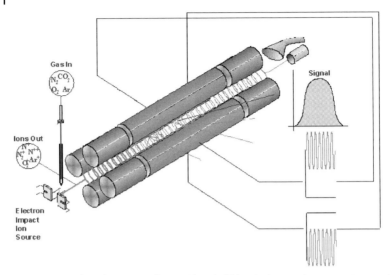

Fig. 19.4 Quadrupole mass analyzer, with m/z 32 impinging on detector; note gaussian peak profile. (Used with permission of ThermoOnix Inc.)

19.2.4.3 Choice of Analyzer

The choice between magnetic sector and quadrupole mass analyzers depends upon the application. Magnetic sectors are inherently more stable for a number of reasons. The resulting peak shape has a flat top, which is less susceptible to small fluctuations in mass alignment. In contrast, the peak shape is gaussian with a quadrupole, which results in greater signal drift with only slight mass alignment fluctuations. This stability is particularly important in the case of overlapping fragmentation patterns, since matrix deconvolution algorithms can propagate significant error into the final result if stability is poor. Furthermore, with a magnetic sector, ion injection into the magnet field is typically at 1000 eV, compared to 3–5 eV ion injection into a quadrupole. In operation, both systems will experience a degree of contamination that will slightly change the electrostatics of the order of 0.1–0.5 eV within the analyzer; this effect is relatively insignificant to the 1000 eV of the magnetic sector, but is much more significant to the lower-energy quadrupole, requiring more frequent cleaning and more frequent calibration to compensate. This can be particularly important for contamination-prone processes, such as with high concentrations of hydrocarbons, halogenated compounds or sulfur compounds.

Quadrupoles are generally faster, less expensive, more compact and have higher mass ranges for analysis (up to 600 m/z, compared with 200 m/z for a magnetic sector). This last aspect makes them particularly well suited for analysis of trace VOCs in ambient air. Although the gaussian peak shape is more prone to drift, this is often not a problem for sample streams that do not contain components with overlapping fragmentation patterns. Likewise, a quadrupole may be ideal for applications that do not involve significant amounts of contamination-inducing components, such as with air monitoring or many fermentation processes.

Other mass analyzer designs, such as time-of-flight, ion trap and ion cyclotron resonance, hold promise for higher resolution, faster analysis speed and increased mass range, but these designs are currently seldom used in on-line applications due to their complexity, higher cost and increased requirements for operator expertise. Other novel designs are on the horizon that utilize further advances in miniaturization, signal processing, and more stable electronics. As with all process analysis technologies, developments in laboratory instrumentation are often eventually transferred to process analyzers.

19.2.5
Detectors

Most process analyzers utilize either a Faraday cup or a secondary electron multiplier (SEM) for detection. The Faraday cup is the simpler and more rugged and stable of the two, but is generally useful for detection of species at higher concentrations (100 ppm to 100%). The SEM is much more sensitive, capable of measurements in the ppb range. It is quite common to configure a process MS with both detectors, along with a set of electrostatic lenses to switch the mass-filtered ion beam between the two detectors. This results in a single process analyzer that is capable of quantitation from 1 ppb to 100%!

19.2.6
Vacuum System

As shown in Fig. 19.1, the ion source, mass analyzer and detector are incorporated into a vacuum system. This vacuum must be sufficient (i.e., the mean free path must be sufficiently long) to prevent collisions between particles prior to analysis. A roughing pump is first used to provide vacuum for the sample inlet, as well as to provide the backing pumping for the turbomolecular pump, which provides the vacuum needed (between 10^{-5} and 10^{-7} Torr) for the ion source and mass filter. Specialized pumps and oils are available for corrosive processes. The fact that process MS requires vacuum pumps is often cited as a reason for avoiding the technique, but modern pumps are sufficiently rugged that they are rarely a cause of analyzer failure. Ion, getter and diffusion pumps are not routinely used for process monitoring, although getter pumps may become more common as process MS instrumentation is miniaturized.

19.2.7
Data Analysis and Output

Figure 19.5 is an EI mass spectrum of a typical sample stream that might be associated with a fermentation process; for simplicity, only nitrogen, oxygen, argon and carbon dioxide will be considered. In the case of a particular m/z to which only one

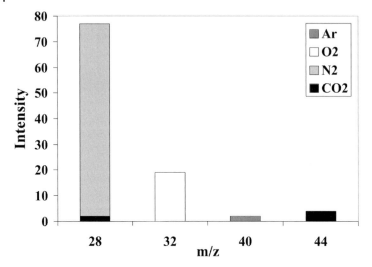

Fig. 19.5 Mass spectrum of air, showing only relevant mass-to-charge (m/z) ratios.

compound contributes a fragment (e.g., for mass 32, only oxygen contributes a fragment), signal intensity (i) is simply a product of the sensitivity factor (s) and the concentration (c) of that one compound:

$$i = sc$$

In the case of oxygen, quantitation is then simply a matter of monitoring the intensity at m/z 32 and dividing by the sensitivity factor.

The situation is more complex when two or more compounds contribute ion fragments to the same m/z. In our fermentation example, m/z 28 is predominantly due to nitrogen, but carbon dioxide has a minor fragment at that same m/z (due to cleavage of an oxygen atom from the molecule), and thus contributes to the total signal at m/z 28, as seen at the bottom of this peak in Fig. 19.5. Carbon dioxide has a major fragment at m/z 44, to which no other compounds in the sample contribute fragments. Since fragmentation patterns are very consistent, one could measure the intensity at m/z 44, determine the concentration of carbon dioxide, and then calculate the relative contribution of carbon dioxide to m/z 28, thus correcting for the potential error in nitrogen quantitation due to the presence of carbon dioxide. In practice, this correction is performed in real-time via matrix deconvolution using least squares techniques, according to the simplified matrix:

$$C = S^T I$$

in which C is the concentration vector, S is the sensitivity × fragmentation matrix, and I is the intensity vector. In practice, other factors are included in the deconvolution, such as detector gain weightings.

Resulting concentrations are often reported as relative concentrations, with all specified gases totalized to 100%, rather than as absolute concentrations. This further enhances the stability of the measurements, since small changes in flow and pressure in the sample line are effectively normalized out. However, this can lead to measurement error if unknown or unexpected compounds are present in significant quantity. Close review with process engineers will usually assist in determining which mode is most appropriate for a given process.

Modern deconvolution algorithms and fast computers permit deconvolution of very complex matrices, and minor components can be measured in the presence of components that at first glance might seem to overlap too much to permit quantitation. However, instrument noise eventually places a limit on how well this will work in practice. A common rule of thumb is that a minor component cannot be deconvolved and accurately measured if the major component contributes more than 30 times the signal of the minor component for a given m/z. With this rough rule, one can use library spectra and historical or modeled stream composition information to determine if MS is a feasible solution for the application. One caution with using library spectra: these often do not include the contribution of minor elemental isotopes that result in trace peaks in the spectrum, and thus do not indicate the degree to which these trace peaks of major components will interfere with minor components. For example, formaldehyde has a major peak at m/z 29, but nitrogen also has a fragment here due to the $^{14}N^{15}N$ dimer; although minor, in air this dimer contributes much more to m/z 29 than do ppm levels of formaldehyde, and thus prevents the measurement of formaldehyde at trace levels in air.

Modern process MS analyzers are controlled by microcontrollers and PC computers. Many incorporate internal processors that permit stand-alone operation, with a PC only required for initial configuration. The processor then handles all measurement and quantitation, as well as data interfacing, fault diagnosis and alarming, and calibration. Process mass spectrometers can be directly interfaced to plant distributed control systems, programmable loop controllers, or other process control systems.

19.2.8
Calibration System

As might be deduced from the above discussion of data analysis, accurate quantitation requires accurate determination of fragmentation patterns and sensitivity factors for calibration. Several options are described below for calibrating a process MS, but all have some common issues. It is common for calibration software to permit one to calibrate the MS to one gas, and then establish the sensitivity of all other components relative to that first gas. For example, one might establish the absolute sensitivity factor of nitrogen, and then establish the relative sensitivity factors for oxygen (relative to nitrogen). Drift mechanisms within MS are such that they can usually be corrected by calibrating nitrogen alone more frequently, and only occasionally re-establishing the relative sensitivities of oxygen. Since instrument drift can be caused by a number of factors, some of which might not have an equal effect on all gases, this relative sensitivity calibration scheme should be checked for each application. However, most commercial process MS vendors can provide recommendations on the frequency of calibration for common applications. In many cases, it is necessary to perform a complete recalibration only on a monthly basis. A complete recalibration should also be performed following any maintenance on the analyzer, especially if the ion source has been cleaned or the filaments replaced.

19.2.9
Gas Cylinders

Gas cylinders are commonly used for process MS calibration because of their relative simplicity and economy. Cylinders should be prepared gravimetrically, ideally by a supplier experienced in the preparation of accurate trace gas mixtures. Caution must be exercised in relying completely on gas cylinder suppliers' certificates of analysis, and it is frequently advisable to validate cylinders via another primary technique such as GC, although one should first scrutinize how that primary technique is itself calibrated. Stability of the gas mixture must be checked. PTFE-linings and electropolished cylinder interiors are often used to slow degradation.

In practice, fragmentation patterns are determined with a series of binary mixtures of a specific analyte and an inert balance gas (frequently Ar, He or N_2) that does not produce fragment ions that overlap with the analyte of interest. Background signals (due to trace leaks or minor filament outgassing) can be corrected by calibrating with an additional cylinder containing only the balance gas. Finally, it is common to have an additional cylinder of a representative gas mixture for fine tuning of sensitivities and correction of errors due to differences in gas viscosities.

19.2.10
Permeation Devices

Permeation devices have long been used for calibrating process analyzers, and they are an excellent alternative to gas cylinders, particularly if the analyte is unstable in a cylinder at low concentrations. These devices make use of the fact that most liquid compounds permeate through polymers at a constant rate for a particular temperature and gas flow rate. In practice, a known amount of liquid of a particular compound is sealed within a polymer tube (often PTFE) of known dimensions, which in turn is placed within a thermostatted container. A carrier gas, often dry air or nitrogen, is then swept at a constant rate through the container, thus transporting a constant amount of the compound of interest to the analyzer. Permeation devices can be gravimetrically calibrated to a high degree of accuracy, and in turn provide a very accurate means of calibrating process analyzers. Permeation rates are low, making these devices particularly useful for calibration at trace levels; this also means that the polymer tubes need to be refilled usually only a few times a year.

19.2.11
Sample Loops

Sample loop calibrators are commonly used for low concentration analytes that are difficult to prepare or are unstable in gas bottles. They are also extremely flexible, in that new compounds can be calibrated upon acquisition of a sample of the pure compound in liquid form. Sample loops are basically small heated known-volume vessels into which a small amount of liquid standard is injected. The sample vaporizes, and the entire volume is then pumped past the inlet, into which a small amount of the vapor is drawn for calibration. Drawbacks of sample loops include the lack of automation for unattended calibration, and inaccuracies inherent in making manual injections of nanoliter-scale liquid volumes. For ambient air monitoring of toxic compounds, one should bear in mind the safety implications of handling syringes that contain potentially hazardous analytes.

19.2.12
Maintenance Requirements

Maintenance requirements for process mass spectrometry are generally less than with process gas chromatography, which process MS often replaces. In general, one should check pump oil levels and color on at least a weekly basis. The degree of oil darkening is often an indicator of aging, but even in the absence of color change, the oil should be replaced semi-annually. In certain processes, ion source filaments often remain in service for over a year, but it is generally recommended that active filaments be replaced on the same schedule as the pump oil. Ion sources may need to be cleaned with some regularity in certain processes. Filters for samples and cooling air should be monitored and changed as required. Most commercial process MS analyzers incorporate sophisticated diagnostics capabilities to alert when there is a problem requiring correction or maintenance.

19.2.13
Modes of Operation

Process mass spectrometers are used predominantly for continuous quantitation of compounds that are included in the initial configuration. In the interest of speed, only the *m/z*s that are required for the analysis are measured. However, one of the most powerful aspects of process MS is the ability to do qualitative analysis on process streams. This is particularly useful for pilot plant operation, or for diagnosing process upsets in scaled-up processes. Several commercial vendors provide instruments that can operate routinely in quantitative mode, then occasionally perform a full mass scan to be archived for future scrutiny if a problem is subsequently detected downstream of the MS sample point. These full mass scans can be compared with library spectra to identify new byproducts. Some systems incorporate rather sophisticated pattern recognition and deconvolution algorithms that can achieve a degree of semiquantitation with no a priori knowledge of sample components or concentration. While this can be an extremely powerful application of MS, such automated compound identifications are best confirmed by someone with some training in mass spectral interpretation.

19.3 Applications

Process MS is commonly used for gas analysis in many manufacturing processes, including:

- Fermentation off-gas
- Steel manufacturing (blast furnace gases, etc.)
- Ethylene oxide
- Ethylene cracking
- Ammonia
- Partial oxidation of hydrocarbons

In addition, MS has also proven to be an excellent choice for ambient air monitoring in such processes as:

- Vinyl chloride monomer
- Ethylene oxide
- Acrylonitrile–butadiene–styrene (ABS) polymer processing
- Solvent monitoring (toluene, benzene, acetone, dimethylacetamide, etc.)

Due to its speed and flexibility, process MS is very useful for pilot plant studies, and for troubleshooting process upsets [3]. In many cases, a "roving" MS can be installed temporarily to gather as much process development information as possible, or to solve a vexing upset problem. Subsequently, it may be determined that a simpler, single-component analyzer is sufficient to control the process, but this determination is now based upon comprehensive real-time empirical data rather than on process models alone. Many process analyzer professionals spend much time retrofitting analyzers into new processes that were designed with inadequate or inappropriate monitoring capabilities. Pilot plant installations of powerful analyzers such as mass spectrometers can avoid such retrofits and thus greatly speed time-to-market of new products.

Development of new applications can be a complex exercise for any process analyzer technology, including MS. One must consider what stream component information is required to control or monitor the process, including analysis accuracy and precision, as well as how rapidly such information is required. One should determine the concentration ranges expected in both normal and upset conditions; a process analyzer is frequently of greatest value in helping to recover rapidly from upsets, so one must take care that it will operate accurately at those times. Additionally, data regarding sample stream temperature, pressure, and flow are required to design an appropriate sample conditioning system. The range of ambient conditions will dictate the design of the analyzer housing or shelter. Finally, one should compare the analyzer maintenance requirements to the expertise and availability of local maintenance personnel. In many cases, more than one analyzer technology may appear to be suitable, which can lead to emotions and vendor loyalties playing a disproportionate role in the final analyzer decision. However, the best solution usually appears if one reiterates through the above process, involving both process engineers and line operators in the information gathering.

19.3.1
Example Application: Fermentation Off-gas Analysis

Fermentation processes are commonly monitored with mass spectrometry, and serve as an excellent application to demonstrate the accuracy, flexibility, and economical benefits of the technique. Fermentation processes are common in pharmaceuticals, brewing and ethanol production processes, and are an active area of industrial research as a possible alternative to classical petrochemical processes. Fermentations are usually performed in agitated tanks, known as fermentors, that contain the organism of interest and various nutrients; gases are sparged into the fermentor to provide the required level of oxygen. Although fermentations can be monitored with several other techniques, mass spectrometry offers the advantage that it can measure major air gases (N_2, O_2, Ar, CO_2), as well as trace volatile organic species (ethanol, toluene, acetic acid, etc.) that are present in the headspace of the fermentor. In addition, MS is ideal for fermentation process development and scale-up, since one instrument can be multipointed to several fermentors, with each requiring different measurement methods. Thus, although the initial analyzer expense can be quite substantial, the resultant cost per analysis per sample point frequently makes MS more economical than installing individual discrete (i.e., single-component) analyzers on each fermentor. Often the economics are such that two MSs can be installed in such a way that each monitors half of the total number of fermentors (thereby effectively halving the measurement cycle time); in the event of an analyzer failure, the remaining functioning unit can then switch to monitoring all fermentors while repairs are made.

With most aerobic fermentations, it is important to accurately measure the respiratory quotient (RQ), which is the ratio of CO_2 evolution to O_2 uptake. Although RQ can be determined using discrete oxygen and carbon dioxide monitors (paramagnetic and infrared analyzers, respectively), such an approach could lead to errors due to differing drift characteristics of the separate monitors. With MS, both gases can be measured on the same instrument, thus increasing both accuracy and precision. (This same benefit is important for many other processes such as partial oxidation of hydrocarbons, in which a single analyzer can be used to monitor both oxygen and the hydrocarbon, thus increasing efficiency and safety.) In addition, the MS can also measure N_2 and Ar, which are not usually consumed in the fermentation process, and these measurements can then be used to calculate total flow in and out of the fermentor; in practice, however, it is common to use flow meters for such measurements, and have the MS running in a relative concentration mode.

19.3 Applications

As stated above, volatile organic species can be measured in the headspace with MS. The liquid-phase concentration of these same components can then be inferred using Henry's Law. This assumes that the fermentation media is well mixed, and that none of the volatile species is present in the fermentor gas inlet. Although thermodynamic properties can be used to calibrate for these liquid phase concentrations, it is generally best to inject a known amount of the compound of interest into the media, measure the liquid phase concentration with a laboratory technique such as gas chromatography, and calibrate the MS measurements against these liquid measurements. Camelbeeck et al. [4] have demonstrated that headspace component measurements are not affected by changes in the rate of mixing by aeration and/or agitation.

The flexibility of MS can further be utilized to monitor for microorganism mutation in certain situations. For example, in one proprietary process, the microorganism is known to be prone to a mutation in which the resulting mutant produces a small amount of hydrogen during glucose metabolism. The MS can be configured to monitor for hydrogen occasionally, and thus alert to this condition well before the reduced product yield is detected.

Table 19.1 contains composition information for a typical aerobic fermentation off-gas stream, along with the precision that is achievable with a magnetic sector process MS. Table 19.2 contains the fragmentation pattern matrix for this same sample stream; note that this is similar to the example presented above in the discussion of data analysis, but this particular stream now contains ethanol. This more complicated stream presents further challenges to deconvolution due to more severe fragmentation pattern overlap: ethanol interferes with the direct

Table 19.1 Fermentation off-gas composition, with precision obtained with process MS.

Analyte	Mol %	Precision (% absolute)
Nitrogen	79	0.006
Oxygen	16	0.004
Argon	1	0.001
Carbon dioxide	4	0.004
Ethanol	0.06	0.001

Table 19.2 Fragmentation pattern matrix for fermentation off-gas.

Mass (m/z)	Nitrogen	Oxygen	Argon	Carbon dioxide	Ethanol
28	100			5	
31		0.015			100
32		100			
40			100		
44				100	
45				1.2	45.2

Table 19.3 Recommended calibration gases for aerobic fermentation process, with ethanol.

Component	Cylinder 1	Cylinder 2	Cylinder 3	Cylinder 4	Cylinder 5
Nitrogen		78.08			79
Oxygen		20.95			16
Argon		0.934			1
Carbon dioxide		0.033	5		4
Ethanol				0.08	0.06
Helium	99.999		balance	balance	

simple measurement of oxygen, carbon dioxide interferes with nitrogen, and ethanol and carbon dioxide interfere with each other. Note, however, from the precision values shown in Tab. 19.1 that the deconvolution algorithm adequately accounts for these overlapping fragmentation patterns.

Table 19.3 describes the set of gas standards that would be used for this fermentation monitoring application. Gas cylinders are the recommended calibration method for this application because accurate and stable mixtures can be procured from many gas suppliers. As described above in the calibration section, initially a complete calibration must be performed with all calibration gases, to be repeated monthly, with a weekly or daily subset calibration with nitrogen alone to correct for most normal instrument drift. Frequency of calibration is very application

Table 19.4 Process mass spectrometry.

Characterized parameter:	Surface specificity:	
Mass-to-charge ratio (Molecular mass of ions)	Information depth: n/a	Detectability: ppb to 100 %
Type of information:	**Resolution:**	
Molecular composition (quantitative, qualitative)	Depth: Lateral: Other: Not used for spatially resolved analaysis.	
Measurement environment: Difficulties	**Time needed for analysis:**	
Vacuum	Fragmentation pattern overlap	Total: 5 s/analyte/stream (excluding sample transport time)
Equipment:	**Cost [ECU]:**	**No. of facilities:**
Magnetic sector or Quadrupole Mass Spectrometer, Electron Impact ionization	60,000–150,000	Fairly common
Type of laboratory: **User skill needed:**	**Sample**	
n/a Unskilled	Form Type: Gas, occasionally for dissolved VOCs with use of membrane inlet	
Techniques yielding similar information:		
FTIR, GC, IMS		

dependent, and the above schedule may be adjusted as dictated by the required accuracy and precision for process control. In this example, Cylinder 1 is pure He, used to establish background levels for all analytes. Cylinder 2 is essentially dry air, used to establish sensitivities for nitrogen, oxygen and argon. Cylinders 3 and 4 are used to establish the fragmentation patterns and sensitivities for carbon dioxide and ethanol, respectively. (Although ethanol has the major peak at m/z 31, the minor peak here from a major component, oxygen, usually dictates that m/z 45 is required for the analysis of ethanol.) Finally, Cylinder 5 contains a mixture that closely resembles the stream composition, and this is used to fine-tune the sensitivities of all gases and correct for viscosity differences. Cylinder 5 also constitutes what is commonly known as a check blend: if operators have reason to suspect that the analyzer is reporting incorrect values, they can switch to monitor this check blend for validation of the instrument.

19.4
Summary

As will have become tediously apparent to the reader of this chapter, any potential installation of process MS application requires a thorough review of the process, including the key information that is needed to control it. This is true for any process analysis technology. As described above, process MS offers several options for each part of the overall system, and therein lies much of the unique flexibility of the technique. Again, a thorough process application review will narrow the choices for components. In addition, several process MS vendors and consultants are very experienced at designing systems to meet the needs of the process.

Table 19.4 summarises the main relevant characteristics of process mass spectrometry.

Process MS is a reliable technique that can provide rapid and precise multicomponent analysis on process streams. The ability to monitor multiple sample points with a single analyzer makes MS very economical for many applications, even with the high cost of the analyzer itself. Maintenance requirements on modern MS analyzers are on a par with or lower than most other analyzer technologies. In addition to permanent installations for routine process control, these attributes also make process mass spectrometry extremely useful for process development and troubleshooting.

References

1 J. Workman, D. J. Veltkamp, S. Doherty et al., *Anal. Chem.*, **1999**, *71*, R121–R180.
2 M. A. Lapack, J. C. Tou, C. G. Enke, *Anal. Chem.*, **1991**, *63*, 875A.
3 M. A. DesJardin, S. J. Doherty, J. R. Gilbert et al., *Process Control Qual.*, **1995**, *6*, 219.
4 J. P. Camelbeeck, D. M. Comberbach, M. Orval et al., *Biotechnol. Tech.*, **1991**, *5*, 443.

20
Elemental Analysis

J. S. Crighton

20.1
Applications of Atomic Spectrometry in Process Analysis

Elemental analysis in general, and atomic spectrometry in particular, play a key role in all aspects of process development, optimisation and control. The most obvious aspects of this contribution are in the control of the composition of the product itself. For example, in the cement industry, X-ray fluorescence (XRF) has long been used to measure the elemental composition of kiln feed (particularly Ca, Si, Fe and Al concentrations), since this can affect the efficiency of operation of the kiln as well as the quality of the clinker produced [1, 2]. Similarly, in the metallurgical industry, XRF has been used alongside atomic emission spectrometry (AES), particularly utilising arc/spark sample introduction systems, to control alloy composition during the production process (see e.g. [3–12]). In other cases, atomic spectrometry may be used on a quality control basis to control the content of specific additives in the final products. For example, inductively coupled plasma atomic emission spectrometry (ICPAES) is often used to measure Ca, Mg, P and Zn in lubricating oils as a means of checking that the correct concentrations of additives have been added to the products [13]. Other applications, which are perhaps less obviously directly related to process control, can still play an equally important role in terms of overall operation of the process and some of these are described briefly below.

Detailed description of all aspects of the application of atomic spectrometry to analysis of the huge variety of sample matrices associated with industrial processes is clearly outside the scope of this chapter and so only a brief overview is given of each application area. For detailed discussion of the application of atomic spectrometry in industrial analysis, the reader should refer to the annual reviews of this topic published in *Atomic Spectrometry Updates* in conjunction with *J.Anal. At. Spectrom.*, [3–12]. In Section 20.2, specific applications of atomic spectrometry to on-line/at-line analysis are discussed in more detail.

20.1.1
Catalyst Control

In order to achieve optimum performance from any catalytic process, it is essential that the concentrations of active components and any associated promoters are maintained within relatively narrow boundaries. For homogeneous catalysis processes, it is relatively easy to take samples and analyse using ICPAES or XRF in order to ensure that optimum concentrations are maintained and this application also lends itself well to on-line or at-line approaches (see Section 20.2). For heterogeneous catalysts however, the taking of regular samples is generally much more difficult and the contribution of atomic spectrometry may be restricted to process development at the pilot plant stage or analysis of post-mortem samples. Again, ICPAES or XRF may be used for analysis of heterogeneous catalysts, depending on the support matrix and active elements which require to be determined. ICPAES suffers from the obvious disadvantage that samples normally require to be digested prior to analysis but XRF generally has poorer limits of detection and can suffer from particle size effects and other matrix effects arising for example from the presence of coke or corrosion metals in used catalysts. Atomic spectrometry can also be used as a means of agreeing settlement prices (based on precious metal concentrations) between catalyst suppliers and users, although due to the high values involved, the analytical precision required (often agreement within 1 % for analyses carried out by each party), can necessitate use of more labour intensive, time consuming analytical approaches (e.g. fire assay).

Performance of heterogeneous catalysts can be adversely affected by the presence of poisons in feedstocks, even at very low (e.g. ng g^{-1}) concentrations since these can become more concentrated on the catalysts. Similarly, low concentrations of catalyst elements can migrate into downstream parts of the process, where they can cause problems with downstream catalysts. For example, low (ng g^{-1}) concentrations of iodine in acetic acid (originating from the catalyst), can cause poisoning of the vinyl acetate catalyst. For precious metals, even very low concentrations migrating into the products can have a significant economic impact on the process by the time they are multiplied up by high flow rates and extended operating times. Although ICPAES can be used for some of these applications, limits of detection can often be inadequate and more sensitive analytical techniques such as inductively coupled plasma – mass spectrometry (ICP-MS) must be used. For the determination of poisons on heterogeneous catalysts, the nature of the potential poisons is generally unknown and it can be difficult to choose a sample digestion approach which can be guaranteed to retain all possible elements. In these cases, laser ablation ICP-MS can be extremely useful in view of the low limits of detection which can be achieved and broad elemental coverage without the necessity of digesting the sample (and danger of losing elements through precipitation or volatilisation).

Although in most cases it is sufficient to determine total concentrations of elements in the samples, in a few instances, it is necessary also to determine the chemical species present. ICP based techniques lend themselves well to these requirements since they can be readily combined with chromatographic techniques

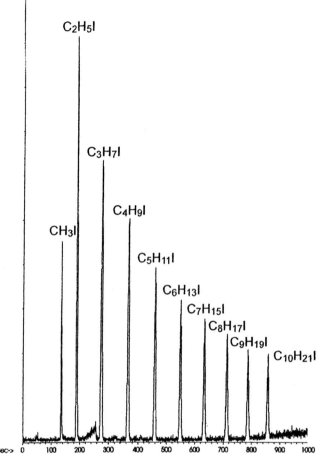

Fig. 20.1 a

Fig. 20.1 (a) GC-ICPMS trace of standard containing 350 ppb alkyl iodides in acetic Acid. and (b) similar trace (to Figure 20.1 (a)) obtained from an intermediate stream within a mixed acid/anhydride carbonylation process.

(see [3–12]). For example, Fig. 20.1(b) shows a chromatogram obtained from an intermediate stream within a mixed acid/anhydride carbonylation process, analysed using capillary gas chromatography coupled with ICP-MS (the corresponding chromatogram obtained from a standard containing 350 ng g^{-1} of mixed alkyl iodides in acetic acid is shown in Fig. 20.1(a) for comparison). The ICP-MS was used to monitor mass 127, providing a completely iodine specific detection for the GC system, thus allowing measurement of organo-iodine compounds in the process sample down to ng g^{-1} levels. A full knowledge of the iodo compounds present at various stages of the process can allow the process to be optimised to ensure low concentrations of residual iodine in the products.

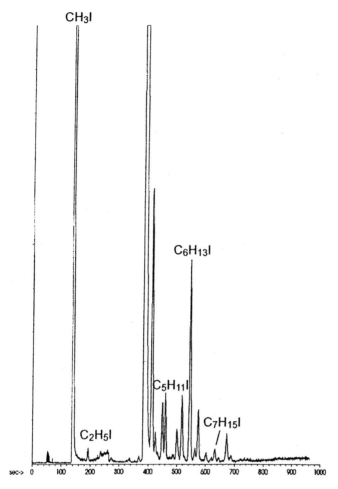

Fig. 20.1 b

20.1.2
Corrosion Monitoring

Corrosion of component parts of the process plant can be readily monitored by measuring the concentrations of associated "corrosion metals" (e.g. Fe, Ni, Cr, Mn, Mo) in downstream process streams or products. In view of its rapid, multi-element capability, ICPAES lends itself well to these applications, although ICP-MS with its inherently lower detection limits, can permit measurement of lower rates of corrosion and of minor alloy components which can aid in identifying which plant component is corroding.

In many cases, the alloy composition of the major plant components will be known and the one responsible for the corrosion can often therefore be identified from the ratios of the concentrations of the "corrosion metals" measured in the downstream process streams or products. In cases where the alloy compositions are not known. however, obtaining this information can be difficult, since it is often impossible to analyse the material without destructive sampling and accessibility restrictions may prohibit analysis using portable analytical equipment. In these instances, a technique which has regularly been used within the BP Group is the XRF "Rubbing" technique. This technique, which was developed at the BP Research Centre in Sunbury involves gently abrading/polishing the surface of the alloy of interest with a flexible polymer disk impregnated on one side with 15 μm diamond particles (661X; 3M, St Paul, MN, USA). During this process, approximately 1 mg of sample is transferred onto the surface of the disk. Since it is only the tips of the diamond particles which abrade the alloy surface, the metal transferred onto the rubbing disk is in the form of very small particles of less than 1 μm. Thus, in XRF terms, the sample is in the form of a "thin film" resulting in no absorption and/or enhancement effects. This therefore gives rise to universal, linear calibrations for the elements of interest, which are independent of alloy type. The resulting rubbing disks can be analysed by wavelength dispersive (WD) or energy dispersive (ED) XRF using relative element response factors (determined either empirically or calculated using fundamental parameters) and results normalised to 100%. The results are thus independent of the weight of sample transferred to the rubbing disk and analysis can be accomplished with few or no calibration standards. The method can be used for any alloy type and gives results which are typically within 5% relative (95% confidence) for elements in the concentration range 0.1 to 100% (lower concentrations can be measured by analysing the disks using laser ablation ICP-MS [14].

Figure 20.2 shows results obtained for a number of elements in a variety of alloy types including aluminum, copper, iron, nickel and titanium based alloys using an

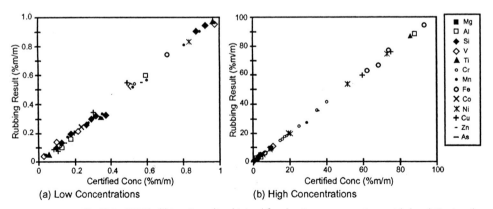

Fig. 20.2 XRF "rubbings" results obtained for aluminum, copper, iron, nickel and titanium based alloys: (a) Low concentrations; (b) high concentrations.

EDXRF spectrometer with calibration based on element response factors calculated using fundamental parameters (no standards). The advantages of the technique are that it is fast, simple, essentially non-destructive and can be used to sample an item of almost any shape or size, in almost any location without requiring any special tools or equipment. The sampling can even be carried out by plant personnel and posted to a remote laboratory for analysis. The only real precaution required is that since only 1 mg of material is transferred to the rubbing disk, the sample surface should be clean and representative of the component as a whole. This can usually be accomplished by cleaning the components with solvent and then abrasive paper prior to sampling, although in cases of severe corrosion, it may be necessary to remove surface contamination with a scurfing tool prior to final clean-up and sampling. The method has even been used for analysis of welds in order to check that correct welding rods had been used by contractors.

20.1.3
Reducing Environmental Impact

With the increasing public awareness regarding the environmental impact of industrial processes and products, there is increasing pressure on companies to ensure not only compliance with any existing legislation but also that all possible steps are taken to minimise any risks of adverse effects Again, atomic spectrometry plays a pivotal role in this assurance process. This can range from ensuring low concentrations of harmful components in the products produced to ensuring that any potentially hazardous components in waste streams and products are eliminated or minimised and that any residual hazardous waste is treated and/or disposed of in a responsible manner.

The pressure to reduce environmental damage generally results in an ever increasing drive to measure harmful components at lower and lower levels and this can put pressure on analytical techniques used to make the measurements. For example, EDXRF and WDXRF are widely used throughout the industry for determination of sulfur in petroleum products. However, the continuous tightening of legislation and lowering of specification levels for sulfur in road transport fuels over recent years has resulted in some international standard methods and some instrumentation being inadequate for use at the lower sulfur specification levels. A report detailing which methods are adequate for the European fuel specifications for 2005 (50 mg kg^{-1} S max.) and beyond has recently been published [15].

For measurement of heavy metals and other harmful trace elements in effluents and solid waste, standard methods are available which generally utilise digestion and/or concentration followed by analysis using atomic absorption spectrometry (AAS), ICPAES, or increasingly ICP-MS. A comprehensive range of such methods has been produced by the US Environmental Protection Agency [16]. For a more detailed discussion of the use of atomic spectrometry in environmental analysis, the reader is referred to the comprehensive annual reviews of this topic produced as *Atomic Spectrometry Updates* [17–26].

20.1.4
Troubleshooting Process Problems

Many process problems fall into one of the categories already described above. However, often problems can be associated with formation of deposits of unknown origin and composition in various parts of the plant. For a completely unknown deposit, XRF is generally more suitable than solution techniques such as ICPAES or AAS since only small amounts of material are often available and it can be very difficult to decide on a digestion approach if the nature of the material is unknown. Qualitative analysis of deposits using XRF is generally straightforward but obtaining quantitative results can be difficult since, by the very nature of the sample, appropriate calibration standards are not available. One approach to this problem which has been used at the BP Research Centre for many years is the XRF 'SMEAR' technique [27]. In this approach, the sample (down to 10 mg) is ground and mixed with zirconium oxide and a small amount of the mixture (about 1 mg) is slurried with a few drops of ethanol and smeared on the outside of the polymer film window of an X-ray cell. This results in production of a thin film sample which can be analysed by XRF with virtually no absorption/enhancement effects using universal calibrations produced using simple oxide standards. By ratioing the net element intensities to that of zirconium, effects of sample thickness (weight) and film distribution/homogeneity can be compensated. Accuracy/precision of the 'SMEAR' technique has been shown to be $\pm 10\%$ relative and the technique can be used to measure element concentrations in the range 0.1% m/m to 100% m/m.

Another approach to semi-quantitative analysis of unknown samples such as deposits is to produce theoretical calibrations based on fundamental parameters. The problem with this approach is that it relies on carrying out a complete analysis of the sample in order to correctly calculate corrections for matrix effects. Unfortunately, many deposits contain high concentrations of light elements such as carbon which cannot be reliably determined using XRF. Unless this information is obtained in another way (e.g. using an alternative analytical technique such as microanalysis), concentrations calculated using fundamental parameters can be seriously in error (for example, many programmes will assume that all of the elements which can be measured represents 100% of the sample, even although in reality they may only comprise 10% of the sample with the rest being carbon). Another problem with some fundamental parameter approaches is that they assume that the sample is a bulk sample and no thickness corrections are applied. Again, this can lead to serious errors if only a very small amount of sample is available. Some XRF instruments utilising fundamental parameter approaches can get round these problems by using measurements of scattered X-ray tube lines to gain information about the concentration of low mass elements which cannot be measured directly and this information can be fed into the fundamental parameter calculations.

Figure 20.3 shows results obtained with a Spectro X-Lab EDXRF spectrometer utilising this approach for samples with light element concentrations (loss on ignition) up 57% m/m. Accuracy for the major elements is generally similar to that

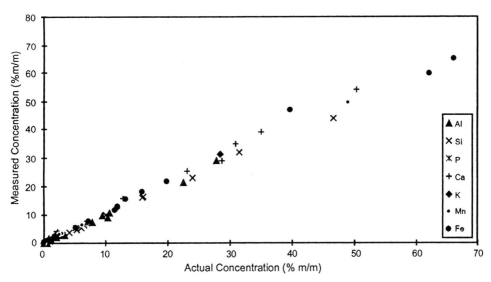

Fig. 20.3 EDXRF results for a wide variety of reference materials with loss on ignition up to 57% m/m obtained using fundamental parameter calculations with correction for unmeasured light elements using scattered radiation.

obtained using the 'SMEAR' approach, but the X-Lab method has the advantages that it can measure down to much lower concentrations (mg kg^{-1} levels) and is less operator dependent (obtaining the correct film thickness for the 'SMEAR' technique and producing well dispersed films without agglomeration is somewhat of an art which must be perfected by the operator). Utilising one of these approaches for semi-quantitative analysis using XRF, the nature of most deposits comprising inorganic components can normally be established, particularly when used in conjunction with X-ray diffraction (XRD).

20.2
On-stream/at-line Analysis

At the present time, most elemental analysis for process control is still based on laboratory analysis as described above. With currently available analytical technology, in many cases, this is the only viable approach. However, where measurements can be carried out on-stream or at-plant, the benefits can be substantial. With traditional laboratory based analysis, sampling frequency is limited and labour costs can be substantial if continuous round-the-clock process monitoring is required. On-stream analyses can provide the capability of closed loop control of the process allowing optimum operating conditions to be maintained, minimising costs and resulting in a more stable product with fewer off-specification products produced. On-stream analysers can also be used to control product blending to specification limits with less 'give-away' from production of over specified product. However, one must bear in mind that typical costs for a fully installed on-line system, complete with sample conditioning and safety systems are about an order of magnitude higher than that of a similar at-plant system. These additional costs can only be justified where a very high sampling frequency is required and where the closed loop control loop can make major improvements to the operation of the process. In many cases, particularly with continuous processes, the high frequency sampling may only be justified during the period where process conditions are changing rapidly (e.g. start-ups or shut-downs). During steady-state operations, the benefits of the high sampling frequency may be marginal. In these cases, an at-plant approach may be more appropriate from a cost-benefit point of view. In this section, on-stream applications of atomic spectrometry for process analysis/control are discussed, however, virtually all of the approaches described could equally be applied on an at-stream basis.

20.2.1
X-ray Fluorescence (XRF)

For on-stream elemental analysis, by far the most suitable and widely used technique is XRF. The technique is particularly suitable for on-stream applications since it is robust, non-destructive, uses no hazardous reagents, requires minimal maintenance and can be easily mounted in an inherently safe enclosure. The technique is capable of simultaneous measurement of elements from Al to U in the concentration range from ppm levels up to high percentages and in many cases can achieve precisions of 1% RSD or better. Analysis times vary depending on the particular application and type of equipment used but a typical cycle time would be a few minutes, thus lending itself to closed loop control systems requiring a high sampling frequency.

Figure 20.4 shows a schematic diagram of a typical on-stream XRF analyser. The figure shows a flow cell for on-stream liquid analyses, but this could equally well be a specialised cell designed for powder/slurry analysis, a conveyer belt containing solid material, or even a continuous flowing solid material such as a roll of paper (further details of these specialist applications are given in Sections 20.2.1.3 and 20.2.1.4). The flowing sample is irradiated with a source of X-rays (normally either an X-ray tube or a radioisotope source) and the fluorescent X-rays produced in the sample are measured using a detector appropriate to the particular application involved.

For efficiency of excitation, the energy of the X-ray source should be close to, but higher than the absorption edge for the element(s) to be measured. Traditionally radioisotopes were used as the source of X-rays for on-stream XRF analysers. Table 20.1 shows typical isotope sources used and the range of elements which they can be used for. These sources have the advantages that they are small, have good stability and require no external power. However, as shown in Tab. 20.1, the sources decay at vastly different rates. Some, such as ^{109}Cd and ^{55}Fe will decay and require replacement every few years (during which time performance will be degrad-

20 Elemental Analysis | 345

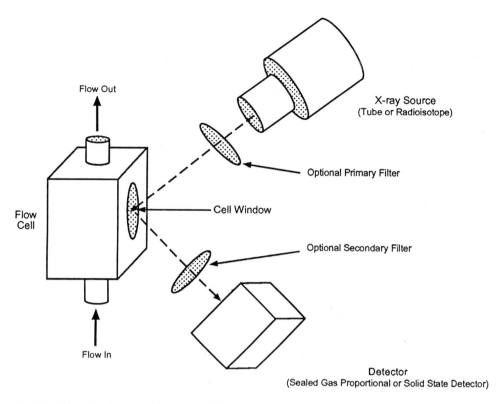

Fig. 20.4 Schematic diagram of on-stream EDXRF analyser.

ing all the time), whereas others such as ^{241}Am have half lives measured in several hundred years which can present disposal problems. Furthermore, the flux produced by radioisotope sources is generally too low for use with high resolution (WDXRF) detectors and the excitation efficiency is not high for all elements. A further complication is that the X-ray source cannot be switched off, which can hamper maintenance operations and can present a potential safety hazard. For these reasons, most modern systems now use an X-ray tube as the source. Excitation efficiency of the X-ray tube is generally higher than with radioisotope sources and can be optimised through use of an appropriate tube anode material, operating vol-

Table 20.1 Common isotope sources for XRF analysis.

Source	^{55}Fe	^{241}Am	^{109}Cd	^{244}Cm
Half Life (years)	2.7	433	1.3	17.8
Energy Range (kV)	1.5 to 5.0	8.5 to 40	5.0 to 18	4.5 to 11.5
Analytes (K lines)	Si to V	Zn to Nd	Cr to Mo	Ti to Se
(L lines)	Nb to Ce		Tb to U	La to Pb

tage and filter material. The flux from the X-ray tube can also be high enough to permit use of WDXRF detectors or secondary targets for optimal excitation of the analyte elements in the sample. Unlike radioisotope sources, X-ray tubes can be switched off when not in use, simplifying maintenance operations. The major disadvantage of the tubes is that they need an external high voltage source and occasionally, in the case of high power tubes, cooling water.

For on-stream/at-plant applications, the detectors which are used are generally of the energy dispersive (EDXRF) type. In the normal range of operation (< 20 keV), these have poorer resolution than wavelength dispersive (WDXRF) detectors but have the advantages that they can measure multiple elements simultaneously and so are generally lower cost than an equivalent WDXRF system (effectively requiring separate crystal and detector for each element to be determined). The smaller collection angle for a WDXRF system, compared to EDXRF also generally necessitates a higher power source, thus again adding to the cost differential. The EDXRF detectors used in XRF spectrometers fall into two distinct categories. The lowest cost systems are based on sealed gas proportional detectors. In addition to the low cost, these have the advantage that they can be operated at room temperature. The disadvantage is that the resolution of these detectors is very poor (typically between 10 and 20% of the X-ray photon energy measured). Thus, for all but the simplest applications/matrices, peak overlap can be a severe problem. The second type of EDXRF detector is based on solid state devices (usually lithium drifted silicon or germanium). These detectors have much better resolution than the sealed gas proportional detectors, but in the normal operation range (< 20 keV) they are not as good as WDXRF detectors. The main disadvantage of the solid state detectors is that they must be cooled down (with liquid nitrogen or Peltier cooling), in order to operate. Typical resolution for a Peltier cooled Si(Li) detector is around 195 eV. Figure 20.5 shows a typical EDXRF spectrum of a process solution containing high levels of corrosion metals compared with the single peak which would be observed for Fe and Ni if a sealed gas proportional detector had been used (inset). One method of compensating for the poor resolution of the sealed gas proportional detectors is the use of filters. These can either be used to remove contributions from interfering lines, to reduce background, or can even be used to perform "non-dispersive" XRF using "balanced filters". In the latter approach, two filters are used sequentially, one which absorbs the analyte line well and the other which does not. The difference between the two measurements gives information on the analyte element of interest.

One aspect of XRF analysis in general, and on-stream XRF in particular, which is often poorly understood is that of sampling depth. There is a general tendency for people to think of X-rays as being a very penetrating form of radiation and to infer that the XRF system is carrying out an analysis of the entire sample which is presented to it. While in some cases, this is a good assumption, in others XRF is only measuring the concentration of analyte in a very shallow region of the sample near the cell window. Table 20.2 shows the "critical depths" (i.e. depth below which greater than 99% of the fluorescent X-rays would be absorbed before they reached the sample surface (cell window)) for a selection of materials as a function of X-ray

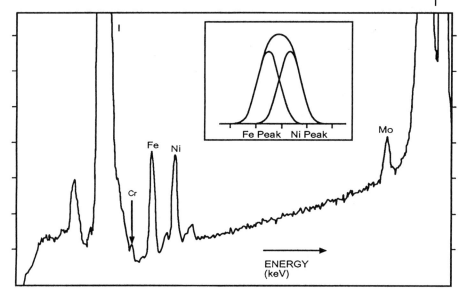

Fig. 20.5 EDXRF spectrum of a reactor sample containing corrosion metals obtained using a Si(Li) detector and (inset) a typical spectrum obtained using a sealed gas proportional counter showing unresolved Fe and Ni peaks.

line energy (analyte element). From this, we can see that, whereas if we were to measure a heavy element such as Pd (K_α line) in a hydrocarbon matrix such as gasoline, the sampling depth would be over 9 cm, if we were to measure sulfur in the same matrix, the sampling depth would be less than 0.3 mm. Furthermore, if we were to attempt to measure sulfur in a relatively heavy matrix such as steel, the analytical measurement would be restricted to a depth of less than 3 µm from

Table 20.2 XRF critical depths[a] for a selection of sample types.

Element	Line	Energy (kV)	Critical Depth (cm)			
			Gasoline	Water	Quarz	Iron
Na	K	1.04	0.0022	0.0008	0.0004	0.00005
S	K	2.31	0.021	0.0068	0.0009	0.0003
Sn	L	3.44	0.074	0.023	0.0028	0.0010
Ca	K	3.69	0.091	0.028	0.0034	0.0012
Fe	K	6.40	0.48	0.14	0.016	0.0051
Zn	K	8.63	1.2	0.36	0.039	0.0014
Au	L	9.67	1.6	0.50	0.055	0.0020
PB	L	10.50	2.0	0.63	0.068	0.0024
Mo	K	17.40	6.8	2.6	0.31	0.010
Pd	K	21.10	9.2	4.0	0.51	0.017

[a] depth is defined as depth below which more than 99% of the fluorescent radiation would be absorbed before it reached the detector (for a conventional XRF geometry with take-off angle approximately equal to 45°).

the sample surface. It is therefore clear that for any successful XRF measurement (on-stream or at-line), we must ensure that the sample depth which is sampled is representative of the sample as a whole.

Limits of detection depend very much on the analyte element, the matrix, the excitation conditions, the detector, the presence of any special filters and the measurement time. Under optimal conditions, limits of detection down to single ppm levels can be achieved using counting times of a few minutes.

20.2.1.1 Liquid Process Streams

The main limitation associated with on-line analysis of liquid process streams using XRF is that the stream must be contained within a window material. This material must be strong and robust enough to withstand extended exposure to the temperature and pressure of the stream, be resistant to chemical attack or abrasion and yet be sufficiently transparent to the fluorescent X-rays to enable measurements to be made down to low enough concentrations to allow accurate measurements to be made. There are very few materials which will satisfy these requirements and in practice, the only suitable materials are beryllium or polymer films of thickness between about 10 and 50 μm. The choice of material and thickness for the window of the flow cell represents a trade-off between analytical sensitivity and maximum temperature, pressure and chemical/mechanical resistance. Figure 20.6 shows the transmission of various window materials as a function of X-ray energy (analyte element). It is clear, that for measurement of light elements such as sulfur, the window material should be as thin as possible (subject to temperature and pressure limitations). Normally where the sample is not corrosive, beryllium windows are used since these are generally more robust and less liable to stretching and distortion than polymer films and require less frequent replacement. For corrosive streams, the beryllium windows can be protected with a thin plastic coating, but for abrasive

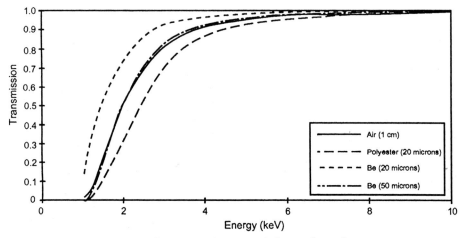

Fig. 20.6 Transmission of various window materials for XRF flow cells.

or very corrosive materials, this is not robust enough and a replaceable polymer film must be used inside the beryllium window. For acidic streams, any of the commonly used polymer films (e.g. polyester, polyimide or polypropylene) can be used, but for caustic streams, polypropylene films are generally best. None of the films are particularly resilient for use with strongly alkaline streams and so in some instances, the flow cell may need to be bypassed during caustic flush cycles to avoid window rupture. The limitation on stream temperature and pressure is normally due to the plastic window material. For polymer film windows, the limits of temperature and pressure are generally in the range 30 to 90 °C and 5 to 10 psi depending on the polymer used and the film thickness. For beryllium windows however, temperature and pressure can be up to 90 °C and 30 psi depending on the window thickness. In extreme cases, very robust windows made of polyvinylidene fluoride (PVDF) can be used for applications with streams at up to 200 °C and 800 psi, but in these cases, the absorption of the X-rays by the window can be so large that only X-ray transmission measurements (rather than XRF) can be made and measurements are restricted to high concentrations of analytes (e.g. measurement of percent levels of sulfur in crude oil and fuel oil pipelines).

Figure 20.6 also shows the absorption of X-rays as a function of energy (analyte element) for air. It is clear that for light elements absorption of fluorescent X-rays due to air in the optical path can be significant. For this reason, the optics used for these measurements should be as closely coupled as possible. Furthermore, the optical path should be purged with helium or argon-free nitrogen (argon significantly interferes with sulfur measurements when sealed proportional detectors are used). For on-stream applications, nitrogen is generally preferred on cost grounds.

As discussed in Section 20.2.1, for measurement of light elements, the sampling depth can be as low as a few microns. Thus any deposition on the sample window can significantly affect the measured results. For this reason, windows normally require regular cleaning and/or replacement. The frequency of cleaning/replacement will depend very much on the particular sample streams being measured but could typically be every few weeks. Furthermore, the presence of any particulate matter can cause severely erroneous and erratic results in cases where sampling depth is limited. In order to reduce susceptibility to such problems, many liquid streams require a sample conditioning system to be installed upstream of the analyser to remove particulates and water (in the case of hydrocarbon streams). Since XRF results are also influenced by sample density, the sample conditioning system may also require facilities for temperature and pressure regulation. This obviously adds to the cost of an on-stream system compared to an equivalent at-plant or laboratory based system.

Commercial on-stream XRF analysers for liquid process streams are produced by a number of manufacturers (see Appendix). These can either be configured for monitoring single or multiple streams. In most cases, the X-ray monitoring head (source and detector) moves sequentially from one flow cell to another for sequential monitoring. Solid reference samples are normally also incorporated in one or more of the positions monitored by the measurement head and used to calibrate and correct for instrumental drift. In many instances, however, plumbing multiple streams to the same location may be impractical and, in these cases, installing multiple single stream monitors may be more appropriate. The commercial units available are generally mounted in fully interlocked and X-ray shielded NEMA 12, 4X or cast aluminum cabinets which can be equipped with a variety of CENELEC and NFPA purge systems for use in potentially explosive environments. Most systems incorporate leak sensors in the base of the cabinet which will completely shut down the unit in the event of a cell window rupture or other leak. In addition, a combustible gas monitor can be incorporated to monitor the purge gas from the enclosure.

Use of on-stream XRF analysis for monitoring liquid process streams has been reported for a number of applications including measurement of Fe, Cu, Co, Ni and Mo from five different points in a solution purification process of a cobalt refinery [28]; analysis of Cu, As and S in copper electrolyte purification solutions [29]; control of a solvent extraction process for La and Nd [30, 31]; continuous monitoring of catalyst elements (Mn, Co and Br) in terephthalic acid process solutions [32]; and measurement of various elements (particularly sulfur) in petroleum product and refinery streams [33, 34].

The latter application area has seen a great deal of effort and development expended over the past few years in view of the increasing global concern regarding the impact of the burning of fossil fuels on the environment. Sulfur concentration is one of the critical components responsible for damaging emissions to the environment, since it is a major contributor to the formation of acid rain. Oil companies are currently making enormous efforts to modify refinery processes to produce a range of low sulfur products (particularly road transport fuels), ahead of ever more stringent government specification limits. In order to understand and control the processes involved, on-stream X-ray analysis systems can be used to measure the sulfur concentrations in various process streams including the crude oil feed and the various streams used in the blending of gasoline and diesel road transport fuels. Sulfur concentrations of crude oil feedstocks are generally very high (percent concentrations), so these can be measured using X-ray transmission systems capable of withstanding pressures up to 800 psi and temperatures of 100 °C. For the streams used in the production of road transport fuels however, measurements must be made at ever decreasing levels. For example, within the European Union, the current specifications (2000), for sulfur in road transport fuels are maximum 350 mg kg^{-1} for diesel and 150 mg kg^{-1} for gasoline. The proposed limits for 2005 are currently maximum 50 mg kg^{-1} sulfur for both gasoline and diesel. Although on-line XRF systems are available which claim to be able to achieve precisions of better than 1 mg kg^{-1} standard deviation at 20 mg kg^{-1} sulfur concentration, in reality, the reproducibility is likely to be much higher than this once all sources of error are included (e.g. calibration errors, drift etc.). In a recent Europe-wide round-robin including over 30 laboratories for example, reproducibilities for sulfur measurements at the 50 mg kg^{-1} level were found to be 17 mg kg^{-1} and 24 mg kg^{-1} for diesel and gasoline respectively, even using the latest generation of laboratory based EDXRF spectrometers specifically optimised for sulfur determination. The EDXRF technique was not considered suitable for use at sulfur concentrations below 50 mg kg^{-1} [15]. Thus it seems likely that, as sulfur concentration specifications continue to reduce, on-stream XRF analysers may be replaced with techniques which are inherently more suited to low level measurements (e.g. UV fluorescence) for this application area.

20.2.1.2 Trace Analysis and Corrosion Monitoring

With conventional on-line XRF systems such as those described above, limits of detection are restricted to, at best, ppm concentrations and often much higher levels. However, many applications, such as corrosion monitoring in pure water streams, require that metals are measured at much lower concentrations (down to ppb levels). These metal contaminants can be present either in the form of particulate matter or as fully dissolved components of the stream. Generally, measurement of metals at these levels requires the use of techniques such as ICPAES, AAS or, for very low concentrations, ICP-MS. However, these techniques are not well suited to on-line applications and in the case of particulate matter require that the samples are digested with acids prior to analysis. This therefore restricts the monitoring process to time consuming and labour intensive laboratory based analyses, resulting in a relatively poor sampling frequency. An elegant solution to this problem has been described by Connolly and Walker [35]. They have developed an on-line XRF system specifically designed for measurement of corrosion metals and other trace elements down to sub-ppb levels in water streams (although the approach could equally be applied to other flowing process streams). The system, which is now available commercially (see Appendix), utilises one or two flow cells equipped with holders capable of holding either conventional membrane filters for particulate analyses or ion exchange filters for dissolved metals. With a two-cell system, the stream can be passed first through the membrane filter for particulate removal/analysis and then through the cell equipped with an ion exchange filter for dissolved metal analysis. Metals on the filters are continuously monitored using an XRF system equipped with either an X-ray tube or radioisotope source and a sealed gas proportional detector or solid state Si(Li) detector. The systems also incorporate a Coriolis based flow sensor and software to permit the total mass of analyte metals on the filter to be plotted as a function of total flow through the cell(s). The gradient of this plot effectively gives the concentration of the metal of interest in the stream.

The detection limits for concentration of elements in the stream are a function of flow rate through the cell and sampling time. However as an example, the absolute detection limit in terms of the smallest amount of analyte on the filter which can be detected is of the order of 5 µg. Thus for a flow rate of 0.4 L min^{-1} and a sampling time of 20 min, the limit of detection in terms of concentration in the solution stream is 0.6 ppb. Even lower limits could be achieved with higher flow rates and/or sampling time, although obviously the latter would result in a poorer sampling frequency. The systems can cope with sample flow rates of between 0.2 and 0.4 L min^{-1} with temperatures in the range 5 to 32 °C and pressures up to 35 psi. Since the filter samples are essentially thin films (from an XRF calibration point of view), calibrations are linear over the normal operation range and can be accomplished using simple thin film standards for particulates analyses or can be easily prepared by passing laboratory prepared solution standards through the ion exchange filters. Care should be taken with the latter approach, however, to ensure that flow rates used are compatible with the kinetics of the absorption process. A new standard test method for on-line measurement of low level particulate and dissolved metals in water by X-ray fluorescence based on the above approach has recently been approved [36].

20.2.1.3 Analysis of Slurries and Powders

Flow cells for on-stream XRF analysis of slurries are generally similar to those for liquid streams (Section 20.2.1.1), although the abrasive nature of some slurries can place additional constraints on the window material and more frequent replacement of the windows may be required. Furthermore, since in general, solid materials absorb X-rays more strongly than liquid streams, detection limits are usually not as good for solid materials as for liquid streams and are rarely better than 50 to 200 ppm. Also, to avoid settling of particles during the measurement process, flow rates for slurry samples must be maintained relatively high (typically 20 to 45 L min^{-1}).

For analysis of powder samples, a variety of approaches have been adopted. Fig. 20.7 shows some typical examples. The gravity sampler is generally similar to the flow cells used for liquid streams but can include a vibrating sample packing device for reproducible packing density. A novel variation of this approach which has been developed for on-stream cement analysis utilises a highly polished metal closure over the viewing port which retains the sample during the packing process, producing a smooth flat surface which can then be analysed directly during the measurement phase without requiring use of a window between the sample and the measuring head [2]. Other approaches which do not require use of a window material include the reciprocating arm sampler and the rotating disc powder sampler (Fig. 20.7). In the former approach, a moving arm extends into the product chute and collects a powder sample which is automatically compacted into a pellet and then moved under the measuring head for analysis. With the rotating disc sampler, the measuring head is placed over a turntable that rotates into the sample chute to collect the sample. The latter approach can allow larger particles to be analysed but is less likely to give precise and accurate results. This results from one of the main problems associated with any analysis of powder or slurry samples using XRF: particle size effects. These effects can occur when the penetration depth of the X-rays is similar to or smaller than the particle size of the sample to be measured. In this case the XRF intensity observed becomes a function of the size and composition of the individual particles rather than of the concentration of the analyte element in the sample as a whole. In these cases, the XRF intensity observed will even depend on the chemical form of the analyte as well as its concentration, even after application of interelement corrections (this effect is often referred to as the "heterogeneity" or "mineralogical" effect [37]). This effect is particularly severe in cases where

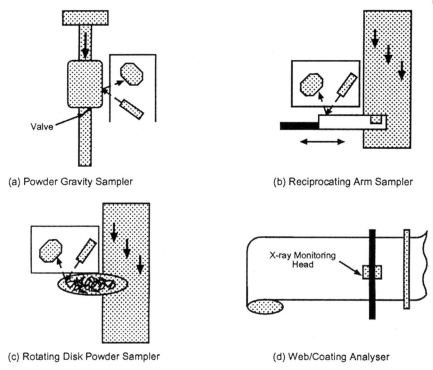

Fig. 20.7 Configurations for on-stream XRF analysis of solids.

strongly absorbing materials are mixed with a light matrix. Figure 20.8 shows how the XRF intensity would vary as a function of particle size for iron in petroleum coke. Curves are shown for iron present in metallic form (e.g. steel particles) and if it were present as the oxide (e.g. rust particles). It is clear that for particle sizes even as small as a few microns, completely different results would be obtained depending on which form of iron particles were present in the sample, even for exactly the same overall iron concentration. Since penetration depth is a function of the energy of the fluorescent X-rays (Tab. 20.2), this problem is generally worse for light element measurements.

Although one must be aware of the potential problems posed by particle size effects in slurry and powder analysis, in many cases, the range of particle size and particle composition for a process stream is sufficiently constant that reliable measurements can be made using on-line XRF systems. Successful applications which have been reported include: an on-stream XRF measuring system for ore slurry analysis [38]; a system for direct XRF analysis of pulverised coal streams [39]; and on-stream analysis of cement using XRF [2, 40]. The latter is available as a commercial on-stream cement analyser which it is claimed can measure lime saturation factors (a function of Ca, Si, Al and Fe concentrations) with a relative standard deviation of better than 1% in 100 s measurement time. Obtaining such results with a sampling frequency of over 10 measurements per hour can allow effi-

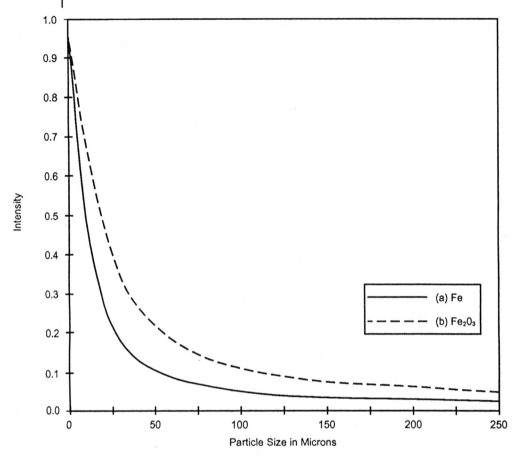

Fig. 20.8 Plot showing variation in XRF intensity for iron in coke as a function of particle size: (a) Iron present as metallic iron particles; (b) iron present as iron oxide (rust) particles.

cient control of the raw mix chemical composition, permitting optimum operation of the kiln and ensuring the quality of the finished product.

20.2.1.4 Direct Analysis

Commercial systems are available which permit direct analysis of solid material on a conveyor belt (see Appendix). Typically the XRF monitoring system comprises a low power X-ray tube and sealed gas proportional detector monitoring head which is mounted about 100 to 300 mm above the conveyor belt, in some cases using shock absorbing mountings. A material height limiter/regulator is often installed upstream of the measuring head to control the material surface to probe distance. Systems of this type are commonly employed in the mineral industry and can cope with tonnage up to 2000 t h^{-1} using a conveyor system with width up to 1500 mm.

A system of this type has been described for automatic sorting of waste glass [41]. In the latter system, the XRF spectrum of a glass fragment is measured using the monitoring head mounted above the conveyor belt and the spectrum is compared with a library of stored spectra for known types of glass. Once identified, the system automatically diverts the glass fragment to a container of like composition using rams. In this way, it is claimed that a substantially continuous stream of glass fragments on the conveyor belt can be sorted according to glass type (composition).

A similar approach can also be used for web/coating applications. The monitoring can be carried out either on a discrete strip on the web, or by using a computer controlled scanning pattern over the width of the web This approach can be used either to determine composition or sample/coating thickness. Typical applications are: silicone on release paper; coatings on steel or other metals; polycarbonate coatings on fibers/textiles, silver emulsion on photo film and so forth. For relatively straightforward applications, simpler systems based on X-ray transmission can be used to measure sample/coating thickness or density [42]. Such a system could typically comprise a radioisotope source and simple, low cost NaI detector mounted at the opposite side of the sample. The intensity (I) of the transmitted X-rays is given by:

$$I = I_0 e^{-\mu \rho l}$$

where, I_0 is the incident X-ray intensity, μ is the mass attenuation coefficient, ρ is the sample density, and l is the thickness of the sample along the X-ray path. If the mass attenuation coefficient is known (or effectively determined from calibration samples), then the measurement of I effectively gives a measurement of the areal density (ρl). Thus, if the density of the sample is known (or constant), the measurement gives a direct measure of the thickness of the sample. Conversely, if the thickness is known (or constant), the measurement of transmitted intensity gives a measure of the sample density. The range of thickness which can be measured depends on the absorption properties of the material being measured and the energy of the X-ray source, but is typically in the micron range. In some cases, the method can be extended for use with layer materials and when a conventional XRF geometry probe is used, attenuation of the XRF line from an element in the substrate can be used to measure the thickness of a coating above the substrate.

A particularly novel application of on-line XRF using direct analysis has been described by Creasy [43]. The author described development of a wavelength dispersive X-ray spectrometer, specifically designed to analyse the composition of molten metal in an electron beam furnace (effectively similar to an extremely large electron microscope!). Specially designed twin wavelength monitor systems were used to permit simultaneous measurements of the analyte element and a proportion of the Bremsstrahlung which was used to compensate for variation of beam current in the furnace. The monitor systems were protected from the furnace environment by water cooling and through the use of thin foil windows. A flow of argon was maintained from the spectrometer to the furnace to help reduce condensation on the windows. It was found that the system was stable for an entire melt (which could last over 10 h) and that in the course of the actual melting, the system could be used for control purposes, providing instant feedback to the operators. Attempts at using EDXRF for this application were unsuccessful since the very high electron flux from the furnace destroyed the Si(Li) detectors.

20.2.2
Atomic Emission Spectrometry

20.2.2.1 Plasma Spectrometry

Plasma spectrometry, particularly ICPAES and ICP-MS, is widely used for laboratory based elemental analysis of process samples in view of its inherent sensitivity, relative ease of use and rapid multi-element capability (see Section 20.1). However, these techniques are inherently unsuitable for on-stream applications since the sample introduction systems are not robust and are prone to blockage and drift problems caused by changes in sample physical properties, temperature or degradation of pump tubing. Furthermore, conventional plasma based systems (ICP or microwave plasmas) cannot cope with particulates larger than a few microns and can rarely tolerate significant flow rates of volatile solvents or gaseous samples. The systems also generally require large quantities of expensive gases (Ar or He), are not well suited to operation in a hostile environment (e.g. large temperature fluctuations or vibration) and due to the very high temperature of the source (up to 10,000 K), are difficult to make inherently safe for use in potentially explosive atmospheres. Nevertheless, some authors have reported successful application of specially modified plasma based systems to on-stream analysis for particular applications.

Liquid streams

Frederici et al. [44] have modified a charge injection device (CID) based ICPAES spectrometer for on-line continuous process monitoring of aqueous industrial waste streams. As discussed in Section 20.2.2.1, the weak point of ICPAES instruments in terms of on-stream analysis is usually the sample introduction system. The authors modified the sample introduction system to facilitate low maintenance and continuous unattended operation. The normal peristaltic pump was replaced by dual metering pumps to eliminate the need for frequent tube replacement and the autosampler was replaced by a valving system to permit switching between the waste streams to be monitored and rinse/calibration and quality control check streams. A separate stream containing internal standards was mixed with the sample/calibration streams on-line resulting in an approximate 20 % dilution. Internal standards were carefully chosen to compensate for changes in the composition of the waste streams (particularly pH) and multiple wavelengths were monitored for each analyte to check for any potential spectral interference. The normal Meinhard nebuliser was replaced by a Burgener high solids nebuliser to improve the tolerance of the system to the presence of particulates in the waste streams and to reduce the incidence of blockages, (although large particles still had to be removed from the streams using an up-stream filter). Axial viewing of the plasma was selected to provide better limits of detection, which were in the range 0.0008 to 0.031 mg L^{-1} for the 15 elements measured. The observed limits of detection were typically about three orders of magnitude lower than the prescribed action levels for this application. The system was found to perform well for most elements although some memory problems were experienced with mercury. Results obtained were compared with those obtained using the traditional laboratory approach (including sample digestion) and were found to give good agreement for a sample resembling tap water. However, the system is unlikely to perform well for more complex waste streams, which normally contain a significant proportion of particulate matter and which would definitely require digestion prior to analysis.

Gaseous effluents
An ICPAES system has also been designed for analysis of high temperature and high pressure fossil fuel process streams, such as those encountered in coal combustors and gasifiers [45]. Such streams can vary enormously in composition and particle loading on a timescale of seconds and any sampling systems must maintain the sample stream at high temperature (typically 650 °C) and flow rates above 2 L min^{-1} in order to prevent condensation or loss of particulate matter in the sample transport system. Conventional ICP excitation sources, typically operating with argon as plasma gas and with rf power between 1 and 1.5 kW will not tolerate such high flow rates of gases containing combustion products and particulates. The authors therefore used an argon/helium plasma operating at very high power (5 kW), together with a specially designed torch with ceramic injector and heated sample transfer line to cope with the high temperature gaseous process stream. Elements in the sample stream were measured using a battery of 0.1 m monochromators, viewing the plasma through a bundle of fiber optics. Calibration was affected using an ultrasonic nebuliser equipped with a heater and condenser for aerosol desolvation. Once calibrated, the system was claimed to be stable for 8 h or longer.

One of the problems with the above approach is that due to the very high power required, a very large, bulky rf generator must be used. Trassy and Diemiaszonek [46] have modified a conventional 1.6 kW argon ICP system for analysis of the gaseous effluents from incineration and other industrial plants. Injection of air into the central channel of an ICP causes the channel to increase in width from its normal value of 2–3 mm to 8–10 mm. This results in a corresponding reduction in the depth of the torroidal region of the plasma, resulting in a reduced rf coupling efficiency. The authors found that this could be compensated using a larger diameter torch (23 mm cf. 19 mm for a conventional torch). Using a 1.5 mm diameter alumina injector, the authors were able to maintain a stable plasma at 1.6 kW rf power with flow rates of up to 0.7 L min^{-1} of air introduced to the plasma via the injector tube. This flow rate is however, still much too low to allow isokinetic sampling of a gaseous effluent stream and so a two-stage sampling device was used (Fig. 20.9). Sample gas from the flue was sampled at high flow rate using a diaphragm pump (protected from particulates using a filter). This gas stream was sub-sampled (either before or after the particulate filter) for injection into the ICP using a two-headed peristaltic pump. The two pump heads were exactly out of phase to reduce modulation of the flow (reduced from 80% with single pump head to 10% for the double head arrangement as described.). Thermal conductivity of the plasma was found to vary substantially with water vapor loading and so the gaseous effluents and aerosol produced from the ultrasonic nebuliser used for calibration were passed through a desolvation system. Limits of detection obtained by the authors for various metals are shown in Tab. 20.3. These are typically well below the threshold limit values for flue gases of industrial plants which are typically in the 0.050 to 1.0 mg m^{-3} range [47]. Precision was found to be around 2 to 4% RSD using a 2 s integration time, although some memory problems were experienced with mercury.

The main problems associated with systems such as those described above which require sample streams to be transported to the plasma is that unless the stream is maintained at a high temperature, condensation can occur in the pipework and/or injector tube; and unless a very high sample flow rate is maintained

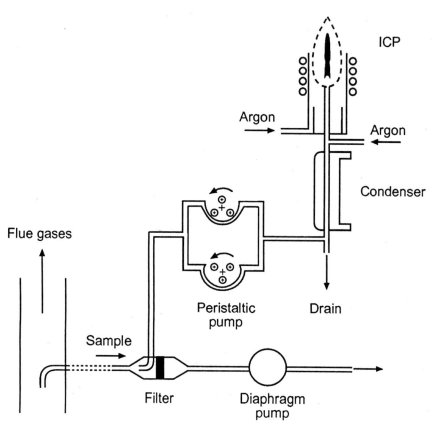

Fig. 20.9 Two-stage sampling system used by Trassy and Diemiaszonek [46] for on-line analysis of gaseous effluents using ICPAES.

Table 20.3 Detection limits[a] for on-line analysis of gaseous effluents by ICPAES.

Element	Line (nm)	Detection Limit ($\mu g/m^{-3}$)
Al	308.215	0.5
As	193.696	25
Ca	317.933	0.3
Cd	228.502	0.5
Co	228.616	0.5
Cr	267.716	0.2
Cu	324.754	0.3
Fe	259.940	0.3
Hg	253.652	5
K	766.491	2
Li	670.784	1
Mg	279.079	3
Mn	257.610	0.05
Na	589.592	5
Ni	231.604	3
Pb	220.351	6
Ti	334.941	0.1
Zn	213.856	1

[a] Limits of detection quoted are based on 3 standard deviations in air, according to Trassy and Diemiaszonek [46].

(> 2 L min^{-1}), particulate matter is unlikely to be transported representatively into the plasma. Both of these problems could potentially be eliminated if the plasma could be mounted directly in the off-gas stream itself. Woskov et al. [48] have described a continuous, real time, microwave plasma element sensor which may be suitable for this application. A schematic diagram of the device is shown in Fig. 20.10. The device comprises a tapered, shorted, microwave waveguide made of refractory material which is mounted in the hot flue gas stream. A passageway in the waveguide allows flue gas to flow freely through the waveguide and a microwave plasma initiated in this region of the waveguide excites contaminant atoms present in the flue gas. Atomic emission from the contaminant elements excited in this way is measured using a spectrometer viewing the plasma through a robust fiber optic such as unclad quartz which is capable of withstanding long term exposure to the hot environment of the flue gas. Calibration of the device may be affected by ablating an alloy plate mounted below the gas inlet to the waveguide, using a laser system. In addition to eliminating the problems of condensation and particulate drop-out associated with sample transport systems such as those described above, the in-stream microwave sensor should also eliminate the memory problems often encountered with mercury measurements.

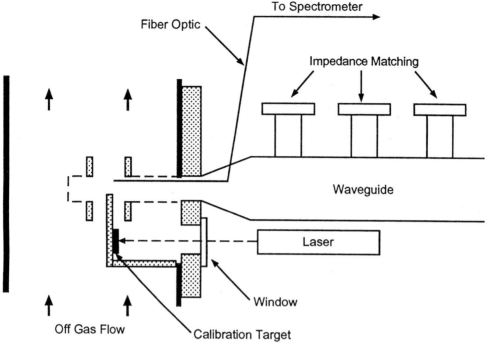

Fig. 20.10 Continuous real time microwave plasma element sensor for flue gas monitoring [48].

Reactive gases

In the semiconductor industry, there is an increasing requirement to measure elemental contamination in reactive gases down to extremely low (sub ppb) concentrations since these can have a major influence on the quality of the devices produced. In particular, devices produced using chemical vapor deposition and reactive ion etching are particularly sensitive to impurities in the process gases used. The elemental contaminants can be present in the gases as particulates or vapors or a combination of both forms [49]. A direct analytical system is therefore preferred which will allow determination of the total contaminant concentrations independent of their chemical and physical forms. Calibration of direct plasma based systems for analysis of gases is more difficult than for solutions since the former generally involves carefully adding a controlled flow rate of a known concentration of volatile analyte contained in a carrier stream into a carefully controlled flow of the gas stream to be analysed. The volatile analyte stream can be obtained using a gaseous form of the analyte (if available), using diffusion tubes, or by passing the carrier gas stream through a liquid analyte for which the vapor pressure is accurately known. All of these approaches have been used, but unfortunately, for many analytes it is very difficult to obtain compounds suitable for use in this way to calibrate the system. A further complication is that for corrosive gases (e.g. HCl), any valves used to control the gas flow rates are likely to be a significant source of contamination. A solution to this problem which was proposed by

Schram [50], used a bypass-backflush balancing system in which peristaltic pumps were used instead of valves to control the gas flow rates for introduction of HCl gas into an ICPAES instrument. The author obtained a limit of detection of around 2 ng g^{-1} for Fe in gaseous HCl. A similar system has also been used for analysis of gaseous HCl using a microwave induced plasma (MIP) [51]. A typical example of the bypass-backflush system used by these authors to introduce the HCl gas to the MIP and to calibrate the system by standard additions with iron pentacarbonyl is shown in Fig. 20.11. This system permits continuous sampling of a gaseous HCl stream and achieves a limit of detection of 0.25 µg L^{-1} of Fe in argon with a reproducibility of 6 %. The use of peristaltic pumps permitted gas flow rates to be changed without moving any valves. This was important since it was found that the iron

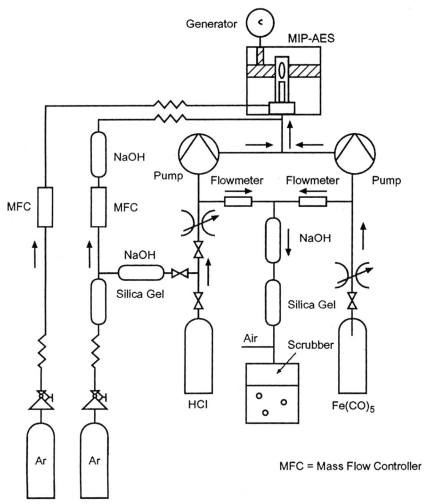

Fig. 20.11 Bypass-backflush balancing system For direct sampling of reactive gas (HCl) for plasma emission spectrometry [51].

emission intensity dropped for about 10 min before reaching a steady state when the hydrogen chloride cylinder was opened (due to corrosion of the valve). A similar approach has been used together with a modified electrothermal atomisation atomic absorption spectrometer (ETA-AAS) for determination of Fe, Ni and Cu in HCl, Cl_2 and BCl_3 [49]. The ETA-AAS approach afforded limits of detection which were similar to those obtained using plasma emission spectrometry but had the advantage that solution standards could be used to perform the standard additions calibrations.

In order to minimise the amounts of toxic gases used, Barnes and co-workers have advocated the use of a sealed ICP source for the analysis of arsine [52,53], silane [54], HCl [55], and Cl_2 [56]. However limits of detection obtained with this system were relatively high compared with the alternative approaches described above and long stabilisation times were often required in order to obtain stable plasma conditions. For example, under flowing conditions, the limit of detection for Sn in HCl was found to be around 300 ng g^{-1}, and the authors found that it was necessary to add chlorine gas to the plasma in equal volumes to the HCl to prevent deposition of tin. The limit of detection reported by the authors for arsenic in silane was equally unimpressive (200 ppb v/v) and compares unfavourably with the limit of detection of 0.5 ppb (v/v) reported by Hutton et al. for direct analysis of silane using ICP-MS [57]. The latter authors also reported a detection limit of 0.65 ppb (v/v) for iodine in the same application with precisions between 2 and 5% RSD.

20.2.2.2 Laser Based Techniques

Conventional plasma spectrometry (e.g. ICPAES, ICP-MS) can also be used for direct analysis of solid samples by using a laser to ablate the surface of the sample to be analysed, and transporting the ablated material to the plasma using a flow of gas (e.g. argon) contained in a transfer tube. This approach has been successfully used for analysis of a wide variety of materials using laboratory based instruments. However, the approach is not well suited to process analysis (particularly on-line analysis), since in addition to the problems associated with the plasma source itself (high cost, high temperature, requirement for large amounts of expensive gases, e.g. Ar or He), the sample transport system also introduces a number of additional problems associated with sample deposition/condensation, memory effects and non-representative sample transport.

A more suitable approach for on-line analysis is to view atomic emission directly from the small plasma formed above the sample surface when the sample is ablated using the laser. This approach, which is shown schematically in Fig. 20.12 has been applied to analyses of a variety of materials (e.g. [58–74]) and is known under a variety of names such as laser induced plasma spectrometry (LIPS), laser induced breakdown spectrometry (LIBS) and laser spark emission spectrometry (LASS). The laser used is typically a Nd:YAG laser operating either at its fundamental, doubled or quadrupled frequency, although excimer and CO_2 lasers have also been used. The important parameter of the laser is that it must be capable of producing focused pulses with power density of at least 10^9 W cm^{-2}. Spot sizes for the focused laser are usually between 0.1 and 1 mm diameter and a common configuration may use a Nd:YAG laser capable of delivering 250 mJ in 7 ns at a rate of 10 Hz.. The spectrometer used often incorporates a diode array or CCD detector so that multiple wavelengths can be monitored simultaneously and so that full peak shapes can be observed to check for broadening or self-absorption effects which are more common in LIPS than in cases where the plasma is independent of the ablation event (e.g. LA-ICPAES). This latter problem is particularly prevalent during the early stages of plasma formation when temperature and electron density are high and recombination events give rise to broad continua emission (high backgrounds). For this reason, the LIPS system must incorporate a rapid time-gated detection system as illustrated in Fig. 20.12. A typical experiment may involve a delay time of 0.5 µs to 1 ms and an integration time of 1 to 50 µs depending on the laser conditions, sample material and concentration of analyte element to be measured. Figure 20.13 shows how the signal to noise ratio changes with delay time for a typical ablation of a glass sample (taken from [69]).

With any laser ablation process, the interaction of the laser with the sample, and hence the amount of material ablated can vary significantly, depending on the sample matrix, color and condition of the surface and the thickness of any coating. This normally necessitates use of an internal standard approach to the analysis, where the intensity of the analyte element is ratioed to the line from another element in the sample whose concentration is known. One potential problem with this approach however, is that one element may be preferentially ablated in preference

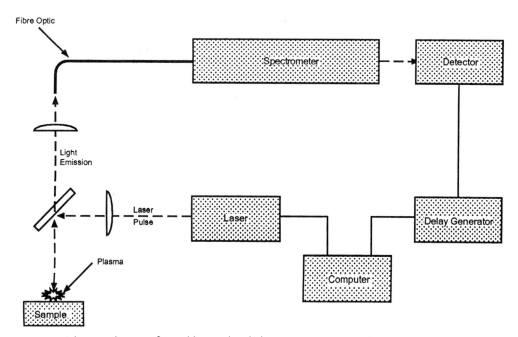

Fig. 20.12 Schematic diagram of typical laser induced plasma spectrometry (LIPS) system.

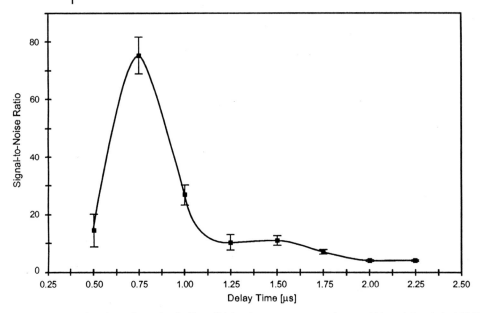

Fig. 20.13 Example of effect of delay time on signal-to-noise ratio for laser induced plasma spectroscopic (LIPS) analysis. (Example shows Si: 288.158 nm in molten glass with 0.1 microsecond gate width and Q-switched Nd:YAG laser with 75 mJ laser pulse energy. Reprinted from [69] with permission from Elsevier Science.

to the other (e.g. it may be present in a more volatile form in the sample matrix). In some cases, this can necessitate that the entire sample is ablated. For example, in the LIPS analysis of coal ash deposits collected on copper substrates, Ottesen [70] found that consistent results could only be obtained after several successive laser shots had ablated most of the outer material and copper lines from the underlying substrate began to appear in the emission spectra (indicating total ablation of the remaining thin coating deposit).

A particular problem associated with LIPS which does not occur in systems with separate ablation and excitation plasmas (e.g. LA-ICPAES) is that not only can the amount of sample ablated vary from laser shot-to shot, but the plasma temperature itself can also vary [69]. Figure 20.14 (taken from [69]) shows how the plasma electron temperature can vary from shot-to shot for a homogeneous glass sample. In such cases, unless both analyte lines and their associated internal standard lines originate from excited electronic states of similar energies, the internal standard approach will not work. A potential solution to this problem has been described by Panne et al [69] in which results were normalised using Saha–Boltzman equilibrium relationships calculated using electronic excitation temperatures and densities obtained from measurements of several atom and ion line ratios. This approach was found to reduce errors associated with plasma temperature variations in the LIPS analysis of major elements in glass samples, resulting in improved measurement precision and highly linear calibrations for Si, Ca and Al in more than 20 different glass samples.

Despite the problems discussed above, LIPS has been successfully applied in a number of process analysis applications. In some cases, the LIPS system has been used off-line as in the application described by Ottesen in which air cooled metallic substrates were used to collect fly ash deposits from a pulverised coal combustion for subsequent analysis off-line [70]. Calibration standards were prepared by spraying aqueous solutions onto heated substrates using an air brush and the method was found to work well provided that the deposits were sufficiently thin to permit complete ablation. Other workers have proposed on-line LIPS systems for process control. An example is the apparatus proposed by Sabsabi [71] for in situ analysis of pre-selected components of homogeneous solid compositions. In particular, the author proposed that the system could be used for measurement of the concentration of active ingredients (e.g. drugs) in pharmaceutical products such as tablets, by monitoring an element present in the active component (e.g. P, Na or S). In the example quoted, phosphorus was measured using a carbon line as internal standard.

One of the advantages of a laser based technique is that it affords the possibility of making spatially resolved measurements with a spatial resolution typically around 0.1 mm. In addition, since each laser pulse typically removes about 1 to 2 µm, repeated ablation/analysis at the same spot allows information to be obtained regard-

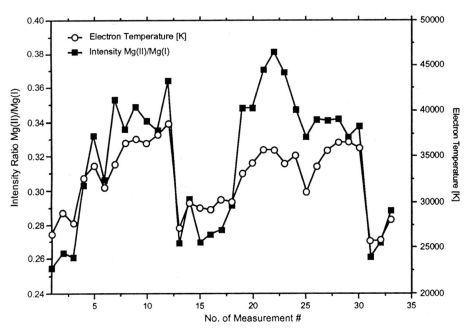

Fig. 20.14 Variation in Mg(II)/Mg(I) line intensity ratio and electronic excitation temperature for LIPS analysis of homogeneous glass sample NIST SRM 1830. Reprinted from [69] with permission from Elsevier Science.

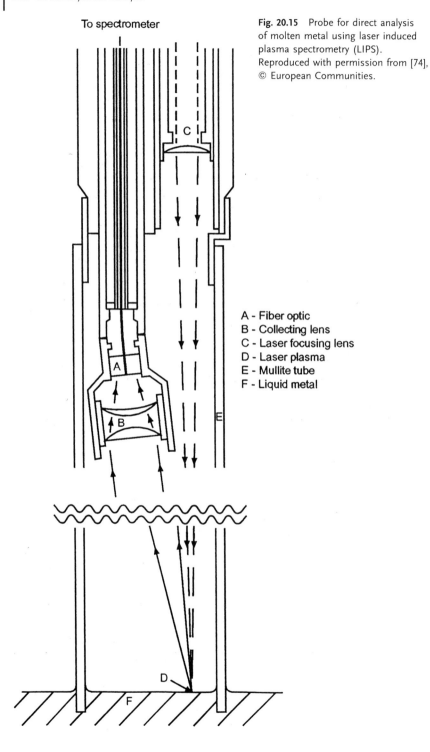

Fig. 20.15 Probe for direct analysis of molten metal using laser induced plasma spectrometry (LIPS). Reproduced with permission from [74], © European Communities.

A - Fiber optic
B - Collecting lens
C - Laser focusing lens
D - Laser plasma
E - Mullite tube
F - Liquid metal

ing concentration distributions as a function of depth from the surface (i.e. elemental depth profiles). Hakkanen and Korppi-Tommola [72] have demonstrated that these features of LIPS can be useful for measurement of the weight and composition of paper coatings. Lines originating from the major elements present in the coating pigments (Al, Si and Ca) were used to measure coating weight with corrections for plasma temperature accomplished using Mg ion-atom line ratios. In addition, the ratios of carbon line intensity to that of carbon plus silicon provided a measure of organic binder content. Using the latter approach, a good correlation with known coating composition was obtained and the authors were able to demonstrate an enhancement in organic binder content in the top few microns of the coatings studied. One problem identified by the authors however, was that the results obtained for kaolin based coatings were influenced by the particle size of the filler. The method could therefore only be applied in cases where the particle size distribution of the filler materials was known and constant.

In addition to its use with solid samples, LIPS can also be applied to analysis of liquids. Panne et al [73] have used the technique for in situ, on-line process analysis of major constituents in glass melts during a vitrification process for fly and bottom ashes resulting from waste incineration. The system used employed a vertical, single axis observation geometry in which the laser was directed through a pierced 45° mirror then focused onto the sample surface using a single lens. Light emission from the plasma was collimated using the same lens and directed into a fiber optic bundle using the mirror. The use of the fiber optic bundle permitted the spectrometer to be located remotely from the hot melt. The single axis geometry permitted the glass melt to be observed through a small (30 mm) opening in the oven, thereby minimising thermal losses during the melting process. Changes in plasma temperature and electron density were corrected using Saha–Boltzman equilibrium relationships calculated using measurements of several atom and ion line ratios. Good agreement was obtained between LIPS results and those obtained using conventional analytical techniques for concentration ratios of Si, Al and Ca.

As with all techniques for on-line production control analysis, very often as much effort must be expended to make the equipment robust enough to operate in a hostile environment as in the basics of the analytical technique itself. Jowitt and Whiteside [74] have described development of such a system for laser analysis of liquid steels. The laser, spectrometer and associated accessories were built into a specially designed mobile unit to allow easy access to and from the furnaces to be studied. The laser was mounted on a lift mast which allowed the laser unit to be placed above the furnace. Special measures had to be taken to protect the laser and optical components from heat and to reduce the effects of vibration on the optical alignment. Analysis was accomplished using a probe consisting of a single refractory tube which entered the molten metal (Fig. 20.15). A flow of argon was used to protect the lenses from dust and fumes and this flowed down through the probe and out through the molten metal, holding the surface of the metal at the focus of the laser focusing lens. The lens unit had to be mounted 1.8 m away from the molten metal surface to prevent optical effects due to lens heating and damage to the fiber optic caused by the intensity of the infrared radiation focused on its tip. The system was successfully applied for monitoring silicon (ratioed to iron) in a blast furnace. However, use of a fiber optic restricts the system to measurements above 200 nm and therefore prohibits measurement of important elements in the steel making process such as carbon, sulfur and phosphorus (whose main emission lines lie in this spectral region). The authors have carried out some preliminary measurements on solid samples using a modified probe incorporating a direct spectral path in place of the fiber optic but this has yet to be incorporated in a system for direct, on-line analysis of molten metal.

20.3 Conclusions

As discussed above, atomic spectrometry plays a key role in the development, optimisation and efficient operation of most industrial processes. At the present time most of these applications are carried out using laboratory based analyses. In some cases the benefits of on-line analysis can be substantial, affording a much higher sampling frequency, permitting tighter control of the process. Some progress has

Table 20.4 X-ray fluorescence (XRF).

Characterized parameter:		Surface specificity:		
Energy/wavelength of X-ray emission		Information depth: Microns to centimeters		Detectability: ppm to percent
Type of information:		**Resolution:**		
Elemental composition		Depth:	Lateral: Not normally used for spatially resolved analysis but can be down to sub mm	Other:
Measurement environment:	**Difficulties**	**Time needed for analysis:**		
X-ray tube or radioisotope excitation Air, vacuum, nitrogen or helium optical path	Matrix/particle size effects	Prep. 0.5 to 5 min	Measurement 1 to 5 min	Evaluation 1 to 5 min
Equipment:		**Cost [ECU]:**		**No. of facilities:**
Energy Dispersive (ED) or Wavelength Dispersive (WD) XRF spectrometer		15,000–150,000		Very common
Type of laboratory:	**User skill needed:**	**Sample:**		
Small	Unskilled	Form type Solid, liquid or slurry		Size: 1 cm to several m
Techniques yielding similar information				
ICPAES, AAS, LIPS				

Table 20.5 Plasma emission spectrometry (ICPAES or MIPAES).

Characterized parameter:		Surface specificity		
Atomic emission lines		Information depth:		Detectability: ppb to percent
Type of information:		**Resolution**		
Elemental composition		Depth:	Lateral:	Other:
		Not normally used for spatially resolved analysis		
Measurement environment:	**Difficulties:**	**Time needed for analysis:**		
High temperature plasma	Spectral interferences Sample introduction	Preparation Up to 1 h	Measurement 1 to 5 min	Evaluation 1 to 5 min
Equipment:		**Cost [ECU]:**		**No. of facilities:**
Inductively coupled plasma or microwave induced plasma, High resolution uv/vis spectrometer		30,000–100,000		Very common
Type of laboratory:	**User skill needed:**	**Sample:**		
Small medium	Moderate	Form type Liquid or gas Solids after digestion		Size: ml
Techniques yielding similar information:				
ICPMS, AAS, LIPS, XRF				

been made in developing atomic spectrometric techniques for on-stream or at-line analysis, particularly using XRF, which is inherently well suited to these applications in view of its stability and robustness. However, all of the techniques employed can suffer from severe limitations which restrict their use to specific niche applications. Furthermore, a high cost and substantial development time is often associated with making the systems robust enough for operation in hostile and potentially hazardous environments, and in many cases, this is not completely successful. Thus, with currently available technology, although on-line elemental analysis can provide substantial benefits for a few applications, there seems little prospect of this leading to a substantially reduced requirement for laboratory based analysis in the near future.

Tables 20.4 to 20.8 summarize the use of atomic spectrometry in process control.

20.3 Conclusions

Table 20.6 Laser induced plasma spectrometry (LIPS (or LIBS)).

Characterized parameter:	Surface specificity:	
Atomic emission lines	Information depth: 1 to 2 microns	Detectability: ppb to percent

Type of information:	Resolution:		
Elemental composition	Depth: 1 to 2 microns	Lateral: 0.1–1 mm	Other:

Measurement environment:	Difficulties	Time needed for analysis:		
Laser induced plasma	Spectral interferences Changes in plasma temp.	Prep. None	Measurement 1 to 5 min	Evaluation 1 to 5 min

Equipment:	Cost [ECU]:	No. of facilities:
Laser, high resolution uv/vis spectrometer, time gated detection system	50,000–100,000	Limited

Type of laboratory:	User skill needed:	Sample:	
Large	Moderate to high	Form type Solid	Size: >1 mm

Techniques yielding similar information:

Laser ablation ICPMS or ICPAES, XRF

Table 20.7 Inductively coupled plasma mass spectrometry (ICPMS).

Characterized parameter:	Surface specificity:	
Atomic mass of ions	Information depth:	Detectability: ppt to percent

Type of information:	Resolution:
Elemental composition	Depth: Lateral: Other: Not normally used for spatially resolved analysis but can be down to 10 microns if used with laser ablation

Measurement environment:	Difficulties:	Time needed for analysis:		
High temperature plasma	Spectral interferences Sample introduction	Prep. Up to 1 h	Measurement 1 to 5 min	Evaluation 1 to 5 min

Equipment:	Cost [ECU]:	No. of facilities:
Inductively Coupled Plasma, quadrupole or sector mass spectrometer	100,000–250,000	Fairly common

Type of laboratory:	User skill needed:	Sample:	
Medium Large	Moderate	Form type Liquid or gas Solid (with laser ablation) or after digestion)	Size: ml

Techniques yielding similar information:

ICPAES, AAS, LIPS, XRF

Table 20.8 Atomic absorption spectrometry (AAS).

Characterized parameter:	Surface specificity:	
Absorption of atomic lines	Information depth:	Detectability: ppb to percent
Type of information:	**Resolution:**	
Elemental composition	Depth: Lateral: Other: Not normally used for spatially resolved analysis	
Measurement environment: Difficulties:	**Time needed for analysis:**	
Flame or graphite furnace — Matrix effects, Sample introduction	Prep. Up to 1 h	Measurement 1 to 5 min — Evaluation 1 min
Equipment:	**Cost [ECU]:**	**No. of facilities:**
Atomic absorption spectrometer	10,000–30,000	Very common
Type of laboratory: User skill needed:	**Sample:**	
Small — Low	Form type Liquid or gas Solids after digestion	Size: ml
Techniques yielding similar information:		
ICPAES, ICPMS, AAS, XRF		

Acknowledgements

The author gratefully acknowledges information provided by Hobré Instruments, Oxford Instruments and Spectro Analytical Instruments which greatly assisted in the production of the section of this chapter relating to on-line XRF analysis. The author would also like to thank Carole Hampton of the BP Information Centre, Sunbury for obtaining copies of most of the references quoted in the text.

Appendix: Suppliers of On-line XRF Equipment

General On-line XRF
Kevex Spectrace
1275 Hammerwood Ave.
Sunnyvale
CA 94089
USA
www.spectrace.com

Metorex International OY
Nihtisillankuja 5
PO Box 85
FIN-02631 ESPOO
Finland
www.metorex.fi/default.htm

Spectro Analytical Instruments
Boschstr. 10
47533 Kleve
Germany
www.spectro-ai.com

Trace Analysis/Corrosion Monitoring
Detora Analytical Inc.
PO Box 2747
Alliance
Ohio 44601-0747
USA
www.detora.com/default.htm

On-line Cement Analysis
Oxford Instruments
Wyndyk furlong
Abingdon Business Park
Abingdon
Oxon OX14 1UJ
UK
www.oxinst.com/analytical/

References

1 R. Duer, *World Cement*, **1999**, 63–65.
2 S. T. Pedersen, M. S. Finney, *World Cement*, **1998**, 29, 30–37.
3 J. Marshall, J. Carroll, J. S. Crighton, *J. Anal. Atom. Spectrom.*, **1990**, 5, 323R–360R.
4 J. Marshall, J. Carroll, J. S. Crighton, *J. Anal. Atom. Spectrom.*, **1991**, 6, 283R–321R.
5 J. Marshall, J. Carroll, J. S. Crighton et al., *J. Anal. Atom. Spectrom.*, **1992**, 7, 349R–388R.
6 J. Marshall, J. Carroll, J. S. Crighton et al., *J. Anal. Atom. Spectrom.*, **1993**, 8, 337R–375R.
7 J. Marshall, J. Carroll, J. S. Crighton, et al., *J. Anal. Atom. Spectrom.*, **1994**, 9, 319R–353R.
8 J. Marshall, J. Carroll, J. S. Crighton, *J. Anal. Atom. Spectrom.*, **1995**, 10, 359R–402R.
9 J. S. Crighton, J. Carroll, B. Fairman et al., *J. Anal. Atom. Spectrom.*, **1996**, 11, 461R–508R.
10 J. S. Crighton, B. Fairman, J. Haines et al., *J. Anal. Atom. Spectrom.*, **1997**, 12, 509R–542R.
11 B. Fairman, M. W. Hinds, S. M. Nelms et al., *J. Anal. Atom. Spectrom.*, **1998**, 13, 233R–267R.
12 B. Fairman, M. W. Hinds, S. M. Nelms et al., *J. Anal. Atom. Spectrom.*, **1999**, 14, 1937–1969.
13 Method number IP 437/98, Standard Methods for Analysis and Testing of Petroleum and Related Products, Institute of Petroleum, London, 2000.
14 A. Raith, R. C. Hutton, I. D. Abell et al., *J. Anal. Atom. Spectrom.*, **1995**, 10, 591–594.
15 *Sulphur Methods for EN228 and EN590 Fuel Specifications*, CEN/TC19 WG27 Report on Round Robin Exercise, May 2000.
16 *Methods for the Determination of Metals in Environmental Samples*, US Environmental Protection Agency, Cincinnati, 1991 (see also www.epa.gov/epahome/index/key.htm).
17 M. S. Cresser, J. Armstrong, J. Dean et al., *J. Anal. Atom. Spectrom.*, **1991**, 6, 1R–68R.
18 M. S. Cresser, J. Armstrong, J. Dean et al., *J. Anal. Atom. Spectrom.*, **1992**, 7, 1R–66R.
19 M. S. Cresser, J. Armstrong, J. Cook et al., *J. Anal. Atom. Spectrom.*, **1993**, 8, 1R–78R.
20 M. S. Cresser, J. Armstrong, J. Cook et al., *J. Anal. Atom. Spectrom.*, **1994**, 9, 25R–85R.
21 M. S. Cresser, J. Armstrong, J. Cook et al., *J. Anal. Atom. Spectrom.*, **1995**, 10, 9R–60R.
22 M. S. Cresser, L. M. Garden, J. Armstrong et al., *J. Anal. Atom. Spectrom.*, **1996**, 11, 19R–86R.
23 J. Dean, L. M. Garden, J. Armstrong et al., *J. Anal. Atom. Spectrom.*, **1997**, 12, 19R–87R.
24 J. R. Dean, O. Butler, A. Fisher et al., *J. Anal. Atom. Spectrom.*, **1998**, 13, 1R–56R.
25 M. R. Cave, O. Butler, J. M. Cook et al., *J. Anal. Atom. Spectrom.*, **1999**, 14, 279–352.
26 M. R. Cave, O. Butler, J. M. Cook et al., *J. Anal. Atom. Spectrom.*, **2000**, 15, 181–235.

27 G. E. Purdue, R. W. Williams, *X-ray Spectrom.*, **1985**, *14*, 102–108.

28 M. Hietela., D. J. Kalnicky, *Adv. X-ray Anal.*, **1989**, *32*, 49–57.

29 R. A. Davidson, E. B. Walker, C. R. Barrow et al., *Appl. Spectrosc.*, **1994**, *48*, 796–800.

30 M. Casarci, F. Bellisario, G. M. Gasparini et al., *Value Adding Solvent Extr: [Pap ISEC '96]*, **1996**, *2*, 1121–1126.

31 Li Wenli, G. P. Ascenzo, R. Curini et al., *Anal. Chem.*, **1998**, *362*, 253–260.

32 Kevex Spectrace Application Note (http://www.spectrace.com/Applications/analysis°g_terephthalic^{a}cid.htm)

33 R. J. Fredericks, R. R. Comtois, W. Holtman, *Proc. Annu. Symp. Instrum. Process Ind.*, **1995**, *50*, 33–38.

34 S. Yamashita, T. Shimizu, *Adv. Instrum. Control*, **1996**, *51*, 23–32.

35 D. Connolly, C. Walker, *Ultrapure Water*, **1998**, *15*, 53–58.

36 ASTM Method No. D6502-99, American Society for Testing and Materials, Philadelphia, **1999**.

37 R. Tertian, F. Claisse, *Principles of Quantitative X-ray Fluorescence Analysis*, Heyden, London 1982, Ch. 17.

38 B. Holynńska, M. Lankosz, J. Ostachowicz et al., *Adv. X-ray Anal.*, **1989**, *32*, 45–47.

39 D. J. Connolly, R. W. Dye, M. J. Mravich et al., *US Pat.*, 5,818,899, Oct. 6 1998.

40 R. E. Collins, R. J. Blue, M. C. Mound, *ZKG Int.*, **1995**, *48*, 540–549

41 R. Dejaiffe, J. E. Willis, A. L. Heilveil, *PCT Int. Appl.*, WO96 23212 (Cl.G01N23/223), 1 Aug. 1996, *US Appl.*, 379696, 26 Jan 1995.

42 W. D. Drotning, *Anal. Instrum.*, **1988**, *17*, 385–397.

43 L. E. Creasy, *Adv. X-ray Anal.*, **1994**, *37*, 729–733.

44 C. Frederici, S. Doorns, D. Villanueva et al., *At-Process*, **1998**, *3*, 125–131.

45 R. R. Romanosky, A. S. Viscomi, S. S. Miller et al., *Prepr. Pap. Am. Soc., Div. Fuel Chem.*, **1993**, *38*, 272–278.

46 C. C. Trassy, R. C. Diemiaszonek, *J. Anal. Atom. Spectrom.*, **1995**, *10*, 661–669.

47 Council directive 94/67/CE on the Incineration of Hazardous Wastes, *Off. J. Eur. Comm.*, 31 December 1994, L365, 34.

48 P. P. Woskov, D. L. Smatlak, D. R. Cohnet al., *US Pat.*, 5,479,254, 26 Dec 1995.

49 B. Baaske, C. Högel, S. Kirschner et al., *Spectochim. Acta, Part B*, **1997**, *52*, 1459–1467.

50 J. Schram, *Fresenius' J. Anal. Chem.*, **1992**, *343*, 727–732.

51 S. Kirschener, A. Golloch, U. Telgheder, *J. Anal. Atom. Spectrom.*, **1994**, *9*, 971–974.

52 T. Jacksier, R. Barnes, *J. Anal. Atom. Spectrom.*, **1992**, *7*, 839–844.

53 T. Jacksier, R. Barnes, *Spectrochim. Acta, Part B*, **1993**, *48*, 941–945.

54 M. J. Jajl, R. Barnes, *J. Anal. Atom. Spectrom.*, **1992**, *7*, 833–838.

55 T. Jacksier, R. Barnes, *J. Anal. Atom. Spectrom.*, **1994**, *9*, 1299–1303.

56 T. Jacksier, R. Barnes, *Spectrochim. Acta, Part B*, **1994**, *49*, 797–809.

57 R. C. Hutton, M. Bridenne, E. Coffre et al., *J. Anal. Atom. Spectrom.*, **1990**, *5*, 463–466

58 I. Ahmad, B. J. Goddard, *J. Fiz. Malays.*, **1993**, *14*, 43–54.

59 K. Song, Y. I. Lee, J. Sneddon,. *Appl. Spectrosc. Rev.*, **1997**, *32*, 183–235.

60 D. A. Rusak, B. C. Castle, B. W. Smith et al., *CRC Crit. Rev. Anal. Chem.*, **1997**, *27*, 257–290.

61 V. Majidi, M. R. Joseph, *CRC Crit. Rev. Anal. Chem.*, **1992**, *23*, 143–162.

62 L. Moenke-Blankenburg, *Laser Microanalysis*, Wiley, New York 1989.

63 L. J. Radziemski, D. A. Cremers, *Laser Induced Plasmas and Applications*, Marcel Dekker, New York 1989.

64 L. J. Radziemski, *Microchem. J.*, **1994**, *50*, 218–234.

65 C. Haisch, J. Lierman, U. Panne et al., *Anal. Chim. Acta*, **1997**, *346*, 23–35.

66 C. Haisch, R. Niessner, O. I. Matveevet al., *Fresenius' J. Anal. Chem.*, **1996**, *356*, 21–26.

67 R. Wisbrun, I. Schechter, R. Niessner et al., *Anal. Chem.*, **1994**, *66*, 2964–2975.

68 R. E. Neuhauser, U. Panne, R. Niessneret al., *Anal. Chim. Acta*, **1997**, *346*, 37–48.

69 U. Panne, M. Clara, C. Haisch et al., *Spectrochim. Acta, Part B*, **1998**, *53*, 1957–1968.
70 D. K. Ottesen, *Symp. (Int.) Combust. [Proc.]*, **1992**, *24*, 1579–1585.
71 M. Sabsabi,. J. F. Bussiere, *PCT Int. Appl.*, WO98 20325 (G01N 21/71, 33/15), 14 May 1998.
72 H. J. Häkkänen, J. E. I. Korppi-Tommola, *Anal. Chem.*, **1998**, *70*, 4724–4729.
73 U. Panne, M. Clara, C. Haisch et al., *Spectrochim. Acta Part B*, **1998**, *53*, 1969–1981.
74 R. Jowitt, I. Whiteside, *Laser Analysis of Liquid Steels*, Comm. Eur. Communities (Rep. Eor. 13932), 1997.

Section X
Hyphenated Techniques

Introduction

John C. Fetzer

This chapter covers the use of spectrometers as detectors for chromatographic separations. Modern separation methods can often give a chromatogram that is composed of a large number of individually resolved peaks. The use of spectrometers in this fashion yields molecular information about each peak. This can greatly aid in peak identification and quantitation. The identification of peaks can also give the analyst information on contaminants in a product, the occurrence of side-reactions in a synthesis, the distribution of isomers, and the answers to many other specific questions.

Full-spectrum UV absorbance and fluorescence detection, and mass, infrared, nuclear magnetic resonance, and atomic spectrometries are the more common methods used. The basic aspects and limitations of these types of spectrometric detection are described.

UV absorbance and fluorescence detection are only of moderate use as liquid chromatography detectors for organic compounds because most of these do not have very characteristic spectra and many do not even fluoresce. These indistinct spectra are marked by one or two broad bands. For a few classes, however, this is not the case. The polycyclic aromatic hydrocarbons (PAHs), for example, have spectra that contain several sharp bands in a distinct pattern for each PAH. For this class of compounds, these detectors are much more sensitive and give more information on the peak identities than any other type of detector.

Mass spectrometry can be used for peak identification, with fragmentation patterns showing the presence of specific chemical groups. Infrared and NMR detection are useful for this because they also give information about the chemical functionality of the peaks detected. This can be used either to determine the structure of unknown peaks or to monitor specific chemical structures such as methyl groups, a carbonyl group, or an ether linkage.

Atomic detectors monitor the presence of specific atoms that are contained in the components of each eluting peak. This can be extremely useful when the analytes contain less common elements such as the halogens or metals. Simultaneous monitoring of several elements, even of the very common elements carbon, hydrogen, sulfur, oxygen, and sulfur, can also help identify the component molecules.

Several applications of each technique are given to show the capabilities and utilities of each. The discussion will also focus on the complementary nature of the information each detector provides in comparison to some of the others.

21
Hyphenated Techniques for Chromatographic Detection

John C. Fetzer

21.1
Introduction

In all of the many forms of chromatography, detection is an inherently important final step. The type of detection can aid in the analysis by gathering information that can be used to identify the peaks seen. There can be many peaks that elute from the column of a gas, liquid, or supercritical-fluid chromatograph. Certain detectors are in fact spectrometers that examine each peak for specific information on its identity. This chapter deals with this use of spectrometers as the tail-end detector in chromatography. Other separation techniques, such as field-flow fractionation or capillary electrophoresis, differ in their separation mechanisms, but as far as coupling to spectrometers behave like one of these three types of chromatography.

If the analysis is the separation of many common complex mixtures, then many peaks can be either partially or fully resolved. The simplest and least expensive, and therefore the commonest, chromatographic detectors only yield a response for a peak, with little diagnostic power to identify it other than the retention time. The flame-ionization detector in GC and the refractive index or single-wavelength UV absorbance and fluorescence detectors in HPLC or SFC are good examples of simple, widely-used detectors of this type. This is not a limitation for routine analyses, such as those used for product quality control or process monitoring. For complex mixtures or for the situation where a problem has been identified and its causes need to be determined, chromatography with simple detectors is woefully insufficient. The hyphenated techniques, however, are ideal tools in many of these situations.

The hyphenated techniques provide a synergy where the combination far outperforms either technique alone. The spectroscopist often only thinks of the chromatograph as a novel, albeit very useful, sample inlet device. Only moderate thought is given to optimizing the separation in the way a chromatographer might do. Concurrently, many chromatographers think of the detector as only a device to identify peaks, with little concern about resolution, matrix effects, and other factors that are the spectroscopists' major concerns. They do not approach the detection in the way a spectroscopist might. In reality, the strengths of one part compensate for the weaknesses of the other. Chromatograms are inherently complicated, but spectral information can readily simplify this by identifying some (if not all) of the peaks. The spectrometers, on the other hand, often cannot identify very similar compounds, such as isomers, and suffer greatly from matrix effects. The chromatographic retention time, coupled to the spectrometric data, often surmounts these. Additionally, the separation by its fundamental nature reduces the matrix effects because the components are separated and elute individually, making the spectrometric detection more valid.

The use of spectrometers as detectors has become very prevalent in the past decade. More and more of the hyphenated techniques reviewed here have moved from the realm of unique devices found only in an academic or government research facility. Many are now available as off-the-shelf instrument packages readily available from several possible instrument companies. In some cases, these have become very commonplace detectors because they have been commercially available for a decade or more. These are the cases when a mass spectrometer or atomic emission spectrometer is used as a gas chromatography detector or when the mass spectrometer or UV absorbance spectrometer are used as a liquid chromatography detector.

These more common separation–detection combinations have had wide usage for many years and will, therefore, be covered here in what can only be considered as a cursory fashion in relation to all of the work done with them. From the myriad of references to the operation of these detectors and their use, only a few examples have been chosen to illustrate the power of each detector when coupled to a chromatograph.

The reader will be referred to much more extensive review articles or books for details on their wide range of applications. The annual review issue of the journal *Analytical Chemistry* is a good starting point. In these review issues, alternating years cover techniques and applications. The first are a series of reviews focusing on the variety of analytical techniques, principally the various types of chromatography and spectrometry. The second is divided into a series of articles on areas such as pharmaceuticals, polymer analysis, petroleum and other fossil fuels, and environmental analyses. These issues usually appear as the mid-June volume of the journal.

The area of hyphenated techniques is so active that any researcher who wishes to stay abreast of the use of new applications and techniques must read the literature constantly. Each new issue of any of the major analytical chemistry journals contains one or two or more articles in this field. Conversely, by the nature of hyphenated techniques, an article on one particular technique can appear in many possible places. For example, an article on an HPLC separation with full spectrum UV absorbance detection of polycyclic aromatic hydrocarbons (PAHs) could appear in any of the journals that deal with analytical chemistry, chromatography, spectrometry, the chemistry of the PAHs, or materials containing PAHs. This author's own publications list shows several examples of papers appearing in each type of journal, even though each could have appeared in a different one than the one it did.

This leads to the need for table-of-contents and electronic search services to keep up.

In this chapter the various spectrometric detectors will be reviewed. These will be arranged by spectrometer type, as the use of certain detectors is not limited to only one certain mode of chromatography. For each detector, however, the individual problems inherent to coupling with each mode of chromatography will be covered. For example, the issue of how to remove the mobile-phase solvents in LC-MS will be part of the discussion of that technique in the MS section. This will be a separate segment from the problems inherent to GC-MS or SFC-MS.

These techniques generate such a wealth of data that their advent at the same time as the development of computer systems that are of high-capability and capacity, fast, reliable, and inexpensive, cannot be mere coincidence. The use of data systems with the instrumentation systems for hyphenated techniques, however, is beyond the scope of this review and will only be mentioned when it is part of an inherent advantage or disadvantage of a particular technique.

21.2
Electronic Spectral Detection

UV absorbance and fluorescence are useful types of spectrometry for many classes of molecules. They measure changes in the energy levels of the molecular electrons. The wavelengths of light used correspond to energies that send electrons from a ground to an excited state. This usually involves the π bonding or lone pair electrons in the molecule, but higher energy (lower wavelength) transitions involving σ bonding electrons can also be seen. As a rule of thumb, the lower the wavelength, the greater the variety of molecules that will absorb.

UV absorbance spectrometry measures the energy absorbed when the electrons go from the ground to an excited state, while fluorescence measures the full process of that energy change and the one that results when the excited state energy is lost through photon emission. Fetzer and Biggs reviewed the use of these types of detectors for environmental analyses [1]. The specific advantages of full-spectrum UV and fluorescence detection were highlighted by many examples relating to detection of the polycyclic aromatic hydrocarbons (PAHs).

The photodiode-array based UV absorbance detector, usually abbreviated as either the PDA or DAD, but in this chapter as DAD, has been used in a wide range of applications since it was introduced commercially almost two decades ago. It allows collection of the complete UV absorbance spectrum of the HPLC column eluent. The general principles and operation of the DAD have been reviewed by Huber and George in a book on DADs and their application [2]. This book is also a good primer on the use of the DAD in fields such as clinical, pharmaceutical, environmental, polymer, and biotechnology analyses.

In this detector, a beam from a "white light" source is passed through the flow cell. Any wavelength that is absorbed by a compound flowing through the cell is attenuated following normal Beer–Lambert law behavior. The resulting light is diffracted and focused on a row of optically sensitive semiconductor devices (the photo-diode array). Each photo-diode corresponds to a small spectral range and the light intensity on it affects its electronic output. With the collection of the UV spectra of the column eluent, peak identification and quantitation can both be accomplished at the same time. The simple substitution of a high-pressure flow cell for the one routinely used in HPLC makes the DAD amenable for SFC detection. There are no specific issues that arise in SFC-DAD that differ from those described for HPLC.

Certain compound classes, such as the PAHs, are ideal for this type of detection. The UV spectra of the PAHs are very intense and contain many bands that aid in definitive identification. Seemingly small differences in isomeric structure can lead to great differences in the pattern of band locations and intensities. As an example, the spectra of two isomeric PAHs are shown. The difference in the spectra arises from the different arrangements of the π electrons in the different ring structures of the PAH isomers. Figures 21.1 and 21.2 show the UV absorbance spectra for two similar PAHs, each one having only one or two rings more on the core structure of the bottom one in Fig. 21.1, dibenzo[cd,lm]perylene. The second pair consists of isomeric PAHs. Although the two spectra are similar, there are significant differences that characterize each. For other compound classes the differences between isomers and even very structurally different species, will be much less pronounced. This is because the presence of sulfur, oxygen, nitrogen, and other heteroatoms makes the electronic distributions less distinct and individualistic.

A DAD in static mode, when used as a stand-alone spectrometer, has slightly less sensitivity than a conventional scanning-monochromator spectrometer based on grating optics. The DAD can collect usable spectra at around 0.001 absorbance units (AU) full scale. For certain compounds with a number of intense and narrow bands, such as the PAHs, this may be even lower because the pattern of bands is so distinct. Since the spectra of many compounds have only one or two broad bands, the higher value would be the typical limit. In the dynamic mode used with HPLC, the limit for useful diagnostic spectra is around 0.01 AU, with 0.002 AU to 0.004 AU being the limit for most PAHs.

Other molecular structure factors, such as the presence of methyl or other alkyl substitution also have an effect on DAD spectra. A single alkyl-group attachment shifts the absorbance spectrum of the parent compound upwards by 1–2 nm. Although additional alkyl substitution shifts the spectrum even higher, the shift is not strictly additive. The presence of a fused saturated ring usually causes an upward shift of 5–10 nm. These small shifts in spectral location require that the DAD have corresponding resolution to see these changes.

Interestingly, when perdeuterated (every hydrogen atom totally substituted by deuterium) compounds are used as internal standards for MS detection, their presence can also be seen in the HPLC-DAD spectra. These compounds appear as blue-shifted spectra relative to those of the unsubstituted versions. For example, perdeuteroperylene has a downward shift of about 2 nm in its spectrum relative to the unsubstituted perylene, its absorbance maximum is at 432 nm compared with 435 nm for normal perylene. This is due to the electronic nature of deuterium

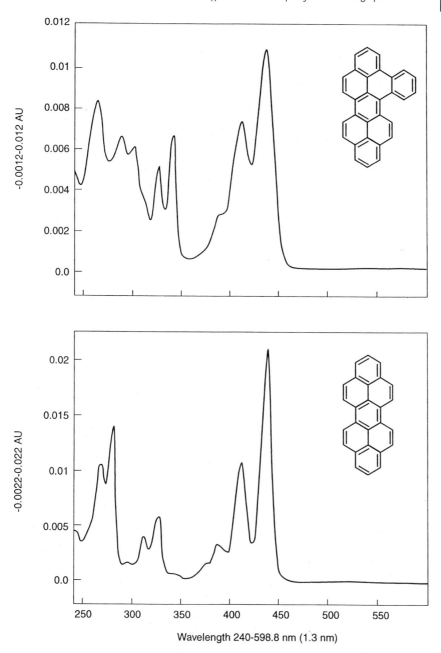

Fig. 21.1 The UV absorbance spectra of tribenzo[a,cd,lm]perylene and dibenzo[cd,lm]perylene, showing the complexity inherent to the spectra of PAHs.

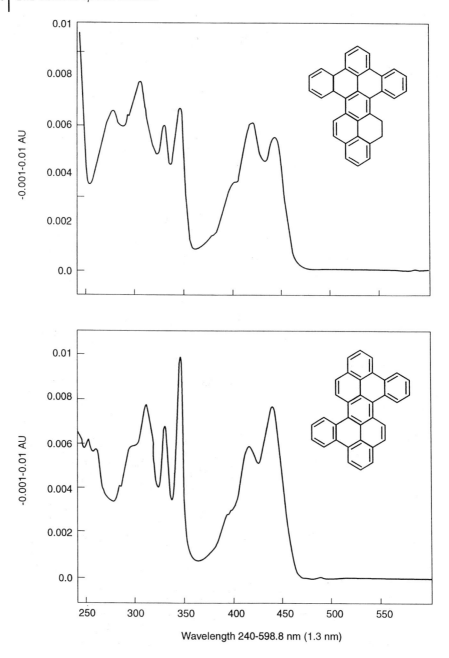

Fig. 21.2 The UV absorbance spectra of the two isomers tetrabenzo[a,cd,f,lm]perylene and tetrabenzo[a,cd,j,lm]perylene. Note that these spectra are not only different from each other, but are very different from those of the two similar structures shown in Fig. 21.1.

as compared to normal hydrogen atoms. This difference also results in the perdeutero versions having slightly shorter retention times than the unsubstituted versions in reversed-phase HPLC.

The DAD allows monitoring of all the wavelengths within a chosen spectral range. This makes it much more of a universal detector than the older single-wavelength designs. In those devices, major components could be assumed to be only minor constituents if the wavelength used was not at one of the compounds' stronger absorbance bands. If the wavelength was in a spectral region where a component did not absorb at all, it could be missed altogether. For example, since benzene absorbs strongly at 254 nm, this wavelength was often used to monitor for alkylbenzenes. High numbers of alkyl or fused saturated ring substituents will shift the spectrum to a very different maximum. The compound 1,2,3,4,5,6,7,8,9, 10,11,12-dodecahydrotriphenylene is such a compound [3]. It consists of a benzene ring with three saturated rings, fused at the 1,2; 3,4; and 5,6 faces. Its analogous most-intense absorbance maximum is at 273 nm, a shift of 19 nm because of the saturated ring substituents (Fig. 21.3). This band has a molar absorptivity of about 325. At 254 nm, the absorptivity is only 110. If equal amounts of benzene and this compound were injected, the chromatogram monitored at 254 nm would have two peaks. Their relative intensities would be 6.3 to 1.

Commercially available DADs usually acquire spectra from a low wavelength of 190 nm to an upper wavelength limit as high as 800 nm. For most applications, this high wavelength range is not needed. Most compounds absorb appreciably in the UV and lower visible range (350 nm to 500 nm, compounds appearing yellow or orange will absorb in this range). Metal complexes, dyes, and certain PAHs, are among the common compounds that absorb in the > 500 nm range (these compounds would appear to be red, green, purple, or blue to the human eye).

The HPLC separation of the higher fullerenes yielded one example of the advantage of monitoring at all absorbance wavelengths. Fetzer and Gallegos [4] initially performed a non-aqueous reversed-phase HPLC separation of a crude fullerene mixture. They also performed concurrent direct-probe MS analyses. All reports, at that time, of the composition of fullerene soot, reported only the major components C-60 (~ 90 %) and C-70 (~ 70 %). A series of small peaks in the DAD output after the expected two much larger ones of the C-60 and C-70 fullerenes was noticed. These small peaks had UV spectra that were characterized by almost monotonically increasing intensities when going to lower wavelengths, with very small shoulders and less intense bands at the high wavelengths. The absorbances, however, were not intense at either of the wavelengths of the two stronger bands of C-60 or for those of the many bands of C-70. Single-wavelength monitoring would either have missed these components or greatly underestimated the concentrations. Figure 21.4 shows the HPLC-DAD chromatogram with these small peaks appearing on the much more intense tail of the C-70 peak.

These small HPLC peaks corresponded to small peaks in the MS caused by C-76, C-78, C-80, C-82, and C-84 fullerenes. The total concentration of all these higher fullerenes is less than 1 %. The DAD easily observed these at wavelengths that were most sensitive for them, making their observation possible. Dieterich and

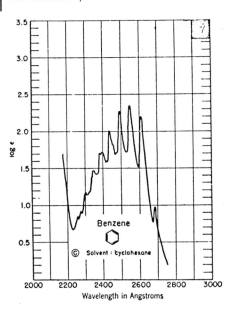

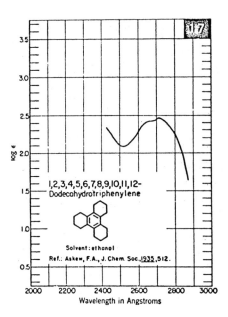

Fig. 21.3 The UV absorbance spectra for benzene and 1,2,3,4,5,6,7,8,9,10,11,12-dodecahydrotriphenylene, with both the loss in sharp energy levels and 20 nm red shift of the latter.

co-workers [5–7] then used this observation and the same separation method to isolate several of these larger fullerene species, which detailed molecular characterization, such as the mass spectrum in Fig. 21.5, showed to be C-76 fullerene, several C-78 isomers, and a wide variety of even larger fullerenes. Although these interesting carbon clusters would have eventually been discovered and isolated, the use of

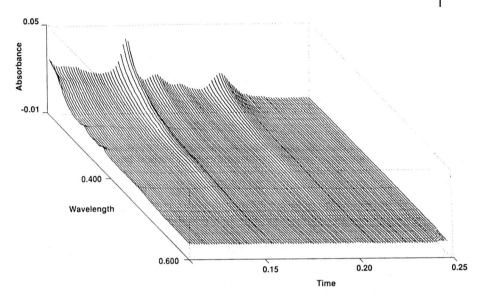

Fig. 21.4 The later-eluting portion of the DAD chromatogram of a fullerene separation, showing several small peaks. These turned out to be heavier fullerenes.

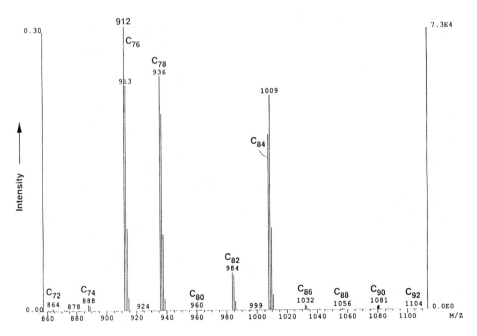

Fig. 21.5 The mass spectrum of the material in Fig. 21.4 which showed that it was fullerenes larger than the C60 and C70 that predominate.

a DAD that enabled the observation of these minor peaks accelerated that discovery by several months.

Another example where using the DAD was advantageous was the unexpected discovery of a new eight-ring PAH [8–10]. A synthesis had been performed to make a certain other eight-ring isomeric PAH. This target PAH was preparatively separated by reversed-phase HPLC and a DAD was used to monitor the separation. After the isolation of the desired compound was completed, a very strong mobile phase of pure dichloromethane was used to clean off the column. After a short period of time, the column effluent showed a peak, which from the spectrum was obviously a PAH (the many bands clustered in packets of increasing intensity is characteristic of many PAH spectra, Fig. 21.6). This turned out to be a strongly retained isomer of the target PAH. It had a capacity factor of around 2 orders of magnitude greater than that of the dominant, but much earlier-eluting isomer (the upper spectrum of Fig. 21.1). The target PAH was a compact, non-planar structure and the unexpected one was elongated and planar, with these gross structural differences causing the retention difference.

The shift to slightly higher wavelengths described for the alkyl substitution of benzene is much more generic. It applies for all compound types and a large number of substituents. The classic Woodward–Hoffman additivity "rules" describe

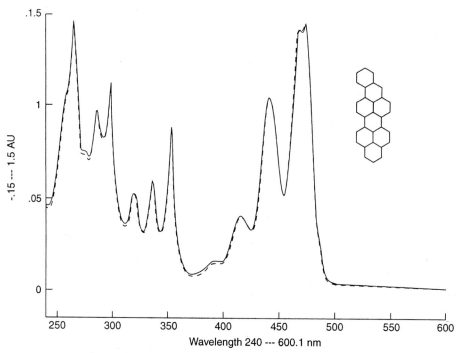

Fig. 21.6 The UV absorbance spectrum of benzo[*rst*]-naphtho8,1,2-[*cde*]pentaphene, a new PAH unexpected discovery through the use of a DAD.

these and have been used to estimate the shifts for a wide variety of compound classes. These orderly trends in UV absorbance relative to structure can aid in identifying unknown peaks.

For simple compounds, such as either the unsubstituted compounds or their monomethyl derivatives, spectral matching programs that are included in the standard DAD data software usually work well. This is not true, on the other hand, for other more substituted compounds. Where the spectral shift due to alkyl-substitution is greater than 1–2 nm, the spectral shift is not recognized as such by the software. The currently available software does not work well [11, 12]. These programs generally work by spectral matching done on a point-to-point comparison, rather than by looking at the patterns of the spectral bands by comparison of the relative intensities of spectral features and the distance of each band from the others. The point-by-point method takes each wavelength intensity and, based on the Beer–Lambert law, normalizes the spectrum for comparison. Each wavelength's intensity is then compared to those in the standard spectra. Matches result from spectra with the smallest total of differences between the unknown peak and the reference spectra. Good matches may result because the unknown and a standard both contain a large absorbance band at the same wavelength, although numerous small bands in each may be very different.

With greater substitution, the common programs instead only see the large differences in location, due to the shift of the spectrum by the substitution, as altogether different peaks. They do not see the similarities in patterns because each band is examined separately. For example, the spectra, shown in Fig. 21.7, for unsubstituted PAH pyrene and for a highly alkyl-substituted pyrene (like the compound, 1,2,2a,3,4,4a,5,6-octahydrocoronene, with the dissimilarities accentuated for the spectral band at 305 nm due to residual impurity coronene) can be compared [3]. The similarity in the peak shapes and relative locations of the absorbance bands is obvious to the naked eye. A spectral matching program, however, cannot recognize the second compound as a pyrene-type species because the spectral shift is too large for the similarity to be found in a point-to-point comparison.

Co-elution of peaks is also not a serious problem with DADs because the spectra obey the Beer–Lambert law. The data computer, therefore, can examine the spectra across a peak for proportionality and yield a peak purity measurement. Since each co-eluting component contributes to the total absorbance at each wavelength, the changes in absorbance with retention time reflect the compositional changes. Software can assess these changes in the increase or decrease at specific wavelengths, and mathematically extract the contributions of each component. When recombined for all wavelengths, individual spectra of each component are obtained [13]. These can then be compared to spectral libraries for identification (similar to the treatment of any of the spectra of any peak).

Since the DAD is a multiple-wavelength detector, data manipulations are possible with it that are impossible for single-wavelength detection. A simple example is co-elution of two components. If the retention times are slightly different, then the spectra from different times across the chromatographic peak will differ and show the co-elution. In the case where there is co-elution of a small number of compo-

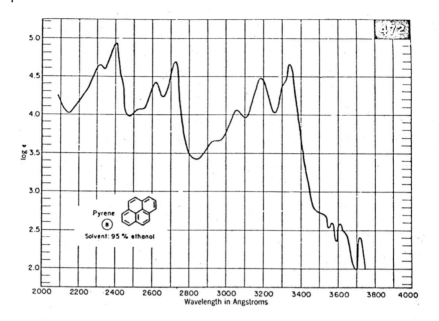

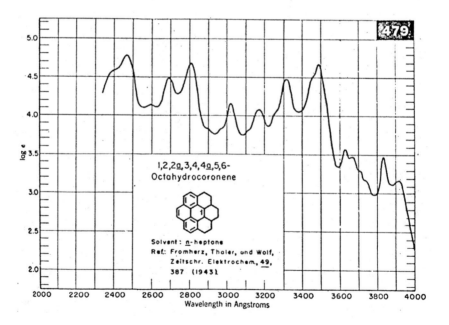

Fig. 21.7 An example of the similar spectral pattern of highly alkyl-substituted PAHs. The peak in the lower spectrum at 305 nm is due to some residual coronene from the starting material.

nents in a retention range, the overlapping peaks can be deconvoluted if there is any difference in retention times. For each wavelength, the intensity changes as the sum of the individual contributions due to each component (the Beer–Lambert Law). Across the peak of a single component, the spectra will all be proportional to each other. If two components with even slightly different retention times are within the peak, then the relative intensities will change. The contribution of the two components to each wavelength will be their individual absorbances based on their concentration at the retention time of the spectrum. Algorithms can separate the contribution from each and recreate the spectra of each component.

In the simplest case of two co-eluting compounds, the molar absortivities need only be very different at two distinct wavelengths for deconvolution to be possible. The relative changes in intensity at these two wavelengths are enough to mathematically separate the spectral contributions of each compound at the different retention times. Both the retention times of the peak maxima of each component and their absorbance spectra can be separated from each other by the computer's algorithms.

Ramos et al. [13] first separated the individual spectra of benzo[b]fluoranthene and benzo[k]fluoranthene or chrysene and benz[a]anthracene from each other in a purposely created co-eluting peak of mixtures of each of the two pairs of PAHs. They were then able to deconvolute the individual spectra from a mixture of the isomers benzo[e]pyrene, benzo[b]fluoranthene, and benzo[k]fluoranthene in a similar fashion. In all cases they purposely generated peaks with severe overlap (greater than 90 % of each peak co-eluting with the other components) to show the power of deconvolution. Tauler et al. described another algorithm to deconvolute the individual spectra in co-eluting peaks and reviewed similar efforts by others. Multivariate curve resolution has been used as an alternative approach for peak deconvolution. It can identify minor impurity peaks and yield the "true" retention times [14, 15].

One problem arises in DAD use that was not as major in single-wavelength detection, the absorbance of the mobile phase. In single-wavelength detection under isocratic (constant composition) conditions, any absorbance due to the mobile-phase solvents only resulted in a constant increase in the baseline. This was easily overcome by normalizing the baseline output to zero. If gradient elution was used, the baseline rose in a continuous, readily accounted for fashion. With DADs, the total absorbance of the mobile-phase interferes with spectral collection in that range. Deconvolution software, however, can extract the spectra.

The advent of high-capability personal computers has fueled the growth in capabilities of the DAD. Early commercial versions of the diode-array absorbance detector (ca. 1982), for example, were operated off a small desktop computer with similar capabilities to today's hand-held calculators and data storage of only a few kilobytes. These instruments could only collect the spectra, determine peak purity, and match spectra from a small set of about 24 standards. Only a similar number of sample spectra could be collected because of the limited memory of such small computers. As the previous paragraphs highlight, the abilities of DAD software and data storage are much greater today. A typical 30 min HPLC-DAD run that collects a spectrum every other second over a 250 nm range, might take several megabytes of storage.

Solvent selection with DADs must take the mobile-phase absorbance into account if the analytes have absorbances in the range of any of the potential choices. The choice of strong solvents for a gradient separation can be severely limited if some of the possibilities absorb in the wavelength regions of interest. For example, in aqueous reversed phase HPLC, methanol and acetonitrile have similar elution strengths and from a chromatographic standpoint one or the other can be used. Acetonitrile is much more favored, however, from a spectrometric standpoint because it does not absorb above 195 nm. Methanol has strong absorbance up to 230 nm and would mask any compounds that absorb below that wavelength.

Because of this much lower wavelength capability, acetonitrile allows detection of compounds containing sulfur, nitrogen, oxygen, and most other heteroatoms. Even the saturated alkanes absorb in the 210 to 220 nm range. Thus, aqueous acetonitrile gradients can be used for the analysis of sugars, amino acids, vitamins, and many other compounds that would be masked by the use of methanol.

Additionally, some solvent choices may require extra steps that must be taken to reduce the absorbing impurities of the solvent. Ethyl acetate and tetrahydrofuran (THF) are such solvents because they are unstable to hydrolysis or oxidation. UV absorbing impurities form when these solvents are exposed to water or oxygen prior to storage. Passage through freshly activated silica removes the impurities from ethyl acetate, while the peroxide that forms in the THF can be removed with sodium metal. These steps lower the UV wavelength cut-off for these solvents by 20 to 25 nm.

There have been many publications on the use of HPLC-DAD. As one set of examples that highlights both the growth and now-current wide application, is the analysis of the larger PAHs. This class of compounds has many, many isomers that have very similar structures and retention times. The collection of spectra by the DAD allows each HPLC peak to be monitored and compared to standard or reference spectra. PAH mixtures are usually very complex, so the separation and identification of these samples by using HPLC-DAD highlights the powerful capability possible with this combination.

One of the earliest works using HPLC-DAD for the identification of the large PAHs involved the analysis of a diesel particulate extract [16–18]. About a dozen LPAHs were found. The identifications made were later correlated to the observed mutagenicity of this diesel particulate. A carbon-black extract, obtained from the same carbon black was examined by HPLC-DAD. The DAD, as well as a much larger collection of standard compounds, allowed the identification of around 20 more LPAHs than in earlier studies of this same material.

McCarry et al. [19–25] performed a similar series of HPLC fractionations to determine the PAHs in sediment samples. They observed $C_{24}H_{14}$ LPAHs similar to those found by Wise and co-workers in a coal-tar SRM. In later work, LPAHs of 26, 28, 30, and 32 carbons were found. Both a DAD and direct atmospheric-pressure chemical-ionization mass spectrometry were used for detection. To make the DAD less specific and more universal, the average response from 250 nm to 370 nm was collected as a total-absorbance chromatogram.

A new eight-ring PAH, phenanthro-5,4,3,2-[efghi]perylene, was observed in the HPLC-DAD analysis of a deposit from the catalytic hydrocracking of a petroleum-based feedstock [26]. This new PAH eluted close to the known isomer, benzo[a]coronene, but even with severe overlap of the two peaks, the DAD could deconvolute the combined spectra and yield one for each of the components. Preparative HPLC-DAD resulted later in the isolation of the pure new compound. A coal-tar pitch was separated and DAD spectra were used to identify the PAHs in it [27]. Several of the large PAHs, such as benzo[a]perylene and dibenzo[a,j]perylene, were found in this sample indicating that the formation mechanism included condensation of smaller PAHs through formation of bridging rings.

Some studies that focus on the formation of PAHs as a route to soots have relied heavily on HPLC-DAD analysis of the products of the pyrolysis of smaller PAHs [28, 29]. The analyses have found numerous larger PAHs that indicate that both condensation reactions and gross molecular rearrangements occur. The DAD allowed identification of several minor components, as well as some major ones that suffered from co-elution. Deep-sea hydrothermal vents, where magma seeps through faults and contacts ocean water and the detritus on the seafloor, have been shown to produce PAH-containing material. Several large PAHs were found by HPLC-DAD analyses in these complex mixtures [30, 31]. The use of on-line UV spectral collection allowed definitive identification through comparison with the retention times and spectra of standards. The identifications then led to ideas about the formation mechanisms. An example is found in the hydrothermal vent work. Several very condensed structures, including benzo[ghi]perylene, coronene, and ovalene, pointed to a formation through a series of one-ring additions. Other species seen could not form in this fashion, but their structures suggest formation through condensation reactions.

There have only been a few reports of gas-phase UV detection. The biggest drawbacks are that many volatile compounds have few characteristic chromophores and that gas-phase spectra differ dramatically from the solution spectra found in most references and in the published literature. In the gas-phase, the electronic transitions are better defined, whereas in solution the molecules have interactions and collisions with the molecules that spread out the energies, thus broadening the spectral bands.

For the sake of brevity we will use the acronym FSFD for full-spectrum fluorescence detector. There have been a few reports of FSFDs. There have been a variety of optical elements, including vidicons (television-type cameras with associated storage devices) and diode-arrays. There have been a few publications describing the use of FSFDs, many of which were used in PAH analyses. The PAHs are generally highly fluorescent and their spectra are very rich with many bands in both the excitation and the emission spectra. Two examples of PAH fluorescence spectral pairs are shown in Fig. 21.8 and 21.9. The first shows the common mirror-image pattern found for many PAHs, but the latter shows the asymmetrical one seen for many common highly condensed PAHs.

Jadamec et al. [32] reported an early FSSD that was based on a fast-scanning fluorometer with a flow cell. The output from the spectrometer was displayed on an oscilloscope. The display was recorded by a vidicon (television-type) camera to collect spectra from an oscilloscope. They separated crude oil fractions from an oil spill sample and identified naphthalene and fluorene. Some of the later-eluting peaks were described as possible "polyphenyls", but a lack of available standards could not preclude that these were not similar larger PAHs or other fluorescing heteroatom polycyclic compounds.

Gluckman and Novotny [33–35] and Cecil and Rutan [36] built two examples of in-lab assembled FSFDs that were based on diode-arrays as the detection element. The first group described a diode-array based emission monitor. They separated a variety of PAH standard mixtures and the carbon-black extract, identical to that used for the earlier GC-MS studies described above. Only a few of the PAHs were identified due to a lack of standard reference compounds or spectra, but two peaks were ascribed to non-alternant structures (containing five-member rings) to the large PAHs rubrene and decacyclene on the basis of similar patterns in their spectra.

Cecil and Rutan examined the corrections that need to be made in the fluorescence spectra gathered by diode arrays when there are differences in mobile-phase solvents. Normally, the changes in the strength of solvation between PAHs and different solvents lead to shifts in the spectra. Non-polar solvents, like

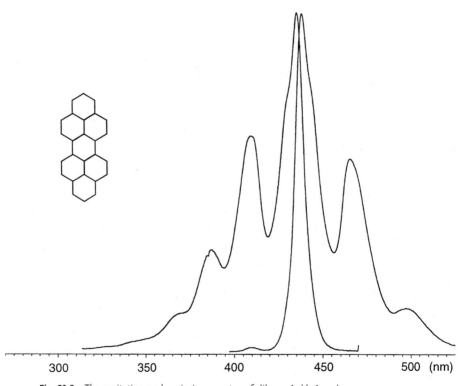

Fig. 21.8 The excitation and emission spectra of dibenzo[cd,lm]perylene.

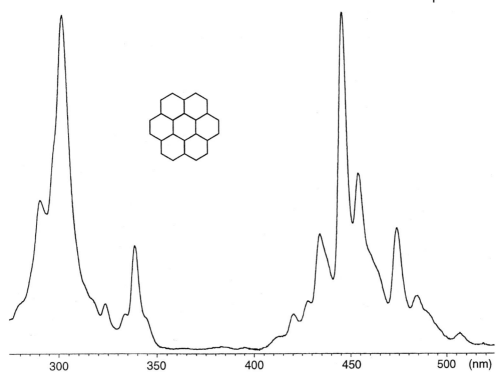

Fig. 21.9 The excitation and emission spectra of coronene, showing their asymmetry.

n-hexane, iso-octane, or supercritical carbon dioxide, exhibit the lowest wavelength spectra because their solvation energies are the smallest. As solvents interact more strongly with the PAH, the fluorescence transition energy goes down. This leads to higher wavelength spectra. They found that both the wavelength shifts and band height changes are significant. Spectral matching to literature spectra (which would be the common mode since large collections of standard compounds are not practical) is made difficult. For example, perylene has a 6 nm higher wavelength when the solvent is changed from pure methanol to 20% water in methanol. The collection of a standard spectral library under set conditions was recommended, with algorithms dealing with the wavelength shifts. The second issue of relative changes in band heights was not addressed.

There is a recently introduced commercial FSFD [37]. Through the use of DAD-based optics, data can be collected as either the excitation spectra at a fixed emission wavelength or the emission spectra at a fixed excitation wavelength. The layout of this FSFD is shown in Fig. 21.10. The introductory brochure for this instrument gives examples of its use for PAHs, aflatoxins, vitamins, carbamates, and gyphosate and its main metabolite (the latter two examples were after appropriate derivatization to form fluorescent compounds). As the use of this detector increases, so too should literature reports of its application. Figures 21.11 and 21.12 are exam-

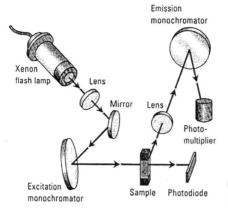

Fig. 21.10 Schematic of a commercial FSFD. (Courtesy of Agilent Technologies Co.)

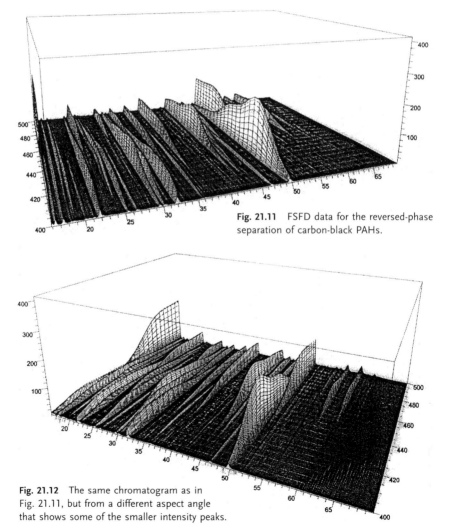

Fig. 21.11 FSFD data for the reversed-phase separation of carbon-black PAHs.

Fig. 21.12 The same chromatogram as in Fig. 21.11, but from a different aspect angle that shows some of the smaller intensity peaks.

ples of the type of data available with this instrument. A carbon-black extract containing numerous large PAHs was separated by non-aqueous reversed-phase HPLC. The FSFD spots even minor components because it sees all of the fluorescence wavelengths. One unfortunate feature of this detector can be seen in these chromatograms. The designers chose to sacrifice spectral resolution for greater sensitivities by building in a wide spectral slitwidth (15 nm). The usually sharp multitude of peaks occurring in PAH spectra are not observed.

Burt et al. [38] coupled an HPLC to detect and identify several PAHs through measurement of their fluorescence decay lifetimes. Several wavelengths were monitored simultaneously to differentiate some closely eluting peaks.

Fogarty and co-workers [39, 40] used a dye-laser source and a videofluorometer, with a diode-array emission detector, to collect the complete three-dimensional excitation-emission map (EEM) of individual PAHs in a mixture of 18 standards. Their chromatographic data could be displayed as individual EEMs of peaks, or as the excitation or emission spectra as functions of the retention times.

Fluorescence detection has many inherent advantages and disadvantages. The major advantages include very high sensitivities, less interference from co-elution or chances for misidentification because only a small number of compounds fluorescence, and more selective detection because both the excitation and emission wavelengths are used [41]. The disadvantages are somewhat similar, by including too much selectivity so that wavelength selection for more than a single compound can be complicated (involving wavelength programs for retention windows), the higher sensitivity often requires more dilutions to get into the working range, quenching due to dissolved oxygen in the mobile phase or other components in the sample may unknowingly reduce the responses, and certain molecules (in particular some of the PAHs) have very solvent dependent responses.

One of the biggest advantages is the very high sensitivity. Since the signal of the sample's emitted light is measured directly (in contrast to UV absorbance measurement which measures small differences in the light beam intensities); there is little interference. This is accentuated by the measurement being at different wavelengths than the excitation and by the viewing optics being positioned at right angles to the incident excitation beam.

With this high sensitivity, the use of a DAD and a FSFD in series does not work well for highly fluorescent compounds. The analyte concentrations needed to yield DAD spectra of greater than 0.01 AU usually result in fluorescence emission peaks that are way off scale with the FSFD. The inherent sensitivities of a FSFD for PAHs, for example, are two or three orders of magnitude lower than those for a DAD. So, at the DAD limit of detection, the FSFD may have a signal a hundred or a thousand times larger in scale. In this case both spectral identification and quantitation with the DAD is more difficult. The regions of the absorbance spectrum that are off scale are, of course, unusable, but there are generally some wavelength ranges where less intense bands absorb. Integration of the chromatogram at the wavelength of one of these bands can also be used for quantitation.

For compounds that are weakly fluorescent, with quantum efficiencies of less than 0.1, the differences are much smaller and it might be possible to use these two detectors in series. In this case, the FSFG would be in the emission spectral mode since the absorbance and excitation spectra are generally similar (because they arise from similar electronic transitions).This combination, however, is advantageous when both fluorescing and non-fluorescing analytes are targeted. One notable example is the EPA 16 priority-pollutant PAHs, where 15 fluorescence intensely. The three-ring acenaphthylene does not fluorescence, but the DAD can readily determine it.

The selectivity of detection is also very high, since few compounds will both elute in the retention range expected for a component of interest and excite and emit at the chosen set of wavelengths. The latter characteristic, however, can also be a disadvantage if the analyst is trying to determine the composition of an unknown mixture because the use of selective wavelengths may lead to missing components that fluoresce at other wavelengths or do not fluoresce at all.

An additional advantage inherent to the FSFDs is that the presence of co-eluting impurities can be observed. These compounds that would normally go undetected and affect quantitation should be readily seen, either as additional fluorescence if the compound does fluoresce or as reduced fluorescence or skewing of the spectrum if the compound only absorbs UV light. When working at typical analytical levels, the fluorescence signals from two compounds are additive, the first problem can be both spotted and corrected for.

The latter effect is known as inner-system filtering and is common in many "real world" samples. There are two types of inner-system filtering that reduce the fluorescence signal. In the first, the co-eluting species absorbs at the chosen excitation wavelength, thus reducing the incident beam intensity. The analyte then has a reduced excitation rate and the signal is diminished. In the second type, the emitted light from the analyte is absorbed by the co-eluting species, which would also diminish the signal. In either case, the effect should be proportional to the absorbance of the co-eluting species. This is extremely unlikely to match the spectral pattern of excitation or emission of the analyte (whichever occurs from one or the other type of inner-system filtering). Thus, the fluorescence spectrum would be altered non-uniformly and appear to be skewed relative to that of a standard injection of the same compound where no filtering occurs.

21.3
MS Detection

GC-MS is the most widely used hyphenated technique and there have been many comprehensive reviews. This description will only be a brief overview and touch on specific issues relevant to the coupling of the GC to the MS. The interfacing of the GC outlet to the MS inlet usually requires some type of selective carrier gas removal. Although direct connection of the GC to the MS is feasible (if large enough vacuum pumps are used), this is rarely done. This is because the vacuum at the outlet of the column can affect the separation efficiency, making most calculations of column retention parameter or efficiency calculations impossible, and the MS system must be shut down for column switching. The large excess of carrier gas is inherently not compatible with the vacuum needed for MS.

Common interfaces include the molecular jet and flow splitters. The first uses the difference in momentum between the low-molecular weight carrier gas and the high-molecular-weight analytes. The column effluent passes into the separator inlet line, which is enclosed in a glass chamber that is under vacuum. A small gap separates this line from the outlet line. Sample molecules move preferentially from one line to the other by inertia, while much of the carrier gas is removed tangentially by the vacuum. The enrichment also increases the sensitivity of GC-MS.

In splitters, a narrow length of connecting tubing restricts the flow into the MS. The remainder of the GC effluent flows out as waste. Two types of splitter designs are used, open and direct. In open splitters, the mechanism is similar in concept to the jet separator. The column outlet butts into the restrictor, which is contained in a sheath. A stream of helium sweeps this area. The carrier gas is preferentially swept away from the restrictor, while the heavier analytes move toward the restrictor opening. In direct splitting, a simple tee connection results in only part of the flow passing through the restrictor.

The molecular jet and both types of splitter suffer from sample discrimination. In the jet separator and open splitter the enrichment varies with molecular weight. In the direct splitter, the ratio of the split changes as the temperature is ramped upwards in the commonly used gradient mode. The relative quantities of different analytes vary, making absolute quantitation impossible without tedious measurements of the enrichment or split flow ratios.

A few applications highlighting the power and limitations of GC-MS will be given, but the reader is directed to the reviews that focus on applications for a broader perspective of the use of GC-MS.

An example of the powerful combination of high-resolution GC with MS detection is shown in Fig. 21.13 to 21.19. Figure 21.13 is the total ion chromatogram of the separation of a commercial lubricating oil additive. Figures 21.14 to 21.19 are some of the individual peaks and the structures assigned to them by mass spectral interpretation of the fragmentations and mass losses. Note that certain peaks are identified as members of an isomeric set, with the other peaks yielding almost identical spectra.

The petroleum industry was one of the more important spawning grounds for GC-MS. The inherent nature of petroleum, and more importantly those of the much more valuable processed products derived from it, made this a natural occurrence. Petroleum products are almost totally composed of the non-polar or low-polarity compound classes of saturated hydrocarbons, olefinic hydrocarbons (those that contain an alkene double bond), and aromatic hydrocarbons. By being non-polar or only slightly polar, the volatilities of these compounds are very high. The inter-molecular interactions are primarily the weak van der Waal's and aromatic π bonding dipole–dipole interactions. Hydrogen bonding and acid–base interactions are only prevalent in the heavier materials such as residual (asphaltic) materials where the nitrogen and oxygen content is high. Processed material has

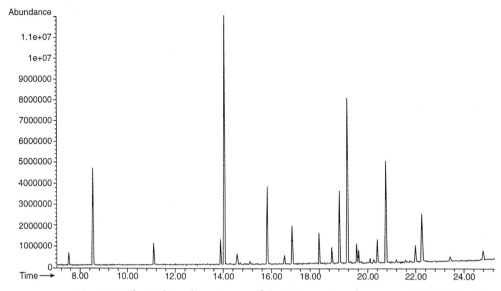

Fig. 21.13 The total ion chromatogram of the GC separation of a commercial lubricating oil additive.

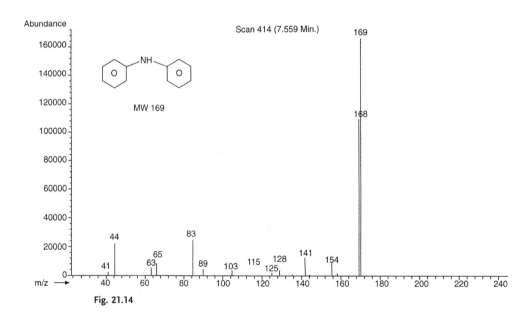

Fig. 21.14

Fig. 21.14–19 The mass spectra of individual peaks in the chromatogram in Fig. 21.13. The numbers refer to the retention times. Courtesy of J. D. Hudson and M. T. Cheng, Chevron Research and Technology Co., Richmond, CA, USA.

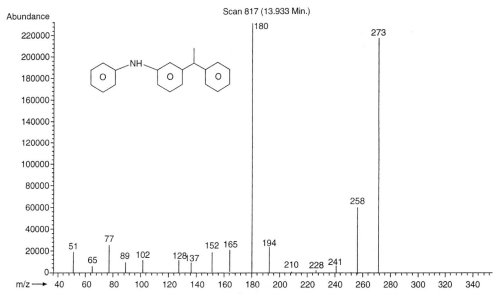

Fig. 21.15

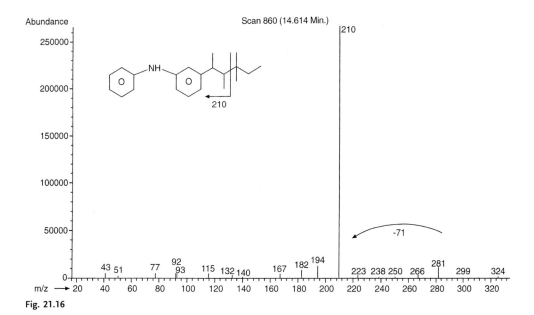

Fig. 21.16

404 | *21.3 MS Detection*

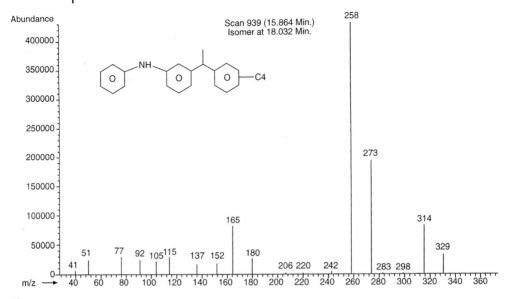

Fig. 21.17

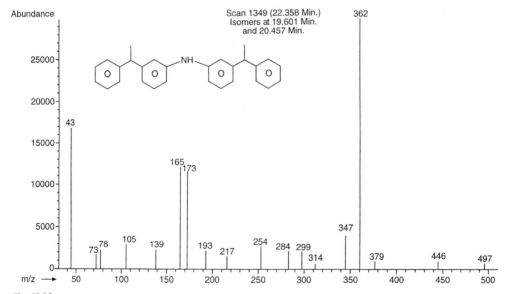

Fig. 21.18

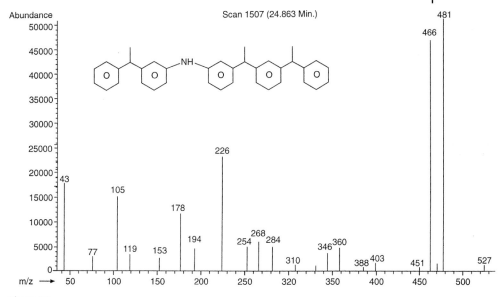

Fig. 21.19

undergone treatment to remove heteroatoms, so even the heavier fractions of lubricating base oil and wax are volatile enough for GC analysis.

The major impurities in these materials are the sulfur-containing analogues of these classes, which are also very non-polar. Due to the lack of polar functionalities, these compound classes are all relatively volatile. The only limitation to volatility is increasing molecular weight, but saturated hydrocarbons of at least 80 carbon atoms can be separated by high-temperature GC methods. The waxes and lubricating oil base stocks that are among the heavier processed products are lower than this carbon number. The most economically important materials, gasoline (petrol), diesel fuel, and aviation jet fuel (kerosene) all range from 5 to 20 carbon atoms.

Concurrently, there are large numbers of hydrocarbon compound isomers because of the presence of chain branching and saturated rings. The physical and chemical properties of each of these as components of the fuels determine the overall fuel properties. The high-resolution of GC, coupled to the carbon number and degree of saturation information from MS, can be used to predict these properties. The high efficiency of capillary GC is needed as the best separator of such complex mixtures, but the complexity is often so great that MS detection is a necessity in order to differentiate peaks. For the heavier products, the carbon number range and the proportions of linear and branched species affects the overall properties. For example, a motor oil base stock must have tight limits on its carbon number range. If it is too high, the oil will be too viscous at low temperatures to flow effectively at engine start-up. If it is too low, volatility will occur at high temperatures that can lead to uneven combustion in the engine. GC-MS is an almost ideal analytical tool for these types of samples.

One whole field of research relating to petroleum is dominated by GC-MS analyses: the use of biomarkers to characterize the location and origin of petroleum. Petroleum arises from a variety of organic material that is geologically aged and degraded [42]. The remnants of the original biological materials are reflected by a multitude of hydrocarbon species, including polyterpenes, steranes and hopanes (polycyclic saturated hydrocarbons whose cores resemble those of steroidal compounds), and porphyrins. These generally result by defunctionalization and reduction of the ring systems to saturated and partially-hydrogenated forms of the ring structures. This field, initially spearheaded by researchers such as Seifert, Moldowan, and Gallegos, has grown to be a core tool in petroleum exploration because of its reliance on GC-MS.

High-resolution capillary GC can separate the myriad of saturated hydrocarbons found in petroleum. Correlations of individual components or classes of compounds have been made to a variety of variables. These included: whether the source material was originally terrestrial, marine, lacustrine (originating in lakes), etc.; the conditions (predominantly the temperature and the age) of the diagenesis, the period of transformation from biological material to the defunctionalized, predominantly hydrocarbon coal, shale oil, or petroleum; and the possible migration of the petroleum away from its source rocks to the ultimate pools where it is found. For example, the ratios of the stereoisomeric forms of a compound are used to determine racemization, which is then correlated to the age of the petroleum. This requires a GC separation capable of separating a multitude of very similar compounds, and the selectivity of MS to differentiate them. The geochemical literature is literally loaded with new uses of GC-MS to extract more information from each petroleum.

One common GC method for petroleum characterization is simulated distillation, sim-dist [43]. The GC retention times on a non-polar phase of a petroleum or processed material are correlated to the boiling points of the n-alkanes. These boiling point markers are then used to determine the amounts of material eluting in boiling point ranges. Sim-dist is a rapid method to estimate the yields from refinery distillation processes and is used to estimate the distribution of the various products from a process. Roussis and Fitzgerald reported a GC-MS method for sim-dist analysis of petroleum. The MS data were used to sub-divide the boiling point fractions into sub-classes of saturates, one-ring aromatics, two-ring aromatics, thiophenes, etc.

Another hydrocarbon class, the larger PAHs, those of 24 or more ring carbons, is also important in the analysis of petroleum and related materials in certain applications. These compounds elute at the upper temperature limit of current commercial capillary GC columns, but their isomeric complexity requires the higher resolution. Their separation and identification by GC-MS, however, has been very useful in a variety of analytical problems that highlight the power of high-resolution separations with highly selective detection. Schmidt et al. identified several 24-carbon PAHs in the effluent from burning hard coal by GC-MS with a methylphenylpolysoloxane column [44, 45]. Many peaks of 26 carbons eluted towards the end of their chromatograms, but were not identified. They later extended their identifications by synthesizing several more $C_{24}H_{14}$ isomers. Wise and co-workers [46] used similar columns (DB-5) to separate the 24-carbon PAHs in extracts from a coal-tar standard reference material. They used GC-MS, and were unable to identify many peaks due to a lack of available standards for retention time comparisons. Simoneit and co-workers [47, 48] used high-temperature GC-MS to examine the PAHs in fractions obtained from alumina absorption chromatography fractions of the tar-like bitumens from deep-sea hydrothermal vent areas. They found PAHs of 24, 26, 28, and 30 carbons. Similarly to the other groups using only GC-MS, they were not able to identify any specific isomers.

One shortcoming of GC-MS for isomer analysis is that the "normal" electron impact (EI) ionization mechanism does not usually differentiate between isomers. The molecular ions, by definition, are the same and the fragmentation patterns through loss of substituent groups are also usually very similar. The ratios of ions are not reproducible enough to definitively distinguish between isomers. This leaves reliance on the separation and the resulting retention times as the only way to tell isomers apart.

For the PAHs, however, Simonsick and Hites [50] showed that special chemical ionization reagent gases could be used so that isomers will appear to have different fragmentation patterns. They used methane as the reagent and then used the M/M+1 ratio to compare to calculated ionization potentials (IPs) to assign structures to several LPAH peaks. They separated the same extract as had been earlier studied by Lee and Hites. They saw eight $C_{28}H_{14}$ isomers and four $C_{30}H_{14}$ isomers. The comparison of calculated IPs to ion ratios and the possession of two isomers from each set (which were used to compare retention times as well as ion ratios) let them assign probable structures to all 12 peaks. There was, however, a note of caution given. Even with the high resolution of capillary GC, they observed an ion ratio for one of their standards, benzo[*pqr*]naphtha-8,1,2-[*bcd*]perylene, that did not match the corresponding peak in the carbon- black separation. They relied on the GC retention time and a separate preparative HPLC separation and UV analysis of the peaks to confirm this component and assumed that there must be co-elution that changed the observed ion ratio.

This is just one example of the use of added reagents to cause selective ionization, commonly called chemical ionization (CI). The use of specific reagent gases for determining certain types of analytes in GC-MS is a rich field of study in itself [51]. Reagents are commonly chosen that aid in the selective ionization of target analytes through acid–base reactions or in the enhancement of the ionization of certain functional groups. For example, in negative ion CI, a reagent gas is chosen so that it has a slightly lower proton affinity than the target group. The target molecular type, as well as all others with a greater proton affinity, will ionize by giving up a proton to the reagent gas. Any molecules with a lower proton affinity than the reagent will not ionize. This specific mode is referred to as NICI, with the opposite approach of generating positive ions being PICI. In PICI, the reagent gas acts on electron affinity differences. For molecular classes with high proton affinities, such as the basic pyrroles, carbazoles, and aza-arenes, ammonia will selectively ionize them and many other common classes of compounds that occur in the same samples (such as the PAHs, thiophenes, and furans) will remain neutral [52].

Suzuki et al. [53] used methanol chemical ionization to differentiate PAHs from similar-sized heteroatom-containing polycyclic aromatic compounds. They assigned example structures to the masses observed. Since neither were their methodologies for the LPAHs able to distinguish isomers nor did they use a large reference compound set, these structures can only be deemed to be possibilities among the huge number of isomers.

One of the many other areas in which GC-MS is widely used is in forensic analysis. Kaye [54] has reviewed many uses of GC-MS in his book dealing with the use of modern analytical methods of analysis in criminal investigations. He highlights GC-MS analyses of opiates, including heroin, codeine, and morphine characterization. Not only is this approach useful in identifying what a suspected substance might be, but if it is an illegal substance the pattern of components and impurities can aid in determining the source of the drugs. There are also examples given showing the widespread use of GC-MS as one of the preferred methods in the testing of athletes for use of performance-enhancing substances. There are a wide variety of compounds that can increase the performance of human (and equine) competitors, as well as a variety of measures and countermeasures described by both the analytical chemist responsible for monitoring any banned substances and athletes and trainers bent on circumventing the rules.

GC-MS has been widely used in environmental analyses. The U. S. Environmental Protection Agency's mandated methods for volatile and "semi-volatile" priority pollutants in effluent water call for GC-MS analyses for a wide variety of acid, base, and neutral compounds. An aqueous sample was sequentially extracted to give the various fractions after appropriate pH adjustments. Selected-ion monitoring of key ions for a pollutant during a range of time around its retention time is the required method. This requirement was one of the major driving forces in the sales of GC-MS equipment during the 1980s. Lacorte and co-workers [55] describe a similar method for a greatly expanded listing of environmental pollutants.

For LC-MS, the chemical ionization mode for the mass spectrometry is the most readily used since the large amounts of mobile-phase solvents naturally act as the chemical ionization reagent. This can limit the utility of LC-MS because the mobile-phase solvent choices are determined by the HPLC conditions. There is usually little flexibility in the choice of solvents. For example, the use of acetonitrile or methanol is a common variable in reversed-phase HPLC, but most other solvent switches lead to gross changes in the chromatographic separation. The most ideal chemical ionization reagents are very likely not usable for this reason. Rosele-Mele et al. used this approached to identify the porphyrins, both free and bound to metals, in a shale oil.

The first interfaces between LC and MS were mechanical devices, such as the moving belt interface. In this interface, the column effluent was deposited onto a surface, which moved to collect the sequence of eluents. Solvent was then removed from the belt by heating, sometimes aided by vacuum. The belt then moved into the MS sample generation area where the deposited compounds were volatilized by further heating and the very-high-vacuum conditions. Although this type of interface helped prove the utility of LC-MS, its cumbersome design and the presence of ghost peaks from previous runs were severe problems. With the advent of direct introduction methods, the mechanical interfaces became obsolete.

The most widely used modern interfaces between HPLC and MS are atmospheric-pressure ionization (API) using spray techniques. Thermospray, electrospray, ionspray are the three common modes. Each of these techniques utilizes the nebulization of the liquid stream with specific modes to increase the efficiency. Thermospray uses the inherent expansion of the solvents when they are exposed to a rough vacuum region (before the high vacuum of the MS), assisted by heating. Ionization is normally attained by the addition of ionic buffers, with the somewhat volatile ammonium acetate being a favorite. Subsequent conventional ionization with electron impact or high-charge fields of the stream may be used in addition, if there are an insufficient number of ions. Thermospray is most effective for HPLC mobile phases where no organic modifier is used (pure water with buffers, as in ion chromatography or an electrophoretic separation).

Electrospray ionization utilizes a high voltage field of several kV to ionize the droplets of the nebulized effluent. The charged droplets are accelerated and focused by ion optics into an area where a countercurrent of inert gas removes the uncharged droplets and vapor. Ionspray is similar, but a pressurized countercurrent of heated inert gas assists in the nebulization and evaporation of the mobile phase. Ionspray is more effective than electrospray with mobile phases with low levels of organic modifiers. Both approaches are relatively mild forms of ionization, so the molecular ions of the peaks are predominant. The slight degree of fragmentation can make the assignment of specific structures more difficult, especially for isomers, which would have different fragmentation patterns in their electron-impact spectra.

In a brochure describing their ionspray LC-MS interface [56], Waters shows examples of the separation and identification of several types of molecules, including polypeptides, bisphenol A polymer additives, and steroidal anti-inflammatory drugs. Hewlett-Packard (now named Agilent Technology) describes similar applications for their commercially available electrospray LC-MS system [57. This design differed from many by using an orthogonal flow stream in which the HPLC flow was dispersed and the ion inlet was perpendicular to it [58, 59]. This gave a better signal-to-noise and a more reproducible peak height. Charlwood and co-workers [60] used a microbore HPLC system with this interface to characterize derivatized oligosaccharides. The N-linked glycans of up to 10 sugars were separated, with the mass spectra of species being above 1800 Da. Peng and co-workers [61] used LC-MS-MS with an ionspray interface to determine candidate anti-arthritic drugs in human plasma and cartilage tissues. Mobile phase gradients of water, acetonitrile, and formic acid provided the chemical ionization reagent for positive ion MS. The target drugs were hydroxamic acid based protease inhibitors. The use of the selective MS-MS mode gave quantitation in plasma of sub-ng mL^{-1} in plasma.

Suzuki and Yasumoto [62] used liquid chromatography-electrospray ionization mass spectrometry to measure the diarrhetic shellfish-poisoning toxins okadaic acid, dinophysistoxin-1 and pectenotoxin-6 in bivalves. Holcapek and co-workers determined extremely low levels of several glycols, including ethylene glycol, using derivatization with benzoyl chloride [63]. HPLC-MS with an electrospray interface had limits of detection of 10 to 25 (g L^{-1}. Boyer [64] used LC-MS with an electrospray interface to determine the pharmaceutical nortriptyline, a tricyclic antidepressant, and its metabolites. A mobile phase of water:methanol with 30 mM of ammonium acetate was used for selective ionization.

To highlight the on-going growth of LS-MS techniques, they were the subject of two chapters in a recent review volume [65, 66]. These dealt with the analysis of oligonucleotides by electrospray MS and the analysis for herbicides in aqueous media. The first, by Deforce and Van den Eeckhout, highlights the advantage of electrospray over other competing MS modes, such as matrix-assisted laser desorption/ ionization, time-of-flight (MALDI-TOF) MS. The easier coupling to an HPLC and the greater mass resolution at higher molecular weights are the main advantages cited. The resolution advantage is especially pronounced for oligonucleotides of more than 60 bases. In the herbicide review, D'Ascenco and co-workers focus mainly on the use of thermospray interfaces and API. Besides giving detail descriptions of the interfaces, they emphasize the advantages of these approaches with aqueous samples.

The dietary flavenoids have been proposed as being beneficial in reducing the risk of contracting colon cancer and of heart attacks. The analysis of these very polar compounds in biological fluids has, therefore, received some attention. Nielsen et al. [67] monitored 12 of these compounds through reversed-phase HPLC analysis of urine. API-MS was used to measure the glycoside and aglycon forms of these polyphenolic compounds.

Hsu et al. [68] and Mao et al. [69] examined the nitrogen-containing heterocyclic compounds in diesel fuel and other processed petroleum products. A combination of normal-phase HPLC and chemical ionization MS showed the presence of indole, carbazole, and benzocarbazole with varying degrees of methylation up to four.

HPLC-MS has been heavily utilized to measure the "true" relationships between molecular size, molecular weight, and retention in size-exclusion chromatography (SEC) in polymer characterization. SEC separates a polymeric material through permeation differences as the sample passes through columns with pores with well-defined diameters. Permeation is based on the size of the molecules, but often the need is to know the molecular weights of the molecules. The relationship between the two is not straightforward, depending on structural factors that control the chain flexibility and intra-molecular interactions. The MS provides the absolute molecular weights. Its use for routine polymer characterization is not practical, however, so SEC with refractive index or evaporative light scattering detectors is used for its speed and simplicity of operation after the columns have been calibrated by SEC-MS.

Asserud et al. [70] used SEC with an electrospray interface to couple to a FTMS. They characterized a variety of poly(methyl methacrylate) polymers. This allowed them to accurately determine not only the molecular weight distribution, but also the end-group functionality and observe secondary polymer distributions due to the formation of cyclic species. In the off-line mode (collection of the SEC effluent for subsequent FTMS), they were able to determine polymers with molecular weights of over 500,000 Da.

MALDI-MS has become increasingly popular as a tool to help calibrate SEC separations. In MALDI, the sample is trapped in a matrix. In this case, as the name implies, this matrix acts not only as a sample trap, but also takes an active role in the laser-induced ionization mechanism. The isolated polymer fractions are analyzed and molecular weight distributions based on the MS data are used to calibrate retention times on the SEC columns.

Pace and Betowski [71] used micro-column HPLC to introduce samples into a particle-beam MS. The separation was with a polymeric octadecylsilane bonded phase and methanol–tetrahydrofuran gradients. They examined a set of standard large PAHs, from 24 to 36 carbons, and compared those results to species seen in two extracts from soils collected at hazardous-waste sites. Generally their detection limits were approximately 1 ng.

Rosenberg et al. [72] used reversed-phase HPLC with atmospheric pressure chemical ionization MS to separate and measure several organotin species that had been extracted from sediments. These species were used as a fuel additive when organolead compounds were replaced. They are environmentally important because of their bioaccumulation and toxicities.

One interesting hybrid of LC-MS has been developed by Hercules and his research group: TLC-MS [73]. The MS approach utilizes MALDI MS. MALDI MS has much lower detection limits than many other MS approaches and can be used for both low and high molecular weight components. The TLC plate surface and the MALDI matrix are coupled. This is accomplished through pressing the TLC plate against a second plate coated with the MALDI matrix. Several cyclic peptides were separated and measured with this set-up.

One unique type of MS, ICP-MS, needs to be discussed separately because it does not deal with molecular species, but with atomic ones. The inductively coupled plasma is a common atomization source for atomic spectrometry. This "sample preparation/ sample introduction" mode has been coupled with an MS to yield an instrument capable of trace level elemental analysis. Each element has a unique set of isotopes in known proportions. These can be used to quantify the element. In the case of elements with overlapping isotopic mass numbers, simple deconvolution can be used to give results for each. ICPMS has very low detection limits.

GC- and LC-ICP-MS have been used as a means of separating species and then identifying individual peaks by their unique elemental mass spectra. Braverman [74] separated the rare-earth elements using HPLC-ICPMS. Schminke and Seubert, for example, used ion chromatography as the separations tool and the ICP-MS detector was used to measure bromate ion at µg L^{-1} levels in the presence of large excesses of sulfate and nitrate ions [75]. Organolead and organotin compounds have been analyzed with GC-ICPMS [76, 77]. The vanadium and nickel metalloporphyrins in a shale oil were examined by both GC-ICPMS and HPLC-ICPMS by Ebdon and co-workers [78]. They found that the HPLC approach gave much more reliable quantitative data.

There are a few issues that arise when a SFC is coupled to an MS that differ from those inherent to GC or HPLC. Since the fluid density in SFC is comparable to the liquid densities of HPLC, similar interfaces can be used to remove the large excess of mobile-phase molecules if there are slight modifications made. The most common SFC mobile phase is carbon dioxide. The task of removal of the excess is made even simpler because this fluid readily converts to the gas phase upon decompression. This, however, leads to one major problem. Expanding carbon dioxide gets very cold due to the Joule–Thompson effect. The interface area must be heated to prevent any sample deposition or co-precipitation on the interface surfaces of the cold carbon dioxide as dry ice. This problem of fluid expansion and rapid cooling is inherent to any type of detection where the detector is not maintained at high pressures. The thermospray interface is the most commonly used coupling device. As with HPLC-MS the vacuum system for SFC-MS must be of a much higher capacity than for GC-MS.

For the separation of very-polar phenolic Mannich bases, Fuchsluefer et al. used the novel fluid ethane with dimethyl ether as the modifier [79]. The normal carbon-dioxide-based mobile phases, even with polar modifiers, could not elute these compounds. They identified the main oligomeric products and several of the byproducts in the manufacture of these compounds, which are used as hardeners and accelerators in epoxy resins. They coupled their SFC outlet to a MS in the atmospheric-pressure chemical ionization mode.

21.4
NMR Detection

Nuclear magnetic resonance (NMR) is routinely used to determine the types of carbon and hydrogen present in molecular structures. NMR requires nuclei with ½ spins, and so is useful for other nuclei like silicon and phosphorus. The carbon-13 isotope is the NMR active one. It is present at about 1% in a bulk of inactive carbon-12. The reverse situation is present for hydrogen. The common hydrogen-1 isotope is most prevalent, and hydrogen-2 (deuterium) is the minor NMR-inactive isotope.

For observation of the hydrogen-1 in analytes, there must not be hydrogen-1 in the mobile phase or the strong solvent peaks will mask the small analyte ones. The requirement of the mobile phase containing only deuterated components is a severe limitation to the use of HPLC-NMR because these solvents are prohibitively expensive in the amounts used as mobile phases. Many perdeutero forms of solvents are hundreds of times more expensive. The more complex a solvent's structure, the more expensive and less available will be its perdeutero form (the same statement is true if carbon-13 versions are needed as NMR active versions of analytes). With the use of special radio-frequency pulse sequences and nuclear pre-saturation techniques, the proton signals due the mobile phases can be damped, but the use of deuterated solvents is still recommended. For carbon-13 NMR, the large signal peaks of the mobile-phase solvents will overwhelm the signals of the chromatographic peaks unless the sample components have been labeled with carbon-13. Labeling of the sample is convenient only in synthetic work and even then it can be very expensive to use labeled starting materials. For more exotic nuclei, like fluorine-19 and phosphorus-31, these limitations are not a problem because these nuclei are the most abundant and the common solvents do not interfere.

The residence time that a component must be in the NMR cavity in order to collect a spectrum with good signal-to-noise ratio (S/N) can be a limit on the chromatographic conditions or the flow rate. An individual NMR scan may only take small fractions of a second, but thousands (or even millions) of scans may be needed. Multiple scanning, even with the Fourier-transform based instruments in current use, can take several seconds or even minutes to acquire. For conventional HPLC, peak widths are of the order of several seconds. For microbore HPLC, peak widths are typically only a few seconds. Stop-flow methods are commonly used, since using slower flow rates increases the peak spreading due to diffusion leading to a loss in peak resolution. Stop-flow operation, however, results in a loss of accurate retention times for comparison (necessitating separate runs for collection of retention times and NMR spectra) and chromatographic resolution is degraded because of diffusion as the analytes sit on the column during the stop-flow periods.

Subramanian and co-workers [80] utilized a microbore HPLC with special connectors for coupling to a proton NMR. When the onset of a peak was detected, the flow was diverted to the NMR flow cell and the flow was stopped. The NMR spectrum of the peak was then collected. Flow was resumed, with repeated stoppages as each peak eluted. Mixtures of small amino acids and polypeptides were separated and identified. An example HPLC-NMR contour-map chromatogram is shown in Fig. 21.20. Figure 21.21 shows the three extracted NMR spectra of the components. A typical spectral collection time, for 5 μg of the peptide phenylalanine-alanine, was 3.5 h. Earlier work by this same group utilized a conventional, larger flow system (and thus larger sample sizes and shorter times needed for spectral acquisition). Typical total run times were 5 to 14 h for complex mixtures, even with the much larger sample volumes. The microbore system, however, was proportionally better because the higher separating power of microbore columns results in sharper peaks that are relatively more concentrated than those obtained with HPLCs of conventional flow rates. All types of spectrometric detection benefit from this advantage of microbore HPLC over conventional HPLC.

Preiss et al. [81] reported the use of stopped-flow HPLC-NMR to identify dyes and other pollutants in the effluent from a textile manufacturing plant. They utilized a mobile phase of gradient of acetonitrile and deuterium oxide. Although they used conventional volume equipment, stop-flow times of 0 min to 2 h were required to achieve good S/N for individual components. They identified 14 dyes, their degradation products, or other compounds (such as long-chain benzenesulfonates).

Levsen et al. [82] reviewed their research on the use of HPLC-NMR in the characterization of environmental samples. They examined a contaminated ground water from near a munitions plant, a leachate from a waste disposal site, and the effluent water from a textile mill. Since NMR detection is non-

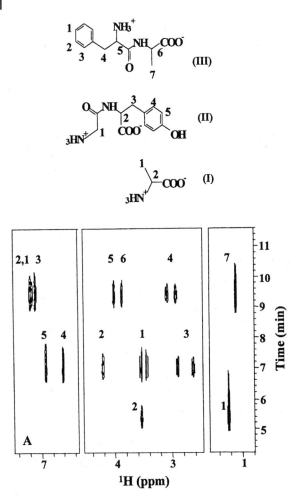

Fig. 21.20 The three-dimensional contour map of a separation of simple polypeptides. From R. Subramanian, W. P. Kelley, P. D. Floyd, Z. J. Tan et al., Anal. Chem., **1999**, 71, 5335–5339.

destructive, they showed through subsequent off-line MS analysis of individually collected peaks that the two methods complement each other in helping to determine the structures of unknown compounds. They found carboxylic acids, aromatic sulfonates, aminoanthraquinones, and other compounds in these water samples.

SEC-NMR of stereo-regular poly(methyl methacrylate) (PEMA) polymers has been used by Kitayama and co-workers [83] to determine their tacticity (the orientation of functional groups relative to the polymer backbone). Since SEC uses much larger sample amounts than HPLC, this analysis could be done in real time. Each PEMA sample was separated on a single mixed-bed SEC column and

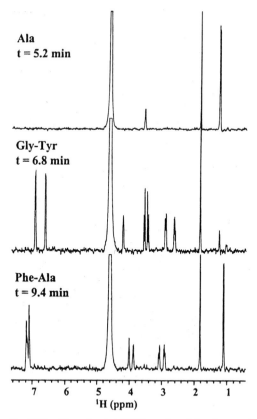

Fig. 21.21 The extracted NMR spectra of the three polypeptides found in Fig. 21.20. From R. Subramanian, W. P. Kelley, P. D. Floyd et al., *Anal. Chem.*, **1999**, *71*, 5335–5339.

the proton NMR data was collected in a 1 h run time. The separation gave four peaks, which the NMR identified as the isotactic, heterotactic, predominantly syndiotactic, and syndiotactic forms.

21.5
FTIR Detection

FTIR detection can be a very useful tool in both the observation of a specific set of target compounds and in the elucidation of unknown ones. Infrared light causes molecules composed of a number of atoms to exhibit vibrational spectra. Some of the molecular motions are due to very complex contortions of the molecular structure, but many can be ascribed to the stretching of specific bond types or to certain motions within a functional group. Infrared radiation distorts the normal molecular bonding framework by stretching individual bonds or causing combinations of bonds to undergo more complicated motions (such as the scissoring or wagging of two bonds linked to the same atom, the bending of a portion of an aromatic ring out of the plane, etc.).

In common practice, the units used are either of frequency, in cm^{-1}, or wavelength, in μm, with the first being more frequently used. When a spectrum is displayed with frequency as the *x*-axis, the low-energy molecular motions are at high cm^{-1} and are displayed to the left. The motions of larger atoms or groups of atoms that require more energy are at lower cm^{-1} values. This lower frequency region is not normally used for spectral interpretation of functionalities, but is useful for fingerprinting purposes when matching sample spectra to those of standards.

The molecular information that FTIR detection offers that MS, the other common hyphenated GC technique, cannot, includes elucidation of aromatic ring substitution (for example, with di-substitution, *ortho*, *meta*, or *para*), *cis* or *trans* or geminal substitution on a carbon–carbon double bond, the arrangement of rings in PAHs (benz[*a*]anthracene versus chrysene or triphenylene or benzo[*c*]phenanthrene or tetracene for the four-ring ortho-fused PAHs), alkyl-chain branching isomers, and alcohols, which often shows in the MS as the easily dehydrated product alkene [84].

There are two main types of FTIR detection for GCs, in the gas-phase using an in-stream optical system and through vapor deposition with detection being away from the GC flow stream. In the first, a light pipe that can transmit IR radiation is positioned on either side of a detection cell. Transparent windows pass the IR radiation into the flow cell. The whole assembly is maintained at temperatures of 250 °C to 350 °C to prevent deposition of sample molecules. Most interfaces for this type of GC-FTIR also have heated transfer lines to and from the flow cell to ensure that no deposition occurs before introduction into the spectrometer.

In the second mode, the eluting GC peaks are trapped and spectra of each component are collected offline. Cryogenic trapping of the GC eluent has become commonly available in recent years. An inert plate, often coated with a zinc selenide or a gold film, is positioned at the column outlet. This surface is cooled with either liquid nitrogen or helium. Stepper motors move it so that the eluting GC peaks are deposited and collected on the cold surface as a series of spots. These can then be individually examined by repositioning the plate in the FTIR optical path.

The spectra that are collected in this fashion have very narrow lines. The cooling reduces the inter-molecular interactions and collisional broadening that are found in KBr or solid phase spectra. These narrow-line spectra have both the bands useful in structural assignments and enough complexity in the patterns to be fingerprints. A broad band, such as the methyl-stretching band around 3000 cm^{-1}, will typically collapse into two, three or more narrow bands when the sample is cooled. The locations and relative intensities of these sets of bands are fingerprints for many very common compounds. For example, each of the methylchrysene isomers gives a characteristic spectrum in the C–H stretching region, so that their separated peaks can be easily identified. Figure 21.22 shows the spectra of the two PAH isomers phenanthrene and anthracene, while Fig. 21.23 shows some of the spectra of the even more complex five-ring isomers set. These spectra were col-

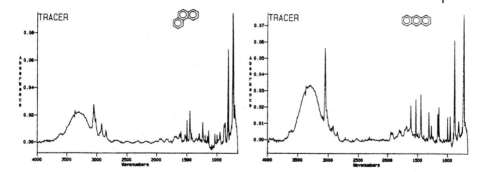

Fig. 21.22 The GC-FTIR spectra of the three-ring PAH isomer phenanthrene and anthracene. Reproduced with permission from Gordon and Breach Science Publishers from A. M. H. Budzinski, J. R. Powell, P. Garrigues, *Polycyclic Aromat. Compd.*, 11, 334.

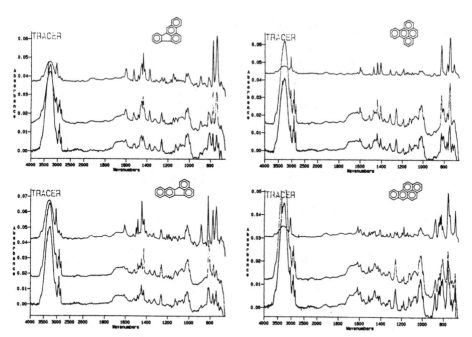

Fig. 21.23 The GC-FTIR spectra of four isomeric 20-carbon PAHs. Reproduced with permission from Gordon and Breach Science Publishers from A. M. H. Budzinski, J. R. Powell, P. Garrigues, *Polycyclic Aromat. Compd.*, 11, 335.

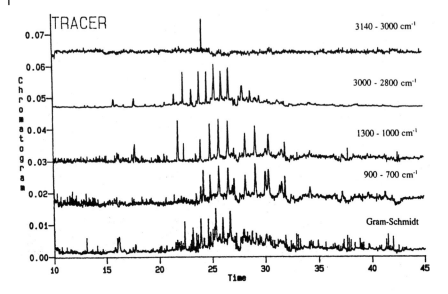

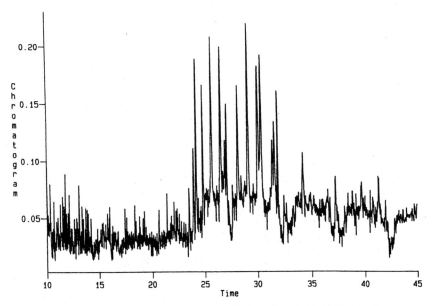

Fig. 21.24 Selected wavenumber chromatograms of a sediment extract. Reproduced with permission from Gordon and Breach Science Publishers from A. M. H. Budzinski, J. R. Powell, P. Garrigues, *Polycyclic Aromat. Compd.*, 11, 335.

lected during the analysis of a marine sediment sample. Figure 21.24 is an example of the use of selected wavenumber monitoring, showing the presence of certain structural features.

One major drawback of GC-FTIR is that the spectra collected are either in the gas phase at high temperatures or trapped at cryogenic temperatures. Most reference spectra in the literature, particularly for older synthetic work, were collected at room temperature as liquid films or as solid solutions in KBr pellets. These reference spectra have much broader bands and so spectral matching is not possible for gas-phase collection. Matrix effects, particularly for the more polar compounds, can occur in vapor deposition samples. These samples are undiluted, so that inter-molecular interactions can occur. Hydrogen bonding or acid–base interactions are minimized in the reference spectra because of the diluting KBr matrix. In spite of these possible problems, in a recent review article Bruno [85] estimated that GC-FTIR spectral matching was accurate around 95 % of the time. This contrasts with his estimate of only 75 % for correct GC-MS spectral identification. He attributed the difference mainly to the similarities in the mass spectra of isomers, which FTIR can differentiate.

As an example of the identifying power of GC-FTIR, a recent report is a good example. The PAHs in a marine harbor sediment were examined by Meyer et al. [86]. The sediment sample was extracted and compared to reference spectra obtained for the 24 PAHs in the reference mixture NIST SRM 1491. Although the PAHs are very similar structurally, being composed of ring aromatic carbons and peripheral hydrogen atoms, the relative arrangement of the rings and the number of hydrogen atoms and their positioning on each ring are the only variables. The overall spectrum, however, is a fingerprint, particularly in the lower frequency range as mentioned earlier. Of the 15 PAHs also detected by HPLC with fluorescence detection, GC-FTIR saw 13. Of these, 11 were identified, with only 2 being "not unequivocally" identified.

In an early report of the use of FTIR detection [87], Gurka and Betowski studied the volatiles from soil samples and from a commercial chemical still bottom sample using GC-FTIR and GC-MS. They found a wide variety of chlorinated compounds, particularly chlorobenzenes. Compton and Stout [88] reported the use of GC-FTIR to monitor the volatile compounds in coffee. Five selective functional group ranges were monitored in one example, the aromatic C–H stretching region of 3000–3140 cm^{-1}, the aliphatic C–H stretching region of 2800–3000 cm^{-1}, the carbonyl stretching region of 1680–1780 cm^{-1}, the C–O region of 1000–1300 cm^{-1}, and the C–N region of 900–1000 cm^{-1}. Using the individual FTIR scans they identified over 40 components, including caffeine, pyridine, carbon disulfide, and a variety of carboxylic acids, ketones, and alcohols.

Wilkins et al. [89] describe the use of GC-FTIR to separate and identify some very similar cyclic alcohols, the pairs of *cis* and *trans*-menth-2-ene-1-ols, sabinene hydrates, and terpineols. The 800–1500 cm^{-1} region gave distinct fingerprints that differentiated each isomer from the other.

HPLC-FTIR suffers greatly from the inherent interference of the large excess of mobile-phase solvent. Since the common solvents are only simple organic molecules, such useful infrared regions as the C–H stretching one around 3000 cm^{-1}, are lost. Other solvents have to be avoided if certain regions are of interest, which limits the choices for chromatographic optimization. For example, acetone or ethyl acetate cannot be used as the strong solvent in a reversed-phase separation if the carbonyl region is of interest. Many approaches have been utilized to surmount this limitation, including sample enrichment or solvent evaporation (as in GC-MS) or computer-generated solvent subtraction. The latter is inherently limited because it involves the, at-times, tenuous approach of subtracting a very large signal to observe a very small residual one.

Microbore HPLC can be coupled without modification to a conventional electrospray GC-MS type of interface for cryogenic trapping FTIR detection. A heated sheathing gas aided in solvent vaporization. A zinc selenide plate was used for sample spot collection.

Kaye [54] describes some examples of the use of HPLC-FTIR in identifying illicit drugs. He describes the characterization of heroin and lysergic acid samples by HPLC followed by identification from the FTIR spectra of isolated peaks. He highlights the advantage of HPLC methods over GC ones in that HPLC generates larger amounts of purified components that are in an easily collectable form. This allows subsequent definitive identification.

In a more general statement Kaye highlights that there is an inherently high need for validity of methods and confidence in the results for both identification of illegal drugs and their concentrations, so that the hyphenated techniques are widely used in drug testing and other forensics applications. Since many of these molecules are polar and water soluble, reversed-phase HPLC is the separation method of choice.

Huang and Sundberg [90] separated polyolefinic polymers by SEC, and FTIR detection was used to determine the variety of structures present in the resulting hydrocarbon polymers. Both the products of polymerization of a pure olefin and blends of two starting olefins were examined.

SFC-FTIR is also possible [91]. The carbon dioxide mobile phase that is most often used is relatively free of strong IR absorbances. It has two strong bands at 2325–2375 cm^{-1} and 3550–3650 cm^{-1}. French and Novotny [92] were the first to use supercritical fluid xenon as the mobile phase to observe throughout the IR range. This inert gas has no IR active bands. Through the use of microbore columns, the amount of xenon needed was relatively small since the flow rates were only a few microliters of fluid per minute. Since xenon gas can be expensive, this translated into only a few milliliters per minute of gaseous xenon being used. The use of the microbore system makes the cost of using xenon amenable. As a mobile phase, xenon is very non-polar and so behaves somewhat like carbon dioxide. Unlike carbon dioxide, however, it does not mix well with polar modifiers like methanol and, therefore, it is not as useful in the separation of polar analytes [93–96]. The use of xenon SFC has increased, being applied to the separation of sesquiterpenes and PAHs [97, 98]. The differences between supercritical fluid xenon spectra and those in the condensed phase have been assessed. Corrections are relatively simple.

Cryogenic trapping with SFC is also possible. Norton and Griffiths [99] report a typical detection limit of 1 to 10 ng injected, with the very strongly absorbing caffeine being seen at 500 pg injected. They found linear response for quantitation either from the FTIR spectrum or from the functional group wavelength chromatograms.

SEC, which is mainly used for high-molecular weight compounds like polymers and proteins, has been coupled to FTIR. Using SEC-FTIR, Liu and Dwyer [100] examined the types of branching in styrene–butadiene copolymers and Jordan and Taylor [101] measured the additives in several commercial polymers.

21.6
Atomic Spectrometric Detection

There are a variety of atomic spectrometers, in addition to the ICP-MS already discussed, that have been coupled to chromatographs to give element specific monitoring. Generally the chromatographic stream is introduced into the plasma area of an atomic spectrometer. In this high-energy environment, all of the molecular bonds are ruptured and only free atoms are present. In this state the atoms can either absorb or emit light at characteristic wavelengths. Each element has a characteristic set of wavelengths, which are based on each element's unique atomic electron orbital energies.

The GC with atomic emission detection (AED) has become widely used during the past few years. The development of optical systems that can monitor several emission wavelengths simultaneously has made these systems both affordable and utilitarian. The interfacing of atomic spectrometers to GC is more straightforward than interfacing to HPLC or SFC. In GC, the helium carrier gas is in much lower amounts and can be an active component in the spectrometer plasma. The other two separation methods have large amounts of carrier that are not inherently compatible to the plasma and must be removed to ensure efficient atomization of the samples.

In GC-AED, the GC eluent stream is directly introduced into the plasma of an atomic emission spectrometer. The helium GC gas readily mixes with the helium or argon plasma that is normally used. This plasma is maintained in the plasma state at high energy through the use of several types of energy. Both AC and DC arcs have been used, as well as radio frequency and microwave radiation. Microwave plasmas are most common in commercial GC-AED instruments. Often volumes of reagent gases or a make-up gas are added to ensure optimum atomization of the molecules by the plasma.

Since the AED can detect carbon quantitatively, it is as universal a GC detector as the commonly used FID. The AED can also readily detect deuterium. This means that the same surrogates or sample spiking compounds useful in GC-MS can be used with GC-AED for estimation of extraction recoveries or sample handling losses. The detection of multiple elements simultaneously can yield approximate empirical formulae for the peaks. The precision and accuracy for all the common

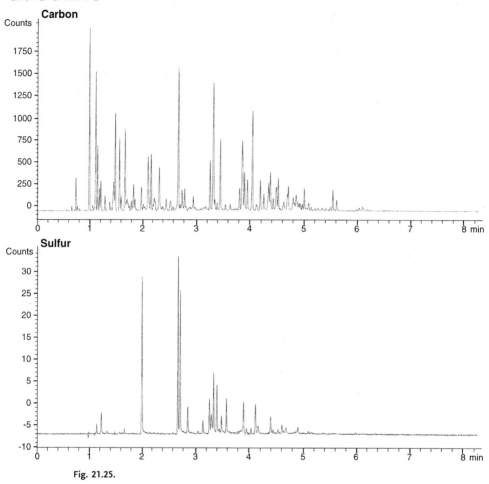

Fig. 21.25.

Fig. 21.25–27 The GC0AED carbon and sulfur profiles of three different light processed petroleum distillates. Courtesy of J. Iwamoto, Chevron Research and Technology Co., Richmond, CA, USA.

elements, C, H, O, N, and S, are not good enough yet to give a true empirical formula for each peak that can then be used for absolute identification. They are, however, adequate to confirm peak identities. Figures 21.25–21.27 show the carbon and sulfur profiles obtained for three light petroleum distillates. While other GC detectors can detect the common heteroatoms nitrogen (the bead therionic detector and the chemiluminescence detector) and sulfur (the flame photometric detector and the chemiluminescence detector), the GC-AED is the only common GC detector that is selective for oxygen. It is also the only detector that can simultaneously de-

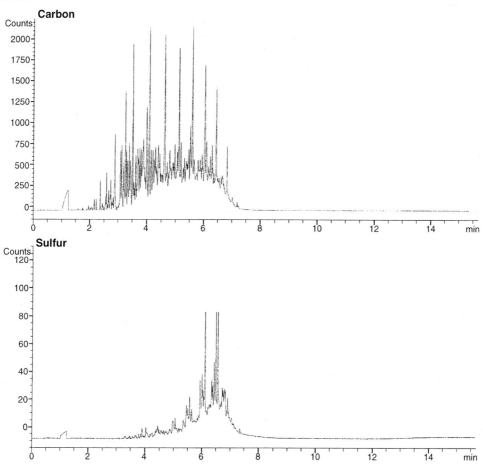

Fig. 21.26

tect more than one type of heteroatom, and commercial instruments commonly can detect 5 to 10 at a time. The capability of the GC-AED to analyze the volatile compounds of almost any element, makes its potential use very wide-ranging and valuable. Figure 21.28 shows such an example, with carbon, sulfur, nickel, and vanadium profiles for a heavy petroleum distillate sample.

Deuterium and carbon-13 have significantly different atomic emission spectra than the predominant normal-abundance isomers hydrogen-1 and carbon-12 so that the AED can be used for selective detection of labeled compounds. Quimby et al. [102] compare emission chromatograms collected for the 171.4 nm line of carbon-12, the analogous 171.0 nm line for carbon-13, and the 174.2 nm line for nitrogen. The chromatograms show that normal-abundance n-dodecane, n-tridecane, and n-tetradecane show at the first line, but at neither of the other two. Car-

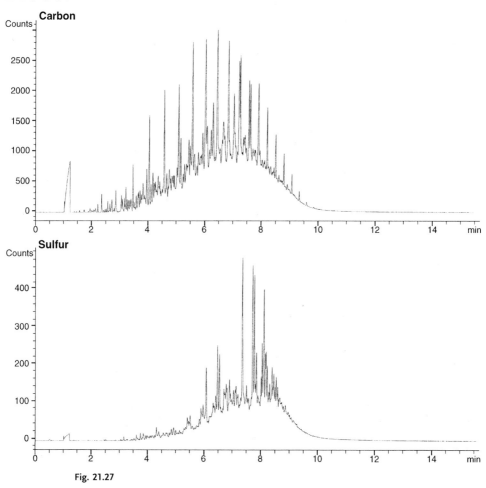

Fig. 21.28 GC-AED profiles for carbon, sulfur, nickel, and vanadium of a vacuum gas oil distillation cut. Courtesy of J. Iwamoto, Chevron Research and Technology Co., Rickmond, CA, USA.

Heavy VGO

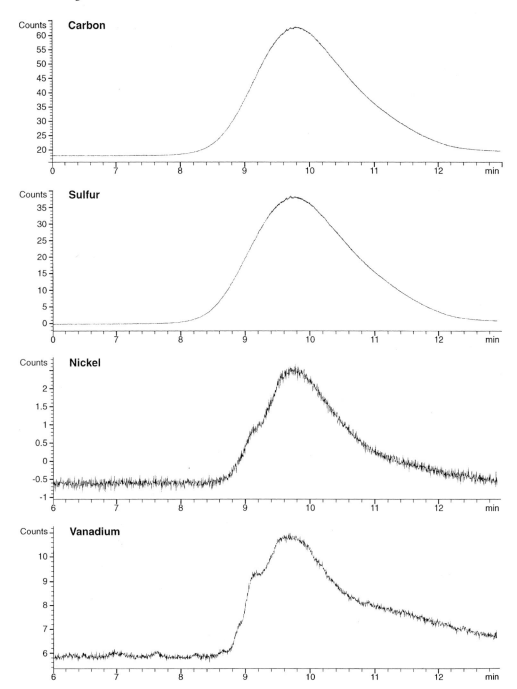

bon-13 labeled nitrobenzene shows a peak at the same retention time in the latter two, but not in the first.

GC-AED has been used to determine organomercury and organotin species in a variety of environmental samples, including marine sediments and several standard reference material samples of codfish oil, sediments, a variety of fish and other marine animal tissues [103–107]. Agilent Technologies highlights this application in their sales literature by showing the organotin compounds in a marine sediment standard reference material [108]. In another environmental analysis, Johnson et al. [109] use the chlorine, sulfur, and phosphorus lines to selectively detect pesticides such as chlorpyrifos. This work was part of a larger program to develop a method for the analysis of several hundred pesticides and suspected endrocrine disrupters. Many pesticides contain two or more types of heteroatoms, so the simultaneous detection of these at the correct retention time and in the proper proportions is definitive for identification. Repetitive GC runs using MS detection [110] gave complementary information and allowed better identifications and comparisons of retention times between the two methods (the GC-MS retention times were perturbed by the outlet pressure being sub-ambient from the MS vacuum system).

Quimby et al. [111], Albro et al. [112], and Quimby et al. [113] used GC-AED to examine the nitrogen and sulfur species found in gasoline, diesel fuel, aviation fuel (kerosene), and heavier processed petroleum products. The species found were mainly carbazoles and thiophenes. A carbon emission line was used for universal detection. The selectivity and sensitivity of various emission lines were compared, as was the effect of reagent and make-up gas flows. The first paper also examined a variety of oxygen-containing species in gasoline. They used a test mixture that contained nine common alcohols and five ethers that are of some importance as gasoline additives (mandated by the US EPA for cleaner combustion). This is an important method in those areas that add alcohols or ethers to gasoline due to environmental regulation, since some of these are suspected carcinogens and their water solubility leads to build-up in the aqueous environment. This method selectively detects the oxygen-containing compounds. Other methods generally rely solely on chromatographic retention times for identification.

Boduszynski and co-workers [114] and Andersson and Sielex [115] have used GC-AED to examine the sulfur-containing molecules in petroleum and its heavier processed products. Both groups found a preponderance of thiophenic sulfur compounds in processed materials, in contrast to petroleum. The unprocessed petroleum had mercaptans, sulfides, thiophenes, as well as other species containing other heteroatoms in addition to the sulfur. These two studies also showed that this detector has the selectivity for sulfur that is attainable with the two older commonly used sulfur detectors, the flame photometric detector (FPD) and the chemiluminescent detectors. The first of these, however, suffers from a very non-linear response. This effect is very compound-class dependent. This is due to the fact that the light-emitting species that is detected is a two-sulfur atom one that results from recombination of the sulfur atoms in the combusted sample peak. The efficiency of combustion and the rate of recombination are both dependent on the number of sulfur atoms and the structure of the sample molecules. This makes the AED a much more favored detection based just on its sulfur detecting capabilities.

The "volatile" porphyrins in crude oils (petroleum) were separated by high-temperature GC [116, 117]. The vanadium, nickel, and iron porphyrins were individually monitored. The optimum detection lines were chosen by looking at the emission spectra at the peaks of vanadyl octaethylporphyrin, nickel octaethylporphyrin, and ferrocene in individual runs of the standard compounds. The individual GC-AED chromatograms differed from each other in petroleum samples, but only a few peaks could be discerned on top of broad peak profiles. In some petroleum samples, no iron was seen.

As is the case with LC-MS, the solvent from an HPLC must be removed before the column output can be introduced into the spectrometer, or plasma conditions must be compatible to large amounts of organic solvents. For microwave or inductively-coupled plasmas (ICPs), the first must be done. Direct-current plasmas (DCPs) can accommodate direct solvent introduction. The high-molecular weight compounds when lead binds to human erythrocytes were separated by size-exclusion chromatography and ICP-AED was used to monitor the eluent.

Biggs and co-workers used a DC arc plasma detector to measure iron, nickel, vanadium, and sulfur profiles from SEC of petroleum and its distillation and adsorption chromatography fractions [118–123]. They found that the metal species were present as three distinct molecular sizes. They identified the low molecular weight ones as metalloporphyrins. They also measured the variety of silicon-containing species in polysiloxane polymers and synthetic alkyl- and arylsulfonates [124]. This combination of separation and elemental detection allowed them to see several oligomeric series of species in the polysiloxanes, which were compared with the retention times of standard compounds. The series proved to be linear chains, branched chains, and cyclic structures. The chromatographic method was able to give a baseline separation of each of the oligomers up to 60 units of trimethoxy-polydimethylsiloxane (silicone oils and greases are this type of polymer).

Slejkovec et al. [125] used HPLC-atomic fluorescence to separate and quantify the anionic arsenic compounds found in urban aerosol samples. The initial sample preparation produced aqueous extracts. These were found to contain only arsenate, although the separation worked for mixtures of arsenite, arsenate, monomethylarsonic acid, and dimethylarsonic acid. The detection limit for arsenate was 80 pg mL^{-1}.

Butcher et al. [126] coupled a diode-array based atomic absorption spectrometer to the output of an HPLC. They used this to determine the fuel additive methylcyclopentadienyl manganese tricarbonyl and the non-methylated species. They found a detection limit of 2 ng(Mn) mL^{-1}. They also examined the photolytic stability of these organomanganese compounds and found that samples should only be exposed to laboratory light during sample introduction.

21.7
Other Types of Detection

There have been a few reports of HPLC with Raman detection. Raman spectroscopy is complementary to FTIR and responds to those molecular motions that do not produce a change in a molecular dipole moment. Molecular motion such as stretching of an olefinic double bond or a sulfur–sulfur bond is not IR active, but is Raman active. Raman signals are inherently much weaker than FTIR ones, requiring laser sources and monitoring of small energy difference between the incident and observed light. Raman detection is usually done through the monitoring for a target analyte at a specific wavelength. Nguyen-Hong and co-workers [127] describe one such system. Cooper et al. [128] describe another which they used to detect nitrobenzene. The use of the more conventional approach of postcolumn trapping for Raman detection does not require cryogenic trapping. Silica plates (like those used in thin-layer chromatography) can collect the sample spots. The technique of surface-enhanced Raman spectrometry can then be readily performed on each spot. Calabin et al. [129] used this approach, with a post-collection application of colloidal silver as the surface-active agent. A commercial HC-Raman system is available as an option to a GC-FTIR one [130].

The optical activities of the eluting HPLC peaks can be monitored by using a circular dichroism spectrometer (CDS). In CDS, the effect that a sample solution has on the angle of a beam of polarized light is measured. Many organic compounds, particularly those involved in biological processes, show optical activity by exhibiting two forms. Light can either be polarized in a right-hand or a left-hand manner. Isolation of the biologically active form from a synthetic, racemic product mixture is a very important step for the organic chemist. In some materials of biological origin, like petroleum, the prevalent biological form racemizes, or starts to convert to an equilibrium mixture of the two forms, as soon as the biological process stops (for example, the organism dying). Measurement of the ratio of optical forms of certain of these pairs is an important tool in geochemistry.

Normally CDS requires the use of pure compounds, since the total change in the angle is measured and each component in a mixture adds to the total. Therefore, it is seldom used for real world analyses because it is useful only in determining the polarization of very pure compounds. The separating power of HPLC, however, makes CDS useful for certain applications. The chromatography separates a mixture into peaks, each of which is essentially a pure component as it flows through the sample cell. Yamamoto and co-workers [131] built a detector that readily measures the enantiomeric purity of separated peaks. They split the column effluent, with one part going to a regular absorbance detector and the other flowing into a cell using polarized light. Peak profiles and intensity ratios were used to assess the presence of any enantiomeric impurities.

Bringmann and co-workers used HPLC-CDS to determine the absolute stereo configurations of several plant metabolites [132]. They used identical chromatographic conditions for complementary HPLC-MS and HPLC-NMR data by coupling the same chromatograph and column to each type of spectrometer. Mistry and co-workers [133] used HPLC-CDS and HPLC-NMR to identify and determine the stereo configuration of components of the neuromuscular blocking agent Atracurium Besylate. They used a chiral column and found 10 isomers, in four enantiomeric pairs and two meso compounds. This combination of complementary detectors helps assign the specific configurations and structural features of complex biological mixtures. Since chiral bonded phases, like the Pirkle columns, are specifically designed to separate optical isomers from each other, their use with subsequent CDS is potentially a very powerful tool in the characterization of natural products and other biological materials.

A variation on HPLC-CDS is based on magneto-optical rotation. In a longitudinal electric field, almost all molecules show optical activity [134]. The electrical field orients the molecules through their permanent dipole moments and/ or through any anisotropic electrical polarizability. Kawazumi et al. [135] used this principle to build a detector for HPLC. Although the device was complicated, and prone to respond to minute changes in the refractive index or temperature of the mobile phase, it showed promise as a universal detector. This universality would only apply to detectability. Compounds with strong dipole moments and those that are easily polarizable would have more intense responses than compounds that do not.

The interesting combination of capillary electrophoresis-X-ray fluorescence detection was recently described as an element-specific detector [136]. A special plastic cell, which was both X-ray transparent and did not produce any interfering emission, was used. Vitamin B-12 (cyanocobalamin) and the cyclohexanediaminotetraacetic acid complexes of iron, zinc, cobalt, and copper were separated. Although the detection limits found were in the nanogram range, the authors estimate that with optimization that 2 or 3 orders of magnitude more sensitivity is likely.

Hill and Tarver [137] review the use of Fourier-transform ion-mobility spectroscopy with SFC. In this technique, sample molecules are places in an atmospheric-pressure ionization chamber. The entrance and exit slits are rapidly and synchronously gated. This causes the drift times of the ions to be either in- or out-of-phase with the gates. The frequency of the gates can be ramped to control which ions are in-phase, and thus pass through to the ion detector. The real utility of this effect is the introduction of reagent gases, such as oxygen or ammonia. These gases selectively react with the sample molecular ions. The resulting arrangement is a selective detector for chemical reactivities.

In theory, almost every type of spectrometer can potentially be used as a chromatographic detector since the samples are most often in either the gas or liquid states. Given this fact, the number of hyphenated techniques is only bound to grow as more spectroscopists couple a chromatograph to their instruments.

21.8
Serial or Parallel Multiple Detection

The combination of GC with FTIR and MS detection has been used for a variety of applications [138–142]. In all of these the GC effluent is split through a tee and each branch goes to one of the two detectors. Examples include a sample from suspected clandestine use of a chemical laboratory showed phenyl-2-propanone and other compounds indicative of methamphetamine production, the concentration of benzene, toluene, the xylene isomers, and 1,2,4-trimethylbenzene in a gasoline, the trace contaminants in solvents used in semiconductor manufacture, the three flavor essence isomers eugenol, and *cis*-and *trans*- isoeugenol (4-allyl-2-methoxyphenol and *cis*- and *trans*- 2-methoxy-4-propenylphenol), and the components in a perfume sample. In the last of these, the sample was previously run by GC-AED with C, H, O, and N detection to augment the other two types of detectors.

Unlike most of the other spectrometric detectors described here, the DAD and FSFD are non-destructive and can be readily used in a true flow-through mode (unlike NMR, which generally requires slower flow rates or stop-flow conditions). Therefore, the chromatographic eluent from them can easily be coupled to another subsequent detector. The commercial FSFD described earlier in fact typically is run in series with a DAD.

Destructve detectors, such as MS or AED, can be coupled to the output of the non-destructive detectors. This gives complementary and simultaneous information that can aid in peak identification. The retention data will show later retention times, but the sets of data can be reconciled for this lag through monitoring of standard compounds.

The series of DAD-MS detection has been the most commonly used. Quilliam and co-workers [143–145] determined polycyclic aromatic compounds and marine ecotoxins in this fashion. Their series of papers describes the coupling of a mass spectrometer to the outlet of the DAD. The MS mode was atmospheric–pressure chemical ionization, where the effluent was volatilized in a heated nebulizer. The mobile phase is preferentially removed, but the remaining solvent also acts as the chemical-ionization reagent. Both reversed-phase HPLC and supercritical fluid chromatography were used as the separation methods. This allowed separation of LPAHs of over 500 Da. This group was only able to identify a few of the LPAH peaks through the use of standards or reference UV absorbance spectra. These included coronene, benzo[*pqr*]naphtho [8,1,2-*bcd*]perylene, naphtho [8,1,2-*abc*]coronene, and ovalene. Bessant et al. used DAD and electrospray MS to measure hydroxypyridine isomers at different pHs. They noted peak shape difference, with the MS data showing much greater tailing. This was ascribed to the increased void volumes and the mechanism of solvent removal. The MS vacuum acts to effectively expand the void volume or increase post-column mixing because of the rapid expansion of the liquid volumes and the turbulence.

With the development of several commercial, commonly-available LC-MS interfaces, the combination of HPLC-DAD-MS is now becoming a routine sequential analytical approach. This should result in a burgeoning number of applications in a wide variety of fields. Additionally, with the FSFD now also being commercially available, the combination of HPLC-FSFD-MS is easily possible. In a classic paper, Paeden et al. [146)] separated the large PAHs from a carbon-black extract, collecting over 50 individual peaks. Fluorescence excitation and emission spectra and the mass spectrum of each fraction were then collected for characterization. Although the equivalent work has not yet been done, this same task of data collection would take less than a day by using the serial hyphenated technique approach in a few repetitive runs (to collect the two modes of fluorescence spectra at difference excitation and emission wavelengths). This would be in contrast to the several weeks required in the original work.

NMR is also a non-destructive technique, and a small number of sequential applications have been published. Wilson and co-workers [147] used HPLC-DAD-NMR-MS to characterize plant extracts. Hanson and co-workers [148] used a similar approach to examine another plant extract of pharmaceutical interest. In both cases, the complementary nature of the data provided quantitation of both major and minor constituents and aided in the structural identification of several of the minor components, including chiral isomers. Lommen et al. [149] describe a similarly configured system. The DAD and MS outputs were used to detect peaks, which were then transferred for NMR. They examined glycosides found in apple peel. They identified six quercetin glycosides and two phloretin glycosides, with the NMR data providing the definitive conformational data to differentiate the isomers.

References

1 J. C. Fetzer, W. R. Biggs, *J. Chromatogr.*, **1993**, *642*, 319–327.
2 L. Huber, S. A. George, *Diode-Array Detection in HPLC*, Marcel Dekker, New York 1993.
3 R. A. Freidel, M. Orchin, *Ultraviolet Spectra of Aromatic Compounds*, John Wiley and Sons, New York, 1951, spectra 7 and 17.
4 J. C. Fetzer, E. J. Gallegos, *Polycyclic Aromat. Compd.*, **1992**, *2*, 245–251.
5 F. Dieterich, R. Ettl, Y. Rubin et al., *Science*, **1991**, *252*, 548–551.
6 R. Ettl, I. Chao, F. Dieterich, et al., *Nature (London)*, **1991**, *353*, 149–153.
7 F. Dieterich, R. L. Whetten, C. Thilgen et al., *Science*, **1991**, *254*, 1768–1770.
8 J. C. Fetzer, W. R. Biggs, *J. Chromatogr.*, **1984**, *295*, 161–169.
9 J. C. Fetzer, W. R. Biggs, *Org. Prep. Proced. Int.*, **1986**, *18*, 290–294.
10 J. C. Fetzer, *Polycyclic Aromat. Cmpd.*, **1999**, *14/15*, 1–10.
11 H. J. Boessenkoof, P. Cleij, C. E. Goewie et al., *Mikrochim. Acta*, 1986, 2, 71–92.
12 D. M. Demorest, J. C. Fetzer, I. S. Lurie et al., *LC/GC Mag.*, **1987**, *5*, 128–142.
13 L. S. Ramos, R. J. Stewart, B. G. Rohrback, *Chromatogr. (LC/GC)*, **1987**, *2*, 95–102.
14 R. Tauler, G. Durand, D. Barcelo, *Chromatographia*, **1992**, *33*, 244–254.
15 R. Gargallo, R. Tauler, F. Cuesta-Sanchez et al., *Trends Anal. Chem.*, **1996**, *15*, 279–286.
16 K. Jinno, J. C. Fetzer, W. R. Biggs, *Chromatographia*, **1986**, *21*, 274–276.
17 J. C. Fetzer, W. R. Biggs, K. Jinno, *Chromatographia*, **1986**, *21*, 439–442.
18 K. Jinno, Y. Miyashita, S. Sasaki et al., *Environ. Monit. Assess.*, **1991**, *19*, 13–25.
19 C. H. Marvin, L. Allan, B. E. McCarry et al., *Environ. Mol. Mutagenesis*, **1993**, *22*, 61–70.
20 C. H. Marvin, M. Tassaro, B. E. McCarry et al., *Sci. Total Environ.*, **1994**, *156*, 119–131.
21 A. E. Legzdins, B. E. McCarry, C. H. Marvin et al., *J. Environ. Anal. Chem.*, **1995**, *60*, 79–94.
22 C. H. Marvin, J. A. Lundrigan, B. E. McCarry et al., *Environ. Toxicol. Chem.*, **1995**, *14*, 2059–2066.
23 C. H. Marvin, B. E. McCarry, J. Villella et al., *Polycyclic Aromat. Cmpd.*, **1996**, *9*, 193–200.
24 C. H. Marvin, B. E. McCarry, J. A. Lundrigan et al., *Sci. Total Environ.*, **1999**, *231*, 135–144.
25 C. H. Marvin, R. W. Smith, D. W. Bryant et al., *J. Chromatogr.*, **1999**, *863*, 13–24.
26 J. C. Fetzer, W. R. Biggs, *Polycyclic Aromat. Compd.*, **1994**, *5*, 193–199.
27 J. C. Fetzer, J. R. Kershaw, *Fuel*, **1995**, *74*, 1533–1536.
28 M. J. Wornat, B. A. Vernaglia, A. L. LaFleur et al., *Twenty-Seventh International Symposium on Combustion*, The Combustion Institute, Pittsburgh, PA 1998, pp. 1677–1686.
29 M. J. Wornat, F. J. J. Vriesendorp, A. L. LaFleur et al., *Polycyclic Aromat. Compd.*, **1999**, *13* 221–240.
30 J. C. Fetzer, B. R. T. Simoneit, P. Garrigues et al., *Polycyclic Aromat. Compd.*, **1996**, *9*, 109–120.

31 B. R. T. Simoneit, J. C. Fetzer, in *Organic Geochememistry: Developments and Applications to Energy, Climate, Environment and Human History*, eds. J. O. Grimalt, C. Dorronsoro, A. I. G. O. A., Donostia, San Sebastian, Spain 1995, pp. 414–417.

32 J. R. Jadamec, W. A. Saner, Y. Talmi, *Anal. Chem.*, **1977**, *49*, 1316–1321.

33 J. C. Gluckman, M. Novotny, *J. High Resol. Chromatogr.*, **1985**, *8*, 672–677.

34 J. C. Gluckman, D. C. Shelly, M. Novotny, *Anal. Chem.*, **1985**, *57*, 1546–1552.

35 J. C. Gluckman, M. V. Novotny, *Chromatogr. Sci.*, **1989**, *45*, 145–173.

36 T. L. Cecil, PhD. Thesis, Virginia Commonwealth University, 1990.

37 R. Schuster, H. Schulenberg-Schell, *A New Approach to Lower Limits of Detection and Easy Spectral Analysis*, Hewlett Packard Instruments (now Agilent Technology Co.), Waldbronn, Germany 1998.

38 J. A. Burt, M. A. Dvorak, G. D. Gillespie et al., *Appl. Spectrosc.*, **1999**, *53*, 1496–1501.

39 M. P. Fogarty, D. C. Shelly, I. M. Warner, *J. High Resol. Chromatogr.*, **1981**, *4*, 561–568.

40 D. C. Shelly, M. P. Fogarty, I. M. Warner, *J. High Resol. Chromatogr.*, **1981**, *4*, 616–626.

41 J. C. Fetzer, *The Chemistry and Analysis of the Large Polycyclic Aromatic Hydrocarbons*, John Wiley and Sons, New York 2000, ch. 4.

42 K. E. Peters, J. M. Moldowan, *The Biomarker Guide*, Prentice Hall, Englewood Cliffs, NJ 1993.

43 S. G. Roussis, W. P. Fitzgerald, *Anal. Chem.*, **2000**, *72*, 1400–1409.

44 W. Schmidt, G. Grimmer, J. Jacob et al., *Toxicol. Environ. Chem.*, **1986**, *13*, 1–16.

45 W. Schmidt, G. Grimmer, J. Jacob et al., *Fresenius J. Anal. Chem.*, **1987**, *326*, 401–413.

46 S. A. Wise, B. A. Benner, H. Liu et al., *Anal. Chem.*, **1998**, *60*, 630–637.

47 J. M. Gieskes, B. R. T. Simoneit, A. J. Magenheim et al., *Appl. Geochem.*, **1990**, *5*, 93–101.

48 B. R. T. Simoneit, W. D. Goodfellow, J. M. Franklin, *Appl. Geochem.*, **1992**, *7*, 257–264.

49 O. E. Kawka, B. R. T. Simoneit, *Org. Geochem.*, **1994**, *22*, 947–978.

50 W. J. Simonsick, R. A. Hites, *Anal. Chem.*, **1986**, *58*, 2114–2121.

51 J. T. Watson, *Introduction to Mass Spectrometry*, Lippincott-Raven Publishers, Philadelphia 1997, p. 219.

52 A. Rosele-Mele, J. F. Carter, J. R. Maxwell, *J. Am. Soc. Mass Spectrom.*, **1996**, *7*, 965–971.

53 S. Suzuki, T. Kaneko, M. Tsuchiya, *Kankyo Kagaku*, **1996**, *6*, 511–520.

54 B. H. Kaye, *Science and the Detective*, Wiley-VCH, Weinheim1995, ch. 8.

55 S. Lacorte, I. Guiffard, D. Fraisse et al., *Anal. Chem.*, **2000**, *72*, 1430–1440.

56 Water Corporation, *Zspray Mass Detector for API LC/MS*, Milford MA 1999.

57 Hewlett-Packard, *Chemical Analysis*, 1999

58 K. Imatani, C. Smith, *Am. Lab.*, **1996**, **Nov.**

59 Hewlett-Packard, *Development of a new Orthogonal Geometry Atmospheric Pressure Ionization Interface for LC-MS*, 1998

60 J. Charlwood, H. Birrell, E. S. P. Bouvier et al., *Anal. Chem.*, **2000**, *72*, 1469–1474.

61 S. X. Peng, S. L. King, D. M. Bornes et al., *Anal. Chem.*, **2000**, *72*, **1913–1917**.

62 T. Suzuki, T. Yasumoto, *J. Chromatogr. A*, **2000**, *874*, 199–206.

63 M. Holcapek, H. Virelizler, J. Chamot-Rooke et al., *Anal. Chem.*, **1999**, *71*, 2288–2293.

64 A. Boyer, in *LCWorldTalk*, Shimadzu, Columbia MD, 2000

65 D. L. Deforce, E. G. Van den Eeckhout, in *Advances in Chromatography: Volume 40*, eds. P. R. Brown, E. Grushka, Marcel Dekker, New York, 2000, pp. 539–566.

66 G. d'Ascenzo, R. Curini, A. Gentili et al., in *Advances in Chromatography: Volume 40*, eds. P. R. Brown, E. Grushka, Marcel Dekker, New York 2000, pp. 567–598.

67 S. E. Nielsen, R. Freese, C. Cernett et al., *Anal. Chem.*, **2000**, *72*, 1503–1509.

68 C. S. Hsu, K. Kian, W. K. Robbins, *J. High Resol. Chromatogr.*, **1994**, *17*, 271–276.

69 J. Mao, C. R. Pancheco, D. D. Traficante et al., *Fuel*, **1995**, *74*, 880–887.

70 D. J. Asserud, L. Prokai, W. J. Simonsick, *Anal. Chem.*, **1999**, *71*, 4793–4799.

71 C. M. Pace, L. D. Betowski, *J. Am. Soc. Mass Spectrom.*, **1995**, *6*, 597–607.

72 E. Rosenberg, V. Kmetov, M. Grasserbauer, *Fresenius' J. Anal. Chem.*, **2000**, *366*, 400–407. (2000)

73 J. T. Mehl, D. M. Hercules, *Anal. Chem.*, **2000**, *72*, 68–73.

74 D. S. Braverman, *J. Anal. At. Spectrom.*, **1992**, *7*, 43–46.

75 G. Schminke, A. Seubert, *Fresenius' J. Anal. Chem.*, **2000**, *366*, 387–391.

76 T. De Smaele, L. Moens, R. Dams et al., *Fresenius' J. Anal. Chem.*, **1996**, *355*, 778–782.

77 M. Heisterkamp, T. De Smaele, J. P. Candelone et al., *J. Anal. At. Spectrosc.*, **1997**, *12*, 1077–1081.

78 L. Ebdon, E. H. Evans, W. G. Pretorius et al., *J. Anal. At. Spectrosc.*, **1996**, *9*, 939–943.

79 U. Fuchsluefer, G. Socher, H.-J. Grether et al., *Anal. Chem.*, **1999**, *71*, 2324–2333.

80 R. Subramanian, W. P. Kelley, P. D. Floyd et al., *Anal. Chem.*, **1999**, *71*, 5335–5339.

81 A. Preiss, N. Karfich, K. Levsen, *Anal. Chem.*, **2000**, *72*, 992–998.

82 K. Levsen, A. Preiss, M. Godejohann, *Trends Anal. Chem.*, **2000**, *19*, 27–48.

83. T. Kitayama, M. Janco, K. Ute et al., *Anal. Chem.*, **2000**, *72*, 1516–1522.

84 Bio Rad Laboratories, Cambridge MA, *Bio Rad's Infrared Detector*, 1999.

85 T. J. Bruno, *Sep. Sci. Technol.*, **1999**, *29*, 63–89.

86 A. Meyer, H. Budzinski, J. R. Powell et al., *Polycyclic Aromat. Compd.*, **1999**, *13*, 329–339.

87 D. F. Gurka, L. D. Betowski, *Anal. Chem.*, **1982**, *54*, 1819–1824.

88 S. Compton, P. Stout, *LC/ GC*, 1990, 12, 920–926.

89 K. Wilkins, I. Marr, B. B. Allen et al., *Differentiation of some Monoterpene alcohols in Organo Aromas*, Hewlett-Packard applications note IRD 91-3, 1991.

90 N. J. Huang, D. C. Sundberg, *Polymer*, **1994**, *35*, 5693–5698.

91 M. W. Raynor, K. D. Bartle, B. W. Cook, *J. High Resol. Chromatogr.*, **1992**, *15*, 361–366.

92 S. French, M. Novotny, *Anal. Chem.*, **1986**, *58*, 164–166.

93 H. Pichard, M. Caude, P. Sassiat et al., *Lab. Analusis*, 1988, 16, 131–133.

94 M. W. Raynor, G. F. Shilstone, K. D. Bartle et al., *J. High Resol. Chromatogr.*, **1989**, *12*, 300–302.

95 M. A. Healy, T. J. Jenkins, M. Polakoff, *Trends Anal. Chem.*, **1991**, *10*, 32–37.

96 M. W. Raynor, G. F. Shilstone, A. A. Clifford et al., *J. Microcolumn Sep.*, 1991, 3, 337–347.

97 T. J. Jenkins, G. Davidson, M. A. Healy et al., *J. High Resol. Chromatogr.*, **1992**, 15, 819–826.

98 C. H. Kirschner, L. T. Taylor, *J. High Resol. Chromatogr.*, **1994**, *17*, 61–67.

99 K. L. Norton, P. R. Griffiths, *J. Chromatogr. A*, **1995**, *703*, 503–522.

100 M. X. Liu, J. L. Dwyer, *Appl. Spectrosc.*, **1996**, *50*, 349–350.

101 S. L. Jordan, L. T. Taylor, *J. Chromatogr. Sci.*, **1997**, *35*, 7–13.

102 B. Quimby, P. C. Dryden, J. J. Sullivan, *Anal. Chem.*, **1990**, *62*, 2509–2512.

103 G. W. Somsen, S. K. Coulter, C. Gooijer et al., *Anal. Chim. Acta*, **1997**, *349*, 189–197.

104 M. S. Jimenez, R. E. Sturgeon, *J. Anal. At. Spectrosc.*, **1997**, *12*, 597–601.

105 Y. Liu, V. Lopez-Avila, M. Alcaraz, *J. High Resol. Chromatogr.*, **1993**, *16*, 106–112.

106 Y. Lin, V. Lopez-Avila, M. Alcaraz et al., *J. AOAC Int.*, **1995**, *78*, 1275–1285.

107 M. K. Behlke, P. C. Uden, M. M. Schrantz et al., *Anal. Chem.*, **1996**, *68*, 3859–3866.

108 Agilent Technologies, *Sensitive and Selective Universal Element Detection for Routine or Research Analyses*, 1999

109 D. Johnson, B. Quimby, J. Sullivan, *Am. Lab.*, **1995**, Oct.

110 P. L. Wylie, B. D. Quimby, *A Method Used to Screen for 557 Pesticides and Suspected Endocrine Disrupters*, Agilent Technology, Wilmington, DE 1998.

111 B. D. Quimby, V. Giarrocco, J. J. Sullivan et al., *J. High Resol. Chromatogr.*, **1993**, *15*, 705–709.

112 T. G. Albro, P. A. Dreifuss, R. F. Wormsbecher, *J. High Resol. Chromatogr.*, **1993**, *16*, 13–17.

113 B. D. Quimby, D. A. Grudoski, V. Giarrocco, *J. Chromatogr. Sci.*, **1998**, *36*, 435–443.

114 M. M. Boduszynski, D. A. Grudoski, C. E. Rechsteiner et al., in *Proceedings of the 6th INITAR International Conference*, 1995, vol. 2, pp.335–244.

115 J. T. Andersson and K. J. Sielex, J. High Resol. Chromatogr., 19 49-53 (1996)

116 L. A. Bergdahl, A. Schuetz, A. J. Grubb, *J. Anal. At. Spectrosc.*, **1996**, *11*, 735–738.

117 B. D. Quimby, P. C. Dryden, J. J. Sullivan, *J. High Resol. Chromatogr.*, **1991**, *14*, 110–116.

118 J. G. Reynolds, W. R. Biggs, J. C. Fetzer et al., *Collect. Colloq. Semin. (Inst. Fr. Pet.)*, **1984**, *40*, 153–159.

119 W. R. Biggs, J. C. Fetzer, R. J. Brown et al., *Liq. Fuels Technol.*, **1985**, *3*, 397–422.

120 J. G. Reynolds, W. R. Biggs, J. C. Fetzer, *Liq. Fuels Technol.*, **1985**, *3*, 423–447.

121 W. R. Biggs, R. J. Brown, J. C. Fetzer, *Energy Fuels*, **1987**, *1*, 257–262.

122 W. R. Biggs, J. G. Reynolds, J. C. Fetzer, Cosmochim. Geochim. Acta, **1988**, *52*, 2337–2341.

123 W. R. Biggs, J. C. Fetzer, R. J. Brown, *Anal. Chem.*, **1987**, *59*, 2798–2802.

124 W. R. Biggs, J. C. Fetzer, *Anal. Chem.*, **1989**, *61*, 236–240.

125 Z. Slejkovec, I. Salma, J. T. van Elteren et al., *Fresenius' J. Anal. Chem.*, **2000**, *366*, 830–834.

126 D. J. Butcher, A. Zybin, M. A. Bolshov et al., *Anal. Chem.*, **1999**, *71*, 5379–5385.

127 T. D. Nguyen Hong, M. Jouan, N. Quy Dao et al., *J. Chromatogr. A*, **1996**, *743*, 323–327.

128 S. D. Cooper, M. M. Robson, D. N. Batchelder et al., *Chromatographia*, **1997**, *44*, 257–262.

129 L. M. Calabin, A. Ruperez, J. J. Laserna, *Anal. Chim. Acta*, **1996**, *318*, 203–210.

130 Bio-Rad Laboratories, Cambridge MA

131 A. Yamamoto, A. Matsunaga, K. Hayakawa et al., *J. Chromatogr. A*, **1996**, *727*, 55–59.

132 G. Bringmann, K. Messer, M. Wohlfarth et al., *Anal. Chem.*, **1999**, *71*, 2678–2686.

133 N. Mistry, A. D. Roberts, G. E. Tranter et al., *Anal. Chem.*, **1999**, *71*, 2838–2843.

134 J. Michl, E. W. Thulstrip, *Spectroscopy with Polarized Light*, Wiley-VCH, Weinheim 1986, p. 45.

135 H. Kawazumi, H. Nishimura, Y. Otsubo et al., *Anal. Sci.*, **1991**, *7*, 1479–1480.

136 S. E. Mann, M. C. Ringo, G. Shea-McCarthy et al., *Anal. Chem.*, 2000, 72, 1754–1758.

137 H. H. Hill, E. E. Tarver, in *Hyphenated Techniques in Supercritical Fluid Chromatography and Extract*, ed. K. Jinno, Elsevier, Amsterdam 1992, pp. 9–24.

138 *Clandestine Lab Reaction Mixture*, Hewlett-Packard GC-IR-MS application note IRD87-2, 1987.

139 *Analysis of Aromatics in Petroleum*, Hewlett-Packard GC-IR-MS application note IRD87-4, 1987.

140 *Semiconductor Solvents by GC/IR/MS*, Hewlett-Packard GC-IR-MS application note IRD86-4, 1986.

141 R. J. Leibrand, *The Power of the IRD: Isomer Differentiation*, Hewlett-Packard application note IRD88-7, 1988.

142 R. J. Leibrand, B. D. Quimby, M. Free, *The Use of Multispectral Analysis in the Characterization of Perfume*, Hewlett-Packard application note IRD 92-2, 1992.

143 P. G. Sim, M. A. Quilliam, S. Pleasance et al., *Polycyclic Aromat. Compd.*, **1993**, *3* (suppl.) 321–328.

144 J. F. Anacleto, L. Ramaley, R. K. Boyd et al., *Can. Rapid Commun. Mass Spectrom.*, **1991**, *5* 149–155.

145 A. L. LaFleur, K. Taghizadeh, J. B. Howard et al., *Am. Soc. Mass Spectrom.*, **1996**, *7*, 276–286.

146 P. A. Peaden, M. L. Lee, Y. Hirata et al., *Anal. Chem.*, **1980**, *52*, 2268–2271.

147 I. Wilson, E. D. Morgan, R. Lafont et al., *Chromatographia*, **1999**, *49*, 374–378.

148 S. H. Hansen, A. G. Jensen, C. Cornett et al., *Anal. Chem.*, **1999**, *71*, 5235–5241.

149 A. Lommen, M. Godejohann, D. P. Venema et al., *Anal. Chem.*, **2000**, *72*, 1793–1797.

Section XI
General Data Treatment: Data Bases/Spectral Libaries

Introduction

Kurt Varmuza

Production, storage, treatment, evaluation, exploitation and interpretation of spectroscopic data is strongly connected with intensive use of computers. Modern spectroscopy – as many other disciplines – would not be possible without the tremendous developments in hardware and software during the last decades. Most of the mentioned aspects are treated in the three parts of this section, although the authors have focuses on different subjects.

Part *Optical Spectroscopy* by S. Thiele and R. Salzer for instance contains the topics basic data treatment, IR- and UV-databases and spectra similarity search methods. Emphasis is given to multivariate calibration mehtods that are now routinely used for quantitative analyses of compounds in complex matrices. Principles of the widely used chemometric methods, such as PCA, PCR, and PLS are explained together with their applications in IR spectroscopy.

Part *Nuclear Magnetic Resonance Spectroscopy* by W. Robien focuses on structure elucidation of organic compounds. Spectra similarity searches, spectrum prediction (from a given chemical structure), recognition of substructures and automatic isomer generation are the main topics; they are still areas of scientific research in computer-assisted structure elucidation.

Part *Mass Spectrometry* by A. N. Davies gives an overwiew from a user's point of view. Commercially available mass spectral databases and software products for library searches are characterized. A statement from section 24.1 is repeated here because it seems to be essential not only for mass spectral database systems: "... let the people who will be working with the systems have a major say in the testing and selection of the product to be purchased ...".

As a supplement to the more than 80 references given in the three parts an overwiew of books is presented here that are relevant to chemometrics [1] and its applications in spectroscopy. Two comprehensive standard books on chemometrics have been published by *D. L. Massart* et al. [2], and *B. G. M. Vandeginste* et al [3]. The predecessor of these books probably was the most used volume in chemometrics for many years [4]. Introductory and smaller books are from *M. J. Adams* (focus on analytical spectroscopy) [5], *K. R. Beebe* et al. (almost no mathematics) [6], *R. G. Brereton* [7], *R. Kramer* (focus on multivariate calibration) [8], and *M. Otto* [9]. The classical book on multivariate calibration in chemistry is from *H. Mar-*

tens and *T. Naes* [10], one on neural networks in chemistry from *J. Zupan* and *J. Gasteiger* [11].

[1] K. Varmuza, in: P. v. R. Schleyer, N. L. Allinger, T. Clark, J. Gasteiger, P. A. Kollman, H. F. Schaefer III, P. R. Schreiner (Eds.), the Encyclopedia of Computation Chemistry, Wiley & Sons, Chichester, 1998, p. 346–366.

[2] D. L. Massart, B. G. M. Bandegiste, L. C. M. Buydens, S. De Jong, J. Smeyers-Verbeke, Handbook of chemometrics and qualimetrics: Part A. Elsevier, Amsterdam, 1997

[3] B. G. M. Vandeginste, D. L. Massart, L. C. M. Buydens, S. De Jong, J. Smeyers-Verbeke, Handbook of chemometrics and qualimetrics: Part B. Elsevier, Amsterdam, 1998.

[4] D. L. Massart, B. G. M. Vandeginste, S. N. Deming, Y. Michotte, L. Kaufmann, Chemometrics: a textbook. Elsevier, Amsterdam, 1988.

[5] M. J. Adams. Chemometrics in analytical spectroscopy. The Royal Society of Chemistry, Cambridge, 1995.

[6] K. R. Beebe, R. J. Pell, M. B. Seasholtz, Chemometrics: A practical guide. Wiley & Sons, New York, 1998.

[7] R. G. Brereton (Ed): Multivariate pattern recognition in chemometrics, illustrated by case studies. Elsevier, Amsterdam, 1992.

[8] R. Kramer, Chemometric techniques for quantitative analysis. Marcel Dekker, New York, 1998.

[9] M. Otto, Chemometrics. Wiley-VCH, New York, 1999.

[10] H. Martens, T. Naes, Multivariate calibration. Wiley, Chichester, 1989.

[11] J. Zupan, J. Gasteiger, Neural networks in chemistry and drug design. Wiley-VCH, Weinheim, 1999.

22
Optical Spectroscopy

Steffen Thiele and Reiner Salzer

22.1
Introduction

The spectra measured by any method of optical spectroscopy may be subject to qualitative (what is it?) or quantitative (how much is it?) evaluation. We assume here that the basic rules described in other chapters for the overall analytical process are obeyed, in particular for sample selection and sample preparation. Errors during sample preparation or simply due to an incorrect positioning of the specimen in the optical beam can never be corrected for in the measured spectra. Restricted quality in the experimental spectra will lead to errors either in qualitative evaluation (e.g. ill-defined results in spectral search) or in quantitative evaluation (e.g. erroneous determination of concentration).

Modern spectrometer software offers a variety of mathematical tools for processing spectra. These tools provide powerful features to the experienced user but cause serious danger in the case of non-critical application. Some types of processing tools do not alter the information content of the spectra (cf. Tab. 22.1, Basic operations), others do (cf. Tab. 22.1, Manipulations). Particular attention has to be paid if the latter have to be used.

The evaluation of spectra will be discussed separately for qualitative and quantitative analysis. Particular emphasis will be laid (i) on state-of-the-art methods for searching spectra in spectral libraries or searching for spectroscopic information in data banks and on (ii) procedures for multivariate data analysis.

Two of the basic operations mentioned in Tab. 22.1 are of particular importance for the evaluation of spectra, centering and standardization. They will be considered first.

Table 22.1 Types and objectives of spectra processing.

Basic Operations	Prettier Display	Qualitative Evaluation	Quantitative Evaluation
conversion transmission T – absorbance A	X		X
conversion wavelength λ – wavenumber k	X	X	X
background reduction	X	X	X
centering		X	X
standardization		X	X
spectra subtraction			X
Manipulations			
background correction	X	X	
smoothing	X	X	X !
derivation	X	X	
deconvolution		X	
curve fitting		X	(X)
peak integration			X
generation of band tables		X	
spike correction (Raman spectra)	X	X	

22.2
Basic Operations

22.2.1
Centering

Centering is an important step in the pretreatment of spectral data prior to multivariate evaluation both for qualitative and for quantitative analysis. At first the mean spectrum a_M^T is calculated from the set of measured spectra:

$$a_M^T = (\bar{a}_1, \bar{a}_2, \ldots, \bar{a}_p), \tag{1}$$

$$\text{where } \bar{a}_j = \frac{1}{n} \sum_{i=1}^{n} a_{ij} \quad j = 1, \ldots, p. \tag{2}$$

The subscript j represents the number p of spectral data points, the subscript i the number n of measured spectra. Subsequently, the mean spectrum is subtracted from each measured spectrum:

$$a_{C;i}^T = a_i^T - a_M^T \quad i = 1, \ldots, n \tag{3}$$

As a result of the above data treatment the mean of the centered spectra $a_{C;i}^T$ amounts to zero, i.e. *the center of the data set of n spectra has been shifted to the origin of the coordinate system* (cf. Fig. 22.1). All subsequent matrix operations in

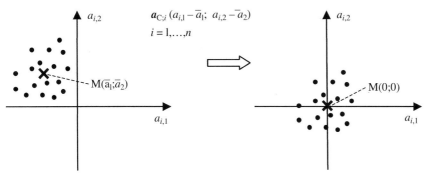

Fig. 22.1 Effect of centering the measured absorbance values $a_{i,1}$ and $a_{i,2}$ illustrated for two spectral data points ($p = 2$).

either qualitative or quantitative evaluation benefit from the centering, because the overlaid offset of the center has been removed and only the significant scattering of the measured spectra around the center is retained.

22.2.2
Standardization (Autoscaling)

Standardization is a method of pretreatment of spectral data prior to multivariate evaluation both for qualitative and for quantitative analysis, just like centering. In the case of standardization, the standard deviation s is calculated at each of the p spectral data points for all n measured spectra:

$$s_j = \sqrt{\frac{\sum_{i=1}^{n}(a_{ij} - \bar{a}_j)^2}{n-1}} \quad j = 1, \ldots, p \quad (4)$$

The standardized absorbance value of each measured spectrum ($i = 1,\ldots,n$) at each spectral wavelength ($j = 1,\ldots,p$) is calculated by using

$$a_{s_{ij}} = \frac{a_{ij} - \bar{a}_j}{s_j} \quad (5)$$

The standardized set of spectral data shows a mean of zero and a variance of one. Standardization may also be useful for gathering and comparing multicolinearities in later evaluation steps (cf. Section 22.3).

One basic rule should be emphasized already here: the set of mathematical operations chosen for the calibration samples has to be applied in an absolutely identical manner to all subsequent test samples. This rule holds as well for the above-mentioned basic operations as for the more advanced operations described later in this chapter.

22.3
Evaluation of Spectra

22.3.1
Introduction

Contemporary spectrometers are able to produce huge amounts of data within a very short time. This development continues due to the introduction of array detectors for spectral imaging. The utilization of as much as possible of the enclosed spectral information can only be achieved by chemometric procedures for data analysis. The most commonly used procedures for evaluation of spectra are systematically arranged in Fig. 22.2 with the main emphasis on application, i.e. the variety of procedures was divided into methods for qualitative and quantitative analysis. Another distinctive feature refers to the mathematical algorithms on which the procedures are based. The dominance of multivariate over univariate methods is clearly discernible from Fig. 22.2.

In the case of qualitative analysis, unsupervised learning procedures are employed for explorative data analysis or for empirical investigation of samples with no additional information available. The analyst wants to sense the correlation within the data or the interrelation between the data and particular features or properties of the sample. Such information qualifies for structuring of data with respect to useful data range, property parameters and arrangement into classes. Excessive data amounts require reduction to a reasonable size by automated procedures prior to structuring. Principle Component Analysis (PCA) is an efficient method for such spectral data reduction. Supervised learning procedures are employed in order to assign new objects to already established classes (available additional knowledge). A simple case is the comparison of features between a new sample and previously characterized samples, e.g. during the search for an analyte spectrum in a spectral data base. The aim of this search is the elucidation of the composition or structure, respectively, of an unknown.

In the case of quantitative analysis, the amount of, or the exact relation between, the constituents of a compound or a mixture have to be established. The direct relation between the properties of a specimen and the concentration of its constituents could also be the aim of the quantitative investigation. In the latter case so-called calibration models have to be established, the corresponding model parameters have to be estimated, and they have to be confirmed by statistical methods. Calibration models established this way may then be used to determine, on a statistically verified basis, the concentration of constituents of an analyte within the calibrated range. All multivariate methods for quantitative analysis mentioned in Fig. 22.2 are employed for the evaluation of spectra.

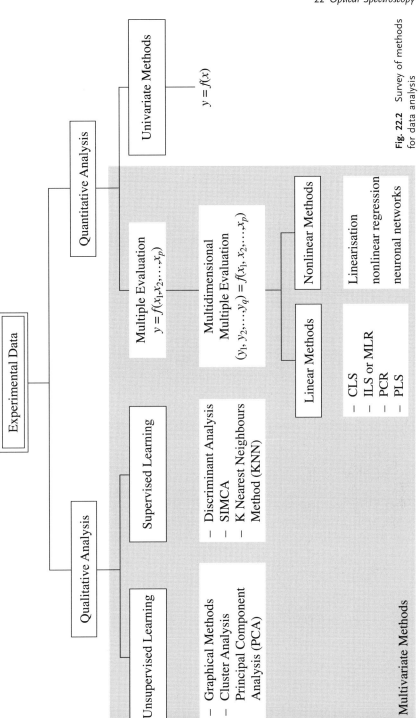

Fig. 22.2 Survey of methods for data analysis

22.3.2
Qualitative Evaluation of Spectra

22.3.2.1 Spectral Data Banks

Many digital spectral libraries have been transformed from printed spectra collections. Well known printed collections are the Aldrich spectra collection [1], the Sadtler spectra collection [2–4], the Schrader–Meyer Atlas of IR and Raman spectra [5], the Hummel collection of IR spectra of polymers [6], the Merck IR Atlas [7] and the Buback collection of NIR spectra [8]. IR spectra have the largest share of digital optical spectra, followed at a clear distance by Raman spectra. Larger collections of UV/VIS spectra have not been established due to their missing fingerprint capability and to the strong sensitivity of the UV/VIS spectra to solvent interactions. A variety of dedicated spectra collections have been created in industrial laboratories without access to the public.

The difference between in-house and on-line versions of spectral libraries consists mainly in the fee, which depends upon the conditions of use. Usually, in-house versions have to be paid only once upon license acquisition and hardware purchase. Afterwards, the actual use is free of charge, regardless of its frequency. In certain cases an annual license fee has to be paid. Follow-up costs may occur if updates are demanded or if hardware service is needed. In the case of in-house systems running on UNIX workstations both initial costs as well as upkeep may be considerable. In contrast, the costs for on-line systems solely depend upon frequency of use. Both the reliability and speed of the internet connection may remarkably influence the cost. If only used occasionally the on-line version is certainly financially more viable than the in-house version. At best one is able to find free-of-charge internet offers, sometimes with restricted access to certain features (cf. Tab. 22.2).

Spectral data banks contain all sorts of information about a particular substance in the form of tables. The requested field of a table can be accessed by the user either via abbreviations (e.g., MF for molecular formula) or via input masks, which place the necessary denominations of the field at the user's disposal. Some fields may contain searchable information only as alphanumerical text (e.g., compound names), others may be searched only numerically (e.g., molecular weight). In the case of numerical fields, some numerical operators (e.g., <; =; > or – for area allocation) may also be applied.

It is often possible to search for bands of particular intensity in selected wavelength ranges. During the search all fields may be interconnected logically (search masks) or by logic operators (and; or; not; proximity operators). A summary of information contained in a spectral data bank for a given compound is given in Tab. 22.3.

The CAS number (Chemical Abstracts Service registry number) is of particular importance. This number went into a wide range of data banks (structural and factual data banks, bibliographic data banks, substance data banks) and may advantageously be used for fast access to information related to a particular substance. Data banks comprising information about several spectroscopic methods are of

Table 22.2 Optical spectral data bases (according to provider specifications, 01.09.2000).

Data Base	Type and Amount of spectra (if Available)	Availability		Charge
		In-house version	On-line version	
Aldrich				
Condensed Phase	FT-IR: 18.500	X		yes
	FT-Raman: 14.000	X		
Vapor Phase	FT-IR: 5.000	X		yes
Environmental Protection Agency (EPA)				
Vapor Phase	FT-IR: 3.300	X		yes
Fiveash Data Manag., Inc.				
Spectra of Drugs/ Canadian Forensic Spectra	FT-IR: 3.750	X		yes
Vapor Phase of Organic Compounds	FT-IR: 5.220	X		yes
Special Spectra	FT-IR: 2.600	X		yes
Galactic Ind. Corp. (URL:www.galactic.com)	IR, MS, NMR,UV/VIS, NIR: 6.000		X	no
Galactic Ind. Corp. and Nicolet Instr. Corp. (URL:FTIRsearch.com)	FT-IR: 71.000		X	yes
	Raman: 16.000		X	yes
Nicolet Instr. Corp.				
Organic Chemical Library	FT-Raman: 1.000	X		yes
Vapor Phase Library (6.543 from Aldrich)	IR: 8.654	X		
Polymer Application Libraries and other special libraries	IR, Raman	X		yes
NIST				
WebBook (URL:webbook.nist.gov)	IR: 7.500	X	X	no
	UV/VIS: 400	X	X	
Sadtler				
Condensed Phase IR Standards	IR: 75.570	X		yes
Vapor Phase IR Standards	IR: 9.200	X		yes
Special Libraries	IR, Raman	X		yes
SDBS				
Organic compounds (URL:www.aist.go.jp/RIODB/db004/menu-e.html)	FT-IR: 47.300		X	no
	Raman: 3.500		X	no
SpecInfo (URL:www.chemicalconcepts.com/products.htm)	IR: 18.500	X	X	yes

22.3 Evaluation of Spectra

Table 22.3 Searchable information in spectral data banks.

Compound Identification	compound name incl. synonyms
	CAS number
	molecular formula
	molecular weight
	graphical structure representation
	additional data possible (e.g., melting point)
Acquisition of Spectra	sample source
	sample purity
	sample preparation
	spectrometer type
	spectral resolution
	additional data possible
	(e.g., special sample conditions; file format)
Spectrum	complete spectrum in numerical and graphical form
	band table

high value for structural elucidation if a combination of methods has to be applied (e.g., NIST, SDBS, SpecInfo).

The spectral search itself is based on computation of similarity by comparison of full spectra or of peak tables. Necessary means are structural editors and spectral editors, which are usually applied off-line before the spectral search is started. Spectral editors may be used to erase existing peaks from or to generate new peaks into spectra. Spectral editors may also be used for the complete generation of new spectra. Structural editors are used to generate chemical 2D structures or to modify structures, which were previously obtained from chemical drawing programs. These structures can be searched in the data bank after they have been encoded by the structural editor. Substructure searches are uncommon in IR data banks because IR spectra are used for fingerprint identification of compounds rather than for assembling full structures from subunits.

In the case of a spectral search by comparison of peak tables, the system first computes the necessary peak table from the experimental spectrum of the analyte. For the subsequent search both the position and the intensity of the peaks in the spectrum of the unknown and in the spectra of the compounds contained in the data bank are compared. After the search has been completed, the top position in the hit list is assigned to the substance with least differences in peak positions and peak intensities.

In the case of a full spectra search, the complete set of spectral features (absorbance values at p wavelength positions) is compared between the spectrum of the unknown and all spectra contained in the data bank. So-called similarity measures are computed for each individual comparison. In the case of the mostly employed similarity measure, the Euclidean distance, the spectrum is regarded as a p-dimensional spectral vector (data points at p wavelength positions). The comparison

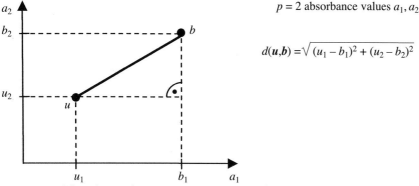

Fig. 22.3 Euclidean distance between vector **u** (spectrum of unknown substance) and vector **b** (library spectrum) illustrated for absorbance values at two wavelengths.

between the spectral vector **u** of the unknown and the vector **b** of the data bank spectrum results in a value $d(u,b)$ (cf. Fig. 22.3):

$$d(\boldsymbol{u},\boldsymbol{b}) = \sqrt{\sum_{i=1}^{p}(u_i - b_i)^2} \qquad (6)$$

The library spectrum showing the minimal Euclidean distance d_{min} from the spectrum of the unknown will be assigned the top position in the resulting hit list. In order to ensure comparability of the distances d each submitted spectrum has to be subjected to a particular normalization (cf. Fig. 22.4). At first, the lowest absorbance value across the spectrum is subtracted from all p data points of the spec-

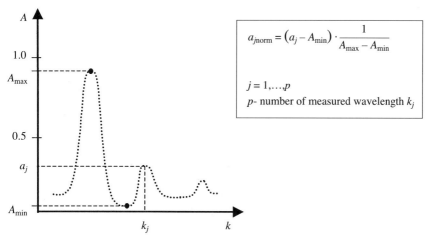

Fig. 22.4 Normalization of a spectrum to the absorbance range between 0 and 1.

trum. Afterwards, the spectrum is normalized to the required absorbance maximum. The necessity of such a normalization constitutes a serious limitation of the search procedure.

In contemporary search algorithms each spectral vector is normalized to unit length (unit vector x_E). The length of a vector x (also called absolute value or norm) is given by

$$\|x\| = \sqrt{x^T x} \tag{7}$$

and the norm vector x_E by

$$x_E = x/\|x\| \quad . \tag{8}$$

By means of the definition of the scalar product of two vectors x and y

$$x^T y = \|x\| \, \|y\| \cdot \cos(x, y) \quad , \tag{9}$$

the library search in the p-dimensional Euclidean space (where p is the number of measured wavelengths) results in the following distance value d_E:

$$d_E = 1 - \cos(u, b) = 1 - \frac{u^T b}{\sqrt{u^T u} \sqrt{b^T b}} \quad . \tag{10}$$

The calculated result corresponds to the cosine of the angle between the spectral vector u of the unknown and the vector b of the library spectrum. In the case of congruency of both vectors u and b we obtain $\cos(u,b) = 1$, thus $d_E = 0$.

A common problem for contemporary search algorithms, caused by varying baseline off-sets, can be overcome by centering the spectra (cf. Section 22.2). Centered spectra u_Z are obtained by calculating the average $\bar{u}$ of a spectral vector u measured at p wavelengths:

$$\bar{u} = \frac{1}{p} \sum_{i=1}^{p} u_i \quad . \tag{11}$$

The average $\bar{u}$ is subsequently subtracted from all components u_i of the vector u:

$$u_z = (u_i - \bar{u})^T \quad \text{with} \quad i = 1, \ldots, p \quad . \tag{12}$$

These centered vectors u_Z are now used for the calculation of a correlated distance d_C as in Eq. (10)

$$d_C = 1 - \cos(u_Z, b_Z) = 1 - \frac{u_Z^T b_Z}{\sqrt{u_Z^T u_Z} \sqrt{b_Z^T b_Z}} \quad . \tag{13}$$

Based on mathematical definitions, Eq. (13) can be reformulated (corr stands for correlation)

$$\mathrm{corr}(\boldsymbol{u},\boldsymbol{b}) = \frac{\boldsymbol{u}_Z^T \boldsymbol{b}_Z}{\|\boldsymbol{u}_Z\| \|\boldsymbol{b}_Z\|} = \frac{\boldsymbol{u}_S^T \boldsymbol{b}_S}{p-1} \quad . \tag{14}$$

After introduction of the standardized vectors $\boldsymbol{u}_S$ and $\boldsymbol{b}_S$

$$\boldsymbol{u}_S = \frac{\boldsymbol{u}_Z}{s(\boldsymbol{u})} \quad \text{and} \quad \boldsymbol{b}_S = \frac{\boldsymbol{b}_Z}{s(\boldsymbol{b})} \quad , \tag{15}$$

where s denotes the standard deviation of the vectors $\boldsymbol{u}$ and $\boldsymbol{b}$ (cf. Section 22.2):

$$s(\boldsymbol{u}) = \frac{\|\boldsymbol{u}_Z\|}{\sqrt{p-1}} \quad \text{and} \quad s(\boldsymbol{b}) = \frac{\|\boldsymbol{b}_Z\|}{\sqrt{p-1}} \quad , \tag{16}$$

we obtain

$$\boldsymbol{u}_S = \frac{\boldsymbol{u}_Z}{\|\boldsymbol{u}_Z\|} \cdot \sqrt{p-1} \quad \text{and} \quad \boldsymbol{b}_S = \frac{\boldsymbol{b}_Z}{\|\boldsymbol{b}_Z\|} \cdot \sqrt{p-1} \quad . \tag{17}$$

Equation (17) corresponds to Eq. (8) with the vectors now standardized to the length $\sqrt{p-1}$. In summary, the improvement in the correlated distance value is due to the removal of the baseline off-set and to the scale-invariance of the pre-treated vectors of the unknown and of the library spectrum. The following problems in searching for an unknown in a spectral library persist:

1. The spectrum cannot be satisfactorily normalized due to a non-horizontal or non-linear baseline.
2. The composition of the spectral library usually does not cover all necessary groups of chemical compounds.
3. The similarity scores which are used to construct the hit list are not necessarily significant measures for the similarity of structures.

A non-horizontal baseline may be corrected for by subtracting an angles straight line. In the case of a non-linear baseline, much experience is needed to minimize possible effects of a mathematical correction on the search results. Here, the computation of the second derivative of the spectrum might be the favorable option.

Even if an unknown belongs to a particular class of compounds, which is not represented in the library, the search will always result in a hit list with a number of entries. In general, entries with small scores are not relevant for an identification. In order to gain a feeling for the significance of an entry, with respect to its scores, one should experiment with the library, and a sound chemical knowledge is necessary.

22.3.2.2 Data Banks Containing Spectroscopic Information

Fields containing spectroscopic information can specifically be accessed in the data banks Beilstein (organic chemistry) and Gmelin (inorganic chemistry, organometallic chemistry) (cf. Tab. 22.4). Covered spectral ranges, measured spectral intensity maxima, and references, may be found. Data concerning the state of aggregation of the sample or the solvent may possibly be included as well. Of course, the full text option may be used in any data bank to search for spectroscopic data or experiments, e.g. in Chemical Abstracts or in Analytical Abstracts (example: raman spectrum AND ...).

If interdisciplinary matters are searched, bibliographic data banks like Biosis (biosciences, biomedicine), Medline (medicine), Inspec (physics, electroengineering, engineering) as well as all data banks containing patent information should be considered.

22.3.2.3 Interpretation of Spectra by Means of Group Frequencies and of Characteristic Bands

This topic has already been dealt with in Chapter 2. Currently there are numerous software products available on the market or under development, which are intended to support the task of spectra interpretation (e.g., IRMentor Pro [9], IR-Tutor [10], SpecTool [11]). Certain tools even permit a spectrum simulation of the structure guessed [12]. Some spectrometer operation systems include tools to aid simple spectra interpretation.

22.3.2.4 PCA (Principal Component Analysis)

PCA is an efficient method for data reduction in particular in spectroscopy (cf. Fig. 22.5). It also has to be performed prior to quantitative evaluations by PCR (principal component regression).

Table 22.4 Search fields for optical spectroscopy in the Beilstein and Gmelin data banks.

Beilstein		Vibrational Spectrum (CTVIB)
IRS	Infrared spectrum	Anisotropy of infrared bands
IRM	Infrared maximum	Degree of depolarization of Raman bands
RAS	Raman spectrum	Fermi resonance
RAM	Raman maximum	Fine structure of infrared bands
CTVIB	Vibrational spectrum	Intensity of infrared bands
		Intensity of Raman bands
Gmelin		Polarization of infrared bands
IRS	Infrared spectrum	Reflection spectrum in the infrared region
UVS	UV/VIS spectrum	Resonance Raman effect

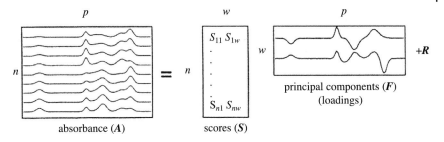

Fig. 22.5 Decomposition of the absorbance matrix A.

n – number of spectra
p – number of measured wavelengths = number of data points
v – number of neglected principal components
w – number of retained principal components $w = p - v$

The mathematical background of PCA consists in the transformation of the initial coordinate system into a new one in order to display the variance of the experimental data much more clearly. To this aim the mathematical algorithms provide that

- the principal components consist of linear combinations (i.e. weighted sums) of the initial variables (i.e. absorbances at the measured wavelengths);
- the principal components are computed in such a way as to cover the largest amount of variance (e.g., the variance in spectral data due to the different properties of the measured samples);
- the linear combinations represent new, so-called latent variables (e.g., variables which cannot be assigned to a particular spectral band) with appropriate properties.

Initially, the n experimental spectra, each comprising p data points, are collected into a n,p-dimensional data matrix A. Any row of the matrix A comprises all p absorbance values of a particular spectrum. Any column consists of all n absorbance values at a particular wavelength. As a first step the data matrix is either centered (A_Z) or standardized (A_S) (cf. Section 22.2). In order to achieve the above stated aims, this pretreated matrix is afterwards split into two matrices by the chosen algorithm:

$$A_Z = S \cdot F^T \tag{18}$$

dimensions $\quad n,p \quad n,p \quad p,p$

Matrix S is called the score matrix. Its rows comprise the scaling coefficients. Matrix F may be called the loading matrix or principal components (PCs) or factors or eigenvectors. Its columns comprise the calculated principal components. By multiplication of the matrices S and F we are able to reconstruct the centered or the standardized spectra.

Since the calculation of the principal components is based on the criterion of covering the largest amount of variance in the experimental data (A_Z or A_S), the first principal component features the maximum variance. Subsequent PCs cover less and less variance. Distant PCs may be omitted for data reduction. In that case Eq. (18) may be rewritten as

$$A_Z = S \cdot F^T + R , \qquad (19)$$
$$\text{dimensions} \quad n,p \quad n,w \quad w,p \quad n,p$$

where w represents the number of retained PCs ($w = p-v$), v is the number of neglected PCs, R is the residual matrix (error matrix) (cf. Fig. 22.5). An analysis of the error matrix R is necessary in order to chose w. Chemical knowledge has strictly to be applied during interpretation of the PCs, which are computed on pure mathematical considerations. Ideally, R merely contains the spectral noise as well as unnecessary information from the experimental spectra.

One method in error analysis is the computation of residual variances. The shares of different PCs in the total variance are sketched in Fig. 22.6.

If 6 PCs are retained for further evaluation, a residual variance in the experimental values remains beyond consideration. Other methods for estimating a reasonable size of w are the eigenvalue-one criterion [13, 14], the Scree-test [13, 14], and cross validation (cf. Section 22.3.3).

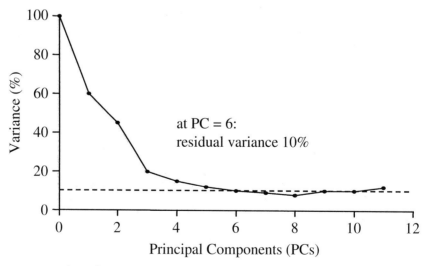

Fig. 22.6 Analysis of residual variance in PCA.

22.3.2.5 Cluster Analysis [13,14]

The aim of cluster analysis is the stepwise merger of objects (spectra) with respect to the similarity of their properties (absorbances at p measured wavelengths). A cluster comprises a group of objects whose similarity is closer than their similarity towards objects outside this group. The similarity of objects is assessed using the same distance measures as described earlier, e.g.

- Euclidean distance (cf. Fig. 22.3) or
- Mahalanobis distance [15, 16].

The calculation of the Mahalanobis distance is based on the interrelations between absorbance values at various wavelengths using their covariances. For this reason the Mahalanobis distance is of particular importance for assessing spectroscopic data.

Cluster analysis is often the preceding step for discriminant analysis.

22.3.2.6 Discriminant analysis [13, 14]

In the simplest case, a discriminant analysis is performed in order to check the affiliation (yes/no decision) of an unknown to a particular class, e.g. in case of a purity/quality check or a substance identification. A sample may equally well be assigned between various classes (e.g., quality levels) if a corresponding series of mathematical models has been established. Models are based on a series of test spectra, which has to completely cover the variations of particular substances in particular chemical classes. From this series of test spectra, classes of similar objects are formed by means of so-called discriminant functions. The model is optimized with respect to the separation among the classes. The evaluation of the assignment of objects to the classes of an established model is performed by statistically backed distance and scattering measures.

22.3.2.7 SIMCA Soft Independent Modeling of Class Analogy (SIMCA) [13,14]

In SIMCA, an independent principal component model (cf. PCA) is established for each individual class of the test data set. The evaluation of the assignment of objects to these classes of an established model is performed by statistically backed distance measures.

22.3.3
Quantitative Evaluation of Spectra

With respect to the applied mathematical algorithms, quantitative evaluation of spectra can be subdivided into univariate and multivariate methods (cf. Fig. 22.2). The independent variables x and x_i, respectively, are denoted regressors, whereas the dependent variables y and y_i, respectively, are denoted regressands. The basic sequence of a quantitative evaluation is always the same:

- Step 1: Choosing a model.
- Step 2: Choosing a training set.
- Step 3: Estimation of model parameters.
- Step 4: Validation of the model by statistical means.
- Step 5: Application of the model for prediction.

In the following sections the above sequence of steps will be discussed for all relevant methods used in evaluating optical spectra quantitatively.

22.3.3.1 Univariate Methods

The only method considered is:

Least squares regression (LSR)

Step 1: choosing a model
The Beer–Lambert law is expressed by

$$A_k = C \cdot K_k \quad , \tag{20}$$

where $\quad K_k = \varepsilon_k \cdot \log e \cdot l \tag{21}$

ε_k is the absorptivity coefficient at wavelength k, and l the thickness of the absorbing medium.

Within its application range, a linear calibration model can be established already by measuring at a single wavelength position

$$A = bC + a \tag{22}$$

The sample concentration C is calculated from the measured absorbance value A at the sensitivity b and the blank value a.

Step 2: choosing a training set
Spectra are measured with the analyte at n different concentrations. The absorbance values A_i are determined right at the maximum of a spectral band or peak, which is unambiguously assigned to the analyte.

The calibration based on peak area evaluation instead of a simple height determination might be advantageous. Likewise, rationing the heights or areas of two peaks might be useful to trace a concentration ratio. An initial baseline correction might be necessary. Preferably, a straight line should be employed for such corrections (cf. Fig. 22.7).

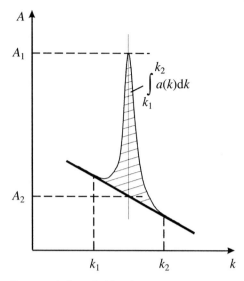

Fig. 22.7 Quantitative evaluation of a baseline corrected spectral band.

Step 3: estimation of model parameters
A least squares regression is performed in order to estimate $\hat{a}$ and $\hat{b}$:

$$\hat{a} = \frac{1}{n}\left(\sum_{i=1}^{n} A_i - \hat{b}\sum_{i=1}^{n} C_i\right); \quad \hat{b} = \frac{n\sum_{i=1}^{n} C_i A_i - \sum_{i=1}^{n} C_i \cdot \sum_{i=1}^{n} A_i}{n\sum_{i=1}^{n} C_i^2 - \left(\sum_{i=1}^{n} C_i\right)^2}. \quad (23)$$

Step 4: validation of the model by statistical means
This can be done in two ways
  (a) Analysis of residuals
  The analysis is performed by calculating

$$r_i = A_i - \hat{A}_i \quad i = 1, \ldots, n. \quad (24)$$

using the absorbance values $\hat{A}_i$, which were calculated according to the model by

$$\hat{A}_i = \hat{b}C_i + \hat{a} \quad (25)$$

and subsequent graphic representation of r_i over C_i (cf. Fig. 22.8)
  (b) Analysis of variance
  The variances s_a and s_b of the estimated parameters $\hat{a}$ and $\hat{b}$ are calculated according to

22.3 Evaluation of Spectra

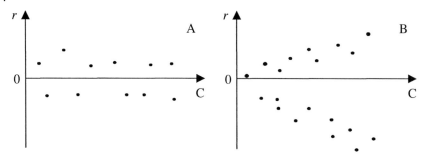

A: r_i shows normal distribution and uniform variance

B: r_i shows uneven variance

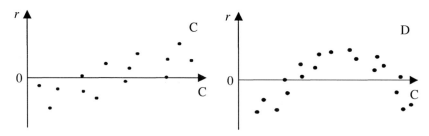

C: Erroneous linear model parameter

D: Non-linear model

Fig. 22.8 Examination of a linear model by analyzing the residuals.

$$s_a^2 = \frac{s_0^2 \sum_{i=1}^{n} C_i^2}{n \sum_{i=1}^{n} (C_i - \bar{C})^2} \; ; \quad s_b^2 = \frac{s_0^2}{\sum_{i=1}^{n} (C_i - \bar{C})^2} \quad , \tag{26}$$

$$\text{where } s_0^2 = \frac{\sum_{i=1}^{n} (A_i - \hat{A}_i)^2}{n - 2} \quad \text{and} \quad \bar{C} = \frac{1}{n} \sum_{i=1}^{n} C_i \quad . \tag{27}$$

The confidence intervals Δa and Δb are calculated according to

$$\Delta a = \pm t(P, f) s_a \quad \text{and} \quad \Delta b = \pm t(P, f) s_b \quad , \tag{28}$$

where t indicates a t-distribution, f the degree of freedom ($f = n-2$) and P the requested probability for the confidence interval.

Table 22.5 Advantages and disadvantages of least squares regression.

Least Squares Regression Advantages	Disadvantages
simple model	analyte has to have an isolated spectral band
fast calculation	no application to complex mixtures with overlaid spectral bands

Step 5: application of the model for prediction (cf. Tab. 22.5)
The predicted concentration C_{pred} may be calculated from measured absorbance values A_{meas} using the developed calibration model according to

$$C_{pred} = (A_{meas} - \hat{a})/\hat{b} \quad . \tag{29}$$

It should be mentioned here, that samples used for the estimation of the model parameters in steps 1–4 must not be re-used for prediction in step 5.

22.3.3.2 Multivariate Methods [13, 14, 17]

From a mathematical point of view, applications of multivariate methods may be subdivided into the multiple case and the multidimensional multiple case (cf. Fig. 22.2). In the former case, several independent variables or features are mapped to merely one dependent variable or target value. In the second case, several independent variables or features are mapped to several dependent variables or target values. As a rule, linear models are used for such problems in optical spectroscopy. In the case of non-linear relations, the calibration range gets restricted, a linearizing data pretreatment is performed in order to get away with linear models, or nonlinear methods (usually neural networks) have to be applied.

All models discussed below belong to the multidimensional multiple case.

(a) CLS Classical least squares (CLS), K-matrix

Step 1: choosing a model
The generalized form of the Beer–Lambert law for mixtures containing m components may be written in matrix form (cf. Fig. 22.9) as

$$\begin{array}{c} A = C \cdot K \\ \text{dimensions} \quad n,p \quad n,m \quad m,p \end{array} \quad . \tag{30}$$

The measured absorbance values of a spectrum (p data points) are contained in one row of the absorbance matrix A. The number of rows n in A corresponds to the number of measured spectra in the training set (cf. Fig. 22.5). In the concentration matrix C, a row contains the concentrations of the individual components of the

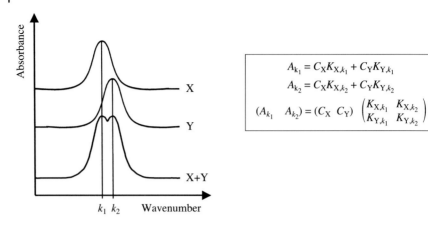

Fig. 22.9 Beer–Lambert law for a two component mixture.

particular training spectrum. K represents the matrix of the absorptivity constants. A row contains these constants of a particular component for all p measured wavelength positions.

Step 2: choosing a training set
Preparation of n mixtures by variation of the concentration of all m components and subsequent measuring of their absorbance spectra.

Step 3: estimation of model parameters
Matrices A (measured) and C (prepared) in Eq. (30) are now known. Estimation of the calibration matrix $\hat{K}$ is performed by the equation

$$\hat{K} = (C^T C)^{-1} C^T A \quad . \tag{31}$$

Step 4: validation of the model by statistical means
This can be done in several ways, e.g.
   (a) Analysis of residuals
   The residual matrix R is obtained by calculating the difference

$$R = A - \hat{A} \tag{32}$$

where $\hat{A} = C\hat{K} \quad .$ (33)

The n rows of R contain difference spectra. Each difference spectrum can be assigned a particular error value

$$R_i = \sum_{j=1}^{p} r_{ij}^2 \quad i = 1, \ldots, n \quad . \tag{34}$$

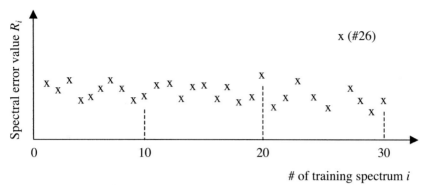

Fig. 22.10 Detection of outlier spectra.

These error values may be depicted graphically with the number n of the corresponding training spectrum on the abscissa (cf. Fig. 22.10). This graph permits an easy detection of outlier spectra.

(b) Cross validation

The following sequence has to be completed (leave-one-out strategy):

- Selection of a subset of n-1 training spectra.
- Estimation of a model.
- Prediction of the concentration of the omitted sample.
- Repetition of the above sequence until each of the n training spectra has been omitted and predicted once.

After completion of the above cycle, all n experimentally prepared concentration values (C_{exp}) are depicted graphically against the predicted concentration values

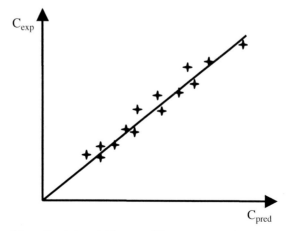

Fig. 22.11 Rating of the quality of the calibrated model for one of the m components.

Table 22.6 Advantages and disadvantages of classical least squares regression.

Classical Least Squares Advantages	Disadvantages
relatively fast calculation	the components of the sample have to be known completely, all concentrations have to be varied during calibration
no wavelength selection necessary, complete spectrum applicable	during prediction all these and only these components have to be present in the sample
large number of data points per spectrum and large number of calibration spectra ensures low noise	fails in case of new impurities or components which were not present in the mixture during calibration
	matrix inversions necessary both during calibration and during prediction

(C_{pred}) for each of the m components (cf. Fig. 22.11). The quality of the calibrated model can be rated from this set of m diagrams.

The data couples (C_{exp}; C_{pred}) in Fig. 22.11 may be evaluated by all statistical procedures. Cross validation is a very powerful method for outlier detection and optimization of calibration models. If the predicted concentrations obey the necessary quality standard for all but one training spectrum, the latter can be regarded as an outlier.

(c) Set validation

The training spectra are randomly separated into two sets. The first set is used for calibration, the second set for validation. Set validation does not demand as much computing power as cross validation but is inferior in the quality of the results, and should be applied only for large data sets.

Step 5: application of the model for predictions (cf. Tab. 22.6)
The calibrated model is finally used to predict the concentrations of the m components (vector c^T comprising m concentration values) from the experimental spectrum (vector a^T comprising p absorbance values):

$$c^T = a^T \cdot (\hat{K}^T \hat{K})^{-1} \hat{K}^T \quad . \tag{35}$$

(b) Inverse least squares (ILS), *P*-matrix or multiple linear regression (MLR, multidimensional)

Step 1: choosing a model
The generalized form of the Beer–Lambert law (cf. Eq. (30)) may be rearranged with respect to the concentration:

$$C = A \cdot P \quad . \tag{36}$$
$$\text{dimensions} \quad n,m \quad n,p \quad p,m$$

The structure of matrices C and A was already described for CLS. The elements of the P matrix are proportional to the reciprocal absorptivity constants. It is the crucial advantage of ILS over CLS, that only the concentrations of the components of interest have to be known during calibration.

Step 2: choosing a training set
Due to the mathematical requirement $n>p$ (dimension of the matrix) a very large number of training spectra should be measured. This might evoke the problem of collinearity among the spectra, i.e. the spectra would no longer be completely independent of each other, as mathematically necessary. Instead, they might be transformed into each other by linear combinations. Linear dependences occur, e.g., if calibration mixtures are simply diluted without changing the concentration ratios between the constituents. Spectra of such diluted samples contain redundant information, which in turn causes mathematical instabilities. In order to keep p small one should restrict the calibration to spectral areas which comprise distinct contributions by the components of the mixture. Such decisions demand chemical knowledge as well as sure instinct.

Step 3: estimation of model parameters
Using the matrices A (measured) and C (prepared) (cf. Eq. (36)) one can now estimate the calibration matrix $\hat{P}$:

$$\hat{P} = (A^T A)^{-1} A^T C \quad . \tag{37}$$

Step 4: validation of the model by statistical means
The validation of the model is performed in the same way as above for CLS.

Step 5: application of the model for prediction (cf. Tab. 22.7)
Based on a measured spectrum of an unknown (vector a^T comprising p absorbance values) and on the calibration matrix, the concentrations of the m components can now be predicted (vector c^T comprising m concentration values):

$$c^T = a^T \cdot (A^T A)^{-1} A^T C \quad . \tag{38}$$

Table 22.7 Advantages and disadvantages of inverse least squares regression.

Inverse Least Squares	
Advantages	Disadvantages
relatively fast calculation	difficulty in choosing the right spectral areas
the calibration model may only be based on knowledge about the interesting components, impurities are not important	Often a larger number of calibration samples necessary ($n>p$ required for mathematical reasons)
only one matrix inversion during calibration	multi-collinearity may cause problems
complex mixtures may be analyzed	time-consuming calibration

(c) Principal component regression (PCR)

Step 1: choosing a model
The main problem in the above described ILS is caused by possible multicollinearities between regressed spectra. This problem can be overcome by using a PCA (cf. Section 22.3.2) not only for data reduction but also to combine the data reduction with elimination of multicollinearities. The principal components computed this way are afterwards used as regressors in the ILS scheme.

Step 2: choosing a training set
- Full spectra or only spectral ranges showing contributions by the components of interest may be selected.
- Concentrations of the components of interest have to be known.
- Estimation of the principal components with respect to the chemical components to be calibrated (chemical knowledge!). This estimation based on chemical grounds ensures the closest relation between the variance represented in the principal components and the change in concentration of the chemical components of interest.

Step 3: estimation of model parameters
At first we introduce into the PCA (Eq. (18)) the equation $F^{-1} = F^T$, which is based on the orthogonality of F:

$$S = A_Z F \tag{39}$$

The second step yields in analogy to ILS (B corresponds to P)

$$C = S \cdot B = A_Z \cdot F \cdot B$$
$$\text{dimensions} \quad n,m \quad n,w \quad w,m \quad n,p \quad p,w \quad w,m \tag{40}$$

B represents the matrix of coefficients of the regression of the PCA scores (regressors) versus the concentrations of the chemical components (regressands). All other symbols are identical to those already described above or in Section 22.3.2.
The estimation of the calibration matrix $\hat{B}$ yields

$$\hat{B} = (S^T S)^{-1} S^T C \tag{41}$$

Step 4: validation of the model by statistical means
The selection of the number of principal components has already been described for PCA (Section 22.3.2). The validation of the model is effected in analogy to CLS.

Table 22.8 Advantages and disadvantages of principal component regression.

Principal Component Regression	
Advantages	Disadvantages
full spectrum or larger parts of a spectrum may be used	PCA demands much chemical knowledge and sure instinct
large number of spectral data points leads to noise reduction	large number of calibration samples necessary
only information about the components of interest necessary	large calibration effort necessary
rather complex mixtures may be analyzed	selection of non-significant principal components or neglect of important principal component possible
data reduction by PCA	

Step 5: application of the model for prediction (cf. Tab. 22.8)
The concentrations of all m components of interest (vector c^T containing m concentration values) can be predicted based on the measured spectrum of an unknown mixture (vector a^T containing p absorbance values), the principal component matrix F and the calibration matrix $\hat{B}$:

$$c^T = a^T \cdot F \hat{B} \quad . \tag{42}$$

(d) Partial least squares (PLS)

Step 1: choosing a model
The starting point is a model analogous to inverse calibration (ILS)

$$C = A \cdot B \quad . \tag{43}$$
$$\text{dimensions} \quad n,m \quad n,p \quad p,m$$

Like in PCA, both the concentration matrix C and the absorbance matrix A are decomposed into score and loading matrices S_C and F_C or S_A and F_A, respectively:

$$C = S_C \cdot F_C^T; \quad A = S_A \cdot F_A^T \quad . \tag{44}$$

The matrices C and A are decomposed interdependently with regard to their score matrices S_C and S_A, which maintains the close coherence between spectral information (A) and the component concentration (C) in the best possible manner.

Step 2: choosing a training set
Measurement of a large number n of training spectra, which cover as much of the concentration range as possible. The concentrations of the different components have to be varied independently.

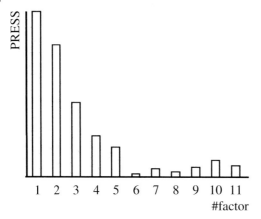

Fig. 22.12 Estimation of the number of significant factors by means of the prediction residual error sum of squares (PRESS).

Step 3: estimation of model parameters
The estimation of the calibration matrix $\hat{B}$ yields

$$\hat{B} = \hat{W}(F_A^T \hat{W})^{-1} F_C^T \quad , \tag{45}$$

where W represents the weight matrix, which is given a value during the decomposition.

Step 4: validation of the model by statistical means
The selection of an optimal number of factors (loadings) is a central point in PCR and PLS. In both methods the so-called prediction residual error sum of squares (PRESS) is calculated

$$\text{PRESS} = \sum_{i=1}^{n} \sum_{j=1}^{m} (C_{ij} - \hat{C}_{ij})^2 \quad , \tag{46}$$

where n is the number of training spectra, m the number of components in the mixture, C_{ij} are the prepared and $\hat{C}_{ij}$ the predicted concentrations. After a cross validation is performed, the prediction error (PRESS) can be depicted versus the number of employed factors in order to select an optimized number of factors.

In Fig. 22.12, the minimum for the prediction error is found at 6 factors. The following factors most render noise, their retention would deteriorate the quality of the model. Other validation is performed analogous to CLS.

Table 22.9 Advantages and disadvantages of partial least squares regression.

Partial Least Squares Advantages	Disadvantages
full spectra may be used	abstract model
most robust calibration	large number of training spectra necessary
only information about components of interest necessary	extended computing time
very complex mixtures may be analyzed	determination of the optimal number of factors difficult
computation of factors and regression in a single step, projection of the features of interest (concentrations) onto the factors superior to PCR	

Step 5: application of the model for prediction (cf. Tab. 22.9)

The concentrations of all m components of interest (vector c^T containing m concentration values) can be predicted based on the measured spectrum of an unknown mixture (vector a^T containing p absorbance values) and the calibration matrix $\hat{B}$:

$$c^T = a^T \hat{B} \quad . \tag{47}$$

A survey of presently available software products for data analysis is given in Table 22.10.

Table 22.10 Software products for data analysis.

Product	Manufacturer	Methods Covered
GRAMS/32, PLSplus/IQ Add-on	Galactic Ind. Corp.	LSR, PCA, PCR, PLS
MATLAB	MathWorks Inc.	
Chemometrics Toolbox	Applied Chemometrics Inc.	CLS, ILS, PCA, PCR, PLS
PLS_Toolbox	Eigenvector Research Inc.	
QuantIR	Nicolet Analytical Instruments	CLS, ILS, PLS
SCAN	Minitab Statistical Inc.	PCA, PCR, PLS and others
Statgraphics	Statistical Graphics Corp.	cluster analysis, discriminant analysis, PCA, linear and non-linear regression
Statistica	Stat Soft	cluster analysis, discriminant analysis, PCA, linear and non-linear regression
Unscrambler	Camo A/S	PCA, PCR, PLS

References

1 C. J. Pouchert, *The Aldrich Library of Infrared Spectra*, Aldrich Chemical Co., Milwaukee, WI 1981.

2 W. W. Simons, *The Sadtler Handbook of Infrared Spectra*, Sadtler Research Laboratory, Philadelphia, PA 1978.

3 *Sadtler Handbook of Ultraviolet Spectra*, Sadtler Research Laboratory, Philadelphia, PA 1979.

4 J. R. Ferraro, *The Sadtler Infrared Spectra Handbook of Minerals and Clays*, Sadtler Research Laboratory, Philadelphia, PA 1982.

5 B. Schrader, *Raman/Infrared Atlas of Organic Compounds*, VCH, Weinheim 1989.

6 D. O. Hummel, *Atlas of Polymer and Plastic Analysis*, VCH, Weinheim 1991.

7 K.G.R. Pachler, F. Matlok, H.-U. Gremlich, *Merck FT-IR Atlas*, VCH, Weinheim 1988.

8 M. Buback, H. P. Vögele, *FT-NIR Atlas*, Wiley-VCH, Weinheim, 1993.

9 Bio-Rad Labor. Inc., Sadtler Group, Philadelphia, USA.

10 C. Abrams, *IR Tutor*, Wiley-VCH, Weinheim 1998

11 *Spec-Tool*, Chemical Concepts, Weinheim, Germany.

12 URL:www2.ccc.uni-erlangen.de/research/ir/index.html.

13 M. Otto, *Chemometrics*, Wiley-VCH, Weinheim 1999.

14 R. Henrion, G. Henrion, *Multivariate Datenanalyse*, Springer-Verlag, Berlin, 1995.

15 P. C. Mahalanobis, *Proc. Natl. Inst. Sci. India*, **1936**, *2*, 49 (original reference).

16 R. De Maesschalck, D. Jouan-Rimbaud, D. L. Massart, *Chemom. Intell. Lab Syst.*, **2000**, *50*(1), 1–18.

17 H. Martens, T. Naes, *Multivariate Calibration*, John Wiley & Sons, Chichester 1991.

23
Nuclear Magnetic Resonance Spectroscopy

Wolfgang Robien

23.1
Introduction

Structure elucidation of organic compounds is mainly based on different types of spectroscopic techniques. Among the methods available NMR spectroscopy contributes a large amount of information; in many cases the spectral data are so rich in structural information content that the constitution, configuration and conformation of the unknown can be derived exclusively by the interpretation of such spectra. NMR spectroscopy is, in comparison to other techniques, an insensitive method leading to a low signal-to-noise ratio of the spectra obtained. This disadvantage has been circumvented by the introduction of the principle of Fourier-transformation leading to much shorter data acquisition times and the possibility of observing more insensitive nuclei than protons and therefore getting more detailed information about the sample under investigation. During the past three decades many sophisticated pulse techniques have been developed, which allow one, together with the use of highfield instrumentation, to investigate extremely small amounts of samples and additionally to get more complete information about them. In particular, the tremendous improvement based on two(three)-dimensional methods opened a new horizon in structure elucidation [1]. For all these reasons the earlier bottleneck of data acquisition as the most time-consuming step has been dramatically shifted to data interpretation, especially with the background of combinatorial chemistry and LC-NMR coupling.

Structure elucidation of a complex molecule from its spectral data without software support is a challenging and time-consuming procedure which can only be done by an expert in this field. In order to supply the expert with the necessary information during the spectrum interpretation and structure generation procedure, a large number of computer programs [2–4] and data collections have been developed [5].

23.2
Comparison of NMR-Spectroscopy with IR and MS

In order to understand the range of possibilities in computer-assisted structure elucidation methods it is essential to compare the different spectroscopic techniques with respect to their information content and their suitability for computer-assisted handling of their data. High resolution mass spectral data can lead to the molecular formula, which is a necessary parameter for most of the interpretation and isomer generation programs. MS is also well-suited to the detection of certain functional groups, as also is IR. The main disadvantage of MS is the extremely high dependence of the spectral pattern on the ionization technique applied; this effect can be used to get some more specific information about the unknown sample. Another reason why MS usually gives only supporting information during the structure elucidation process, is the complex and yet only partially understandable correlation between spectral and structural properties resulting in the absence of reliable methods to predict the MS spectrum for a given structure. Some work on this topic has been published [6], but the techniques developed have never been routinely applied. From IR spectra a large variety of functional groups and substitution patterns can be derived but the correlation between spectral behavior and structural property is not so well-defined as in NMR spectroscopy. For all these reasons MS and IR are not well-suited to rank a few hundred structural proposals as created by an isomer generation program, which is the usual case during computer-assisted structure elucidation, especially when dealing with molecules having a few non-carbon atoms which are only indirectly visible via the chemical shift values of the attached carbons.

NMR spectroscopy is an insensitive method compared to IR and MS, but it has been dramatically improved since its beginning. In organic chemistry ^{1}H- and ^{13}C-NMR-spectroscopies are usually routinely applied for structure elucidation purposes. The total shift range for protons spans roughly 12 ppm in typical organic compounds compared to 250 ppm for carbon. Solvent effects on protons are usually more pronounced, an effect – aromatic solvent induced shift (ASIS) – which is also systematically used in structure elucidation leads to the disadvantage that spectrum prediction by computer programs becomes more complicated. The comparison of a query spectrum against spectra in a reference library is also a more complex task for ^{1}H-NMR spectra because of the field dependence of the spectral patterns. In carbon NMR spectroscopy the couplings between carbons at natural abundance are usually invisible and the couplings to protons are artificially suppressed by decoupling techniques leading to simplified spectral patterns. Furthermore carbon NMR spectroscopy allows direct insight into the skeleton of an organic sample. For all these reasons discussed above it can be concluded that ^{13}C-NMR spectroscopy is the method of choice for computer-assisted structure elucidation, having the disadvantage of dealing with an insensitive nucleus, which is 5760 times more insensitive than a proton.

23.3
Methods in NMR Spectroscopy

The advantage of NMR spectroscopy is the large number of experimental techniques that allow one to derive a very specific piece of information about the unknown chemical structure. The parameters usually extracted from a NMR spectrum are:

- Chemical shift values
- Signal intensities
- Coupling constants

The chemical shift value reflects the electron density and therefore gives information about hybridization and the environment of the corresponding nucleus. The signal intensity is proportional to the number of nuclei and therefore gives information about symmetry. The coupling constants give information about spin systems and therefore insight into the relationship between different nuclei. Application of additional techniques allows one to determine the signal multiplicities in ^{13}C-NMR spectra (for example: attached proton test (APT) and distortionless enhancement by polarization transfer (DEPT)) and to get more insight into the coupling network. The most prominent two-dimensional techniques used are:

- H–H correlation: Correlated spectroscopy (COSY)
- C–H correlation over one bond: Heteronuclear multiple quantum coherence (HMQC), heteronuclear single quantum coherence (HSQC)
- C–H correlation over multiple bonds: Heteronuclear multiple bond correlation HMBC), correlation by long-range couplings (COLOC)
- C–C correlation: Incredible natural abundance double quantum transfer experiment (INADEQUATE)

Chemical shift values are the main information source for spectral similarity searches, whereas the correlation between spectral lines and the corresponding structural environment is the basis for spectrum prediction programs. The information derived from the correlation techniques is used by interpretation programs to prove the presence of certain structural fragments from a given set of correlation signals. These correlation signals are also used as distance constraints, in terms of number of bonds between two coupled nuclei, in the case of ambiguity, which occurs frequently in HMBC-type spectra, later on during the structure generation process.

23.4
Spectral Similarity Search Techniques

One very well-established method used in computer-assisted structure elucidation is the comparison of the spectrum of the unknown against a reference data collection. The largest databases available hold at the moment some 250,000 spectra of the same method, corresponding to roughly 1% of the known chemical structures. The consequence for the design of the algorithms is that the procedure used for comparison must be able not only to detect the identical spectral pattern in any case, but it must also be able to retrieve similar patterns, which usually give a lot of information about partial structures and typical skeletons contained in the unknown.

The basic algorithms for spectral comparison use the Euclidian distance between corresponding data points when comparing curves or a given deviation when comparing peak lists in order to select compounds having similar resonances. The procedure for comparing peak lists can only be applied sequentially in some computer programs, leading to the unwanted effect that similar structures can be excluded when starting with an uncommon chemical shift value. The better approach handles the complete peak list of the unknown at the same time and afterwards allows the selection of lines which must be present in the reference spectrum. Different implementations of this basic algorithm allow the user to control the number of lines present in the reference spectrum compared to the number of lines available in the query spectrum allowing him to analyze the reference structure with respect to structural fragments present in the unknown or to derive components within a mixture.

The disadvantage of such a type of similarity search based on the comparison of a peak list against the reference data collection on a line-by-line basis is that regions without any line are completely neglected. Another approach which also takes into account regions without lines is the SAHO (spectral appearance in hierarchical order) search method [7]. The typical range of the chemical shift values is divided into smaller ranges (typically 10–15 ppm for carbon) and the number of signals within such a region is counted. In order to achieve a higher selectivity different multiplicities (either odd/even or singlet, doublet, triplet and quartet) can be counted separately, leading to an array of numbers describing the spectral pattern very well. The same procedure is applied to the reference data collection and the resulting arrays are stored. The comparison of the pattern of the unknown against the reference patterns is extremely fast, because only a small amount of data must be handled (typically 8 byte per spectrum) leading to a speed of 10^5 to 10^6 comparisons per second. This algorithm is an ideal tool to deduce at least the compound class of the unknown under the assumption that the reference data collection contains examples of similar structures. In any case it should be mentioned that the results from the similarity search based on line-by-line comparison and the SAHO method are usually complementary and the user is strongly advised to apply both methods when available. These spectral similarity search techniques are extremely fast and should therefore be applied first during the structure elucidation process in order to avoid more time-consuming techniques for solving trivial problems.

23.5
Spectrum Estimation, Techniques

For the reasons given in the Introduction ^{13}C-NMR spectrum prediction is more popular than ^{1}H-NMR spectrum prediction, moreover, some programs are known to perform at a reasonable level of precision [8–10]. The early work of collecting chemical shift values was severely influenced by the idea of predicting spectra for a given structure. The first very simple, but still useful approach of increment rules has been implemented into a large variety of computer programs. The central concept of this method is to use a parent structure (for example benzene [11]) and to start with this chemical shift value in the calculation. Increments are derived from the difference between the chemical shift value of the parent compound and the corresponding chemical shift value of the mono-substituted derivatives. Polysubstituted compounds are treated as "overlapping" mono-substituted derivatives and the tabulated increments are simply added to the base value of the parent compound neglecting therefore any substituent interaction. More sophisticated systems allow for additional correction parameters leading to improved results [12, 13]. The advantages of this method are the simple principle behind it and the good results for certain compound classes where other methods tend to fail (for example polysubstituted benzenes). One main disadvantage is that only a limited number of increment tables for a limited range of parent compounds is available in the literature.

A more elegant method has been developed to make use of all compounds in a large reference data collection. The increment table for a certain query structure is generated on-the-fly and therefore all reference spectra contribute to the solution of a particular problem. This method automatically includes substituent interactions if this information is available within the reference data. Furthermore a poly-substituted compound can be generated by formally overlapping only mono-substituted derivatives (as with the basic increment method), but also by selecting for example di- or tri-substitued derivatives, allowing different calculation pathways leading to an expectation range for the signals of the query structure. The disadvantage of this method is the long computing time and the complex algorithms behind using partial structure search technologies [14].

A different approach is called the HOSE (hierarchically ordered spherical description of environments) code [15] that is based on the extremely well pronounced correlation between a ^{13}C chemical shift value and the corresponding carbon-centered substructural unit. The HOSE code starts at a carbon atom (focus atom) and describes its hybridization and multiplicity. The neighboring atoms are described in the same manner taking into account their atom type, hybridizations, number of directly bound hydrogens and the type of bonds between them. This scheme is applied to the first neighbors of the focus atom, then to the second neighbors and so on, leading to a spherical description of the structural environment. This carbon-centered fragment is sorted in a canonical way within each sphere giving a structure descriptor which is correlated to the known chemical shift value of its focus atom. This procedure is repeated for any carbon of any reference structure available and stored on a file during database creation. The query structure is analyzed in the same way and the fragments generated are compared against the corresponding file of the database. Chemical shift values of coincident structure descriptors are taken for the calculation of the mean value, the total shift range and the standard deviation. The number of coincident spheres between the reference structures and the query structure determines the precision of the result obtained. For sp^3 carbons at least three spheres (γ-effects) are necessary, in the case of conjugated systems four spheres usually give reliable results (effects of *para*-substituents). This basic principle has been implemented in a very similar way into several computer programs. Additionally, solvent induced effects can be added, leading to an improved spectrum prediction capability. The HOSE code was designed to handle a two-dimensional structure representation, therefore stereochemical effects were ignored. Consequently it was necessary to introduce a further extension which is able to describe steric interactions leading to an extreme improvement of the precision of the spectral prediction [16, 22].

A totally different approach is the utilization of neural network technologies as proposed by several authors [17–19]. The intellectual challenge of designing a neural network is the selection of the structure descriptors in order to reflect the correlation between structural and spectral properties in an optimal manner. Network optimization is a time-consuming task which must be done only once. The application of the trained network to a given structure is extremely fast and the prediction of the chemical shift values to be expected is performed within milliseconds, therefore giving an excellent tool for spectrum simulation and subsequent ranking of a large list of candidate structures.

23.6
Spectrum Prediction, Quality Consideration

There is a strong relationship between the quality of the database, the level of sophistication of the algorithms used and the results obtained. General databases are usually built from literature data; despite most of the assignments given in the public domain literature being correct, a large number of either wrong structures and/or misassignments are known [20]. Even some very common functional groups (for example tosylate [21]) are known to be systematically assigned in two different ways. The assignment simply depends on the literature used for reference purpose, thus propagating the wrong assignment when using the wrong reference without checking. These data may appear later in some databases and will be used for further prediction leading to unreliable results. This type of error can be easily detected by analyzing a large reference collection by means of statistical methods and should be frequently applied by a database administrator.

As stated above, spectrum prediction is based on the correlation between structural environments and their corresponding chemical shift values. In order to simulate spectra by the methods described adequate reference material is necessary, which is not given in some journals, frequently, instead of assignments only peak-lists (even without multiplicity information) as given by the spectrometer software have been published, diluting the basis for spectrum prediction of new classes of compounds.

In order to cover the enormous structural diversity it is necessary to have the most diverse database available. On the other hand it is necessary to fill gaps within the area of interest with ones own data. Usually a very small, but specifically dedicated database is much more powerful for solving a limited range of problems than a general database.

It is absolutely essential to have access to all parameters influencing the spectral simulation process; furthermore the typical parameters obtained as a result (mean value, range, deviation) are not sufficient to evaluate the result, because in many cases a visual inspection of the distribution is necessary in order to detect outliers or to understand stereochemical effects. Access to the original data contributing to a specific result must be possible in order to clarify any ambiguity.

Spectrum prediction is a frequently used technique during the structure elucidation process, but a detailed inspection of the results is necessary. Some programs offer the possibility to use different algorithms for spectrum prediction (usually HOSE code technology and neural networks). In such a situation both methods should be applied and the results obtained should be carefully compared [22]. At least in the case of different predictions a further critical evaluation of the result should be an obligation.

23.7
Spectrum Prediction and Quality Control, Examples

Spectrum prediction is an extremely decisive tool for verification of structural proposals and therefore the implementation of the algorithms and the quality of the reference database used cannot be discussed independently, because the results obtained are strongly connected to both parameters. The most severe limitation of the HOSE code is the availability of sufficient and correct reference material within the database, because a description of the query structure at a lower number of spheres dramatically influences the predicted chemical shift values. In Fig. 23.1 an assignment error on carbons 11 and 12 has been *artificially introduced* into the correct assignment as given in the literature in order to demonstrate the effect of accessing wrong entries within the reference database. The predicted spectrum as given in Fig. 23.2 using HOSE code technology with five coincident spheres (i.e. all neighbors up to five bonds) estimates 93.7 ppm and 94.3 ppm respectively for carbons 11 and 12 together with an expectation range of some 60 ppm which is totally useless for the spectroscopist. Estimation with two coincident spheres gives in this case the better mean values for carbon 11 and 12, again with extremely large expectation ranges as shown in Fig. 23.3. A comparison of the estimated values at one to five coincident spheres, as given in Fig. 23.4 demonstrates that there is only a small change in the mean values when increasing the number of coincident spheres, except for carbons 11 and 12 for which wrong reference material contributes heavily

23.7 Spectrum Prediction and Quality Control, Examples

12-HYDROXYCHILOSCYPHA-2,7-DIONE; COMPOUND-#4

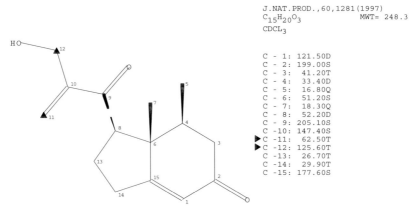

J.NAT.PROD.,60,1281(1997)
$C_{15}H_{20}O_3$ MWT= 248.3
$CDCL_3$

```
C -  1: 121.50D
C -  2: 199.00S
C -  3:  41.20T
C -  4:  33.40D
C -  5:  16.80Q
C -  6:  51.20S
C -  7:  18.30Q
C -  8:  52.20D
C -  9: 205.10S
C -10: 147.40S
▶C -11:  62.50T
▶C -12: 125.60T
C -13:  26.70T
C -14:  29.90T
C -15: 177.60S
```

Fig. 23.1 An entry from the reference database [22, 39] with an *artificially introduced* assignment error at carbons 11 and 12.

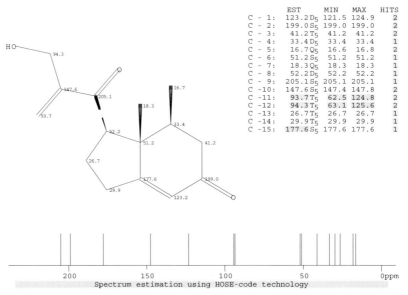

```
         EST      MIN   MAX   HITS
C -  1: 123.2 D5 121.5 124.9   2
C -  2: 199.0 S5 199.0 199.0   2
C -  3:  41.2 T5  41.2  41.2   2
C -  4:  33.4 D5  33.4  33.4   1
C -  5:  16.7 Q5  16.6  16.8   2
C -  6:  51.2 S5  51.2  51.2   1
C -  7:  18.3 Q5  18.3  18.3   1
C -  8:  52.2 D5  52.2  52.2   1
C -  9: 205.1 S5 205.1 205.1   1
C -10: 147.6 S5 147.4 147.8    2
C -11:  93.7 T5  62.5 124.8    2
C -12:  94.3 T5  63.1 125.6    2
C -13:  26.7 T5  26.7  26.7    1
C -14:  29.9 T5  29.9  29.9    1
C -15: 177.6 S5 177.6 177.6    1
```

Spectrum estimation using HOSE-code technology

Fig. 23.2 Spectrum estimation using 5 coincident spheres generating wrong predictions for carbons 11 and 12 at 93.7 and 94.3 ppm respectively, because of wrong reference material.

23 Nuclear Magnetic Resonance Spectroscopy | 477

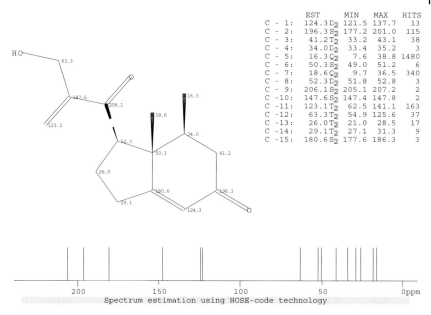

	EST	MIN	MAX	HITS
C - 1:	124.3 D_2	121.5	137.7	13
C - 2:	196.3 S_2	177.2	201.0	115
C - 3:	41.2 T_2	33.2	43.1	38
C - 4:	34.0 D_2	33.4	35.2	3
C - 5:	16.3 Q_2	7.6	38.8	1480
C - 6:	50.3 S_2	49.0	51.2	6
C - 7:	18.6 Q_2	9.7	36.5	340
C - 8:	52.3 D_2	51.8	52.8	3
C - 9:	206.1 S_2	205.1	207.2	2
C -10:	147.6 S_2	147.4	147.8	2
C -11:	123.1 T_2	62.5	141.1	163
C -12:	63.3 T_2	54.9	125.6	37
C -13:	26.0 T_2	21.0	28.5	17
C -14:	29.1 T_2	27.1	31.3	9
C -15:	180.6 S_2	177.6	186.3	3

Spectrum estimation using HOSE-code technology

Fig. 23.3 Spectrum estimation using two coincident spheres generating better predictions [123.1 ppm for C_{11} and 63.3 ppm for C_{12} respectively] than with five coincident spheres (see Fig. 23.4)

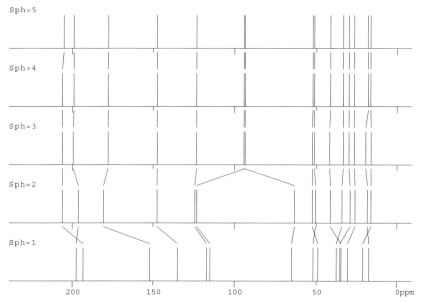

Fig. 23.4 Dependence of the predicted chemical shift values on the number of coincident spheres, given on the left-hand side. The large change when increasing the number of coincident spheres from two to three is worth a more detailed investigation of the reference material used.

to the mean values. From this example it can be concluded that the HOSE code method is very sensitive to wrong assignments, leading to low-quality spectrum predictions in the case of wrong reference material. Neural network technology is far less sensitive to assignment errors, despite the wrong dataset being used during training reasonable chemical shift values are predicted, as given in Fig. 23.5.

Another important feature of spectrum prediction is the utilization of stereochemical information. Stereochemical effects may induce effects of up to *ca.* 30 ppm in typical organic compounds. The four isomeric tricyclo-octane derivatives as given in Fig. 23.6 demonstrate this clearly, especially at C_8 with a chemical shift range starting at 23 ppm in the exo,exo-derivative and going up to 53 ppm in the endo,endo-configurated isomer. The comparison of the corresponding carbon NMR spectra in Fig. 23.7 shows these large increments caused by steric interaction.

Implementation of steric interactions into the HOSE code and into neural networks improves spectrum prediction dramatically, as can be seen from the podocarpane example in Fig. 23.8. The diastereotopic methyl groups located at C_4 can be well predicted – 21.7 ppm for the axial and 33.4 ppm for the equatorial methyl group, which is in excellent agreement with the values found in the literature. The separation can be easily understood when inspecting the distribution of the entries contributing to the predicted values as given in Fig. 23.9. A total number of 34 entries selected from the reference database predicts a useless mean value at 28 ppm for *both methyl groups* when using the structural descriptors from the

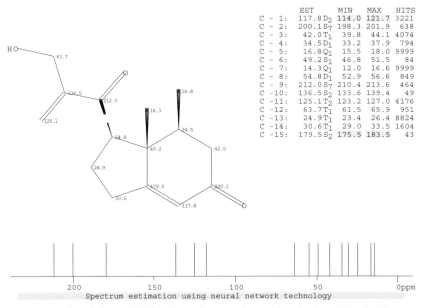

Fig. 23.5 Predicted chemical shift values using neural network technology giving 125.1 ppm for C_{11} and 63.7 ppm for C_{12}, respectively.

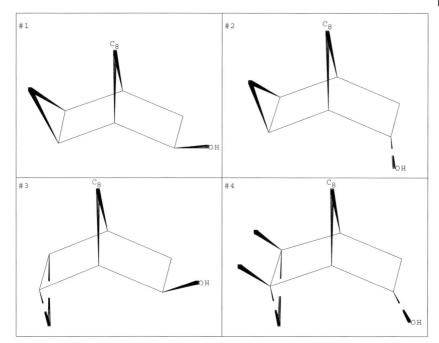

Fig. 23.6 Four isomeric tricyclo-octane derivatives showing extremely pronounced effects of stereochemistry on chemical shift values (see Fig. 23.7).

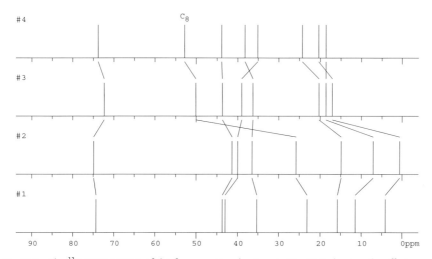

Fig. 23.7 The ^{13}C-NMR spectra of the four compounds given in Fig. 23.6 showing the effect of steric interactions on the chemical shift values.

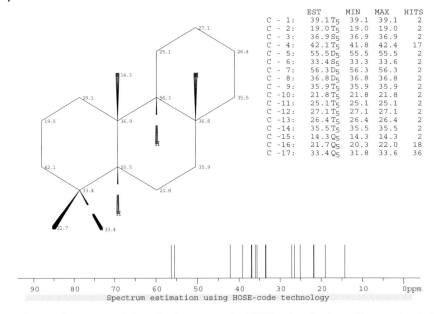

Fig. 23.8 Spectrum prediction of podocarpane using HOSE-code technology with stereochemical information [22]. The chemical shift values of the methyl groups located at position 4 are predicted in good agreement with the literature data.

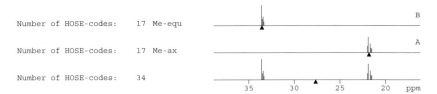

Fig. 23.9 Distribution analysis of the entries contributing to the estimation of the methyl groups at position 4 of podocarpane. Bottom trace: stereochemical effects are ignored leading to a useless mean value around 28 ppm (small triangle); a total number of 34 reference data contribute to this mean value. Trace A and B: Utilization of stereochemical interactions separates the 34 reference data into two distinct sets of chemical shift values leading to a correct prediction for the axial and equatorial methyl group.

original HOSE code disregarding stereochemical effects. The introduction of the number of 1,3-diaxial interactions divides these 34 reference data into two groups of shift values estimating a chemical shift value of 21.7 ppm for the axial methyl group and 33.4 ppm for the equatorial one, which is shown in traces A and B of Fig. 23.9.

23.8
Spectrum Interpretation and Isomer Generation

Isomer generation is the well-defined task to generate, exhaustively and without redundancy, all possible chemical structures that are consistent with a given set of constraints [23, 24]. This combinatorial problem itself is optimally suited for computer application and this definition seems to be ready for easy implementation into a piece of computer software. A more detailed look into the background of isomer generation shows immediately the problems associated with this task. The enormous number of possible candidate structures when starting from only the molecular formula can be seen in Tab. 23.1.

Even a small molecular formula around C_{10} produces a large number of possible chemical structures. The introduction of one heteroatom and/or a degree of unsaturation increases the size of the problem dramatically. The molecular formulae chosen in Tab. 23.1 represent comparably small compounds, far away from typical applications in modern organic chemistry. The main problem for making structure generation programs a common routine tool is the necessity to implement all available pieces of information from the most important spectroscopic techniques at the earliest possible step in order to avoid the "combinatorial explosion" and therefore

Table 23.1 Number of possible isomers (without stereochemistry) depending on the molecular formula and constraints derived from NMR experiments.

Molecular formula	Additional information	Number of possible isomers [25]
$C_{10}H_{22}$		75
$C_{15}H_{32}$		4,347
$C_{20}H_{42}$		366,319
$C_{10}H_{22}$		75
$C_{10}H_{20}$		852
$C_{10}H_{18}$		5,568
$C_{10}H_{14}$		81,909
$C_{10}H_{12}$		201,578
$C_{10}H_{10}$		369,067
$C_{10}H_{10}$		369,067
$C_{10}H_{10}O$		7,288,733
$C_{10}H_{10}O_2$		79,402,190
$C_{10}H_{10}O_2$		79,402,190
$C_{10}H_{10}O_2$	Signal multiplicity (4S, 3D, 2T, 1Q)	10,370,392
$C_{10}H_{10}O_2$	Carbon hybridization ($3sp^3$, $5sp^2$, $2sp$)	240,751
$C_{10}H_{10}O_2$	Signal multiplicity and hybridization as above	38,353
$C_{10}H_{16}$		24,938

long computing times during the structure generation process itself and during the necessary post-processing of the list of candidate structures. From Tab. 23.1 it can be seen that a powerful spectrum interpretation step is necessary even for a small molecular formula. The example $C_{10}H_{10}$ gives 369,067 isomers when using only the molecular formula without further constraints. The introduction of one oxygen increases the size of the problem by a factor of 20, a further oxygen leads to roughly 80 million possible isomers. The utilization of multiplicity information which can be easily derived from DEPT spectra reduces the size of this particular problem by nearly one order of magnitude. Additionally the use of hybridization information as selected in Tab. 23.1 [26] reduces the number of possible isomers to 38,353 candidate structures. The determination of the multiplicity can be done by experimental techniques (usually DEPT) and is therefore information which is very secure. On the other hand the determination of the hybridization is based on the interpretation of chemical shift value together with further supporting information (for example $^1J_{CH}$ couplings) therefore some ambiguity might remain, especially in the region around 100 ppm where sp^3, sp^2 and sp-carbon atoms have resonance lines.

The process of structure generation is a well-defined task based on graph theory, whereas spectrum interpretation, the translation from the spectral information into structural constraints, is based on heuristics. Even the selection of the hybridization state of a carbon atom cannot be performed with absolute security from its chemical shift value and some further information. A much more complex situation is encountered when deriving substructural fragments from spectral data, the interpretation of the data leads to a large amount of alternative possibilities for substructural units. This feature must be taken into account during the isomer generation process, making the programs more complex and slower.

There are two possible extreme situations which should be analyzed in further detail: The first approach uses a very detailed interpretation process based on a library of larger fragments as done for example by SPECSOLV [27], which makes structure generation working fast but with the disadvantage that only a part of the possible candidate structure compatible with the given constraints will be obtained, a situation which is less useful for natural product chemistry dealing with new classes of compounds. The other extreme approach is to take into account only absolutely safe pieces of information, for example signal multiplicities which can be determined experimentally. This situation uses only a very limited set of constraints for the structure generation process leading to an incredible number of candidates for further processing. In any case, ambiguity (for example with valencies and hybridizations) must be handled by the isomer generator which makes this tool much more complex. Usually there is a close connection between the interpretation and the isomer generation part of such systems.

A new horizon was introduced with the popularity of two-dimensional NMR methods giving information about coupled spin systems and therefore distance information (in terms of intervening bonds between coupled nuclei). The most useful experiments for organic structure elucidation are the methods for HH-, CH- and CC-correlation, the detailed experimental conditions (for example gradient-enhanced, normal-reverse detection) do not change the basic principle of these methods for computer application. The general information is always that correlation signals are translated into distance information between two atoms, which will be used either by the interpretation process leading directly to substructural units or, in the case of ambiguity (distance over 2 to n bonds) as a constraint during the structure generation process in order to exclude certain combinations of the substructural fragments. The COSY-type spectra give information about HH-correlations over two(geminal), three(vicinal) and four(allylic, W-coupling) bonds, but in many systems long-range couplings over more than four bonds are known, leading again to some ambiguity. The case of a $^2J_{HH}$ coupling can be easily excluded by a heteronuclear correlation experiment. CH-correlation experiments over one bond (HMQC, HSQC) are able to detect exclusively this type of connection, because $^1J_{CH}$-couplings are much larger than any $^nJ_{CH}$-coupling. Long range correlation techniques have to deal with two problems. The first is the detection of artefacts from $^1J_{CH}$-correlations, which can be easily excluded from HMQC-type spectra. The second problem is a much more severe one, based on the fact that long-range J_{CH}-couplings are of similar size, in particular $^2J_{CH}$ and $^3J_{CH}$ cannot be distinguished in many cases. Prohibiting the differentiation between these two distances leads to an increased computing time during the structure generation process and in many cases to a larger set of candidate structures being produced. Frequently, signals corresponding to correlations over four to six bonds have also been observed adding a further order of magnitude to the complexity of structure generation, whereas ignoring this possibility leads to wrong structure proposals. Additionally, severe signal overlap may also increase the difficulties during the spectrum interpretation process [28]. CC-correlation techniques are well-suited for computer-assisted structure elucidation but they have the disadvantage of low sensitivity. Another well-known concept to restrict the number of generated candidate structures is to use a list of forbidden substructures within the candidates. This option should also be used with extreme care, because exotic structures, like highly strained or unexpected heterocyclic systems, will always be excluded [29].

From the scope of the problem of structure generation it can be seen that there is a demand for spectrum interpretation programs which work in a very carefully designed way, allowing the user access to all constraints and therefore selecting the correct set of constraints from the given experimental data. The basic idea of spectrum interpretation of NMR data is to correlate a certain chemical shift value with a set of functional groups used as building blocks for subsequent structure generation. In order to improve this basic idea of using a single line, a complete subspectrum can be correlated with a set of larger substructural fragments, leading to higher selectivity during the interpretation process and therefore faster post-processing. The approaches known work either with rule-based correlation tables (for example CHEMICS [30], DENDRAL [31], DARC [32]) or HOSE-code technology (SPECSOLV[27]); atom-centered fragments (ACFs) are used in SESAMI [33, 34] and EPIOS [35], three-atom fragments are selected in CSEARCH [36, 37]. The basic principle behind these technologies remains always the efficient correlation between spectral and structural property; in any case these methods become slower when using smaller fragments or they generate only partial solutions when starting from a too specific set of fragments. The latter effect is also extremely dependent on the database used for the generation of the subspectra–substructure correlation tables. In any case a high structural diversity within the reference database is obligatory.

23.9
Ranking of Candidate Structures

The usual case when applying an interpretation/isomer generation program to a real-world structure elucidation problem is not to obtain a single structural proposal to a specific set of constraints. Frequently a list of possible candidate structures will be generated, consisting either of only a few proposals or maybe a few thousands of proposals, depending on the number of constraints given. Therefore a ranking process based on spectrum prediction [37] is necessary to select proposals of higher probability. For ^{1}H- and ^{13}C-NMR-spectra well-established methods are available, which allow one to select a set of most probable candidate structures. It is strongly recommended not only to use a single best solution, but also to include useful alternatives for further investigation. Isomer generator programs also produce very uncommon structures, which are not well-represented by databases used for spectrum prediction purposes, leading to a biased evaluation of the hit list with the result that more common structures might be ranked better than uncommon ones. Spectrum prediction based on a database having a high degree of structural diversity is a fast and reliable method which can be applied to a hit list of a few thousand chemical structures within a reasonable time. In the case that more structures have been generated, the set of constraints should be refined and/or additional experimental data should be collected.

23.10
Conclusions

For the synthetic organic chemist spectrum simulation in order to verify a structural proposal is the most decisive task; in natural product chemistry the starting point is usually a spectral similarity search using the experimental spectrum of the unknown. Both methods are extremely fast and can be done automatically immediately after the measurement of the one-dimensional routine spectra. When these basic methods give no solution further spectroscopic experiments are usually performed, giving more detailed information about the unknown sample. Spectrum interpretation and subsequent isomer generation is mainly successful when a large amount of additional information is available. The most important pieces of information are the presence/absence of certain functional groups and the efficient use of distance constraints as derived from two-dimensional correlation spectroscopy. This information usually allows one to deduce the constitution of an organic compound, in order to determine the configuration and conformation a more specialized set of tools is available. Many of these techniques, either experimental or computational methods, have been developed to deal with specific classes of compounds, especially with biopolymers [38]. The typical application field of the computer-techniques described here is structure elucidation of organic compounds up to C_{50}. Many steps during the structure elucidation process can be performed, or at least simplified and accelerated, by appropriate computational technologies, but the results must be critically evaluated by the expert. A large variety of algorithms and more or less integrated systems have been described in the literature and the field of computer-assisted structure elucidation is evolving dynamically. Numerous tools are commercially available and can support the chemist during the structure elucidation process, a critical evaluation of these tools with specific examples from his own field of application is strongly advised.

References

1 Kessler H., Gehrke M., Griesinger C., *Angew. Chem. Int. Ed. Engl.*, **1988**, *27*, 490–536; Parella T., *Magn. Reson. Chem.*, **1998**, *36*, 467–495
2 Gray, N. A. B., *Computer-Assisted Structure Elucidation*, John Wiley and Sons, New York 1986.
3 Munk, M. E., *J. Chem. Inf. Comput. Sci.*, **1998**, *38*, 997–1009 and references cited therein.
4 Jaspars M., *Nat. Prod. Rep.*, **1999**, *16*, 241–248 and references cited therein.
5 http://www.lohninger.com/spectroscopy/dball.html.
6 Gasteiger J., Hanebeck W., Schulz K. P., *J. Chem. Inf. Comput. Sci.*, **1992**, *32*, 264–271.
7 Bremser W., Wagner H., Franke B., *Org. Magn. Reson.*, **1981**, *15*, 178–187.
8 http://www.biorad.com.
9 http://specinfo.wiley-vch.de.
10 http://www.acdlabs.com.
11 Ewing D. F., *Org. Magn. Reson.*, **1979**, *12*, 499–524.
12 Hönig H., *Magn. Reson. Chem.*, **1996**, *34*, 395–406.
13 Thomas S., Ströhl D., Kleinpeter E., *J. Chem. Inf. Comput. Sci.*, **1994**, *34*, 725–729; http://www.chem.uni-potsdam.de/arosim/index.html.
14 Chen L., Robien W., *Anal. Chem.*, **1993**, *65*, 2282–2287.
15 Bremser W., *Anal. Chim. Acta*, **1978**, *103*, 355–363.
16 http://www.univie.ac.at/orgchem/csearch_server_info.html.
17 http://www.univie.ac.at/orgchem/wralpha.html.
18 Klamt A., Hoever P., Bärmann F. et al., in *Software-Development in Chemistry 7*, ed. D. Ziessow, Gesellschaft Deutscher Chemiker, Frankfurt am Main 1993, pp. 39–44.
19 Meiler J., Meusinger R., Will M., *Mh. Chem.*, **1999**, *130*, 1089–1095.
20 Badertscher M., Bischofberger K., Pretsch E., *Trends Anal. Chem.*, **1980**, *16*, 234–241.
21 http://www.acdlabs.co.uk/publish/nmr_485.html and references cited therein.
22 Robien W., Purtuc V., Schütz V. et al., Lecture at 13th CIC-Workshop, 15 – 17. 11. 1998, Bad Dürkheim/Germany; http://www2.chemie.uni-erlangen.de/external/cic/tagungen/workshop98/paper3.html.
23 Shelley C. A., Hays T. R., Munk M. E. et al., *Anal. Chim. Acta*, **1978**, *103*, 121–132.
24 Masinter L. M., Sridharan N. S., Lederberg J., Smith D. H., *J. Am. Chem. Soc.*, **1974**, *96*, 7702–7714.
25 All calculations have been performed using MOLGEN-3.1, Benecke C., Grund R., Hohberger R. et al., *Anal. Chim. Acta*, **1995**, *314*, 141–147.
26 Rivera A. P., Arancibia L., Castillo M., *J. Nat. Prod.*, **1989**, *52*, 433–435.
27 Will M., Fachinger W., Richert J. R., *J. Chem. Inf. Comput. Sci.*, **1996**, *36*, 221–227.
28 Mukhopadhyay T., Nadkarni S. R., Bhat R. G. et al., *J. Nat. Prod.*, **1999**, *62*, 889–890.
29 Varmuza K., Jordis U., Wolf G., *ECHET96 - Electronic Conference on*

30 Funatsu K., Sasaki S. I., *J. Chem. Inf. Comput. Sci.*, **1996**, *36*, 190–204.
31 Mitchell T. M., Schwenzer G. M., *Org. Magn. Reson.*, **1978**, *11*, 378–384.
32 Dubois J. E., Carabedian M., Ancian B., *C. R. Acad. Sci. (Paris)*, **1980**, *290*, 369–372, 383–386.
33 Razinger M., Balasubramanian K., Perdih M. et al., *J. Chem. Inf. Comput. Sci.*, **1993**, *33*, 812-825
34 Munk M. E., Madison M. S., Schulz K. P. et al., Lecture at 13th CIC-Workshop, 15–17.11.1998, Bad Dürkheim/Germany, http://www2.chemie.uni-erlangen.de/external/cic/tagungen/workshop98/paper8.html.
35 Carabedian M., Dagane I., Dubois J.E., *Anal. Chem.*, **1988**, *60*, 2186–2192.
36 Robien W., *Mikrochim. Acta[Wien]*, **1986**, *II*, 271–279.
37 Seger C., Jandl B., Brader G. et al., *Fresenius' J. Anal. Chem.*, **1997**, *359*, 42–45.
38 Zhu F. Q., Donne D. G., Gozansky E. K. et al., *Magn. Reson. Chem.*, **1996**, *34*, S125–S135.
39 All Figures have been generated using the CSEARCH-NMR-database system Kalchhauser H., Robien W., *J. Chem. Inf. Comput. Sci.*, **1985**, *25*, 103–108.

Heterocyclic Chemistry, 24.6.–22.7.1996, http://www.ch.ic.ac.uk/ectoc/echet96

24
Mass spectrometry

Antony N. Davies

24.1
Introduction

The world of mass spectrometry is blessed with many advantages over the other fields of spectroscopy when it comes to data handling. Not only are mass spectroscopists able to call on the largest collections of high quality reference spectroscopic data of any technique but they are also required to pay out the least amount of money per data set for access! Furthermore, some of the most powerful data analysis tools available for the study of hyphenated data sets from, for example gas-chromatography/mass spectrometry are available for free over the Internet! Finally, most new mass spectrometers come delivered with a large reference data library with advanced competent search software installed as standard. Most scientists from other fields of spectroscopy would regard themselves as having an excellently equipped spectroscopic data handling software if they had access to the same high level of tools and well thought through data analysis packages as are available in a standard mass spectrometry laboratory.

In this chapter the various types of data to be found in mass spectrometry will be detailed. Some of the most common data analysis packages will be described and their strengths and weaknesses probed. The different spectroscopic data packages will be explained along with their differences.

As with all works of this nature it is essential to always go to the original source reference given to find out what the latest situation is for any given software package or reference library. Where a need or deficiency is identified here it might well be the case that the need has been satisfied or the deficiency been corrected by the time the handbook goes into print. Experience has however shown in recent years that this is probably not the case!

Finally, a note of warning: Where a particular opinion is expressed in this chapter it can only have been gained during the work of the author and more correctly his colleagues. This experiences is, by its nature and the nature of all research laboratories, restricted to work carried out in certain analytical problem areas and deals with a limited amount of sample matrices. It is very important for any reader intending to invest in the area of mass spectrometry data handling to test the systems on real samples out of the day-to-day work of their laboratory. Finally, let the people who will be working with the systems have a major say in the testing and selection of the product to be purchased as they will have to be happy working with the new software systems and often possess many insights into the day-to-day running of the laboratory hidden to the average laboratory manager.

24.2
Mass Spectrometry Databases

There are two major collections of mass spectrometry reference data in common use which will be described more fully below. Each has it own unique strategy for data collection and quality control as well as different spectral search options. However, many smaller but very high quality specialised collections can also be found [1]. In July 2000 the 25 millionth CAS registry number was assigned to a mutant epoxide hydrolyase. Figure 24.1 shows the dramatic increase in the number of substances registered. By the time this article appears in print this will of course be an old figure but anyone who looks at the Chemical Abstracts Service Web site can easily find the current numbers of compounds registered under http://www.cas.org/cgi-bin/regreport.pl [2].

Depressingly this number has increasingly outstripped the development of even the largest collection of reference spectroscopic data so that it is often the case that, even though an unknown substance may well have been fully documented in the scientific literature 20 or more years ago, if you rely on commercial reference data collections you may well not be able to identify the compound.

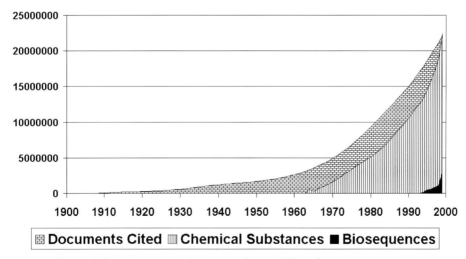

Fig. 24.1 Chemical Abstracts Service registry count changes 1907 to date.

This applies just as well when you are using reference data to carry out structure elucidation of completely new products such as new pharmaceutical agents. It is often the case that, regardless of the high level of intelligence now built into even the most modern of analytical spectroscopy tools, it is important to have added your own well characterised reference data to the knowledge base to ensure that for your particular chemistry the solution space is well populated. More simply put there is no point in trying to elucidate aromatic structures if your reference database only contains aliphatic compounds!

A superb example of such a strategy in action can be found in the description of the AMDIS program from NIST later in this chapter.

Finally, if your spectrometer comes with one of the commercial databases pre-installed it is well worthwhile checking which version has been installed as it could quite possible be older than the current release and may well be easy to upgrade.

24.2.1
NIST/EPA/NIH Mass Spectral Library

The smaller of the two largest collections is that administered by the US National Institute of Standards and Technology (NIST) more commonly known as NIST 02. This collection has been built up over many years from a number of mainly government backed initiatives in the UK and USA and is claimed to be the most widely used [3]. The US government agencies included the old National Bureau of Standards (NBS), the Environmental Protection Agency (EPA) and the National Institute of Health (NIH) who built on the initial work of the UK Atomic Weapons Research Establishment (AWRE) in Aldermaston [4]. Initially some of the data was only stored as an 8 peak data per compound which although saving storage space in an age when this was of primary importance gave spectra that are not useful in a modern environment. Since 1988 the responsibility for this collection has been with the NIST Mass Spectrometry Data Center. These spectra are now updated regularly and the most recent software release has been selected from over 175,000 spectra in the NIST archive. The latest release (January 2001) is known as the NIST 02 Mass Spectral Database and contains some 175,214 spectra of 147,350 compounds for which 107,105 have CAS numbers and virtually all (147,350) have chemical structures stored [3]. In this collection no radio-labelled substances were included and only a single representative spectrum per compound. A further 27,844 spectra of replicate (alternative) spectra for compounds in the main collection are also supplied. Considerable emphasis is placed on the quality of the spectra stored with each spectrum being critically evaluated [5].

Some of the data are available in the useful web-based reference collection of the National Institute of Standards called the WebBook (http://webbook.nist.gov). As with all such services any comment on the size of the collection will be out-of-date before this article goes to print but with over 10,000 mass spectra available on-line linked to a large amount of spectroscopic chemical and physical subsidiary data for the compounds of interest this is a reference data source which all analysts should have close to the top of their bookmarks!

24.2.2
Wiley Registry of Mass Spectral Data

This data collection started out as the Mass Spectrometry Registry under the auspices of Stenhagen and Abrahamsson in Sweden and has grown in recent years out of the work of McLafferty and co-workers. This is the largest of the reference databases of mass spectrometric data which also makes it the largest collection of reference data available in any single field of spectroscopy [6]. The ethos of data inclusion is different in this reference library where multiple entries per compound were included if the spectra were significantly different from one another. Radio-labelled substances were also included.

The usefulness of any particular reference library is greatly enhanced by the availability of subsidiary data on the spectra stored and the Wiley Registry boasted Chemical Abstracts Service (CAS) registry numbers and chemical structure information at an early stage in its development. The Quality Index is in use here to assess candidate spectra for their fitness for use as reference data and this version is slightly different from that adopted by NIST. Unfortunately this has not been 100 percent successful in keeping out incorrect reference data sets. If you should spot what you regard as false data whilst using any collection you should report it to the collators as the best way to quality control any large reference library is to have as many people as possible with as wide an interest base as possible using the collections on a regular basis!

In 1991 this collection was already almost 140,000 spectra (139,859) but from only 118,114 compounds [7]. The fifth edition is available in book form and CD-ROM either with or without structures [8]. The current version is the 6th containing some 229,119 reference spectra from 200,500 compounds which can be combined with a version of the NIST database, increasing the total size to 275,821 spectra of 226,334 compounds. The version marketed by the Palisade Corporation is available with a useful data format conversion utility MASSTransit (http://www.wileyregistry.com). There is also the PC BenchTop/PBM search system described below which can be purchased for off-spectrometer data processing and analysis of GC/MS runs for example.

The 7th Edition contains over 390,000 electron impact mass spectra including the older NIST '98 database. Through use of the AccessPak it may be loaded into many Mass Spectrometer data systems.

24.2.3
SpecInfo/SpecData

The SpecInfo databases from Chemical Concepts GmbH marketed as SpecData have the two databases mentioned above and also their own collection 'CC Mass Spectra 4th Edition' of some 40,000 spectra, although there will be some doubling up of the data available between the collections. The Chemical Concepts web site specifically indicates that, for example, the Industrial Chemicals Collection from Henneberg at the Max Planck Institut für Kohlenforschung, Mülheim, Germany is also available in the NIST '98 database (http://www.chemicalconcepts.com/p11343.htm).

These collections have recently become available for searching through a web based client/server system called SpecSurf written by the company LabControl in Cologne, Germany. 'SpecInfo on the Internet' is hosted by the Wiley New York server at http://specinfo.wiley.com/ [9]. In order to make proper use of the Java programming a small additional free program needs to be installed on your PC to enhance your browser but this is simple to carry out and well documented.

A further version of the SpecInfo database is also available through STN where the mass spectra along with other spectroscopic and chemical substance information can be found through the usual CAS type search options as well as single peak data such as base peak, base peak intensity, nominal mass, peak position etc. The 9th September 1999 upgrade reported over 65,900 mass spectra stored. For more details see: http://www.cas.org/ONLINE/DBSS/specinfoss.html.

24.2.4
SDBS, Integrated Spectra Data Base System for Organic Compounds

This Japanese web-based database is freely available at http://www.aist.go.jp/RIODB/SDBS/menu-e.html and at the time of writing had last been updated on August 18 1999. The database is run by the NIMS National Institute of Materials and Chemical Research in Tsukuba.

The collection has NMR, IR, Raman and EPRdata as well as MS data (approximately 19,600). Unfortunately the search options request only mass numbers and relative intensities (with a 50% error!) see Fig. 24.2.

A trial search carried out by inputting m/z values refused to show the initial hit list as it was deemed too long and a second m/z value was required to restrict the hit list.

The only active link in the results window is the SDBS registry number which will lead the user to the mass spectrum display window shown in Fig. 24.3.

24.2.5
Other Smaller Collections

Depending on the instrument manufacturer a number of smaller specialised mass spectra databases are being made available, either as standard or as optional extras with the spectrometer data systems. A few of the more frequently found libraries are given below as well as collections of more unusual mass spectrometry techniques. Also included are a few references to data collection activities by some concerned organisations.

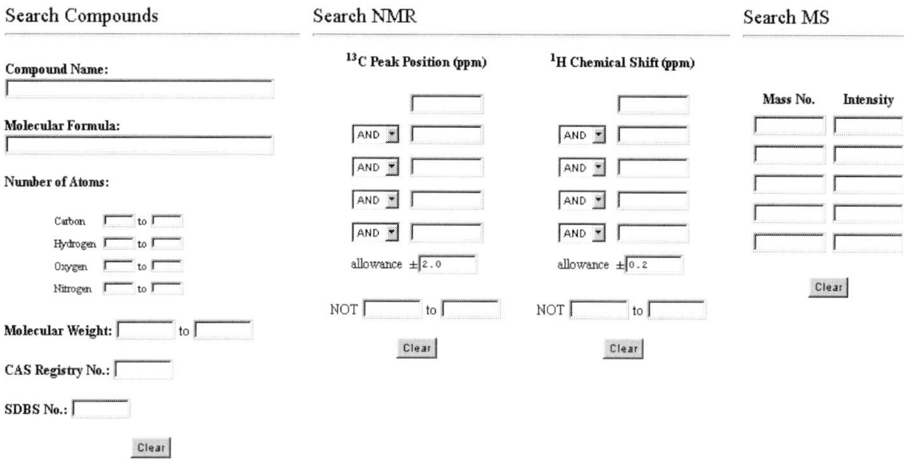

24 Mass spectrometry

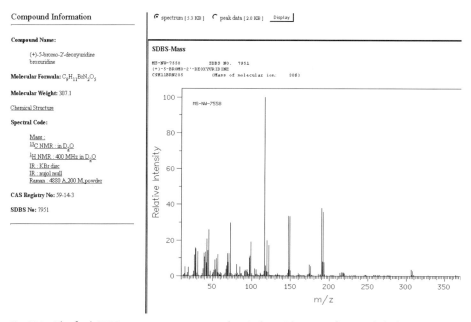

Fig. 24.2 The SDMS database search mask.

Fig. 24.3 The final SDBS mass spectrometry results window with cross-references linked.

24.2.5.1 Pfleger/Maurer/Weber: Mass Spectral and GC Data of Drugs, Poisons, Pesticides, Pollutants and Their Metabolites

One of the smaller databases commonly delivered as standard with many GC/MS spectrometers is the PMW or Pfleger/Maurer/Weber. This database of about 4300 spectra is also published in book form with the 4th part bringing the database up to some 6300 spectra [10]. This is a good collection for those specialising in, for example, clinical toxicology, pharmacology, environmental chemistry and food analysis.

24.2.5.2 Ehrenstorfer

Dr. Ehrenstorfer GmbH in Augsburg, Germany has been supplying analytical standards for many years and they have produced a mass spectra library of over 1450 spectra of pesticides and their metabolites, PCBs, PAHs explosives and other environmentally important compounds [11].

24.2.5.3 Wiley-SIMS

A small CD-ROM based library of around 300 secondary ion mass spectra known as the Wiley-SIMS was created by Henderson and co-workers in Manchester UK. (see http://www.surfacespectra.com).

24.2.5.4 American Academy of Forensic Sciences, Toxicology Section, Mass Spectrometry Database Committee

This group works to co-ordinate the generation of reliable mass spectra of new drugs and their metabolites see: http://www.ualberta.ca/~gjones/msmlib.html

The 1997 version of the data is available on-line for free as a demo library at the Galactic Industries web site. http://www.galactic.com/ although to view the spectra you will need to have the spectra viewing program (also available for free) or another software program supporting Galactic's SPC format installed on your system.

This group is very interested in expanding the collection and would greet any approaches aimed at supplying standards for measurement.

24.2.5.5 The International Association of Forensic Toxicologists (TIAFT)

Another effort to collect reference EI spectra to assist toxicologists in unknown substance identification has been started by this organisation. This small collection is available for free over the internet and specifically aims at gathering spectra of new upcoming or uncommon substances or less frequent derivatives of drugs. This group is interested in receiving new data and can be contacted through their web site at http://www.tiaft.org

24.3
Mass Spectrometry Search Software

Mass spectra are in themselves relatively simple data sets containing essentially simple lists of intensity against mass/charge ratio (m/z). Unfortunately this hides the enormous complexity of the processes which the sample has undergone to yield such a signal. This can cause effects during the recording of mass spectra which, under often quite normal conditions, can deliver substantially different results for the same analyte. You could think that this would make the use of reference data collections for analyte identification effectively impossible unless representative data from all possible experimental conditions were available but fortunately there have been some rather clever algorithms developed. The reference databases themselves are standardised around low resolution electron ionisation mass spectra with a primary electron energy of 70 eV.

There have been several programs available for analysing multidimensional experiments in mass spectrometry such as the MassLib program developed by Henneberg and co-workers at the Max Planck Institute für Kohlenforschung in Mülheim, Germany or the more recent AMDIS package written by Stein and co-workers at the US National Institute of Standards and Technology at Gaithersburg [12].

Three different mass spectrometry search algorithms dominate the database searching systems commercially available today. The Cornell University 'Probability Based Matching' (PBM) software, The Integrated Control System (INCOS) and the MassLib system (see below) with the SISCOM search software.

These systems stem from original publications in the early 1970s but have been extensively copied, improved and adapted. They all use well thought out powerful algorithms and are surprisingly fast considering the size of the reference libraries through which they have to search. These systems initially reduce the size of the result space by the use of pre-searches, often using only a small fraction of the available data in the unknown spectrum and the reference libraries [13]. As only the pre-filtered reference data will be presented for more extensive comparison with the unknown spectrum the pre-search is a critical phase in the library search process and will be discussed again below.

The use of neural networks for mass spectral searching has also been successfully tested and reported recently [14].

Finally a word of warning, even though modern search software can produce superb results from what often looks like very poor starting data, it is also possible to produce hit lists of candidate solutions to a particular problem which are completely wrong. It is important not to rely solely on a computer generated hit list for your analysis but to approach the whole spectral searching task with a reasonable amount of healthy scepticism! On a slightly more positive note, if the first hit does not look right look further down the hit list your software presents you as often the correct hit may not have achieved the top ranking position. Nothing beats asking a friendly mass spectroscopist who knows your chemistry!

24.3.1
INCOS

The INCOS search algorithm originally developed by Finnigan involves the dot product calculation between the unknown spectrum and the library data [15]. An initial pre-search is carried out on a reduced number of peaks out of the unknown spectrum against a reduced peak number reference library. Having reduced the number of candidate hits through the pre-search a second more complete search is carried out through the surviving candidates to produce the final hit ranking. The search itself is weighted by the root of the product of the mass and intensity values giving preference to the higher m/z peaks due to their greater specificity in identifying the unknown. Sparkman has recently further qualified his assessment of the various current implementations of the INCOS algorithm amongst different spectrometer manufacturers by a criticism of the different pre-search algorithms which are a critical part of the complete software package [16].

Data reduction is carried out using a windowing technique. The main search under INCOS is similar in nature to how a scientist would compare two mass spectra in that, for a particular m/z region, the data are locally normalised, thereby ensuring that the search locates similar fragments in the reference library, irrespective of the varying intensity ratios between peaks of different fragments in the unknown. This makes the search system very robust against changes in the recorded data arising from slight changes in experimental set-up or conditions.

24.3.2
Probability Based Matching (PBM)

The probability based matching algorithm [17] and the self-training interpretive and retrieval system (STIRS) have been continuously improved over the last 20 years (see for example [18, 19]). The PBM system can carry out forward and reverse searches for pure analyte and mixture analysis. A forward search looks for the unknown spectrum in the reference database and the reverse search looks to find the best reference spectrum in the unknown data set. In a different strategy to the original INCOS algorithm the search is weighted linearly in favour of the peaks from molecular fragment ions.

The hit quality index (HQI) ranks the candidate reference spectra by their similarity to the unknown spectrum. As warned above it is not wise to take the hit list ranked by HQI at face value but to look down the hit list where it may well be possible to find what a human expert would regard as a better candidate solution to the search than the algorithm has picked out.

The forward search is the most rapid but demands unknown spectra of pure compounds to produce good results. If however the unknown spectrum includes peaks from unwanted impurities or of a mixture then it will not work correctly. The alternative reverse search strategy although slower is now required, whereby the resulting hit list shows the best reference spectrum in the library as found in the unknown data. Peaks in the unknown which do not appear in the library spectrum do not downgrade the HQI as they would in the forward search as they may come from impurities or other substances in a mixture. Once the most significant component has been identified the reference spectrum of this component can be subtracted from the unknown mixture and the reverse search started again to identify, if possible, further compounds in the unknown sample.

24.3.3
MassLib/SISCOM

The MassLib software package has been around in various forms for many years migrating from workstations to PCs as windowing became popular [13]. It was developed by Henneberg and co-workers at the Max Planck Institute in Mülheim, Germany for the analysis of GC/MS runs and is now into MassLib/PC version 8.5 [20].

The package uses fragment as well as neutral loss searching yielding hit lists useful for determining possible chemical structures for an unknown compound even when the compound is clearly not in the reference database. A limit of 500 spectra from the users own mass spectrometry database is currently in place for the basic package but this can be overcome with an upgrade.

For all such third party off-spectrometer data analysis packages it is important to be able to handle as many manufacturers' formats and international data transfer formats as possible and Tab. 24.1 typifies this with those formats currently supported by MassLib.

The full GC/MS trace is analysed to locate the various components and candidate unknown spectra generated for searching in the reference databases using SISCOM. The mass spectrum database search system which comes with the MassLib package detailed below is called SISCOM (Search for Identical and Similar COMponents) [21]. The Identity search option looks for the unknown in the reference libraries and can handle binary mixtures as unknowns provided both substances are represented in the reference database. The identity search swaps to the similarity search mode if no identical hits have been found in the reference databases where characteristic ions in the unknown are searched in the reference libraries. If this option is unsuccessful a related search option is available matching the library to the unknown.

Neutral loss masses from the M+ peak are also used via a special tool for spectral searching.

Table 24.1 Data Formats currently supported by MassLib.

Balzers, Bear Instruments, DA5000, EPA I + II, EZ-Scan, GCQ, HP-Chemstation, HP-RTE, ICIS 1 + 2, INCOS, ITDS, ITS40, JCAMP-DX, Mach3, MassLab, MassLynx, MSD, MSS, NetCDF, Saturn, Shimadzu, Shrader, SSX, VG 11-250.

24.3.4
AMDIS

NIST have produced a freely available package called AMDIS (automated mass spectrometry deconvolution and identification system) for the analysis of GC/MS data sets. Developed to assist in the task of verifying the international Chemical Weapons Convention (http://www.opcw.org/) financially supported by the US Defense Special Weapons Agency (DSWA, US Department of Defense) the AMDIS program is also distributed with the NIST 02 Mass Spectral Library (see above).

The system comes with six speciality target libraries installed which consist of selected spectra from the 175,214 in the full NIST/EPA/NIH library (see Tab. 24.2).

These or user generated databases are the libraries in which the initial identification of the chromatographic peaks is carried out. In our work into pesticides in drinking and ground water for example we generated our own AMDIS library containing specifically the pesticides and their metabolites that we were working on (http://www.spectroscopyeurope.com/td_col.html).

In the latest release of the AMDIS package (v.2.1) it has been decided that computer speeds are now such that it is reasonable to allow searching of the whole of the NIST 02 database during the chromatography analysis phase.

Each single mass chromatogram is studied to identify the locations of the chromatographic peaks and all masses which have the same peak position are then extracted from the total ion chromatogram as a candidate spectrum for identification [22, 23]. If a candidate unknown spectrum has not been clearly identified it is possible to add this to the clipboard for transfer to the NIST MS Search package for individual searching through the NIST '98 database. If the automatic peak picking does not work to your satisfaction then it is possible to opt for manual mode in which the area of background and signal can be selected by use of the mouse and the candidate unknown spectrum generated in this way.

Unfortunately the requirement for only black and white images in this chapter makes it difficult to show the main features of AMDIS v2.1 which allows two different GC/MS runs to be displayed in the same window with their time axes tied. In Fig. 24.4 we have been studying PCBs in sewage sludge and have found the use of the top trace for the sludge sample analysed by flash-thermodesorption/GC/MS and the lower trace from our calibration runs using the DIN PCB mixture extremely useful due to the vast number of peaks seen in our complicated matrix.

Table 24.2 Speciality target libraries installed in AMDIS.

- NISTTOX 1251 spectra toxicological library,
- NISTEPA is 1106 Environmental Protection Agency (EPA) target compounds,
- NISTFF 993 flavours and fragrances
- NISTDRUG is 778 drug analysis spectra,
- NISTFDA 419 for Food and Drug Administration (FDA) analyses,
- NISTCW is 62 chemical-weapons related data sets.

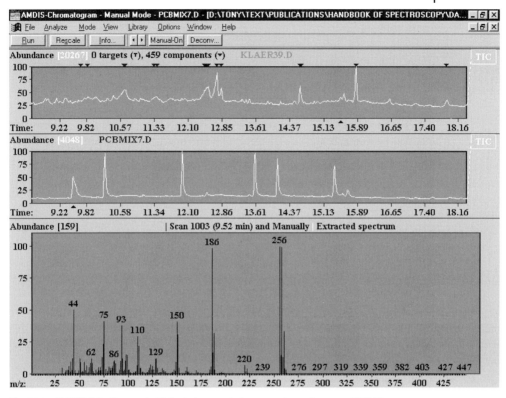

Fig. 24.4 AMDIS 2.0 allows twin TIC windows to help comparisons between GC/MS runs. This figure shows a sewage sludge analysis (upper trace) compared to a reference PCB run (centre trace).

Since its introduction this package has become one of our major GC/MS data analysis tools finding favor with both scientists and laboratory technicians alike [12].

24.3.5
Mass Frontier

A company established in 1997 in Bratislava, Slovakia has produced an interesting mass spectrometry data handling package called Mass Frontier (http://www.highchem.com). The package is PC based and is modular containing a structure editor, spectra manager for spectra/structure databases (also user). The database search algorithm is that developed by NIST and a substructure search is also present. A fragments and mechanisms module can be used for consistency checking and analysis of mass spectra and as a help in MS/MS experiments. The new spectra to be analysed can come from the GC/LC/MS viewer where the TIC and single ion chromatograms can be used for peak location and candidate spectra generation.

24.3.6
The WebBook

One of the more useful free resources on the Internet for analytical chemists is the NIST WebBook http://webbook.nist.gov. The WebBook can be searched through a number of keys but unfortunately spectral searching is not one of them!

One of the most useful features of the NIST WebBook we have found is the name search, where a large selection of alternative names for a particular compound are available and searchable, greatly increasing the chance of finding the information you want on a particular substance of you do not know the IUPAC or CAS name or the CAS registry number and are not really too sure of the chemical structure. However it is possible to upload chemical structures for example (Fig. 24.5).

Amongst the various databases linked to the WebBook results page is that for mass spectrometry which will cause the spectrum to be presented to the user via a Java Applet which now allows direct printing from the browser (Fig. 24.6).

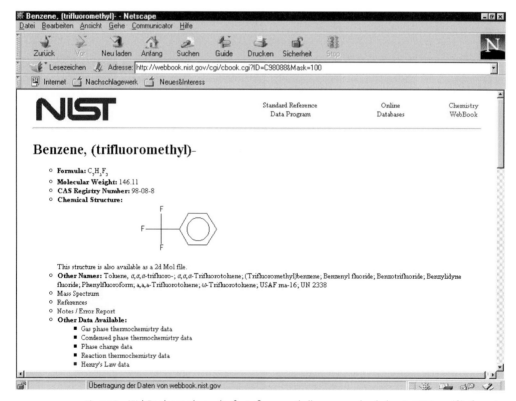

Fig. 24.5 WebBook search results for trifluoromethylbenzene uploaded as in MDL molfile format.

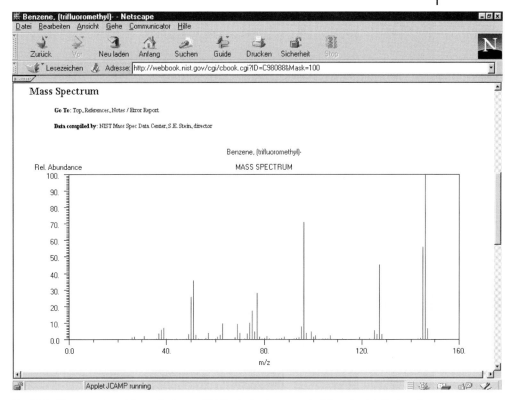

Fig. 24.6 Mass spectrum hit results from a WebBook structure search. A Java Applet displays the JCAMP-DX file from NIST.

A static graphic file is also available in order to print off the spectrum in higher resolution than a screen capture if this is required. Additional information associated with this measurement is also displayed.

24.3.7
General Spectroscopy Packages

The section would be incomplete if we were to leave out the general spectroscopy packages which are not MS specific but which have been receiving steady improvements over the years extending their capabilities to include mass spectrometry. Many of these packages originated in the optical spectroscopy or chromatography fields and have grasped the need to not only address but also be competent in other techniques (as have their users as this change has been demand led!).

Galactic Industries Corporation, famous for their ability to read a vast number of spectrometer manufacturer's data formats, has recently added a GC/MS Application Pack to its long-running GRAMS package (http://www.galactic.com).

The company creon LabControl mentioned above for their SpecSurf product also produces the versatile Spectacle package for many different types of spectroscopy including MS (http://www.creonlabcontrol.com). As with Galactic their software can be found as an OEM product on several spectrometer control packages. They have also tackled the electronic records/electronic signatures rule and have a spectroscopic archiving product known as Q~DIS/R.

BioRad Sadtler have MS functionality built into their ChemWindow Spectroscopy package (http://www.biorad.com). This package is a reporting software with chemical/spectroscopic ability. The MS Tools are aimed at assisting mass spectra analysis by providing modules like producing lists of possible formulae or substructures for a given mass and fragmentation tool for studying fragmentation in proposed structures.

Last but not least ACDLabs have MS capability in their SpecManager product (http://www.acdlabs.com). Their MS module reads single spectra or GC/(LC)/MS data files and can be used for the subsequent analyses including ESI LC/MS minor components recovery, GC/MS separation of co-eluting components as well as structure/substructure formula generation from mass and ion data [24, 25].

24.4
Biological Mass Spectrometry and General Works

Even within the scope of this chapter a quick discussion is required of the newest exciting field which has 'discovered' the power of linking mass spectrometric techniques with intelligent use of computerised data analysis which is biological mass spectrometry. Almost all of the systems above were developed with the analysis and characterisation of relatively uncomplicated chemical moieties in mind but rapid developments have been made especially in the application of mass spectrometric methods in the up-and-coming field of proteomics. For a review of the use of mass spectrometry and database searching in protein analysis for the identification of tryptic peptides see [26]. Additionally a review of electrophoretically separated enzyme digested protein analysis by peptide MS using under 1 pmol of sample and MALDI of electron ionisation methods has also been published [27].

Early attempts at applying chemometric methods to mass spectrometry data sets have been reported see for example [28, 29]. The potential and limitations of using multivariate classification methods for substructure analysis of low resolution mass spectra have also been published [30].

An interesting new reference work for practising mass spectroscopists is the new 'Desk Reference' by David Sparkman [16] which includes a substantial reference section entitled 'Correct and Incorrect Terms'. The information supplied here takes the form of sometimes quite extensive explanations on the correct and incorrect use of many of the phrases and terms in common use in mass spectrometry.

References

1 A. N. Davies, P. S. McIntyre, Spectroscopic Databases, in *Computing Applications in Molecular Spectroscopy*, eds.W.O. George and D. Steele, 1995, The Royal Society of Chemistry, Cambridge 1995, pp. 41–59.
2 CAS, *Chemical Abstracts Service*, 2000, Columbus, OH 43210-0012, 2000.
3 NIST, *NIST '98 Mass Spectral Database*, National Institute of Standards and Technology, Gaithersburg, MD 1998.
4 S. R. Heller, Mass Spectrometry Databases and Search Systems, in *Computer Supported Spectroscopic Databases*, ed. J. Zupan, Ellis Horwood, Chichester 1986, pp. 118–132.
5 P. Ausloos, C. L. Clifton, S. G. Lias et al., *J. Am. Soc. Mass Spectrom.*, **1998**, *10*, 287–299.
6 Wiley, *Wiley Registry of Mass Spectral Data*, John Wiley & Sons, Chichester 1989.
7 F. W. McLafferty, D. B. Stauffer, A. B. Twiss-Brook et al., *J. Am. Soc. Mass Spectrom.*, **1991**, *2(5)*, 432.
8 F. W. McLafferty, *Registry of Mass Spectral Data*, 5th edition, 1989, John Wiley & Sons, Chichester, 1989.
9 A. N. Davies,. *Spectrosc. Eur.*, **2000**, *12(1)*, 26–29.
10 K. Pfleger, H. Mauer, A. Weber, *Mass Spectral and GC Data of Drugs, Poisons, Pesticides, Pollutants and Their Metabolites*, 2nd edition,Wiley-VCH, Weinheim 1992, Vol. Parts 1–3 (Set).
11 D. Ehrenstorfer, *Library of Mass Spectra*, Dr. Ehrenstorfer GmbH: Augsburg, Germany 2000.
12 A. N. Davies, *Spectrosc. Eur.*, **1998**, *10(3)*, 22–26.
13 A. N. Davies, *Spectrosc. Eur.*, **1993**, *5(1)*, 34–38.
14 C. S. Tong, K. C. Cheng, *Chemometrics Intell. Lab. Syst.*, **1999**, *49*, 135–150.
15 S. Sokolow, J. Karnofsky, P. Gustafson, *Finnigan Application Report No. 2.*, Finnigan, San Jose 1978.
16 O. D. Sparkman, *Mass Spectrometry Desk Reference*, 1st edition, Global View Publishing, Pittsburgh 2000, p.106.
17 F. W. McLafferty, R. H. Hertel, R.D. Villwock, *Org. Mass Spectrom.*, **1994**, *8*, 690–702.
18 K. S. Haraki, R. Venkataraghavan, F. W. McLafferty, *Anal. Chem.*, **1981**, *53*, 386–392.
19 F. W. McLafferty, S. Y. Loh, D. B. Stauffer, Computer Identification of Mass Spectra, in *Computer Enhanced Analytical Spectroscopy*, ed. H. L. C. Meuzelaar, Vol. 2. 1990, Plenum Press, New York 1990, Vol. 2, pp. 161–181.
20 F. Friedli, *MassLib. PC*, 2000, MSP KOFEL, Bindenhausstr. 46, CH-3098 Koeniz, Switzerland.
21 H. Damen, D. Henneberg, B. Weimann, *Anal. Chim. Acta*, **1978**, *103*, 289–302.
22 S. E. Stein, *J. Am. Soc. Mass Spectrom.*, **1999**, *10*, 770–781.
23 J. M. Halket, A. Przyborowska, S. E. Stein et al.,*Rapid Commun. Mass Spectrom.*, **1999**, *13*, 279–284.
24 J. E. Biller, K. Biemann, *Anal. Lett.*, **1974**, *7*, 515–528.
25 A. Williams, *An Introduction to CODA: Integration into ACD/MS Manager*, Application Note, Advanced Chemistry Development, Toronto, Ontario 2000.

26 B. T. Chait, *Nat. Biotechnol.*, **1996**, *14*(11), 1544.
27 J. S. Cottrell, C. W. Sutton, *Methods Mol. Biol.*, **1996**, *61*, 67–82.
28 K. Varmuza, A. N. Davies, *Spectrosc. Int.*, **1990**, *3*(4), 14–17.
29 K. Varmuza, *Int. J. Mass Spectrom. Ion Processes*, **1992**, *118/119*, 811–823.
30 K. Varmuza, W. Werther, in *Advances in Mass Spectrometry*, eds. E. J. Karjalainen et al., Elsevier, Amsterdam 1998, Vol. 14, pp. 611–626.

Index

a

7-AAD 2/17
absolute detection limit 1/378
absorption 1/427
absorption coefficient 1/40
absorption index 1/74
acceptor-donor-acceptor (A-D-A) dyes 2/58
ACCORD-HMBC 1/244 ff
accordion-optimized long-range heteronuclear shift correlation methods 1/244
acetylcholine receptors 2/25
acetylcholinesterase 2/21
acetyltransferase 2/21
ACFs 2/483
achievements 2/163 ff
– atmospheric pressure chemical ionisation (APCI) 2/163
– capillary electrophoresis (CE) 2/163
– capillary zone electrophoresis (CZE) 2/163
– electrospray ionisation (ESI) 2/163
– obstacles 2/163 ff
– TSP 2/163
acquired immunodeficiency syndrome (AIDS) 2/47
acridine orange 2/17
acridinium ester 2/65
acrylamide quenching 1/141
actin
– detection of cytoskeletal proteins 2/17
activation energy
– motional 1/278
– rate constants 1/278
active air collection methods 1/8
active vs. passive sampling 1/8
adenylate cyclase 2/24
adrenergic receptors 2/25
adsorbaste 1/553, 1/557
adsorbate 1/508, 1/526 f, 1/560, 1/587, 1/591
adsorption 1/519, 1/529, 1/531, 1/543, 1/545, 1/552, 1/580, 1/582

adsorption geometry 1/535
advantage
– Connes 52
– Felgett 1/52
– and limitations 1/417
– of TXRF 1/399
AEAPS (Auger electron appearance potential spectroscopy) 1/508
Aequorea victoria 2/58
aequorin 2/12
AES (Auger electron spectroscopy) 1/512, 1/591
affinity-proteomics 1/358
AIDS (acquired immunodeficiency syndrome) 2/47
Al^{3+} 2/12
Aldrich 2/447
Alexa Fluor 2/15
Alexa Fluor dyes 2/51
alkaline phosphatase 2/50
– enzyme labels 2/50
Alkemade 1/436
Alzheimer's disease 2/11, 2/13
amalgamation trap 1/451
AMDIS 2/495, 2/498
amino acid analysis 2/16
– solid-phase synthesis 2/12
9-amino-6-chloro-2-methoxyacridine 2/11
7-aminocoumarins 2/20
4-amino-5-methylamino-2',7'-difluorofluorescein (DAF) 2/14
aminopropylsilane 2/84
Amplex Red/resorufin 2/14
amplitude noise 1/405
amylase 2/20
analog-to-digital converter 1/387
analysis
– of binding constants 2/128
– environmental 2/152 ff
– structural 1/98

analytical sensitivity   1/376
ANEP   2/19
angiotensin II receptor   2/25
angular dispersion   1/390
anilides
– groundwater   2/215
– LC-MS/MS   2/215
– natural waters   2/215
– surface water   2/215
anion receptors   2/69
anisotropy   1/149, 1/151, 1/272
– chemical shielding   1/271
– dipolar coupling   1/271
annexin V   2/26
Anthozoa species   2/58
9-anthroyloxy   2/19
n-anthroyloxy fatty acids   1/147
antibody screening   2/6
anticoagulant activity   2/24
antifouling pesticides   2/225
– LC-ICP-MS (inductive coupled plasma)   2/225
– organotin   2/225
antracene   2/70
APCI (atmospheric pressure chemical ionisation)   2/183
– anilides   2/192
– antifouling   2/199
– AEO   2/189
– APEO
  – halogenated   2/191
– biodegradation   2/195, 2/199
– biological fluids   2/194
– capillary zone electrophoresis (CZE)   2/186
– carbamates   2/193
– coconut diethanol amide (CDEA)   2/189
– drugs
  – metabolites   2/184
– drugs and diagnostc agents   2/184
– dyes   2/186
– estrogenic compounds   2/186
– explosives
  – metabolites   2/187
– FIA-MS   2/190
– FIA-MS/MS   2/190
– fruits   2/193, 2/197
– fruit drinks   2/197
– groundwater   2/192, 2/194, 2/196, 2/199
– haloacetic acids   2/187
– halogenated APEOs   2/191
– herbicides   2/192
– heterocyclic compounds   2/188
– interlaboratory study   2/194
– LAS   2/189, 2/191
– LC-MS   2/190
– anilides   2/185
– antifouling pesticides   2/185
– carbamates   2/185
– diagnostic agents   2/185
– drugs   2/185
– dyes   2/185
– estrogenic compounds   2/185
– explosives   2/185
– fungicides   2/185
– general   2/185
– haloacetic acids   2/185
– herbicides   2/185
– industrial effluents   2/185
– municipal wastewaters   2/185
– organophosphorous compounds   2/185
– pesticides   2/185
– pharmaceuticals   2/185
– phenoxycarboxylic acids   2/185
– phenolic compounds   2/185
– phenols   2/185
– phenylureas   2/185
– polycyclic aromatic hydrocarbons   2/185
– quaternary amines   2/185
– sulfonic acids   2/185
– sulfonylureas   2/185
– surfactants   2/185
– thiocyanates   2/185
– thioureas   2/185
– toluidines   2/185
– triazines   2/185
– LC-MS/MS   2/190
– N-methylcarbamate pesticides   2/193
– miscellaneous   2/199
– MS/MS   2/195 f
– MS/MS library   2/194
– natural waters   2/198
– NPEO   2/189
– NPEO-sulfate   2/189
– organophosphorus   2/194
  – biodegradation products   2/194
– phenolic compounds   2/197
– photodegradation   2/195
– quantification   2/191 ff
– pesticides   2/192
– phenols   2/188
  – quantification   2/189
– phenoxy carboxylic acids   2/195
– phenylureas   2/196
– polycyclic aromatic hydrocarbons (PAH)   2/187 f
– quaternary amines   2/192
– review
  – dyes   2/185

– general 2/185
– pesticides 2/185
– surfactants 2/185
– rivers 2/198
– river water 2/192 f, 2/199
– secondary alkane sulfonate (SAS) 2/189
– sediment 2/197
– sulfonylureas 2/196
– surface water 2/196, 2/198
– surfactants 2/189 f
– thiocyanates 2/192
– thioureas 2/196
– tin-containing pesticide 2/201
– toluidines 2/192
– triazines 2/197 ff
– triazole herbicides 2/199
– ungicides 2/192
– vegetables 2/193
– water 2/196 f, 2/198
APCI-LC-MS/MS
– heterocyclic compounds 2/188
– phenols 2/188
APD (avalanche photodiodes) 2/27
APECS (Auger photoelectron coincidence spectroscopy) 1/557
APFIM (atom probe field ion microscopy) 1/510
API (atmospheric pressure ionization) 2/183 ff
– reviews 2/184
apoptosis 2/11, 2/26
applications
– special 1/103
APS (appearance potential spectroscopy) 1/508
aqueous matrices
– eluates of soil samples 2/153
– groundwater 2/153
– leachates 2/153
– surface waters 2/153
– wastewaters 2/153
ARAES (angle resolved AES) 1/514
Arc 1/477, 1/484
ARC procedures 2/251
argon ion laser 2/40, 2/27
Argonne Protein Mapping Group 2/8
ARIPES (angle resolved inverse photoemission spectroscopy) 1/536
aromatic sulfonates 2/154
ARPES (angle resolved photoelectron spectroscopy) 1/506
artificial enzymes
– by molecular imprinting 2/71

ARUPS (angle resolved ultraviolet photoemission spectroscopy) 1/506, 1/537, 1/584
ARXPS (angle resolved X-ray photoelectron spectroscopy) 1/514, 1/587
ascorbic acid 2/15
atmospheric pressure chemical ionization (APCI) 2/183 ff
atmospheric pressure ionization (API) 2/183 ff
atom reservoir 1/434
atomic absorption spectrometry 1/421 ff
– atomisation efficiency 1/441
– atomiser 1/440
– block diagram 1/437
– cold vapor generation technique 1/451 f
– direct sample introduction 1/452
– double beam spectrophotometer 1/453
– dynamic range 1/465
– electrothermal atomisation 1/443 ff
– external calibration 1/465
– flame atomisation 1/441
– flow injection 1/449
– instrumentation 1/436 ff
– laser ablation 1/452
– optical set-up 1/453
– quantitative analysis 1/465 ff
– quartz furnace 1/467
– quartz tube atomiser 1/449
– radiation sources 1/437 ff
– sample transfer efficiency 1/440
– single beam spectrophotometer 1/453
– solid sample introduction 1/452, 1/470
– standard addition technique 1/466
– transport efficiency 1/469
– vapour generation techniques 1/447 ff
atomic spectroscopy, theory 1/421 ff
atomisation 1/471
atomiser 1/440
ATP 2/20
– AT-Pases 2/20
– bioluminescent determination 2/20
– determination 2/65
– DNA 2/20
– GTPases 2/20
– in mitochondria 2/65
– RNA polymerases 2/20
ATR (attenuated total reflection) 1/511
– correction 1/76
– technique 1/95
– -FTIR 2/92 ff
– in vivo monitoring of glucose 2/92
attached proton test (APT) 2/471
attenuated total internal reflection (ATR) 1/75, 2/75

attenuation 1/402
Auger electron 1/369
Auger electron appearance potential spectroscopy (AEAPS) 1/508
Auger electron spectroscopy (AES) 1/512, 1/591
automatic gain control (AGC) 2/247
autoradiography 2/39
auxochrome 1/125
AV 2/29
avalanche photodiodes (APD) 2/27
avidin (AV) 2/29
azophenol 2/70

b

background correction 1/455 ff
– by wavelength modulation 1/462
– continuum source 1/456 f
– pulsed lamp 1/460
– self reversal 1/460 f
– Smith-Hieftje method 1/460 f
– two line 1/456 f
– Zeeman 1/458 ff
background emission 1/434
background interferences 2/252
bacterial luciferase genes ("Lux genes") 2/68
bacterial luciferases 2/65
bags 1/11
Balmer series 1/421
band shifts 1/109
bands
– characteristic 1/102
baseline 2/451
baseline correction 2/457
– batch processes 2/272
– continous processes 2/272
– control charts 2/272
– feed-back control 2/269, 2/280
– feed-forward control 2/269, 2/280
– operational benefits 2/277
– statistical process control (SPC) 2/272
bathochromic effect 1/125
bathochromic shift 2/28
beenaker cavity 1/476
Beer-Lambert 1/89
Beer-Lambert law 1/40, 1/74, 2/75
bending modes 1/42
benz[e]indolium Sq660 2/35
benzindolium heptamethine cyanine dyes 2/34
benzofuranyl fluorophore 2/11
BHHCT 2/56
BIA see real-time BioInteraction Analysis
BiaCore instrument 2/127

BiaCore system 2/87
bioanalytical applications 1/344
bioinformatics 2/6
biological mass spectrometry 2/502
biological wastewater treatment
– alcohol ethoxylates (AEO) 2/175
– alkyl polyglucosides 2/175
– fluorine-containing surfactants 2/175
biolumines cent determination of ATP 2/20
bioluminescence 2/65
BioMEMS 2/131
biomolecular databases 2/8
biopolymer analysis 1/349
– sequence mutations 1/349
biosensor 2/3 ff
– calorimetric 2/5
– electrochemical 2/5
– optical 2/5
– piezoelectric 2/5
– technologies 2/5
biotin 2/29
BIS (Bremsstrahlung isochromat spectroscopy) 1/536
bis(alkylsiloxysilyl) complexes of naphthalocynines (SiNPcs) 2/36
bisindolylmaleimides 2/24
bisoxonols 2/19
BLE (bombardment-induced light emission) 1/533
blood coagulation 2/4
blotting
– fluorescent strains for 2/15
blue-fluorescent N-methylanthraniloyl (MANT) 2/24
BODIPY 2/15 f, 2/16
BODIPY dyes 2/51
Bohr energy level 1/422
Boltzmann's law 1/426
boosted hollow-cathode lamp 1/438
Born-Oppenheimer approximation 1/41
Bremsstrahlung 1/372
Brewster angle 1/74, 2/83
bromacil 2/53
bromide
– ion, indicators for 2/13
Brookhaven protein data bank 2/7 f
bubblers 1/12
Bunsen 1/421, 1/436

c

Ca^{2+} 2/12
Ca^{2+} channel 2/25
CAE (capillary array electrophoresis) 2/39
calcein 2/12

Calcium Green 2/12
calcium regulation 2/12, 2/23
calcium transport 2/12
calibration models 2/444
calibration tests 2/254
calibration
– Raman shift scale 1/60
– wavelength scale 1/65
calixarenes 2/69 f
calmodulin 2/23
– fluorescently labeled 2/23
Cambridge Crystallographic Database 2/8
candidate structures 2/481, 2/484
canisters 1/11
capacitively coupled microwave plasma (CMP) 1/476, 1/477
capillary array electrophoresis (CAE) 2/39
capillary column technology 2/255
capillary electrophoresis 1/346
– affinity chromatography 1/346
– atomic emission spectroscopy 1/492 ff
– capillary electrochromatography 1/346
– capillary isotachophoresis 1/346
– detection (CE-LIFP)
  – laser-induced fluorescence polarisation 2/55
– mass spectrometry 1/348
– micellar electrokinetik chromatography 1/346
capillary gel electrophoresis 2/39
carbamates 2/177 ff, 2/193
– biological fluids 2/194
– drinking water 2/217
– estuarine waters 27217
– groundwater 2/217
– hydrolytic degradation 2/217
– leaches of soil 2/217
– microbial degradation 2/217
– quantification 2/216 f
– natural waters 2/216
– river water 2/203, 2/217
– surface water 2/217
carbohydrate structures 1/351
carbon-centered fragment 2/473
carbonic anhydrase 2/13, 2/21
carboxyfluorescein 2/11
carriers 2/25
cartesian geometry 1/395
CAS 2/446
caspase protease activity 2/6
catalase 2/21
cation receptors 2/69
CBQCA 2/16
CCD (charge coupled device) 1/481

CCSD 2/9
CE see electrophoresis capillary
CEA see electrophoresis capillary array
CE-AES 1/492 ff
cell
– Alzheimer's disease 2/11
– apoptosis 2/11
– endocytosis 2/11
– gas 1/90
– homeostasis 2/11
– ion transport 2/11
– malignancy 2/11
– membrane-impermeant dyes, incl. stains for dead cells (SYTOX Dyes) 2/17
– muscle contraction 2/11
– multidrug resistance 2/11
– -permeant nucleic acid dye 2/17
– proliferation 2/11
– tracers
  – microinjectable 2/57
cellular
– uptake of lipids 2/19
– diagnostics 2/15
cellulase 2/20
centering 2/442
CF-FAB 2/160 ff
– benzo[a]pyrene conjugates 2/161
– brominated surfactants 2/160
– carbamate 2/162
– collision-induced dissociation (CID) 2/162
– DNA adducts 2/161
– drinking water 2/160
– dyes 2/162
– explosives 2/161, 2/163
– flow injection analysis (FIA) 2/161
– metabolites of surfactants 2/160
– N-containing pesticides 2/162
– ozination 2/161
– PAH metabolites 2/162
– PAHs 2/163
– P-containing pesticides 2/162
– phenylurea 2/162
– products of surfactants 2/161
– raw 2/160
– seawater 2/160
– sulfonated azo dyes 2/161
– sulfonates 2/161
– surface water 2/160
– surfactants 2/160
– wastewater 2/160
chain conformation statistics 1/297
channeling-RBS 1/567
characteristic bands 2/452
charge coupled device (CCD) 1/481

charge injection device (CID) 1/481
charge structures 1/334
charge transfer 2/251
chemical abstracts service 2/489
Chemical Concepts 2/491
chemical exchange 2/111
– in NMR 2/111
chemical interferences 1/454, 1/462
– in AES 1/487
chemical ionization 1/331, 2/249, 2/261, 2/407
chemical ionization process 2/262
chemical modification 1/353
chemical shift (CSA) 1/174, 1/272, 1/275, 1/281, 2/471
chemical shift anisotropy (CSA) 1/174, 1/190, 1/270, 2/99
– aromatic 1/190
– dependence on bonding 1/190
– interaction 1/187, 2/116
  – of ^{15}N 2/116
– methyl sites 1/190
– olefinic 1/190
CHEMICS 2/483
chemiluminescence 2/65
Chemometrics Toolbox 2/467
ChemWindow Spectroscopy 2/502
chitinase 2/20
chloride
– ion, indicators for 2/13
Cholera Toxin 2/21
cholesterols 2/15, 2/69
choline glutamate 2/21
ChromaTide nucleotides 2/18
chromoionophore 2/70
chromophores 2/10, 1/125
– UV/VIS 2/10
– near-IR 2/10
CI conditions 2/249 ff, 2/255
CI ion source 2/261
CI ionization processes 2/265
CI mass spectrometry 2/260, 2/262
CI process 2/250, 2/261
CI techniques 2/260
CID (charge injection device) 1/481
CID procedure 2/257
CID process 2/253 f
CIGAR-HMBC 1/247
circular dichroism (CD) 1/82, 2/6, 2/93 ff
circulatory serine proteases 2/24
classical least squares (CLS) 2/459
clinical diagnostics 2/4
CLS 2/459
cluster analysis 2/455

CMP 1/476
CN^- 2/14
^{13}C-NMR spectrum prediction 2/473
CODEX ^{13}C NMR 1/292
– correlation time 1/292
– mobile segments 1/292
– reorientation angle 1/292
cold vapor generation technique 1/451 f
collection and preparation of gaseous samples 1/4
collection and preparation of liquid and solids 1/17
collision induced dissociation (CID) 2/253
collisional deactivation 2/250
collisional quenching 1/45, 1/141, 1/157
collisional-induced dissociation (CID) 1/337, 2/126
colloidal gold 2/16, 2/89
COLOC 2/471
combinatorial chemistry 2/69
combined rotation and multiple pulse decoupling (CRAMPS) 1/175, 1/287
complexing agents 2/154
compounds
– atmospheric-pressure chemical ionisation 2/152
– non-voltile 2/152
– polar 2/152
– thermolabile 2/152
– volatile 2/153
Compton equation 1/371
Concanavalin A 2/21
conclusions 2/226
confocal 1/138, 1/155
confocal microscopes 1/85
conformation
– molecular 1/578
conformational states 1/349
α-conotoxin probes 2/25
continuous flow FAB (CF-FAB) 2/160
continuum source background correction 1/456 f
continuum sources 1/439
conventional mass spectrometers 2/251
Coomassie Blue 2/16
CoroNa Red 2/11
correlated spectroscopy (COSY) 2/105, 2/471
correlation 2/451
correlation by long-range couplings (COLOC) 2/471
COSY (homonuclear correlated spectroscopy) 1/173, 1/223 ff, 2/471
COSY/GCCOSY 1/224
COSY-type spectra 2/483

coumarin's 2/15
coumarine 2/70
counting rate 1/405
counting statistics 1/404
coupling
– constants 2/471
– dipolar 1/276
– ^{2}H quadrupolar 1/278
– quadrupolar 1/276
CP 1/281 f
– ramped 1/282
CP MAS 1/284
CPAA (charge particle activation analysis) 1/516
CRAMPS 1/175, 1/287
creatinine 2/69
critical angle of total reflection $\emptyset_{crit}$ 1/398
critical penetration depth 1/402
critical thickness 1/402, 2/79
croconine dyes 2/34
cross polarisation (CP) 1/281
cross validation 2/461
cross-relaxation experiments 2/110
cryogenic trapping 1/13
cryotrapping-AAS 1/468
crystal field 1/133
crystal spectrometer 1/391
crystallography
– chemical species specific 1/592
– chemical state specific 1/587
– element specific 1/587, 1/592
CSA (chemical shift) 1/272, 1/174, 1/275, 1/281, 2/471
– carboxyl group 1/272
– CODEX 1/291
– isotropic chemical shift 1/281
– isotropic spectra 1/281
– slow dynamic processes 1/291
– tensors 1/272
CSEARCH 2/483
cut-off thickness 2/79
Cy™dyes 2/31
Cy3™ 2/31
Cy3NOS 2/31
Cy5™ 2/31
cyanide
– determination of 2/14
– in blood 2/14
cyanine dimers 2/17
cyanine dyes 2/17
cyclic AMP 2/71
cyclic nucleotides 2/24
cynine dyes 2/15
cysteine 2/15

cystic fibrosis 2/13
cysticerocosis 2/47
cytochrome c
– electron-transfer reaction in 2/86
cytochrome P-450 2/22
cytological staining 2/15

d

Danish Center for Human Genome Research 2/9
dansyl chloride 2/16
dansyl fluorophores 2/19
DAPS (disappearance potential spectroscopy) 1/508
DARC 2/483
data analysis 2/445
data extraction in EDXRF 1/407
data extraction in WDXRF 1/406
data processing
– Fourier transformation 1/187
– phasing 1/187
– smoothing function 1/187
data treatment 1/404
databases
– spectral 1/99
DC arc 1/477, 1/484
DCIP 2/14
DCP 1/475
debsyl chloride 2/16
decomposition of organics 1/28
– ashing 1/28
deconvolution 2/391 f
decoupling 1/283
– CW 1/283
– TPPM 1/283
deformation vibrations 1/42
degeneration factor 1/428
DELFIA system 2/55
DENDRAL 2/483
density of states 1/537, 1/551
DEPT (distortionless enhancement polarization transfer) 1/215, 2/471, 2/482
depth profiling 1/512, 1/531, 1/546 f, 1/563, 1/565, 1/572, 1/589
descriptive proteomics 2/126
detection and determination limits 1/377
detection limits 1/418
detection of human IgG 2/53
detectors 1/59, 1/53, 1/389
– array 1/53
– DTGS 1/53
– MCT 1/53
– multichannel 1/61, 1/67
– single channel 1/57

determination limit 1/378
deuteration 2/98
– in NMR 2/98
deuterium effect 1/143
deuterium lamp 1/439, 1/456 f
2D exchange 1/303
– isotactic polypropylene 1/303
– tropolone 1/303
DFT 2/28
2D-gel electrophoresis 1/357
diabetes 2/6
diabetes mellitus 2/69
diagnostic agents 2/154
diagnostics
– clinical 2/4
– drug discovery 2/4
2,3-diaminonaphthalene 2/13, 2/14
dichlorodihydrofluorescein diacetate 2/26
dichlorofluorescein 2/11
2,4-dichlorophenoxyacetic acid (2,4-D) luminol/HRP system 2/65
dielectric interface 2/72
3,3'-diethyloxadicarbocyanine (DODC) 2/34
difference interferometer 2/83
diffuse reflection 1/78
diffuse reflection infrared Fourier transform spectroscopy (DRIFTS) 1/518
difluorofluoresceins (Oregon Green) 2/11
dilute liquid crystals 2/99
diode array UV see Photodiode array detection
diode laser source 1/440
diode lasers 2/10, 2/27
1,2-dioxetane derivatives 2/65
dioxins 2/257
dipolar coupling (D) 1/174
dipolar interaction 1/188 f
– definition 1/188
– Pake doublet 1/188
– powder pattern 1/189
direct (or second-element) enhancement 1/404
direct current plasma 1/475
direct-excitation configuration 1/393
direct insertion 1/483
direct liquid introduction (DLI) 2/153, 2/156
direct sample introduction 1/452
disappearance potential spectroscopy (DAPS) 1/508
discoidal bilayered structures 2/99
– in NMR 2/99
Discosoma 2/58
discriminant analysis 2/455
dissociation energy 1/434
dissociation equilibrium 1/463

dissociation of molecules in a plasma 1/434
distortionless enhancement polarization transfer (DEPT) 1/215, 2/471, 2/482
5,5'-dithiobis-(2-nitrobenzoic acid) 2/14
2D *J*-resolved NMR 1/219
DLI
– herbicides 2/156
– pesticides 2/156
2D MQMAS 1/317
– glasses 1/317
– isotropic shifts 1/317
– microporous materials 1/317
– minerals 1/317
2D multiple quantum magic angle spinning (MQ/MAS) 1/176
DNA 2/10
– analysis 2/7
– arrays and microarrays 2/18
– polymerase 2/38
– sequencing 2/10, 2/39
2D NMR
– homonuclear ^{13}C-^{13}C 2D correlation experiments 1/293
3D NMR 1/173, 1/204
– connectivity information 1/206
– HNCA pulse sequence 1/204
DODC 2/33
domain structure 1/569
1D (one-dimensional) NMR 1/210
donor/acceptor 1/152
donor-acceptor-donor (D-A-D) dyes 2/58
dopamines 2/15, 2/71
doppler line broadening 1/430
double beam atomic absorption spectrophotometer 1/453
double bond
– conjugated 1/129
double rotation (DOR) 1/176
double-focusing 1/336
double-quantum filtered (2QF) COSY experiment 2/107
double-quantum spectroscopy 1/295
– BABA 1/295
2D PASS 1/303
DPH 1/151
DQ spectroscopy 1/297
– ^{31}P-^{31}P DQ MAS 1/297
DRIFTS (diffuse-reflectance (or reflection) infrared Fourier transform spectroscopy) 1/518
drinking water 2/256
drug discovery 2/5
drugs 2/154
– ESI-LC-MS/MS 2/203

– MS/MS   2/204
– quantification   2/203 f
– wastewater   2/204
– X-ray contrast media   2/204
DsRed   2/58
DTNB   2/14
dyes   2/154, 2/173, 2/205
– CZE   2/205
– MS/MS   2/206
– wastewater   2/206
dynamic   1/140 f, 1/198
– -angle spinning (DAS)   1/176
– proteomics   2/126
– quenching   1/141
– range of motional   1/198
– solid-state ^{2}H NMR   1/198

e
easily ionisable element (EIE)   1/465
echelle grating   1/439, 1/480
echelle spectrograph   1/481 ff
E-COSY   2/107
edited experiments   2/117
EDX (energy dispersive X-ray analysis)   1/524, 1/567
ED-XRF instrumental configurations   1/393
EEAES (electron excited Auger electron spectroscopy)   1/512
EELFS, EXELFS (extended electron energy loss fine structure)   1/529, 1/535, 1/562
EELS see HREELS
effective path length   2/76
EI conditions   2/253
EI interferences   2/249
EI scan program   2/248
EI spectra   2/248
EI/CI ion source   2/252
Einstein coefficients   1/427 ff
Einstein transition probability   1/427
ejection techniques   2/252
elastic or Rayleigh scattering   1/371
electroanalysis   2/3
electrochemically generated chemiluminescence (ECL)   2/65
electrochemiluminescence   2/65
electrodeless discharge lamp (EDL)   1/439
electromagnetic spectrum   1/367
electromagnetic wave   1/39 f
electron ionisation   1/331
electron pressure in a plasma   1/433
electron temperature   1/435 f
electronic states   1/508, 1/584
electronic structure   1/539, 1/543, 1/571, 1/577

electrophoresis
– capillary   2/4
– capillary array   2/4
– gel   2/4
electroreflectance   1/561
electrospray (ESI) MS   2/122
electrospray ionization (ESI)   1/329, 1/333, 2/201 ff
electrospray ionization mass spectroscopy (ESI-MS)   2/123
electrothermal atomisation   1/443 ff
electrothermal vaporisation   1/483, 1/484
element analysis see process analysis
elemental composition   1/509 f, 1/513, 1/516, 1/524, 1/540, 1/543, 1/548, 1/556, 1/565 ff, 1/574, 1/589,
elemental distribution   1/548, 1/554, 1/569
ELF see enzyme-labeled fluorescence
ELISA (enzyme-linked immunosorbent assay)   2/50
ellipsometric sensors   2/82
ellipsometry   1/528, 2/82
Ellman's reagent   2/14
emission spectrum   1/382
EMPA (electron microprobe analysis)   1/524
EMS (electron momentum spectroscopy)   1/522
end-fire coupling   2/82
endocrine disrupting chemicals (EDC)   2/87
endocytosis   2/11
endogenous glycosidase activity   2/20
endoplasmic reticulum   2/22
– probes for   2/22
energy level diagram   1/425
energy noise   1/405
energy-dispersive XRF   1/393
energy-gap law   1/143
enhanced Raman scattering   1/119
enhancement   1/403
environmental analysis   2/254
environmental monitoring   2/5
Environmental Protection Agency   2/447
environmental waters
– antibiotic   2/203
– sulfonamides   2/203
– tetracyclines   2/202
enzymatic exopeptidolytic cleavage   1/351
enzyme-labeled fluorescence (ELF)   2/15
enzyme-linked immunosorbent assay (ELISA)   2/50
EPA methods   2/256
EPECS (Auger photoelectron coincidence spectroscopy)   1/514
epifluorescence   1/155

epilepsy 2/13
EPIOS 2/483
epitope analysis 2/128
epitope mapping 2/6
EPMA (electron probe microanalysis) 1/524, 1/567
EPXMA (electron probe X-ray Microanalysis) 1/524
ERCS (elastic ecoil coincidence spectrometry) 1/522
ERDA or ERD (elastic recoil detection (analysis)) 1/520
ESCA [electron spectroscopy for chemical applications (originally analysis)] 1/587
ESD (electron stimulated desorption) 1/525
ESDIAD (electron stimulated desorption ion angular distributions) 1/526
ESI (electrospray ionization) 1/329, 1/333, 2/200 ff
– AEO 2/209 f
– agricultural soil 2/222
– alcohol ethoxylate (AEO) 2/213
– alkyl etoxysulfates (AES) 2/211
– alkylphenols 2/208
– alkyl polyglucoside 2/212
– alkyl sulfates (AS) 2/211
– anilides 2/215
– antibiotic 2/203
– azo dyes 2/206
– biodegradation 2/212
– bisphenol A 2/208
– CE-MS 2/213
– chemical degradation 2/222
– coastel waters 2/211
– coconut diethanol amide (CDEA) 2/213
– complexing agents 2/203
– crop 2/222
– CZE 2/205
– degradation products 2/220, 2/222 f
– diagnostic agents 2/203
– disinfection byproducts 2/207
– drinking water 2/203, 2/217 ff, 2/223
– drugs 2/203
– ditallow-dimethylammonium chloride (DTDMAC) 2/212
– dyes 2/205
  – metabolites 2/205
– EDTA 2/203
– effluents 2/211
– electropherogram 2/200
– estrogenic compounds 2/206
– estuaries 2/212
– estuarine waters 2/217, 2/220 f
– explosives 2/206
– fruits 2/222
– fungicides 2/215
– German Bight 2/212
– groundwater 2/217 ff, 2/223
– haloacetic acids 2/207
– halogenated APEO 2/211
– halogenated NPEO 2/210
– herbicides 2/215
– hydrolytic degradation 2/217
– imidazolinone herbicides 2/225
– insource-CID 2/211
– ion chromatograph 2/214
– LAS 2/212 f
– LC-MS 2/202, 2/213
  – anilides 2/202
  – antifouling pesticides 2/202
  – carbamates 2/202
  – complexing agents 2/202
  – drugs and diagnostic agents 2/202
  – dyes 2/202
  – estrogenic compound 2/202
  – explosives 2/202
  – fungicides 2/202
  – haloacetic acids and desinfection byproducts 2/202
  – herbicides 2/202
  – organoarsenic compounds 2/202
  – organophosphorus compounds 2/202
  – pesticides 2/202
  – phenols 2/202
  – phenolic pesticides 2/202
  – phenoxycarboxylic acids 2/202
  – phenylureas 2/202
  – polycyclic aromatic hydrocarbons 2/202
  – quaternary amines 2/202
  – sulfonic acids 2/202
  – sulfonylureas 2/202
  – surfactants 2/202
  – thiocyanate compounds 2/202
  – thioureas 2/202
  – toluidines 2/202
  – toxins 2/202
  – triazines 2/202
– leaches of soil 2/217
– metabolites 2/209 f, 2/222, 2/224
– N-methylglucamides 2/212
– microbial degradation 2/217
– miscellaneous 2/225
– MS/MS 2/204, 2/206 f, 2/209 ff, 2/216, 2/219, 2/221, 2/224 f
– natural waters 2/221
– neutral loss (NL) 2/209 f
– NPEO sulfates 2/213
– organoarsenic compounds 2/207

- perfluorooctanesulfonate (PFOS) 2/212
- perfluorooctanoic acid (PFOA) 2/212
- pesticides 2/215
- phenols 2/208
- photolysis products 2/222
- quantification 2/203 ff
- quaternary amines 2/215
- quaternary ammonium compounds 2/214
- neutral loss (NL) 2/209
- nonylphenolpolyether carboxylate (NPEC) 2/210
- North Sea 2/212
- NPEO 2/209 ff
- OPEO 2/211
- photolysis 2/224
- physicochemical degradation 2/224
- review
  - dyes 2/202
  - general 2/202
  - pesticides 2/202
  - surfactants 2/202
  - sulfonates 2/202
  - toxines 2/202
- river water 2/203, 2/217, 2/221, 2/223
- sea water 2/212
- secondary alkane sulfonates (SAS) 2/213
- sediment 2/212
- SFC-MS 2/223
- sulfonamides 2/203
- sulfonates 2/200
- sulfonic acids 2/208
- surface water 2/217 ff
- surfactants 2/209
- tetracyclines 2/202
- thiocyanate compounds 2/215
- toluidines 2/215
- toxins 2/213
- Waddensea marinas 2/212
- wastewater 2/203, 2/204, 2/206, 2/210 f, 2/218
- wastewater inflows 2/211
- water 2/220
- X-ray contrast media 2/204

ESI-LC-TOFMS
- aromatic sulfonamides 2/204
- sulfonates 2/204
- textile wastewater 2/204

ESI-MS (electrospray ionization mass spectroscopy) 2/123
esters 2/11
estrone-3-glucuronide (E3G) 2/87
ethidium bromide 2/17
Euclidean distance 2/449, 2/472
europium chelates 2/56

evaluation of spectra 2/444
evanescent field 1/75 f
- penetration depth 1/76
evanescent wave spectroscopy 2/71
evanescent wave-based techniques 2/69
EWCRDS (evanescent wave cavity ring-down spectroscopy) 1/530
EXAFS (extended X-ray absorption fine structure) 1/529, 1/535, 1/562, 1/584
excimer 1/154
excitation shift 1/146
excitation temperature 1/429, 1/435 f
EXELFS (extended energy loss fine structure) 1/529
ExPASy see also expert protein analysis system 2/8
expert protein analysis system 2/7
explorative data analysis 2/444
explosives 2/154
- degradation 2/207
- degradation products 2/206
- groundwater 2/206
- quantification 2/207
EXSY (EXchange SpectroscopY) 1/231
external ion sources 2/252 f
extraction and preparation of samples 1/14

f

FAB (fast-atom bombardment mass spectrometry) 1/333, 1/574, 2/160 ff
- benzo[a]pyrene conjugates 2/161
- brominated surfactants 2/160
- carbamate 2/162
- collision-induced dissociation (CID) 2/162
- DNA adducts 2/161
- drinking water 2/160
- dyes 2/162
- explosives 2/161, 2/163
- flow injection analysis (FIA) 2/161
- metabolites of surfactants 2/160
- N-containing pesticides 2/162
- ozination 2/161
- PAH metabolites 2/162
- PAHs 2/163
- P-containing pesticides 2/162
- phenylurea 2/162
- products of surfactants 2/161
- raw 2/160
- seawater 2/160
- sulfonated azo dyes 2/161
- sulfonates 2/161
- surface water 2/160
- surfactants 2/160
- wastewater 2/160

FAD 2/63
FAM 2/40
far-red 2/10
fassel 1/473
fast atom bombardment (FAB) 1/333, 1/574, 2/160 ff
fatty acids 2/15
FCS 2/19
α-fetoprotein (AFP) 2/61
FIA (flow injection analysis) 2/65
FIA-AES 1/492 ff
FIA-MS
– surfactants 2/175
FIA-MS/MS
– surfactants 2/175
fiber and waveguide SPR 2/88
fiber optics 2/82
fibronectin 2/24
field desorption 1/332
field effect transitor 1/387
figures-of-merit 1/376
films
– Langmuir-Blodgett films 1/529, 1/560, 2/91
filter
– notch 1/62
– Rayleigh 1/61
filtered experiments (in NMR) 2/117
FIM (field ion microscope) 1/541
fingerprinting capabilities 2/255
firefly luciferase 2/65
FISH (fluorescence in situ hybridization) 2/18
FITC see fluorescein isothiocyanate
Fiveash Data 2/447
flame AAS 1/441
– burning velocity 1/443
– gas mixtures 1/443
– oxidising flame 1/443
– reducing flame 1/443
flame atomisation 1/441
flame atomiser 1/471
flavinmononucleotide (FMNH2) 2/65
flow cytometry 1/135, 1/138, 1/153, 2/15
– standardization reagents
flow injection analysis (FIA) 2/65
flow injection analysis and atomic emission spectroscopy 1/492 ff
fluorescamine 2/16
fluorescein casein 2/20
fluorescein diacetate 2/11
fluorescein isothiocyanate 2/29
fluorescein's cyanines 2/15
fluoresceins 2/28

fluorescence 1/138
– correlation spectroscopy (FCS) 1/155, 2/19
– detection in HPLC 2/395 ff
– enzyme-labeled 2/15
– laser-induced 2/4, 2/39
– lifetime 1/139, 2/28
– polarisation 1/148, 1/151
– polarisation immunoassay (FPIA) 2/55
– polarisation spectroscopy 2/54
– quencher 1/139
– recovery after photobleaching (FRAP) 1/155, 2/19
– resonance energy transfer (FRET) 1/152 f, 2/35, 2/56
– sensors 1/156
– spectroscopy 2/7
– time-resolved 2/10
fluorescent
– dye loaded micro- and nanoparticles 2/15
– dyes 2/7
– enzymes 2/7
– isothiocyanates 2/16
– latex particles 2/15
– polymixin B analogs 2/24
– probes 2/10 f
– proteins 2/7
fluorescently labeled calmodulin 2/23
fluoride
– ion, indicators for 2/13
fluorinated fluoresceins 2/51
fluorophores 2/10
fluorophores
– near-IR 2/10
– visible 2/10
FluoZin 2/13
FMIR (frustrated multiple internal reflection) 1/511
food analysis 2/5
foot-and-mouth disease virus 2/87
forbidden transition 1/375
forward search 2/496
Fourier transform infrared spectrometry (FTIR) 2/92 ff
– conformational changes in proteins 2/92
– protein unfolding 2/92
– secondary structure content 2/92
Fourier transform ion cyclotron resonance 1/341
Fourier transform ion cyclotron resonance (FTICR) detector 2/124
Fourier transform ion cyclotron resonance instruments 2/124
Fourier transform ion cyclotron resonance spectrometer (FT-ICR) 2/127

FPA 1/53
fragmentations 1/350
Franck-Condon factor 1/45
Franck-Condon state 1/45, 1/144
FRAP 2/19
free induction decay (FID) 1/186
frequencies
– characteristic 1/99
– group 1/99
Fresnell equations 2/72
FRET 2/35
frustrated total internal reflection (FTR) 2/81
^{19}F solid-state NMR 1/287
– biomembranes 1/287
– fluoropolymers 1/287
FT RAIRS (Fourier transform reflection-absorption infrared spectroscopy) 1/559
FTIR (Fourier transform infrared spectrometry) 2/92 ff
– in vivo monitoring of glucose 2/92
FTIR microscopy 2/92 ff
– in vivo monitoring of glucose 2/92
FTIRRAS see IRRAS
full scan monitoring 2/251
full spectra search 2/448
full width at half of the maximum peak height (FWHM) 1/182
Fullerenes 2/387
functional genomics 2/3 ff
fundamental parameter method 1/414
fundamental parameter technique 1/410
fura-2 2/12
furans 2/257
Fura-Zin 2/13

g
GABA$_A$ receptor 2/25
galactic 2/447
galactose 2/21
gallium-aluminium-arsenide laser diode 2/27
gas chromatography and atomic emission spectroscopy 1/491 ff
gas chromatography-atomic absorption spectrometry (GC-AAS) 1/467 ff
gas chromatography coupled with mass spectrometric detection (GC-MS) 1/34, 2/251 f
gas chromatography/ion trap mass spectrometry (GC/ITMS) 2/244 ff, 2/251 ff, 2/255 ff, 2/262, 2/265
gas flow proportional counters 1/384
gas phase ionisation 1/331
gas temperature 1/435 f
GC ion trap mass spectrometer 2/245

GC/chemical ionization-ITMS 2/260
GC/CI MS 2/251
GC/EI MS 2/251
GC/EI-ITMS analyses 2/252
GC/ITMS 2/245, 2/247, 2/251 ff, 2/255 ff, 2/259, 2/262, 2/265
GC/MS 1/344, 2/251 f
GC/MS acquisition 2/247
GC/MS experiments 2/248
GC/MS quadrupole-based systems 2/244
GC/MS/MS 2/253
GC/MS/MS procedures 2/251
GC/MS/MS ion traps 2/254
GC-AES 1/491 ff
GC-MS (gas chromatography coupled with mass spectrometric detection) 2/152
– analysis 2/153
GCOSY 1/173
GDMS (glow discharge mass spectrometry) 1/533
GDOES (glow discharge optical emission spectrometry) 1/531
GE see electrophoresis gel
gel electrophoresis 2/39
gene expression 2/13
gene probes 2/10
genome 2/8
– map of the human 2/8
– project, human 2/3
genomics 2/3 ff
– functional 2/4
– polymorphism 2/4
gentamicin 2/87
GHMBC 1/173
GHMQC 1/173
GIS (grazing incidence spectroscopy) 1/559
GIXFR (grazing incidence X-ray fluorescence) 1/581
GIXFR (grazing-exit X-ray fluorescence) 1/581
glowbar 1/50
glow discharge 1/479
glucose 2/69
glucose oxidase 2/14
glucose-6-phosphate dehydrogenase 2/14
glucuronidase 2/20
β-glucuronidase 2/20
glutathione 2/14
glutathione transferase 2/14
gluthathione 2/15
glycosidase 2/20
– endogenous activity 2/20
glycosylations 1/351
Golgi apparatus 2/22

– probes for 2/22
gradient 1/232 f
gradient 1D NOESY 1/255
gradient experiments 1/233
– GCOSY 1/233
– GNOESY 1/233
– GTOCSY 1/233
GRAMS 2/467, 2/501
graph theory 2/482
graphite furnace (atomiser) 1/443 ff
graphite furnace, L'vov platform 1/445
graphite furnace, temperature profile 1/446
graphite furnace, temperature program 1/447
grating couplers 2/81, 2/83
green fluorescent proteins 2/15, 2/58
Greenfield 1/473
Grimm 1/479
Grotrian diagram 1/425
group frequencies 1/44, 1/100, 2/452

h

Hahn echo 2/104
half width of atomic lines 1/437 f
half-integer quadrupole nuclei 1/315
– fourth-rank anisotropic broadening 1/315
– second-order quadrupolar broadening 1/315
haloacetic acids 2/154
hard ionisation 1/331
^{1}H decoupling 1/282
– TPPM 1/282
^{1}H DQ MAS
– hydrogen-bonded protons 1/306
– kinetics of hydrogen bond breaking and formation 1/306
– order parameter 1/306
– proton-proton distances 1/306
HEIS (high energy ion scattering) 1/543, 1/565
Helicobacter Pylori 2/54
helium-neon laser 2/40
hemicyanine dyes 1/143
hemoglobin 2/69
HeNe laser 2/31
heparin 2/24
heparin-binding growth factors 2/24
4,4'-bis (1',1',1',2',2',3',3',-heptafluoro-4',6'-hexanedion-6'-yl)-chlorosulfo-o-terphenyl (BHHCT) 2/56
herbicides
– benzidines 2/154
– carbamates 2/154
– chlorinated 2/154

– quaternary ammonium 2/154
– phenoxyacetic acid 2/154
– triazine 2/154–
heterogeneity 1/417
heterogeneous catalysis 1/582
heteronuclear 2D correlation (HETCOR) 1/176
heteronuclear correlation (HETCOR) 1/307
– ^{1}H-^{13}C WISE (wideline separation) 1/307
– heteronuclear MQC (HMQC) 1/307
– heteronuclear SQC (HSQC) 1/307
– homonuclear decoupling in t_1 1/307
– recoupled polarisation transfer (REPT) 307
– rigid and mobile chemical moieties 1/307
heteronuclear correlation spectroscopy 2/96 ff
heteronuclear dipolar couplings 1/310
– dipolar couplings 1/310
– internuclear distances 1/310
– REPT 1/310
heteronuclear multiple bond correlation (HMBC) 2/471
heteronuclear multiple quantum coherence (HMQC) 2/114 ff, 2/471
heteronuclear NMR experiments 2/94 ff, 2/113
heteronuclear shift correlation 1/234
heteronuclear single quantum coherence (HSQC) 2/471
^{1}H-^{1}H DQ MAS
– BABA recoupling sequence 1/305
– dipolar coupling constant 1/305
– spinning-sideband patterns 1/305
HIAA (high energy ion activation analysis) 1/516
high-performance liquid chromatography (HPLC) 2/10
high-resolution gas chromatography (HRGC) 2/257
high-resolution mass spectrometry (HRMS) 2/257
high-resolution spectra
– double rotation (DOR) 1/315
– dynamic-angle spinning (DAS) 1/315
hindered rotors 1/151
histidine 2/15
histochemistry 2/15
hit list 2/451
Hitachi Ltd. 2/266
HIV-1 protease inhibitor 2/87
^{1}H MAS NMR 1/287
HMBC (heteronuclear multiple bond correlation) 1/173, 1/242, 2/471
HMQC (heteronuclear multiple quantum coherence) 1/173, 1/234, 2/471, 2/483

hollow cathode discharge 1/479
hollow-cathode lamp (HCL) 1/437, 1/460 f
homeostasis 2/11
homonuclear 2D NMR 1/223
homonuclear dipolar coupling 1/285, 1/290
– BABA 1/290
– C7 1/290
– DRAMA 1/290
– DRAWS 1/290
– DREAM 1/290
– HORROR 1/290
– RFDR 1/290
homonuclear dipolar-coupled spins 1/290
– internuclear distance 1/290
homonuclear Hartmann-Hahn (HOHAHA) 2/109
homonuclear TOCSY, total correlated spectroscopy 1/226 ff
homonuclear two-dimensional
– double-quantum (DQ) coherence 1/294
– INADEQUATE 1/294 ff
HOMSTRAD (HOMologous STRucture Alignment Database) 2/9
horseradish peroxidase (HRP) 2/16, 2/50, 2/65
– labelin immuno assay 2/50
HOSE code 2/473, 2/478, 2/483
HREELS, HEELS (high resolution electron energy loss spectroscopy) 1/533
HRMS 2/258
HRP 2/15 f
^{1}H solid state NMR
– CRAMPS 1/299
– high-resolution 1/298
– windowless homonuclear decoupling 1/299
HSQC (heteronuclear single quantum coherence) 1/173, 1/236, 2/471
human chorionic gonadotropin (hCG) 2/56
– β-subunit of 2/56
human creatine kinase MB (CK-MB) 2/87
human genome project 2/3
human IgG 2/53
human phenylalanine hydroylase 2/87
human serun albumin (HSA) 2/56
Hybrid Q-TOF MS 2/124
hybrid time-of-flight mass spectrometers 1/340
hydride generation technique 1/448 ff
hydridization detection 2/18
hydrocarbons 2/261
hydroxy carbonyls 2/261
hydroxyl number 1/110
8-hydroxypyrene-1,3,6-trisulfonic acid 2/11

hydroxystilbamidine 2/17
hyperchromic effect 1/125
hyperfine structure line broadening 1/431
hypericin 2/24
hyphenated 2D NMR 1/174
hyphenated techniques 1/466 ff
hyphenated-2D NMR experiments 1/252
– GHSQC-TOCSY 1/252
– HC-RELAY 1/252
– HMQC-TOCSY 1/252
– HXQC-COSY 1/252
hypochromic effect 1/125
hypocrellins 2/24
hypsochromic effect 1/125
hypsochromic shift 2/28

i

IASys system 2/85
IBIS biosensors 2/87
IBSCA (ion beam spectrochemical anylysis) 1/533
IC/MS 1/344
ICP (inductively coupled plasma) 1/473 ff
ICR mass spectrometry 2/252
identity search 2/497
IETS (inelastic electron tunneling spectroscopy) 1/535
(IGF)-binding protein-2 2/87
illumination
– sample 1/113
ILS (inverse least square) 2/462
immunoaffinity extraction 1/345
immunoassay 2/7, 2/10, 2/15, 2/47
– competitive 2/47
– non-competitive 2/47
– sandwhich 2/47
immunochemistry
– with NIR fluorophores 2/51
– with visible fluorophores 2/50
immunochromatography 2/15
immunohistochemistry 2/15
– stains for 2/15
immunosensor 2/53
IMPEACH-MBC 1/246
imprinted polymers 2/71
in situ hybridization 2/16
in vivo dynamics
– cytoskeleton 2/17
in vivo glucose monitoring 2/69
INADEQUATE 1/232, 2/471
INCOS 2/496
indirect (or third-element) enhancement 1/404
INDO 2/28

indo-carbocyanines 2/19
indolium heptamethine cyanine dyes 2/32
indolium Sq635 2/35
indolium-squaraine dyes 2/34
inductively coupled plasma (ICP) 1/473 ff
inelastic or Compton scattering 1/371
INEPT (insensitive nuclei enhanced by polarization transfer) sequence 1/214, 2/114
infinitely thick or massive samples 1/402
influence coefficient method 1/410, 1/413
infrared and Raman spectroscopy 2/92 ff
– in bioanalysis 2/92 ff
infrared interfaces 2/416
infrared microscopes 1/85
infrared spectroscopy 1/41, 2/6
in-house database 2/446
in-plane 1/42
INS (ion neutralisation spectroscopy) 1/538, 1/552
instrumental tune-up tests 2/254
instruments
– single beam 1/64
insulin 2/47
intercalating dyes 2/18
interface 1/529, 1/579, 2/152 ff
– atmospheric pressure 2/155
– atmospheric-pressure chemical ionisation (APCI) 2/152 f
– continuous flow FAB (CF-FAB) 2/153, 2/160
– direct liquid introduction (DLI) 2/152, 2/153, 2/156
– electrospray (ESI) 2/153
– environmental analyses 2/155
– fast atom bombardment (FAB) 2/153, 2/160
– hermospray (TSP) 2/153
– ion spray 2/155
– moving-belt (MBI) 2/152 f, 2/156
– particle beam (PBI) 2/153, 2/157
– soft ionisation 2/172
interference 2/89
– chemical 1/454
– fringes 1/96
– spectral 1/454 ff
interferogram 1/51
interferometer 1/50
interferometry 2/83
intermolecular ring current 1/297
internal conversion 1/45, 1/138 f, 1/143
internal energy 2/249
internal standardization 1/412
internal standards 1/410
intersystem crossing 1/45, 1/138 f

intracellular ion activity
– chloride 2/13
intracellular pH 2/11 f
– estimating 2/11
intramolecular processes 1/45
intrinsic zone 1/387
inverse 1/174
inverse least squares (ILS) 2/462
iodide
– ion, indicators for 2/13
ion association reactions 2/262
ion attachment mass spectrometry (IAMS) 2/262
ion attachment reactions 2/263
ion current 2/247
ion cyclotron resonance (ICR) spectrometers 2/249
ion detection 1/340
ion mobility spectrometry 2/123
ion transport 2/11
ion trap 2/247 ff, 2/250 f, 2/254 ff
– mass spectrometry 2/244, 2/250, 2/265
ionisation 1/432
ionisation buffer 1/465
ionisation in flames 1/464
ionisation interference 1/433
ionisation temperature 1/435 f
ionisation, degree of in a plasma 1/432
ionization method 2/249
ionization modes 2/247
ionization time 2/247 f
ion-molecule processes 2/248
ion-molecule reactions 2/247, 2/249, 2/251, 2/253
ion-selective electrodes 2/70
IPMA, SIMP (scanning ion microprobe) 1/539
IPS, IPES (inverse photoelectron spectroscopy) 1/536
IR spectroscopy 1/41, 2/6
IRD™ dyes 2/32
IRMentor Pro 2/452
IRRAS (infrared reflection absorption spectroscopy) 1/75
IRRAS, IRAS (infrared reflection-absorption spectroscopy) 1/559
IR-Tutor 2/452
isoluminol 2/65
isomer generation 2/481
isothiocyanates 2/16
isotope labeling
– in NMR 2/97
Isotopic labeled compounds by atomic emission detection 2/423

ISS (ion scattering spectrometry)   1/565
ITMS   2/261
IUPAC name of the X-ray line   1/373

j
^2J, ^3J-HMBC   1/248
J-modulated spin echo experiments   1/213
– APT   1/213
– DEPT   1/213
– INEPT   1/213
JOE   2/40
jump ratio   1/371

k
K^+
– indicators for   2/10
Karplus equation   2/106
kinetic energy   1/336
Kirchhoff   1/421, 1/436
K-matrix   2/459
Kramers-Kronig relation   1/74
Kramers-Kronig transformation   1/81
Kretschmann configuration   2/85
KRIPES (k-resolved inverse photoemission spectroscopy)   1/536
Kr laser   2/31
KRS-5 2/78
Kubelka-Munk relation   1/79

l
L'vov platform   1/445
labeling
– isotopic   2/38
– labelling   2/17
– detection   2/17
– quantitation   2/17
lactate   2/69
Lambert-Beer law   1/367, 1/429 f, 1/465, 2/75
laminin   2/24
laminar flow burner   1/441 ff
LAMMA, or LAMMS, or LMMS (laser microprobe mass analysis or spectroscopy)   1/533
Langmuir-Blodgett   1/512
Langmuir-Blodgett films   1/529, 1/560, 2/91
lanthanide chelates   2/55
Larmor frequency   1/172
Larmor relation   1/172
laser ablation   1/452, 1/478, 1/484, 1/485
laser desorption   1/329, 1/334 ff
laser induced fluorescence (LIF)   2/39
laser induced fluorescence polarisation   2/55
– in capillary electrophoresis detection (CE-LIFP)   2/55

laser plasma   1/478
latent variables   2/453
latex particles
– fluorescent   2/15
layered synthetic multilayers   1/391
LC-AES   1/492 ff
LC-MS   2/152 ff, 2/163, 2/226 f
– achievements   2/163 ff
– alkylpolyethersulfate   2/164
– capillary electrophoresis (CE)   2/163
– capillary zone electrophoresis (CZE)   2/163
– conclusions   2/226
– history   2/152 f
– library 27227
– non-ionic polyethylene glycol (PEG) surfactant   2/164
– non-ionic polypropylene glycol (PPG) surfactant   2/164
– obstacles   2/163 ff
least squares regression   2/456
leave-one-out strategy   2/461
lectins   2/21
LEED (low energy electron diffraction)   1/527
legionella pneumophila serogroup 1 (LPS1)   2/53
Leis (low-energy ion scattering spectrometry)   1/542
library spectrum   2/450
Li-COR 4200 fluorescence microscope   2/53
LI-COR DNA sequencer   2/45
LIF see fluorescence, laser-induced
lifetime   1/138
– luminescence   1/67
– of an excited state   1/428, 1/430
ligand field   1/133
limit of detection   1/378
line broadening   1/430
– Doppler   1/430
– hyperfine structure   1/431
– Lorentz   1/430
– pressure   1/430
– Stark   1/431
line profile   1/431, 1/461
linear absorption coefficient 367
linear dependences   2/463
linear models   2/459
linear Raman effect   1/43
lineshapes   1/272
linewidths   1/281, 1/430
– anisotropy   1/281
– asymmetry   1/281
lipase   2/21
lipid peroxidation   2/19

lipids
- HRP assay for   2/15
- oxydation/peroxydation   2/15
- metabolism   2/19
- nalling   2/19
- traffic   2/19
LIPIDAT   2/9
lipoprotein lipase   2/24
liquid chromatography and atomic emission spectroscopy (LC-AES)   1/492 ff
liquid hromatography-atomic absorption spectrometry (HPLC-AAS)   1/469
liquid chromatography-mass spectrometry (LC-MS)   1/347, 2/152, 2/163, 2/226 f
liquid crystals   1/529
liquid samples   1/29
- chromatographic separation   1/31
- complexation   1/30
- extraction   1/29
- extraction/separation and preconcentration   1/29
liquid-liquid extraction   1/344
liquids   1/22
lithium
- determination of   2/70
- in blood   2/70
loading matrix   2/453
local thermal equilibrium   1/433
local thermal equilibrium (LTE)   1/427
long-range heteronuclear chemical shift correlation   1/240
Lorentz line broadening   1/430
low-density lipoproteins   2/24
low-viscosity solvents   1/146
luciferase enzyme   2/65
luciferase system
- from the firefly *photinus pyralis*   2/65
luciferin   2/65
lucigenin   2/13, 2/22
luminescence   1/44, 1/67
luminol   2/65
luminol/HRP system   2/65
Lyman series   1/422
lyotropic liquid crystals   2/101
LysoSensor probes   2/11

m

MAES (metastable atom electron spectroscopy)   1/551
mag-fura-2 2/12
magic angle spinning (MAS)   1/175, 1/94, 1/280
- line narrowing   1/280
magic-angle hopping (MAH)   1/300

magic-angle turning (MAT)   1/300
Magnesium Green   2/12
magnesium
- detection of Green, dye   2/12
magnetic dipolar interaction   1/187
magnetic domain   1/555
magnetic domains   1/554
magnetic field strengths   1/171
magnetic materials   1/523, 1/537, 1/576 f
magnetic moment   1/172
magnetic resonance imaging (MRI)   1/199
- image contrast control   1/199
magnetic sector   1/335
magnetoreflectance   1/561
magnetron   1/476
MALDI   2/502
MALDI-TOF mass spectrometry   2/122
MALDI-TOF-TOF MS   2/124
malignancy   2/11
maltose phosphorylase   2/14
MAS   1/271, 1/280
- anisotropic broadening   1/281
- anisotropic lineshape   1/282
- line narrowing   1/280
MAS-*J*-HMQC   1/308
- one-bond correlations   1/308
- through-bond *J* couplings   1/308
MAS-*J*-HSQC   1/308
- one-bond correlations   1/308
- through-bond *J* couplings   1/308
mass analysis   1/335
mass attenuation coefficient   1/367
mass chromatogram   2/498
mass frontier   2/499
mass spectra quality indices   2/255
mass spectral interfaces   2/401 f, 2/408 f
mass spectrometry *see also* process analysis   1/329, 1/338, 2/488
- databases   2/489
- search software   2/495
mass spectroscopy (MS)   2/4, 2/122
- in bioanalysis   2/122
MassLib   2/495
MassLib/SISCOM   2/497
MASSTransit   2/491
MATLAB   2/467
matrices   2/253
P-matrix   2/462
matrix effects   1/401
matrix modifier   1/463
matrix-assisted laser desorption (MALDI) MS   1/330, 2/122
MBI
- pesticides   2/156

- polar pharmaceutical compounds 2/156
- polycyclic aromatic hydrocarbons 2/154
- surfactants 2/154
measurements
- reflection-absorption 1/94
MEIS (medium energy ion scattering) 1/543, 1/565
melatonin 2/63
membrane chloride transport 2/13
membrane fusion 2/19
membrane potential 2/12
membrane potential-sensitive probes 2/19
membrane transport
- chloride 2/13
MEMS see microelectromechanical systems
merocyanine 540 2/15, 2/19
metal complexes 1/133
metal ion association reactions 2/262
metalloproteinases 2/13
N-methyl-4-hydrazino-7-nitrobenzo-furazan 2/4
6-metoxyquinolinium derivatives 2/13
Mg^{2+} 2/12
Michelson interferometer 1/50
microinjectable cell tracers 2/57
micromechanical systems in bioanalysis 2/131
microparticles
- fluorescent 2/15
microscans 2/247, 2/249
microscopic 1/84
microscopic XRF 1/399
microscopy 2/57
- confocal 2/63
- laser scanning, with MPE 2/63
- standardization reagents 2/57
microseparation methods 1/346
microsomal dealkylase 2/21
microsystems 2/3
- microarray 2/3
- microelectrophoresis 2/3
- microfluidics 2/3
microtubule
- cell cycle-dependent 2/17
- dynamics 2/17
- polymerization 2/13
microwave plasma 1/476
microscan 2/246, 2/251
MIES (metastable impact electron spectroscopy) 1/539, 1/551
minor-groove binders 2/18
MIP (multiple internal reflection) 1/476, 1/511
mist chambers 1/13

mitochondria
- Na^+ gradients 2/11
- probes for 2/22
- sodium gradients in 2/11
MitoFluor Probes 2/22
mitotic spindle morphogenesis 2/17
MitoTracker 2/22
mixing chamber 1/441
mixture rule 1/368
MLR 2/462
mode couplers (interferometers) 2/81
modulation spectroscopy 1/561
moisture 1/110
molar ellipticity 1/82
molecular film 1/536, 1/539
molecular imprinting 2/69
molecular interactions 1/580
molecular orientation 1/535, 1/551
molecular recognition structures 1/349
moment
- dipole 1/271
- quadrupole 1/271
monochromator
- Czerny-Turner 1/58
monolayer 1/501, 1/574
- Langmuir-Blodgett 1/512
- self-organised 1/536
Moseley's law 1/374
motion
- rate constant 1/279
- three site jump 1/279
moving belt (MBI) 2/153, 2/156
MPE laser scanning microscopy 2/63
MQC 1/295
MRI pulse sequence
- echo planar sequence 202
- spin-warp 1/200
MS see mass spectroscopy
MST see microsystems
mulls 1/92
multichannel analyzer 1/387
multichannel detection 1/470
multichannel instruments 1/481 ff
multichannel spectrometers 1/392
multichannel wavelength-dispersive instruments 1/393
multicollinearities 2/464
multidrug resistance 2/11
multi-element technique 1/470
multi-frequency irradiation methods 2/254
multilayer 1/501
multi-photon 1/155
multi-photon excitation 1/149

multi-photon fluorescence excitation (MPE) 2/59
multiphoton microscopy 1/138
multiple linear regression (MLR) 2/462
multiple magnetization transfers (spin-diffusion) 2/111
multiple quantum spectoscopy 2/108
multiple-element techniques 1/413
multiple-frequency resonance ejection methods 2/252
multistep elution 2/127
– as sample 2/127
– prep in MS 2/127
multivariate calibration in AES 1/489
multivariate calibration methods 2/69
multivariate methods 2/459
muscle contraction 2/11

n
Na^+
– indicators for 2/10
– channel
  – probes for the 2/25
– efflux in 2/11
NAA (neutron activation analysis) 1/518
NADH 2/15, 2/63
NADPH 2/15
Na^+/H^+ antiporter 2/25
Na^+/K^+-ATPase 2/25
NanoOrange 2/16
nanoparticles
– fluorescent 2/15
naphthalene-2,3-dicarboxaldehyde 2/14
naphthalocyanine dyes
– bis(alkylsiloxysilyl) complexes of naphthalocynines 2/36
naphthalocyanines 2/35
native-like structure 1/354
natural lifetime 1/138
NBD 2/19
NBT 2/14
near-field microscopes 1/86
near-field scanning optical microscopy 2/19
near-infrared (NIR) 1/42, 1/104, 2/10
nebuliser 1/441 f
negative ion chemical ionization (NICI) experiments 2/253
neural network 2/474, 2/478
neuraminidase 2/20
neurokinin receptors 2/25
neuromedin C receptors 2/25
neutral loss searching 2/497
Newport Green 2/13

NEXAFS (near edge X-ray absorption spectroscopy) 1/544
NHS ester *see* N-hydroxysuccinimidyl ester
N-hydroxysuccinimidyl (NHS) ester 2/28, 2/31
Nicolet 2/447
nicotinic acetylcholine receptors 2/91
Nile Blue 2/70
NIR 1/42, 1/104, 2/10
– absorbing chromophores 2/32
– absorption spectroscopy 2/68
– agriculture 1/110
– dye NN 382
– environmental monitoring 1/110
– fiber optic immunosensor 2/53
– food industry 1/110
– pharmaceutical industry 1/111
– polymer industry 1/111
– spectrometers
  – miniaturised 2/68
NIST 2/447
NIST Mass Spectral Library 2/490
nitrate 2/70
nitric oxides 2/14
nitrobenzoxadiazole (NBD) 2/19
NIXSW (normal incidence X-ray standing wave) 1/591
NMR *see also* nuclear magnetic resonance spectroscopy 1/171, 2/297
– dynamic processes 1/277
– ^{2}H 1/278
– parameters
  – chemical shift 1/181 ff
  – detection frequency 1/181 ff
  – gyromagnetic ratio 1/181 ff
  – J-coupling constants 1/181 ff
  – magnetic field 1/181 ff
  – nuclear spin 1/181 ff
– resonance frequency 1/272
  – orientational dependence 1/272
  – single-crystal 1/273 f
– solid-state 1/275
– spectroscopy 2/6, 2/94 ff
  – of proteins 2/94
– active nuclei 1/270
  – magnetogyric ratios 1/270
  – natural abundances 1/270
  – nuclear spin quantum numbers 1/270
NOE (nuclear Overhauser effect) 1/173, 1/212
NOESY (nuclear Overhauser enhancement spectroscopy) 1/228, 1/173, 2/110
noise levels
– in the NIR in visible regions 2/27

non-covalent biopolymer complexes 1/349
noncovalent supramolecular complexes 1/354
non-invasine monitoring
– of glucose 2/69
non-linear least squares strategy 1/408
non-linear methods 2/459
non-linear Raman effect 1/43
Non-RBS or n-RBS (Non-Rutherford backscattering spectrometry) 1/567
Nonylphenolpolyglycolether (NPEO) 2/169 ff
– in-source-CID 2/169
– MS/MS CID 2/169
normalization 2/449
NRA (nuclear reaction analysis) 1/541, 1/563
NSOM 1/86
nuclear magnetic resonance (NMR) 1/171, 2/297
nuclear magnetic resonance spectroscopy 2/469
nuclear Overhauser effect (NOE) 1/173, 1/212
– difference spectroscopy 1/212
– distances between a pair of protons 1/212
– stereochemical relationship 1/212
nuclear Overhauser enhancements (NOE) 2/98
nuclear Overhauser enhancement spectroscopy (NOESY) 1/288, 1/173, 2/110
nuclear spin quantum number 1/172
nuclei
– half-integer 1/271
– integer 1/271
– magnetogyric ratio 1/271
– natural abundance 1/271
nucleic acid analysis
– by MS 2/130
horseradish peroxidase (HRP) 2/16
nucleotide-binding proteins 2/24
number density (of absorbing atoms) 1/429
number density (of excited particles) 1/428

o

obstacles 2/163 ff
– atmospheric pressure chemical ionisation (APCI) 2/163
– capillary electrophoresis (CE) 2/163
– capillary zone electrophoresis (CZE) 2/163
– electrospray ionisation (ESI) 2/163
– TSP 2/163
off-resonance decoupling 1/216
OliGreen 2/17
one-dimensional NMR experiment 2/103
operation
– continous-scan 1/52
ophthaldialdehyde 2/16
opioid receptors 2/25
optical activity 2/428
optical density
– in ATR 2/77
optical rotatory dispersion (ORD) 1/81, 2/93
optical spectral data bases 2/447
optical spectroscopy see also process analysis 2/441
optimization of ICP 1/490
orbital angular momentum 1/423
orbital quantum number 1/423
organoarsenic compounds 2/154
organophosphorus compounds 2/217 ff
– biodegradation 2/195
– drinking water 2/217
– groundwater 2/194, 2/217
– interlaboratory study 2/194
– ion chromatography 2/217
– MS/MS 2/217 f
– photodegradation 2/195
– quantification 2/218
– stability 2/218
– surface water 2/217
– wastewater 2/218
organophosphorus pesticides 2/179
organotin compounds 2/258
orientation
– molecular 1/578
orientational dependence 1/272
– NMR resonance frequency 1/272
oriented matrices 2/102
– in NMR 2/102
oriented samples 1/313
orthogonal injection
– in ESI-MS 2/123
outlier spectra 2/461
out-of-plane 1/42
overview
– aromatic sulfonates 2/154
– complexing agents 2/154
– diagnostic agents 2/154
– drugs 2/154
– dyes 2/154
– explosives 2/154
– haloacetic acids 2/154
– organoarsenic compounds 2/154
– PAHs 2/154
– pesticides 2/154
– phenols 2/154
– surfactants 2/154
– toxins 2/154
– xenoestrogens 2/154

oxa-carbocyanines 2/19
oxazolium pentamethine cyanine dye (DODC) 2/33
oxidation state 1/585
oxidoreductase 2/21
OxyBURST technology 2/26
oxygen 2/15, 2/69

p

PAES (positron annihilation auger electron spectroscopy) 1/555
PAH 2/154
– quantification 2/208
Pake doublet 1/192
parent structure 2/473
partial least squares (PLS) 2/465
particle beam (PBI) 2/153, 2/157
particle size 1/417
partition function 1/429, 1/433
PAS (photoacoustic spectroscopy) 1/552
Paschen series 1/422
passive sampling 1/13
pattern recognition
– FIA-MS 2/175
– FIA-MS/MS 2/175
PBI 2/157 ff
– alkylphenol carboxylates (APECs) 2/159
– alkylphenol ethoxylates (APEOs) 2/159
– anilides 2/158
– biochemical 2/157
– carbamate 2/158
– chlorinated phenoxy acid 2/158
– degradation products 2/157
– dyes 2/159
– herbicides 2/157
– isocyanates 2/158
– library-searchable EI spectra 2/159
– organo-phosphorus 2/158
– PAHs 2/159
– PAH metabolites 2/160
– pesticides 2/157
– phenylurea 2/158
– physiochemical 2/157
– quaternary ammonium 2/158
– triazines 2/158
P-COSY experiment 2/107
PCR 2/18, 2/464
PDT 2/36
2PE cross section 2/60
PED or PhD (photoelectron diffraction) 1/586
PEELS (parallel electron energy loss spectroscopy) 1/530, 1/535
PEEM (photemission electron microscopy) 1/553

2PE fluorescence polarization measurements 2/64
pelletized 1/417
pellets 1/92
penetration depth 1/402, 2/74
penicillin 2/87
peptidases 2/20
peptide
– analysis 2/16
– MS 2/502
perfluoro compounds 2/262
permeability
– of the dielectric media 2/72
permitivity 1/562, 2/72
peroxide 2/15
Perrin plots 1/150
perylene 2/19
PESIS (photoelectron spectroscopy of inner shell) 1/587
pesticides 2/154, 2/176 ff, 2/258
– anilides 1/176, 2/158, 2/192, 2/215
– antifouling 2/199
– benzidines 2/154
– biodegradation 2/199
– carbamate 2/154, 2/158, 2/177 ff, 2/193
– chlorinated 2/154
– phenoxy acid 2/158
– degradation pathways 2/182
– degradation products 2/182
– ESI-CZE-MS 2/215
– ESI-FIA-MS 2/215
– estuarine waters 2/182
– fruits 2/193, 2/197
– fruit drinks 2/197
– fungicides 2/215
– glyphosate 2/161
– groundwater 2/192, 2/196 f, 2/199
– herbicides 2/215
– isocyanates 2/158
– library 2/218
– N-methylcarbamate pesticides 2/193
– MS/MS 2/195 f, 2/216
– MS/MS library 2/194
– natural waters 2/198
– organophosphorus 2/158, 2/194
– biodegradation products 2/194
– organophosphorus compounds 2/179
– phenolic compounds 2/197
– phenoxyacetic acid 2/154
– phenoxycarboxylic acids 2/179, 2/195
– phenylurea 2/158, 2/180, 2/196
– quantification 2/181, 2/192 ff, 2/215
– quaternary amines 2/176, 2/192, 2/215
– quaternary ammonium 2/154, 2/158

- rivers 2/198
- river water 2/192 f, 2/199
- sediment 2/197
- soil samples 2/182
- sulfonylureas 2/161, 2/180, 2/196
- surface water 2/196, 2/198
- thiocyanate 2/176, 2/192
  - compounds 2/215
- thioureas 2/180, 2/196
- tin-containing pesticides 2/201
- toluidines 2/176, 2/192, 2/215
- triazines 2/154, 2/158, 2/181, 2/197 ff
- urea pesticides 2/180
  - quantification 2/180
- vegetables 2/193
- water 2/196 f
Pfleger/Maurer/Weber database 2/494
PH or PEH (photoelectron holography) 1/586
pH
- indicators for 2/10
phagocytosis 2/26, 2/68
phallacidin 2/17
phalloidin 2/17
phallotoxins 2/17
phase cycling routines 1/232
- CYCLOPS 1/232
- EXORCYCLE 1/232
phenanthridine dyes 2/17
2-phenetylamine 2/70, 2/71
Phen Green FL 2/13
phenobarbital 2/56
phenols 2/154
- quantification 2/208
phenoxazine 2/70
phenoxycarboxylic acids 2/179
- drinking water 2/219
- groundwater 2/219
- MS/MS 2/219
- quantification 2/219
- surface water 2/219
phenylureas
- agricultural soil 2/222
- chemical degradation 2/222
- crop 2/222
- degradation products 2/220, 2/222
- estuarine waters 2/220
- fruit drinks 2/197
- fruits 2/197, 2/222
- groundwater 2/196 f
- metabolites 2/222
- MS/MS 2/221 f
- natural waters 2/221
- photolysis products 2/222
- quantification 2/220 ff

- river water 2/221
- sediment 2/197
- surface water 2/196, 2/221
- water 2/196 f
phenytoin 2/56
phosphatase-based signal amplification 2/15
phosphate 2/14
phospholipase activity 2/19
phosphorescence 1/45
phosphorylations 1/351
photoacoustic 1/83 f
- sampling depth 1/84
photobleaching 1/155, 1/157, 2/63
photodiode array 2/381 ff
photodiodes 2/10
photodynamic therapy (PDT) 2/36
photoelectric absorption 1/369
photoelectric effect 1/369
photomultiplier 1/386
photomultiplier tubes (PMT) 2/27
o-phthaldialdehyde 2/14
phthalocyanines 2/10, 2/35
phycobiliproteins 2/15, 2/57
PicoGreen 2/17
PIES (penning ionisation electron spectroscopy) 1/539, 1/551
piezoreflectance 1/561
PIGE or PIGME (particle induced gamma ray emission) 1/541, 1/546
PISEMA (polarisation inversion with spin exchange at the magic angle) 1/313
- internuclear dipolar couplings 1/315
- tilt angle of the polypeptide helix 1/315
PIXE (particle induced X-ray emission) 1/541, 1/548
plasma desorption 1/332
PLS (partial least square) 2/465
PLS_Toolbox 2/467
PLSplus 2/467
PM (polarization modulation) 1/561
PMB 2/496
PMP (proton microprobe) 1/539
PMT 2/27
polarization 1/118, 1/383
polarization excitation spectra 1/149
polarized light 1/81
polychromatic flow cytometry (PFC) 2/57
polychromator 1/481 ff
- Raman grating 1/61
Polycyclic aromatic hydrocarbon (PAH) isomers
- by IR 2/417 ff
- by UV 2/384 ff
polyethylene glycols 2/101

Polymer molecular weight, true value by LC-MS 2/410
polymethines 2/10
polymixin B
– analog 2/24
– fluorescent 2/24
population of excited levels 1/426
portable equipment 1/389
post-translational modifications 1/349
post-translational structure modifications 1/349
potential sensitive dyes 2/70
PPP MO 2/28
preamplifier 1/387
pregnancy-associated plasma protein A 2/56
preparation of gaseous samples 1/4
preparation of samples for analysis 1/24
presaturation method (NMR) 2/104
pre-search 2/495
PRESS 2/466
pressure line broadening 1/430
primary absorption 1/403
primary structures 1/349
principal component analysis 2/452
principal component regression (PCR) 2/464
principal quantum number 1/423
principle
– double-beam 1/49 f
Prion proteins 2/6
prism couplers 2/81
Probability Based Matching (PMB) 2/496
probes for Cl$^-$ channels 2/25
probes for K$^+$ channels 2/25
probes for mitochondria 2/22
process analysis 2/336 ff, 2/271 ff
– atomic emission spectrometry (AES) 2/336, 2/356
  – gaseous effluents 2/357
  – laser based techniques 2/362
  – liquid streams 2/356
  – plasma spectrometry 2/356
  – reactive gases 2/360
– acoustic emission spectroscopy 2/276
– atomic spectrometry 2/336
– atomic spectroscopy 2/274
– chemical composition 2/273
– elemental analysis 2/336
  – applications 2/336
  – catalyst control 2/337
  – corrosion monitoring 2/339
  – on-stream/at-line analysis 2/343
  – reducing environmental impact 2/341
  – troubleshooting process problems 2/342

– inductively coupled plasma atomic emission spectrometry (ICPAES) 2/336
– inferential analysis 2/277
– infrared 2/274
– ion mobility spectrometry 2/276
– IR 2/279
– mass spectrometry 2/316 ff
  – applications 2/330
  – attributes 2/316
  – calibration 2/327
  – data analysis 2/325
  – detectors 2/325
  – fermentation off-gas anaylysis 2/331
  – hardware 2/317
  – ionization 2/231
  – limitations 2/316
  – maintenance 2/329
  – mass analyzers 2/322
  – operation 2/329
  – sample collection and conditioning 2/319
  – sample inlet 2/319
  – vacuum system 2/325
– mass spectroscopy 2/274
– microwave 2/279
– microwave spectroscopy 2/274
– NMR (Nuclear Magnetic Resonance) 2/274, 2/279, 2/297 ff
  – broadline NMR 2/301 ff
  – calibration 2/299
  – curing process applications 2/303
  – food industry applications 2/303
  – FT-NMR 2/307
  – gasoline applications 2/309
  – growth factor β_3 2/313
  – manufacturers 2/306
  – petroleum refining 2/313
  – polymer industry applications 2/306
  – polymer production applications 2/303
  – quantitation 2/297, 2/299
  – sample 2/300
  – sulfuric acid alkylation process 2/311
– optical spectroscopy 2/279 ff
  – cavity ringdown spectroscopy 2/294
  – chemiluminescence 2/280, 2/293
  – Far-IR 2/279
  – fluorescence 2/280, 2/293
  – IR 2/279
  – laser techniques 2/280
  – laser diode techniques 2/291
  – Mid-IR 2/279 ff
  – Near-IR 2/279 f
  – near-infrared spectroscopy 2/282 ff
  – non-dispersive infrared analysers 2/280 f

- optical sensors 2/280, 2/294
- Raman spectroscopy 2/280, 27287 ff
- UV 2/279 f
- UV/visible spectroscopy 2/280, 2/286
- visible 2/279 f
- physical characteristics 2/273
- practical considerations 2/272
- Raman spectroscopy 2/280
- REMPI spectroscopy 2/275
- sample 2/272 ff
- spectroscopy 2/273
- ultrasound 2/276
- UV/visible 2/274, 2/279
- X-ray fluorescence (XRF) 2/336, 2/344
  - corrosion monitoring 2/351
  - direct analysis of solids 2/354
  - liquid process streams 2/348
  - powders 2/352
  - slurries 2/352
  - trace analysis 2/351
- X-ray techniques 2/279
process control 2/269 ff
process industry 2/5
process mass spectrometry 2/316
profile function 1/431
propdium iodide 2/17
proportionality 1/384
proteases 2/20
protective agent for AAS 1/463
protein 1/529
- kinases 2/24
- phosphatases 2/24
proteinase inhibitors 2/24
protein kinase 2/23
- activators 2/24
- inhibitors 2/24
protein quantitation 2/16
proteome 2/4
proteome analysis 1/356
proteomic databases 2/5
proteomics 1/356, 2/4
- proteolytic degradation 1/356
- sequence tags 1/356
proton transfer 2/251
pulsed field gradients (PFGs) 1/173, 2/102
pulsed lamp background correction 1/460
pulse-height selection 1/384
pulse methods 1/186
purple membranes (PM)
- of *Halobacterium salinarum* 2/101
pyrene 1/154, 2/19
pyrophosphate 2/14

q

quadrupolar coupling (C_Q) 1/174
quadrupolar interaction 1/191
- asymmetry parameter, η 1/192
- axially symmetric 1/278
- definition 1/192
- electric charge asymmetry of the nucleus 1/191
- moments 1/191
- nuclear electric quadrupole moment 1/192
- quadrupolar coupling constant 1/192
- tensor 1/192
quadrupolar nuclei
- line shapes 1/195
quadrupole instruments 2/246
quadrupole ion trap 2/265
quadrupole ion trap mass spectrometry 2/244
quadrupole mass analyser 1/337
quadrupole mass spectrometry 2/258
qualitative evaluation of spectra, optical 2/446
QuantIR 2/467
quantitative calibration procedures 1/409
quantitative evaluation 2/455
quantitative proteomics 2/126
quantitative reliability 1/418
quantum number, orbital 1/423
quantum number, principal 1/423
quantum number, spin 1/423
quantum number, total 1/423
quantum yield 1/138 ff, 1/143, 2/28
quartz furnace-atomic absorption spectrometry 1/467
quartz T-tube atomiser (for AAS) 1/468
quartz tube atomiser 1/449

r

radiation
- scattered 1/97
radiationless relaxation 1/427
radiation source for AAS 1/434
radiative de-excitation 1/427
radiative transition 1/428
radio frequency (RF) 1/172
radio frequency (RF) generator 1/473
radioactive α-, β-, and γ-sources 1/382
radioactive sources 1/380
radioisotope XRF 1/397
radiotherapy 2/29
RAIRS (reflection absorption infrared spectroscopy) 1/75, 1/534, 1/559
Raman 2/428
- mapping 1/117
- microprobe 1/117

- scattering   2/27, 2/91
- shift   1/43
- spectroscopy   1/43, 1/557, 2/92 ff
  - near-infrared excitation   2/93
random and systematic error   1/415
RBS (Rutherford backscattering spectrometry)   1/541, 1/565
R-COSY   2/108
reactive sites   1/353
real-time BioInteraction Analysis   2/5
receptor binding   2/25
recoupling methods   1/287 ff
- REAPDOR   1/290
- REDOR   1/287
- TRAPDOR   1/290
red-edge   1/146
REDOR   1/310
- dipolar couplings   1/310
- distance determination   1/290
REELS, EELS (reflection electron energy loss spectroscopy)   1/535, 1/561
reflection   2/72
- absorption   1/75
- diffuse   1/97
- measurements   1/73
- off-axis   1/80
- on-axis   1/80
reflectometric interference   2/81
reflectometric interference spectroscopy (RIfS)   2/89
refraction   2/72
refractive index
- complex   1/74
region
- fingerprint   1/102
regulations   2/256
relative detection limits   1/378
relative random counting error   1/405
relaxation   1/195
- correlation times   1/196
- methyl group rotation   1/196
- spin-lattice T_1 1/217
- spin-spin T_2 1/217
- T_1, spin-lattice relaxation   1/196
- $T_{1\rho}$, spin-lattice in the rotating frame relaxation   1/196
- T_2, spin-spin relaxation   1/196
- times
  - ^{13}C T_1 1/285
  - ^1H $T_{1\rho}$ 1/285
relayed COSY (R-COSY)   2/108
representative sample   1/117
- REPT-HMQC   1/310
residual dipolar couplings (RDC)   2/99

residual variance   2/454
residuals   2/458
resolution   1/51
- spectral   1/52
resonance   1/148
resonance energy transfer   1/45
resonance Raman scattering   1/119
resonant ejection frequencies   2/252
resonant mirror (RM)   2/83
reverse phase HPLC   2/45
reverse search   2/496
review
- dyes   2/154
- environmetal analysis   2/154
- environmetal contaminants   2/154
- general   2/154
- surfactants   2/154
- water analysis   2/154
RF level   2/254
RF ramping   2/248
RF voltages   2/252
rhodamine 110 2/20
rhodamines   2/10, 2/15, 2/19, 2/28
RiboGreen   2/17
ribosomes   2/72
RNRA (resonant nuclear reaction analysis)   1/563
rocking modes   1/42
ROESY (rotating frame Overhauser enhanced spectroscopy)   1/173, 1/230
rosamines   2/22
Rose Bengal diacetate   2/15
rotating anode tubes   1/380
rotating-frame nuclear Overhauser effect spectroscopy (ROESY)   2/110
rotating-frame Overhauser effect (ROE)   2/109
rotational correlation time   1/149 f
rotational-echo double resonance (REDOR)   1/175
rotational resonance (RR)   1/290
rotational temperature   1/434, 1/436
Rowland circle   1/482
Rowland spectrometer   1/481 ff
ROX   2/40
RS (recoil spectroscopy)   1/520
ruthenium trisbipyridyl (Ru(bpy)$_3$) complexes   2/68

s
Sadtler   2/447
Saha equation   1/432
SAHO   2/472
SALI (surface analysis by laser ionisation)   1/573

SAM (scanning Auger microscopy) 1/567
sample preparation 1/105, 1/344
sample preparation for inorganic
    analysis 1/25
– acid digestion 1/25
– fusion reactions 1/27
– nonoxidizing acids 1/26
– oxidizing acids 1/26
samples
– liquid crystals 1/275
– neat solid 1/94
– oriented lipid bilayers 1/275
– polymer fibres 1/275
– powdered 1/275
sampling considerations 1/5
Sanger 2/37
Sanger method *see* DNA sequencing
SAv 2/29
SCAN 2/467
scan repetition rate 2/248, 2/251
SCANIIR (surface composition by analysis of neutral and ion impact radiation) 1/533
scanning modes 2/247
scattering 1/371
– in near-IR 2/10
– interactions 1/369
– Rayleigh 1/58
– Tyndall 1/58
schistosomiasis 2/47
scintillation counter 386
score matrix 2/453
SDBS 2/447, 2/492
sealed X-ray tubes 1/380
search for bands 2/446
secondary absorption 1/404
secondary ion mass spectrometry (SIMS) 2/125
secondary target EDXRF system 1/395
segmental mobility 1/151
selected-ion monitoring 2/251
selected-ion monitoring procedures 2/255
selection rules 1/373, 1/424
selective population transfer (SPT) 1/173, 1/213
– DEPT 1/213
– INEPT 1/213
– spin population inversion (SPI) 1/213
selenium
– ion, indicators for 2/13
self-absorption 1/431, 1/438
self-assembled biomembranes 2/91
self-chemical ionization 2/248
self-CI processes 2/253
self-decoupling 1/185

self-diffusion of 1/202
self-organizing monolayers 545
self-quenching 1/143
self-reversal 1/431, 1/460 f
self-training interpretive and retrievel system (STIRS) 2/496
SEM (scanning electron microscopy) 1/567
semiconductor detectors 1/386
semiempirical quantum chemical methods 2/28
semiochemistry 2/70
SEMPA (scanning electron microscopy with polarisation analysis) 1/567
sensitivity 2/252
sensitivity AES 1/491
sensors 2/71
sequence data 2/8
sequence determinations 1/350
– sequencing 2/16
– synthesis 2/16
serotonin 2/63
SERS (surface enhanced Raman scattering) 1/557
SESAMI 2/483
SEW (surface electromagnetic waves spectroscopy) 1/574
SEXAFS (surface X-ray absorption fine structure) 1/527, 1/586
SFG (sum-frequency generation) 1/578
SHG (second harmonic generation) 1/579
shift correlation experiments 1/237 ff
– accordion-HMQC experiment 1/239
– accordion-optimized direct correlation experiment ADSQC 1/239
– DEPT-HMQC 1/237
– multiplicity'-edited GHSQC 1/238
sialidase 2/20
Siegbahn nomenclature 1/373
signal/background ratio 2/253
signal-to-background ratio 2/251 f
SIM acquisition 2/251
SIM mode 2/260
SIM procedures 2/253
SIMCA 2/455
similarity measures 2/448
similarity search 2/497
SIMS (secondary ion mass spectrometry) 1/571
– dynamic mode 1/572
– state mode 1/571
simulated emission 1/427
simultaneous wavelength-dispersive spectrometers 1/393
single-channel instruments 1/392

single-element techniques 1/412
single molecule detection 1/155
single nucleotide polymorphism studies (SNP genotyping) 2/130
single quantum coherence (SQC) 1/176
singlet state 1/425
SiNPcs 2/36
SLEELM (scanning low energy electron loss microscopy) 1/562
slew-scan monchromator 1/480
slurry analysis 1/470
slurry nebulisation 1/483
small sample NMR 1/257 ff
– cryogenic NMR probe 1/260
– magic angle, liquid Nano-probe 1/258
– µ-coil NMR probes 1/258
– SMIDG probe 1/258
Smith-Hieftje background correction method 1/460 f
SNMS (secondary neutral mass spectrometry 1/533, 1/573
SNOM (scanning near-field optical microscopy) 1/86, 1/571
sodium efflux
– in cells 2/11
sodium green 2/11
soft ionisation 1/332 ff
– interaces
    – APCI 2/168
    – ESI 2/168
    – FAB 2/168
    – TSP 2/168
– method 1/330
electrospray-ionisation 1/330
solid echo 1/278
solid phase extraction 1/345
solid sample introduction 1/452, 1/483 ff
– for AAS 1/470
solid samples 1/24
solid state NMR 1/174, 1/285
– ^{1}H 1/285
solid-phase microextraction 1/345, 2/127
– as sample 2/127
– prep in MS 2/127
solids 1/23
solid-state drift chamber 389
solid state NMR 1/173, 1/187
solute quenching 1/140 f1/141
solution state ^{1}H NMR 1/171, 1/179
solvent quenching 1/143
solvent relaxation 1/144, 1/147
solvent suppression (in NMR) 2/104
sorbents 1/9 ff
space-charge effects 2/247 f, 2/249

SPAES (spin polarised Auger electron spectroscopy) 1/512
spark 1/477, 1/484
speciation analysis 1/466
specimen preparation 1/416
SpecInfo 2/447, 2/491
SpecManager 2/502
SPECSOLV 2/482 f
SpecSurf 2/502
SpecTool 2/452
spectra library 2/247
spectra processing 2/442
spectra
– powder 1/276
– quantitative evaluation 2/455
– solid-state 1/277
– solution 1/277
– static powder 1/275 f
spectral buffering 1/455
spectral data bases, optical 2/447
spectral editing 1/283
– SS-APT 1/284
spectral interference in AES 1/486
spectral interferences 1/454 ff
spectral range
– MIR 1/89
spectral regions
– dead 1/49
spectral search 2/448
spectral similarity search 2/471
spectral simplification
– chemical modification 1/183
– selecticve decoupling 1/183
– ^{13}C labeling 1/186
– deuteration 1/185
– self-decoupling 1/185
spectrometers
– AOTF 1/56
– diode array 1/56
– dispersive 1/48 f
– filter 1/56
– fluorescence 1/66
– Fouriertransform 1/50
– FT 1/48
– FT-NIR 1/55
– FT-Raman 1/61
– LED 1/56
– luminescence 1/66
– MIR 1/48
– multi-channel 1/63
– NIR 1/54
– Raman 1/57
– Raman grating 1/57
– scanning-grating 1/55

- UV/VIS 1/63
- vacuum 1/49
spectroscopy
- near-infrared 1/104
- NIR 1/105
- photoacoustic 1/97
- Raman 1/112
spectrum
- emission 1/66
- estimation 2/473
- evaluation 1/405
- excitation 1/66
- prediction 2/474
SPEELS (spin polarised electron energy loss spectroscopy) 1/575
SPI (surface Penning ionisation) 1/551
SPIES (surface Penning ionisation spectroscopy) 1/551
spiking 1/412
spin 1/172
spin angular momentum 1/423
spin decoupling 1/211
- difference spectroscopy 1/212
- selective population transfer (SPT) 1/211
- spin ticking 1/211
spin-diffusion 2/111
spin-lattice 1/217
- inversion-recovery 1/217
spin quantum number 1/423
spinning-sideband patterns
- ^{1}H-^{1}H DQ MAS 1/305
SPIPES (spin polarised inverse photoelectron spectroscopy) 1/536 f
spirobenzopyran 2/70
SPMP (scanning proton microprobe) 1/539
spontaneous decay 1/428
spontaneous emission 1/427
SPR biosensors 2/85 f
SPR spectroscopy (surface plasmon resonance spectroscopy) 1/579
Spreeta™ device 2/87
SPUPS (spin polarised ultraviolet photo-electron spectroscopy) 1/508, 1/576, 1/584
squaraine dyes 2/34
- benz[e]indolium Sq660 2/35
squarilium dyes
- signal ransducing 2/71
squaryliums 2/32
SRPES (synchrotron radiation photoelectron spectroscopy) 1/588
SRUPS (spin-resolved ultraviolet photo-emission spectroscopy) 1/537
standard addition 1/410
standard temperature 1/433

standardization (autoscaling) 2/443
standardization reagents
- flow cytometry 2/57
- microscopy 2/57
Staphylococcus Aureus (Cowan-1 strain) 2/84
Stark line broadening 1/431
Statgraphics 2/467
static quenching 1/140
Statistica 2/467
statistics of sampling 1/18 ff
step-scan 1/53
stereochemical effects 2/478
stereochemical interactions 2/480
Stern-Volmer equation 1/141
STIRS 2/496
Stokes lines 1/43
Stokes-shift 1/146, 2/10
Stokes-Einstein 1/141
storage ring 1/383
streptavidin (SAv) 2/29
streptomicyn
- residues in whole milk 2/87
structural diversity 2/475
structure elucidation 2/469
STS (scanning tunneling spectroscopy) 1/570 ff
styryl dyes (ANEP) 2/19
β-subunit of human chorionic gonadotropin (hCG) 2/56
sulfide 2/14
- probing of dynamic changes of red cell membrane 2/14
5-sulfofluorescein diacetate 2/11
sulfonic acids 2/208
- ion-pairing 2/209
- leachates 2/208
- MS/MS 2/208
- plumes of landfills 2/209
- quantification 2/208 f
- textile wastewater 2/209
sulfonylureas
- agricultural soil 2/222
- chemical degradation 2/222
- crop 2/222
- degradation products 2/220, 2/222
- estuarine waters 2/220
- fruits 2/222
- metabolites 2/222
- MS/MS 2/221 f
- natural waters 2/221
- photolysis products 2/222
- quantification 2/220 ff
- river water 2/221
- surface water 2/221

superconducting magnets 1/342
supercritical-fluid extraction 1/346
supervised learning 2/444
surface
– analysis technique
  – acronyms 1/594
  – classification 1/499
  – selection 1/501
  – type of information 1/505
– cleanliness 1/514
– concentration 1/501
– contamination 1/506, 1/581, 1/590
– definition 1/501
– diffusion 1/555
– enhanced fluorescence, SEF 2/91 ff
– enhanced fluoroimmunoassay, SE-FIA 2/91
– enhanced IR absorption, SEIRA 2/91
– enhanced Raman scattering (SERS) 1/120, 2/93 ff
– enhanced Raman spectroscopy (SERS) 2/93 ff
– experimental 1/501
– physical 1/501
– plasmon 2/85
– plasmon resonance (SPR) 2/81, 2/85
  – fiber and waveguide SPR 2/88
– probe technique 1/499
– resolution 1/502
– selection rule 1/560
– specifity 1/502
surfactants 2/154, 2/189 f, 2/209 ff
– alcohol ethoxylate (AEO) 2/189 f, 2/209 f, 2/213
– alkylether carboxylates 2/190
– alkyl etoxysulfates (AES) 2/211
– alkyl polyglucamides 2/190
– alkyl polyglucosides 2/190, 2/212
– alkyl sulfates (AS) 2/211
– APEO
  – halogenated 2/191
– betaine 2/190
– biodegradation 2/211 f
– CDEA 2/190
– coastel waters 2/211
– coconut diethanol amide (CDEA) 2/189, 2/213
– ditallow-dimethylammonium chloride (DTDMAC) 2/212
– effluents 2/211
– EO/PO compounds 2/190
– estuaries 2/212
– fatty acid polyglycol amines 2/190
– gemini 2/190

– German Bight 2/212
– halogenated APEOs 2/191, 2/211
– halogenated NPEO 2/210
– ion chromatograph 2/214
– LAS 2/189, 2/211 ff
– NPEO 2/189 f
– NPEO-sulfate 2/189
– metabolites 2/209 f
– N-methylglucamides 2/212
– MS/MS 2/209 ff
– neutral loss (NL) 2/209 f
– nonylphenolpolyether carboxylate (NPEC) 2/210
– North Sea 2/212
– NPEO 2/209 ff
– NPEO sulfates 2/213
– OPEO 2/211
– perfluorooctanesulfonate (PFOS) 2/212
– perfluorooctanoic acid (PFOA) 2/212
– quantification 2/191, 2/210 ff
– quaternary ammonium compounds 2/214
– quaternary carboxoalkyl ammonium compounds 2/190
– sea water 2/212
– secondary alkane sulfonate (SAS) 2/189, 2/213
– sediment 2/212
– SPE concentrated analytes 2/190
– stability 2/190
– stability of SPE concentrated analytes 2/190
– sulfobetaine 2/190
– sulfosuccinates 2/190
– toxins 2/213
– Waddensea marinas 2/212
– wastewater 2/210 f
– wastewater inflows 2/211
SXAPS (soft X-ray appearance potential spectroscopy) 1/508
SXPS (soft X-ray photoelectron spectroscopy) 1/588
SYBR 2/16
SYBR Green 2/18
synaptic transmission 2/13
synchrotron radiation facilities 1/380
SYPRO 2/16
SYTO 2/16
SYTOX Dyes 2/17

t

TAG 2/33
TAMRA 2/40
tandem mass spectrometry (MS/MS) 2/122, 2/124, 2/168 ff, 2/257

- collision-induced dissociation (CID) 2/168
- discharge-on 2/169
- filament-on 2/169
- in-source-CID 2/169
- MS^n 2/168

TCS (total (or target) current spectroscopy) 1/510

technique
- microsampling 1/98
- sampling 1/94

TEELS (transmission electron energy loss spectroscopy) 1/535, 1/562

TEM (transmission electron microscopes) 1/535, 1/562

temperature of different excitation sources 1/436

temperature, electron 1/435 f
temperature, excitation 1/435 f
temperature, gas 1/435 f
temperature, ionisation 1/435 f
temperature, rotational 1/434, 1/436

tensors
- quadrupolar 1/273

term scheme 1/425
tertiary structure 1/349
tertiary structure characterisation 353
tetracycline antibiotics 2/68
tetraethylrhodamine derivatives 2/29
tetramethylrhodamine isothiocyanata 2/29
Texas Red 2/51
thallium acid phthalate 1/391
thermoelectrically cooled ED detectors 1/389
thermometric probe 1/434
Thermoquest 2/266
thermoreflectance 1/561
thermospray ionization (TSP) 2/172 ff
- fungicides 2/172
- herbicides 2/172
- pesticides 2/172
- sulfonated azo dyes 2/172
- surfactants 2/172

thia-carbocyanines 2/19
thiazole green (TAG) 2/33
thiazole orange 2/29
thin film approach 1/412
thin films 1/77
thin samples 1/96
thin-film samples 1/402
thiol
- indicatory for red cell membrane, probing 2/14

thioureas
- agricultural soil 2/222
- chemical degradation 2/222
- crop 2/222
- degradation products 2/220, 2/222
- estuarine waters 2/220
- fruits 2/222
- metabolites 2/222
- MS/MS 2/221 f
- natural waters 2/221
- photolysis products 2/222
- quantification 2/220 ff
- river water 2/221
- surface water 2/196, 2/221
- water 2/196

three-dimensional arrangement of elements 1/510

three-photon excitation (3PE) 2/58
thrombin 2/24
thyroid-stimulating hormone (TSH) 2/83
TIAFT 2/494
time of flight 1/338
time-resolved emission spectra 1/146
time-resolved fluorescence 2/10, 2/55
time-resolved fluorescence polarization 1/150
time-resolved surface enhanced fluorescence 2/91
Ti-Saphire laser 2/45
titanium dioxide waveguide 2/83
TMA-DPH 1/151
TO 2/29
tobacco mosaic virus (TMV) 2/101
TOCSY 1/173
TOF ERDA (time-of-flight ERDA) 1/522
TOF-SIMS 2/125
TO-PRO 2/17
Torch 1/473, 1/474
torsion modes 1/42
torsional angle 1/298, 1/312
total angular momentum 1/423
total correlation spectroscopy (TOCSY) [homonuclear Hartmann-Hahn (HOHAHA)] 2/109
total internal reflection (TIR) 2/72 f
total internal reflection fluorescence (TIRF) 1/155, 2/91 ff
total quantum number 1/423
total suppression of sidebands 1/282
TOTO 2/17
TOTO dimers
- synthesis of 2/29

toxicological analysis 1/344
toxins 2/154, 2/213
- CE-MS 2/213
- LC-MS 2/213
- MS/MS 2/213
- seafood 2/213

training set   2/456
trans and gauche conformations   1/298
TransFluorSpheres   2/57
transformation
– Fourier   1/51
– Kubelka-Munk   1/97
transition
– charge transfer   1/132, 1/134
– d-d   1/132
– electronic   1/125
– n→π*   1/132
– π→π*   1/132
– semiconductors   1/134
– single-quantum   1/273
transmissible spongiform encephalopathy (TSE)   2/6
transmission measurements   1/71
transmission spectroscopy   1/582
transverse electric (TE) wave   2/72
transverse magnetic (TM) wave   2/72
transverse relaxation   2/98
transverse relaxation optimized spectroscopy (TROSY)   2/95, 2/98
transverse resonance condition   2/78
trapped-ion mass analysers   1/339
– dynamic traps   1/339
– ion cyclotron resonance   1/339
– static traps   1/339
trapping efficiency   2/244
TR-FIA   2/55
triazine   2/88, 2/181 f, 2/223
– APCI-FIA-MS   2/199
– APCI-FIA-MS/MS   2/199
– APCI-LC-MS   2/199
– degradation pathways   2/182
– degradation products   2/182, 2/198, 2/223
– drinking water   2/223
– estuarine waters   2/182
– groundwater   2/223
– metabolites   2/224
– MS/MS   2/224 f
– natural waters   2/198
– photolysis   2/224
– physicochemical degradation   2/224
– quantification   2/223 ff
– rivers   2/198
– river water   2/223
– SFC-MS   2/223
– soil samples   2/182
– stability   2/198
– surface water   2/198
– water   2/198
trifluorofluoresceins (Oregon Green)   2/11
trinitrotoluene (TNT)   2/87

triple resonance experiments (in NMR)   2/119
triplett state   1/425
TRITC see tetramethylrhodamine isothiocyanata
trivial quenching   1/144
TRXRFA (total reflection X-ray fluorescence analysis)   1/550
tryptophans   1/141, 2/15
TSE see transmissible spongiform encephalopathy
TSP
– alcohol ethoxylates (AEO)   2/175
– alkyl polyglucosides   2/175
– anilides   1/176
– carbamates   2/177 ff
  – interlaboratory examination   2/178
  – quantification   2/177
– degradation pathways   2/182
– degradation products   2/182
– dyes   2/173
– EEC Drinking Water Directive   2/178
– estuarine waters   2/182
– explosives   2/174
– fluorine-containing surfactants   2/175
– fungicides   2/176, 2/183
– herbicides   2/176, 2/183
– miscellaneous   2/183
– organophosphorus compounds   2/179
  – quantification   2/179
– pesticides   2/176
– phenoxycarboxylic acids   2/179
– phenylureas   2/180
– polycyclic aromatic hydrocarbons (PAH)   2/174
– quantification   2/180 f
– quaternary amines   2/176
– seafood   2/176
– soil samples   2/182
– sulfonylureas   2/180
– surfactants   2/175
– thiocyanate   2/176
– thioureas   2/180
– toluidines   2/176
– toxins   2/176
– triazines   2/181
– urea
  – interlaboratory study   2/181
– urea pesticides   2/180
– quantification   2/180
TSP, LC-MS   2/172 ff
– anilides   2/173
– carbamates   2/173
– dyes   2/173
– explosives   2/173

- fungicides 2/173
- herbicides 2/173
- organophosphorus compounds 2/173
- pesticides 2/172 f
- phenoxycarboxylic acids 2/173
- phenylureas 2/173
- polycyclic aromatic hydrocarbons 2/173
- quaternary amines 2/173
- sulfonylureas 2/173
- surfactants 2/172
- thiocyanate 2/173
- thioureas 2/173
- toluidines 2/173
- toxins 2/173
- triazines 2/173
TSP, review
- dyes 2/173
- general 2/173
- pesticides 2/172
- surfactants 2/172
tubulin
- conjugates 2/17
- transport in neurons 2/17
twisting 1/42
two line background correction 1/456 f
two-dimensional (2D) NMR 1/292
- heteronuclear 1/292
- homonuclear 1/292
two-dimensional (2D) NMR experiment 2/105
two-dimensional experiments
- anisotropic-isotropic correlation 1/300
- CSAs 1/300
two-dimensional HPLC 2/127
- as sample 2/127
- prep in MS 2/127
two-dimensional NMR 1/218
two-photon excitation (2PE) 2/58
two-photon fluorescence excitation 2/61
- single step bioaffinity assays 2/58
TXRF or TRXRF (total reflection X-ray fluorescence 1/580
type standardization 1/410, 1/412
tyramide signal amplification (TSA) 2/16
tyramide-labeled dyes 2/16
tyrosine 2/15

u

UHV (ultra high vacuum) 1/506
Ulbricht sphere 1/79
ULS 2/18
ungicides 2/192
- groundwater 2/192
- river water 2/192

univariate methods 2/456
Universal Linkage System (ULS) 2/18
Unscrambler 2/467
unsupervised learning 2/444
UPES, UVPES, UPS, UVPS (ultraviolet photoelectron spectroscopy) 1/583
UPS (ultraviolet photoelectron spectroscopy) 1/508, 1/591
urea 2/69
urea pesticides
- quantification 2/180
uric acid 2/21
UV shifts due to structure 2/382
UV/VIS
- absorption 1/44
- spectroscopy 1/125
- structural analysis 1/129

v

vapour generation techniques 1/447 ff
variable angle correlation spectroscopy (VACSY) 1/300
Varian Inc. 2/266
vibration
- normal 1/99
vibrational circular dichroism (VCD) 2/94
vibrational mode 1/558, 1/578
vibrational relaxation 1/45
vibrations
- combination 1/54, 1/104
- overtone 1/54, 1/104
- stretching 1/42
VIS/NIR dyes 2/28
volatile organic compounds (VOCs) 2/256

w

wagging modes 1/42
Walsh 1/436
water analysis 2/154
waveguides 2/78 f
- monomodal 2/79
- multimodal 2/79
wavelength 1/367
wavelength modulation background correction 1/462
wavelength-dispersive XRF 1/390
wavenumber 1/39
weakly aligned systems 2/95
WebBook 2/490, 2/500
wheat germ agglutinin (WGA) 2/21
white optics 1/91
Wilatz 1/436
Wiley Registry of Mass Spectral Data 2/491
Wiley-SIMS database 2/494

window material 1/71
windowless homonuclear decoupling sequences
- DUMBO-1 1/299
- FSLG 1/299
- PMLG 1/299
WISE (wide line separation) 1/176
Wollaston 1/436
Woodward 1/125

x

XAFS (X-ray absorption fine structure) 1/584
XANES (X-ray absorption near edge spectroscopy) 1/584
xanthenes 2/10
xantine 2/21
XAPS (X-ray appearance potential spectroscopy) 1/508
XCORFE 1/248
XEAES (X-ray excited Auger electron spectroscopy) 1/512
xenoestrogens 2/154
xenon arc lamp 1/439
Xenon for IR detection 2/420
XNDO/S 2/28
XPD (X-ray photoelectron diffraction) 1/586, 1/591
XPS or XPES (X-ray photoelectron spectroscopy) 1/587
X-ray crystallography 2/6
X-ray detectors 1/384
X-ray fluorescence 2/429
XRF (X-ray induced fluorescence) 1/550
XSW (X-ray standing wave) 1/591

y

YO-PRO 2/17
YOYO 2/17

z

Zeeman background correction 1/458 ff
Zeeman effect, anomalous 1/458
Zeeman effect, longitudinal 1/460
Zeeman effect, normal 1/458
Zeeman effect, transverse 1/458
zinc selenide 2/78

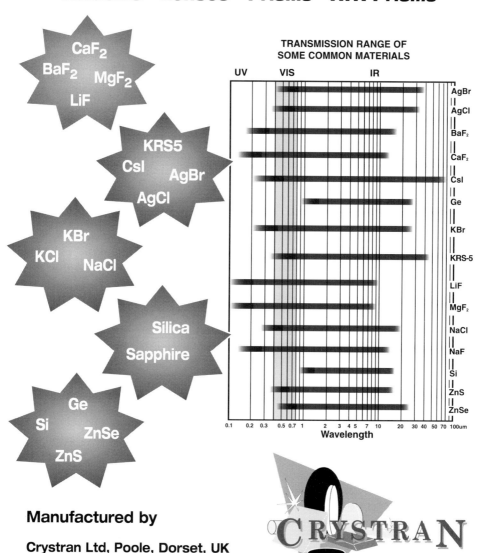

+++ NEW +++ NEW +++

F. GERSON, University of Basel, Switzerland, und W. HUBER, Hoffmann-La Roche & Cie AG, Basel, Switzerland

Electron Spin Resonance Spectroscopy of Organic Radicals

2003. Approx. XVI, 456 pages.
Softcover.
ISBN 3-527-30275-1*

*Prices on request.

Written by Fabian Gerson and Walter Huber, top experts in the field of electron spin resonance spectroscopy, this book offers a compact yet readily comprehensible introduction to the modern world of ESR.
Due to its wide range, from fundamental theory right up to the treatment of all important classes of substances that can be analyzed using ESR spectroscopy, this unique book is ideal for users in both research and industry.
It dismisses with complex physical approaches and uses understandable discussions of example spectra rather than mathematical derivations.

Wiley-VCH Verlag
Postfach 10 11 61 • D-69451 Weinheim
Fax: +49 (0)6201 606 184
e-Mail: service@wiley-vch.de • www.wiley-vch.de